ENGINEER OF REVOLUT

Science, Technology and Culture, 1700–1945

Series Editors

David M. Knight
University of Durham

and

Trevor Levere
University of Toronto

Science, Technology and Culture, 1700–1945 focuses on the social, cultural, industrial and economic contexts of science and technology from the 'scientific revolution' up to the Second World War. It explores the agricultural and industrial revolutions of the eighteenth century, the coffee-house culture of the Enlightenment, the spread of museums, botanic gardens and expositions in the nineteenth century, to the Franco-Prussian war of 1870, seen as a victory for German science. It also addresses the dependence of society on science and technology in the twentieth century.

Science, Technology and Culture, 1700–1945 addresses issues of the interaction of science, technology and culture in the period from 1700 to 1945, at the same time as including new research within the field of the history of science.

Also in the series

Popularizing Science and Technology in the European Periphery, 1800–2000
Edited by Faidra Papanelopoulou, Agustí Nieto-Galan and Enrique Perdiguero

Essays on David Hume, Medical Men and the Scottish Enlightenment
'Industry, Knowledge and Humanity'
Roger L. Emerson

The Language of Mineralogy
John Walker, Chemistry and the Edinburgh Medical School, 1750–1800
Matthew D. Eddy

Engineer of Revolutionary Russia
Iurii V. Lomonosov (1876–1952) and the Railways

ANTHONY HEYWOOD
University of Aberdeen, UK

Routledge
Taylor & Francis Group

LONDON AND NEW YORK

First published 2011 by Ashgate Publishing

2 Park Square, Milton Park, Abingdon, Oxon OX14 4RN

711 Third Avenue, New York, NY 10017, USA

Routledge is an imprint of the Taylor & Francis Group, an informa business

First issued in paperback 2016

British Library Cataloguing in Publication Data
Heywood, Anthony.
  Engineer of revolutionary Russia: Iurii V. Lomonosov (1876–1952) and the railways. – (Science, technology and culture, 1700–1945)
  1. Lomonosov, IU. (IUrii), 1876–1952. 2. Railroad engineers – Russia – Biography. 3. Russia – History – Nicholas II, 1894–1917 – Biography. 4. Engineering and state – Soviet Union.
  I. Title II. Series
  625.1'0092–dc22

Library of Congress Cataloging-in-Publication Data
Heywood, Anthony.
  Engineer of revolutionary Russia : Iurii V. Lomonosov (1876–1952) and the railways / Anthony Heywood.
    p. cm. – (Science, technology, and culture, 1700–1945)
  Includes bibliographical references and index.
  ISBN 978-0-7546-5539-8 (hbk) 1. Lomonosov, IU. (IUrii), 1876–1952. 2. Lomonosov, IU. (IUrii), 1876–1952 – Political and social views. 3. Railroad engineers – Soviet Union – Biography. 4. Railroads – Soviet Union – History. 5. Engineering – Soviet Union – History. 6. Science – Soviet Union – History. 7. Technology – Soviet Union – History. 8. Transportation – Soviet Union – History. 9. Russia – History – Revolution, 1905–1907 – Influence.
  10. Soviet Union – History – Revolution, 1917–1921 – Influence. I. Title.
  TF140.L66H49 2010
  625.10092–dc22
  [B]
                                                                                    2010021616

ISBN 978-0-7546-5539-8 (hbk)
ISBN 978-1-138-25929-4 (pbk)

For Eileen, Susan and John

Contents

List of Illustrations

Source (unless otherwise stated): the G.V. Lomonossoff Collection, Leeds Russian Archive, courtesy Mrs G.M. Browning.

List of Figures

List of Tables

Acknowledgements

All my work on Lomonosov would have been simply impossible without the assistance and encouragement that I have received from my father, who first drew my attention to the engineer's papers at the Leeds Russian Archive; from Richard Davies, the founding archivist of the Leeds archive who has been such a good supervisor, colleague and friend over the years; and from the Lomonossoff family, especially Mrs G.M. Browning and Dr George Lomonossoff. I am very, very deeply grateful to them.

My PhD dissertation and associated monograph *Modernising Lenin's Russia* concerned one aspect of Lomonosov's career in the early 1920s. Working on this present book, I have remained very conscious of the help that I received from so many people for those earlier projects, and I wish to reiterate my gratitude here.

For financial support for the preparation of this book my thanks are due to the Royal Society, the Arts and Humanities Research Council, the British Academy, the British Association for Slavonic and East European Studies, the University of Bradford Research Committee, and the School of Divinity, History and Philosophy, University of Aberdeen.

The staff of the archives, libraries and museums that I have used have been very welcoming and extremely helpful. I owe particular thanks to A.V. Mukhtan, O.V. Krestich and V.V. Bersenev at the Russian State Historical Archive; S.V. Mironenko, E. Chubkova and their colleagues at the State Archive of the Russian Federation; O.M. Redko at the Kyiv City Archive; V.I. Misailova, G.P. Zakrevskaia, L.A. Davidova and S.L. Pogodin at the Central Railway Museum and the Central Museum of the October Railway in St Petersburg; P.E. Rubinin at the Kapitsa museum; Philip Atkins at the National Railway Museum, York; Keith Moore and Katherine Dike at the library of the Institution of Mechanical Engineers; Irina Lukka at Helsinki University Library; Franziska Rogger at Berne University/Berne State Archives; and Marlinde Schwarzenau at the Deutsches Museum, Munich. Martin Maw kindly provided information from the archive of Oxford University Press. The St Petersburg photographic archive and Canadian Pacific Railway archive kindly provided photographs. Particular thanks are due to Mrs J. Slaney at the Inter-Library Loans section of the J.B. Priestley Library, University of Bradford.

I am very grateful to all the numerous people who have helped me in all sorts of ways, including Barbara Allen, Ian Anstruther, Bob Argenbright, Mark Baker, Gennadii Bordiugov, Galya Bradley, Gillian Brown, Ian Button, George Carpenter, Wim Coudenys, Bob Davies, Colin Divall, Paul Dukes, Michael Duffy, V.D. Esakov, Cliff Foust, Robert Frost, V.L. Genis, Isabel Gramer, Mike Hamm, James Harris, D.S. Hellewell, Mike Hickey, Michael Holman, Iurii Il'in,

Polly Jones, Igor Kiselev, J.W. Knowles, B.I. Kolonitskii, Alexandra Korros, Leonid Makarov, Evan Mawdsley, Bruce Menning, Jack Morrell, Aleksandr Nikol'skii, Scott Palmer, Gareth Popkins, Munro Price, Brian Radford, Chris Read, David Saunders, Susanne Schattenberg, Brandon Schneider, V.I. Shishkin, Natal'ia Skachkova, Jon Smele, Jeremy Smith, Tania Sorokina, John Steinberg, A.A. Strutinskii, Jerry Surh, Alex Titov, Katy Turton, Aleksandr Ushakov, Peter Waldron, John Westwood, R. Whittaker, Beryl Williams and Chris Williams. At Ashgate Tom Gray, Emily Yates and their colleagues have been extremely helpful and incredibly patient. A deep regret is that I shall not be able to show this book to A.V. Makurov, latterly of the railway museum at Moscow's Kazan' terminus – a dedicated *parovoznik* and true friend who died in 2007.

I am immensely grateful to John Holroyd for drawing the map, and to Steve Hay for technical assistance with the illustrations. Ben Marsden, Adele Lindenmeyr, my father and my wife Eileen read the draft manuscript, and their many comments and suggestions were immensely helpful.

Special thanks are due to my family and relatives for all their help and support, especially John and Pat Consey, Robert and Mary Sturgill, and my parents. I dedicate this book to my wife Eileen and children Susan and John, with all my love. I hope that they will feel that my many absences and hours at the computer have been worthwhile.

I alone am responsible for any errors of fact and interpretation.

List of Abbreviations and Acronyms

Caltech	California Institute of Technology
Cheka, ChK, VChK	All-Russian Extraordinary Commission for Combating Counter-revolution, Speculation and Misconduct in Office
GOELRO	State Commission for the Electrification of Russia
GOMZA	State Group of Machine-building Factories
Gosplan	State Planning Commission
GPU	State Political Administration
GWR	Great Western Railway (UK)
ILP	Independent Labour Party (UK)
IPS	Institute of Ways of Communication
Kadet	Constitutional Democratic (Party)
Komgosor	Committee for State Civil Construction
KPI	Kiev Polytechnic Institute
LMSR	London, Midland and Scottish Railway (UK)
LNER	London and North Eastern Railway (UK)
MAN	Maschinenfabrik Augsberg-Nurnberg AG
MPS	Ministry of Ways of Communication
NEP	New Economic Policy
NKID	People's Commissariat of Foreign Affairs
NKPS	People's Commissariat of Ways of Communication
NKRKI	People's Commissariat of Workers' and Peasants' Inspection
NKVT	People's Commissariat of Foreign Trade
Nohab	Nydqvist & Holm AB
NTK	NKPS Scientific-Technical Committee
Politburo	Political Bureau of the Bolshevik Party's Central Committee
RSC	Russian Supply Committee in North America
RZhM	Russian Railway Mission Abroad
SD	Social Democratic (Party) (Russian Social Democratic Workers' Party)
SER	South Eastern Railways
Sovnarkom	Council of People's Commissars
SR	Socialist Revolutionary (Party)
STO	Council of Labour and Defence
Tsentrosoiuz	Central Cooperative Society
VChK	All-Russian Cheka

VSNKh	Supreme Council of the National Economy
VTsIK	All-Russian Central Executive Committee
WPI	Warsaw Polytechnic Institute

Note on Renderings

I have used a simplified version of the Library of Congress transliteration system to represent Russian letters with Latin equivalents, except for the letter designation of one steam locomotive class, which is shown as 'Ye' rather than 'E' (see Chapter 6, note 41). All Russian spellings reflect the Soviet reform of the alphabet in 1918. I have normally transliterated the names of people, places and publications – Vitte, Trotskii, Riazan', *Pravda*, etc. However, common English translations have been used for tsars and the cities of Archangel, Moscow and St Petersburg. Non-Russian place names in the tsarist empire/USSR are rendered with the Russian version used by officialdom: for example, Kiev and Khar'kov, not Kyiv and Kharkiv, but note Daugavpils (not Dvinsk) in inter-war Latvia. Likewise, I use the Russian form of the name of any foreigner living and working in Russia who habitually used that form there, notably the Polish-born communist F.E. Dzerzhinskii and engineer A.O. Chechott. Names of Russian/Soviet institutions and organizations are translated. Translations are mine unless otherwise indicated.

The Lomonosov family rendered their name in Latin letters as Lomonossoff (or occasionally Lomonosoff), using a phonetic transliteration system that was widespread in the early twentieth century. I have used the forms Lomonosov and Lomonosova for the engineer and his two wives, following the Library of Congress system, but have retained Lomonossoff where it appears in a quotation or publication reference. Where appropriate the name of Lomonosov's son Iurii is given as George because the family rarely used the Russian form and because he consciously developed a British identity.

For simplicity the term Russia is used for the vast areas that constituted the tsarist empire until February 1917 and the Union of Soviet Socialist Republics (USSR or Soviet Union) from December 1922. The term Soviet Russia is also used for the territory under Soviet control from October 1917.

Dates of events in Russia are recorded with the calendar in official use at the time, which was the Julian calendar until the Soviet authorities adopted the Gregorian calendar in February 1918; the Julian calendar was 12 days behind the Gregorian calendar in the nineteenth century and 13 behind in the twentieth century. Events outside Russia are dated with the Gregorian calendar. For clarity certain dates are shown in both styles or with the abbreviations o.s. (old style) and n.s. (new style).

Metric measurements are used throughout, using the following conversions of Russian and Imperial units where appropriate:

1 pud (pood)	16.30 kilograms
1 verst	1.06 kilometres
1 mile	0.6 kilometres

Currency conversions are, unless otherwise indicated, estimates based on exchange rates shown in an appropriate issue of *The Economist*.

Descriptions of railway steam locomotives usually highlight the arrangement of powered and non-powered axles or wheels. This book uses the Whyte system of wheel notation, which is the norm in the Americas, Britain, Ireland and the British Commonwealth. In this system, for example, a 2-10-0 locomotive has two non-powered wheels at the front, ten powered (driving) wheels and no non-powered wheels at the rear. The Russian railways used the French system of axle notation, where our 2-10-0 becomes a 1-5-0. For a full description of the Whyte system see, for instance, P. Ransome Wallis (ed.), *The Concise Encyclopaedia of World Railway Locomotives* (London, 1959), p. 504.

References from Russian archives give the archive's name in abbreviated form, then the numbers of the collection (*fond*), inventory (*opis'*), file (*delo* or *edinitsa khraneniia*) and folio (*list*). The reverse of a folio is recorded as 'ob.' after the folio number. Ukrainian archive references have the same principle, but different words. A slightly different system is used for files of the tsarist secret police now in f. 102 of the State Archive of the Russian Federation: see Chapter 6, note 55 (p. 147).

Glossary of Technical Terms

armature	rotating coil of an electric motor
automatic train brakes	brake system activated simultaneously on multiple vehicles of a train (as opposed to individual handbrakes on each vehicle)
automatic train couplings	device whereby vehicles couple on impact (as opposed to manual use of hooks and chain)
axle arrangement	see: wheel arrangement
bogie	wheeled undercarriage pivoted below the end of a rail vehicle, with at least four wheels; a bogie-vehicle normally has two bogies (one at each end)
boiler pressure	pressure of the steam in the boiler
calorimeter	instrument to measure the quantity of heat
chassis	the base-frame of a vehicle such as a locomotive, carriage, wagon
common-carrier railway	railway obliged by law to transport any consignment that is presented for transport
compound (steam locomotive)	locomotive in which steam is expanded successively in two (or more) working cylinders (as opposed to single expansion in a 'simple' locomotive)
compression-ignition engine	internal combustion engine in which the source of fuel ignition is the heat generated by compression of air
condensing locomotive	locomotive in which working steam is condensed for reuse
coupled weight (of a locomotive)	the part of the locomotive's weight that rests on the driving wheels
cut-off	the point at which steam is stopped from entering a working cylinder, so that thereafter work is done by expansion of a fixed volume of steam
cycle (of a heat engine)	the sequence of movements (e.g. of the piston in the working cylinder) to complete one stroke of the engine
cycle I, II, III (test or experiment)	Lomonosov's terminology to identify different forms of road test or experiment
cylinder (of heat engine)	chamber in which a piston moves

decapod   locomotive with 10 driving wheels; in Russia, an informal name for the class Ye 2-10-0 locomotive

diesel engine   strictly, an internal combustion engine with blast-air fuel injection, as proposed by Rudolf Diesel, and by extension an internal combustion engine with compression-ignition

diesel locomotive   railway locomotive powered by a diesel engine

**diesel-electric/-mechanical/-   railway locomotive powered by a diesel engine
  hydraulic locomotive**   with electric, mechanical or hydraulic power transmission

direct drive   transmission-less system for powering a vehicle

draughting   process whereby a steam locomotive's exhaust gases and ash are ejected to the atmosphere

drawbar   point on a railway locomotive to which the trailing load is attached

driving wheels (of a locomotive) wheels powered by the engine

dynamometer   self-acting recorder of a locomotive's useful work, expressed as the tractive force exerted at the drawbar

dynamometer car   railway carriage fitted with a dynamometer and other recording equipment

fixed frames (of locomotive)   a rigid assembly through which the axles of the locomotive are threaded (contrast: bogie-chassis)

four-stroke (engine)   engine in which the piston completes four strokes in each cycle

gauge (of railway track)   distance between the two rails; the Russian standard gauge in Lomonosov's lifetime was 1524 mm (now 1520 mm), whereas the standard gauge in much of Europe and North America was 1435 mm (4' 8½")

**gradient profile (of a railway   pattern of variations from the horizontal in a
  line)**   given stretch of railway

heat engine   device for producing motive power from heat

horsepower   imperial (non-metric) unit of power

indicator   device automatically recording changing pressure in the cylinder of an engine with the movement of the piston, to indicate the work done in one cycle

indicator diagram   graph expressing the pressure data obtained with an indicator

internal combustion engine   heat engine where the combustion of the fuel occurs inside the working cylinder(s)

inzhener–puteets	in tsarist Russia: an engineer-graduate of the Institute of Ways of Communication
inzhener–tekhnolog	in tsarist Russia: an engineer-graduate of a technological institute (a production engineer)
locomotive-hours	the time that a locomotive spends away from a depot (in administrative systems for recording the work undertaken by the locomotive)
locomotive passport	initially, a pamphlet that summarized performance data for the given class of locomotive; now, the logbook of an individual locomotive
locomotive testing	the testing of locomotives under power as machines
main-line locomotive	locomotive for the public-service network (as opposed to industrial locomotive on a separate industrial system)
piston	disc or short cylinder, fitting closely within a larger working cylinder, in which it moves to deliver power
pony truck (in a steam locomotive)	sub-assembly with a non-powered axle or axles; can be situated at the front or the rear of the locomotive
regulator (in a steam locomotive)	driver's control to release steam from the boiler to the engine
reversing lever (in a steam locomotive)	driver's control to adjust the cut-off and switch between forward and reverse gear
road test (of a locomotive)	test of a railway locomotive under power on a given section of railway
railway rolling stock	passenger carriages and freight wagons
'scientific road experiments'	Lomonosov's term for his type of road test
shaft (of internal combustion engine)	large power-delivery axle that is rotated by the engine
solenoid	cylindrical coil of wire that acts as a magnet when carrying electric current
stationary test plant/ laboratory/station	apparatus with wheel rollers and measuring equipment to enable a locomotive to be tested under power within a building
superheater (in a steam locomotive)	device to improve the engine's overall efficiency by raising the maximum temperature of the working steam from the boiler, usually to over 600°C; the use of steam at high temperatures promotes efficiencies by allowing, for example, a greater fall of temperature; superheat was seen as a simpler, cheaper alternative to compounding

tender (of steam locomotive)	vehicle for carrying the locomotive's fuel and water supplies
thermal efficiency (of heat engine)	the proportion of a fuel's heat energy that the engine converts to useful work
thermodynamics	science of the relations between heat and other forms of energy
traction (on Russian railways)	pertains mainly to locomotives, but can also refer to passenger carriages and freight wagons
traction calculations	calculations concerning a locomotive class's performance characteristics and use of such data to determine safe speed and haulage limits, traffic schedules, etc.
traction engineer	a specialist in locomotive, carriage and/or wagon technical matters
traction superintendent	the head of the Traction Department of a Russian railway; in charge of the line's locomotive, carriage and wagon affairs
traction theory	scientific theory of the technical processes involved in the operation of a railway locomotive
tractive force	the pulling or haulage power of a locomotive
transmission (of power)	mechanism by which power is transmitted from an engine to an axle so as to move a vehicle; types include electric, gas, hydraulic and mechanical
two-stroke (engine)	the piston completes two strokes in each cycle
useful work (done by a locomotive)	the part of the energy developed by an engine that is deployed usefully, especially for sustaining the motion of a carriage or wagon against resistance, rather than wasted
valve gear (of steam locomotive)	enables the movement of the pistons to move the locomotive
wheel arrangement	the layout of a locomotive's powered and non-powered wheels; this book uses the numerical Whyte system for designating them: for example, a 2-6-0 steam locomotive has two non-powered front wheels (i.e. one axle), six powered wheels (three axles) and no non-powered wheels at the rear
wheelset	pair of wheels connected by an axle

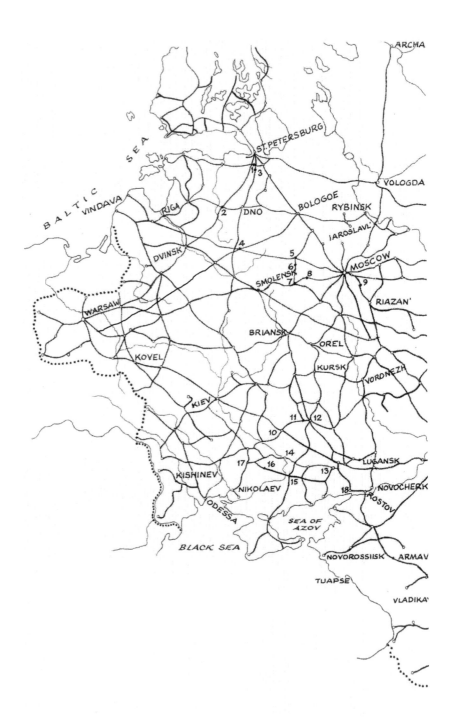

Map 1   Principal railways of European Russia and Western Siberia, circa 1914

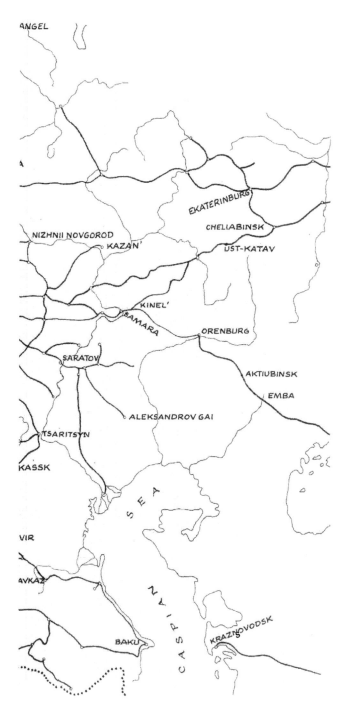

ANGEL

EKATERINBURG

NIZHNII NOVGOROD

CHELIABINSK

KAZAN'

UST-KATAV

KINEL'

SAMARA

ORENBURG

SARATOV

AKTIUBINSK

EMBA

ALEKSANDROV GAI

TSARITSYN

KASSK

SEA

VIR

AVKAZ

CASPIAN

BAKU

KRAZNOVODSK

Key

1. Gatchina
2. Pskov
3. Semrino
4. Velikie Luki
5. Rzhev
6. Sychevka
7. Viaz'ma
8. Gzhatsk
9. Kolomna
10. Poltava
11. Liubotin
12. Khar'kov
13. Iuzovka
14. Ekaterinoslav
15. Aleksandrovsk
16. Nikolpol'
17. Dolinskaia
18. Taganrog

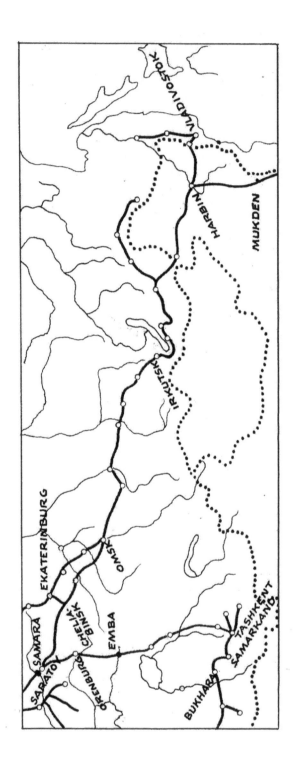

Map 2   Principal railways of Siberia and Russian Central Asia, circa 1914

Introduction

Iurii Vladimirovich Lomonosov (1876–1952) was one of the more conspicuous engineers of a pivotal Russian generation – the generation of Lenin and Stalin that was born in the 1870s and dominated their country's public life for much of the first half of the twentieth century. He became a major figure in the industry that underpinned economic development worldwide in the nineteenth century and that remained vital for the Russian and Soviet economy throughout the twentieth century: railway transport. Ambitious, opinionated and outspoken, he was a controversial character who tended to arouse emphatic reactions. To use J.N. Westwood's neat summary, for some people he was flamboyant, charming and a born raconteur, but for others he was brash, ingratiating and a born liar.[1]

His life was richly varied, eventful and sometimes dramatic, reflecting the turmoil of rapid industrialization, war and revolution in Russia, Europe and North America during the late nineteenth and early twentieth centuries. Born into the rural gentry, he opted for a career in academe and was a full professor by the age of 30 in one of tsarist Russia's few institutes of engineering. Transferring to the Russian state railways in 1908, he became one of the most senior managers within just five years. In the meantime he established himself as an authority in developing and applying scientific techniques for improving railway efficiency and especially for testing steam locomotives under power. As both manager and researcher he helped to lead the all-important work of Russia's railways during the First World War. Yet his privileged background and professional successes did not imply that he accepted Russia's autocratic political system. In fact, he participated in the terrorist activities of the Marxist Russian Social Democratic Workers' Party in Russia's 1905 revolution and made a distinctive contribution to the autocracy's demise in the revolution of February–March 1917. During the following six years he played central roles as an engineer-diplomat in two key aspects of revolutionary Russia's relations with North America and Western Europe, initially for the Provisional government that replaced the Tsar and then for Lenin's Bolshevik regime. From 1923 he concentrated on leading the design and construction of two high-power diesel locomotives for the Soviet railways – work that did much to prove that this technology could replace steam power on the world's railways. Yet despite enjoying Lenin's unstinting support, Lomonosov became one of the many experienced Russian engineers to be denigrated by Bolshevik activists as an anti-Soviet 'bourgeois specialist' in the 1920s. Lucky to avoid arrest and execution, he spent his last 25 years abroad, mostly in England and North America. There he

---

[1]   J.N. Westwood, *Soviet Locomotive Technology During Industrialization, 1928–1952* (London, 1982), p. 12.

became known to his fellow railway engineers as 'the great Russian Locomotive Engineer'.[2] Yet for him personally exile was a tragedy that challenged his deepest notions of his identity and self-worth. Deemed a traitor by Stalin's regime, he became a 'non-person' in his homeland like so many early Soviet public figures, with discussion of him and his work suppressed for decades.[3]

This Soviet censorship, however, never quite eliminated him from historical writing. Within the USSR the most striking exception occurred at the peak of Stalin's power in 1945 in the form of a short, favourable pen-portrait in the memoirs of a former colleague, the shipbuilding expert A.N. Krylov; why the censor permitted these remarks remains a mystery.[4] From the late 1950s, when the regime slightly relaxed its censorship as part of its policy of de-stalinization, Lomonosov figured in a number of Soviet publications that were mostly intended for specialist audiences.[5] These references, however, were almost invariably cursory, and as late as 1984 his name was studiously omitted from a feature article about Lenin and early Soviet diesel locomotives in the Communist Party's newspaper *Pravda*.[6] This censorship naturally restricted awareness of him outside the USSR as well. Lomonosov did acquire many friends and contacts in the railway engineering communities of Western Europe and North America, one of whom published a biographical article in 1973.[7] But these acquaintances could not read Russian, the language of most of his publications. As for foreign historians of Russia, most remained oblivious of him, and the few like Westwood who did

---

[2] R.C. Bond, *A Lifetime with Locomotives* (Cambridge, 1975), p. 81. The phrase is echoed by, for example, the historian M. Rutherford in 'Failure? Part Two', *Backtrack*, 9/12 (December 1995): 657.

[3] For example, Lomonosov was excluded from V.A. Rakov's major survey of Russian and Soviet locomotive development, *Lokomotivy zheleznykh dorog Sovetskogo Soiuza: Ot pervykh parovozov do sovremennykh lokomotivov* (Moscow, 1955).

[4] A.N. Krylov, *Moi vospominaniia* (Moscow, 1945), pp. 277–318, especially 297–8.

[5] For instance: P.V. Iakobson, *Istoriia teplovoza v SSSR* (Moscow, 1960), p. 25; R.F. Karpova, 'Sovetskie otnosheniia so skandinavskimi stranami 1920–1923 gg.', *Skandinavskii sbornik*, 10 (1965): 140. The most extensive comments are in V.A. Shishkin, *Sovetskoe gosudarstvo i strany zapada v 1917–1923 gg.: Ocherki istorii stanovleniia ekonomicheskikh otnoshenii* (Leningrad, 1969), pp. 225–6; L.V. Lovtsov, 'Iz istorii bor'by Kommunisticheskoi partii za sozdanie otechestvennogo teplovozostroeniia', *Stranitsy velikogo puti: Iz istorii bor'by KPSS za pobedu kommunizma: Sbornik statei aspirantov, Chast' III* (Moscow, 1970), pp. 100–33; and L.V. Lovtsov, 'Leninskii printsip edinstva resheniia khoziaistvennykh i politicheskikh zadach v rukovodstve partii vosstanovleniem zheleznodorozhnogo transporta (1921–1925 gg.)', *Iz istorii bor'by KPSS za pobedu sotsializma i kommunizma: Sbornik statei* (Moscow, 1970), pp. 159–83.

[6] A. Erokhin, '"Spetsial'no sledit' za etim delom ...": Rasskaz o knigakh iz lichnoi biblioteki V.I. Lenina', *Pravda*, 4 February 1984, p. 3.

[7] B. Reed, 'Lomonossoff: A Diesel Traction Pioneer', *Railroad History*, 128 (1973): 35–49. But note his absence from J. Marshall, *A Biographical Dictionary of Railway Engineers*, 2nd edn (Oxford, 2003).

perceive his importance could not explore the Soviet archives to learn more.[8] Not until the advent of *glasnost'* in 1985 did this situation begin to change. A modest flowering of interest has since occurred in Russia, mainly with articles in academic journals, popular magazines and newspapers. Unfortunately, almost all of these contributions are blighted by simplistic hagiography or demonization.[9] In the West, meanwhile, the most substantial works have been an article-length biographical survey by Hugh Aplin and my own publications, most of which concern his foreign-trade activity in the early 1920s.[10]

The need for book-length biographies in English of engineers like Lomonosov appears all the more compelling when we consider the existing biographical literature in English for revolutionary Russia (broadly defined as late Imperial and early Soviet Russia). Although we have numerous studies of Lenin, Stalin, Trotskii and certain other familiar individuals, mostly from the worlds of politics and the arts, there are many other notable public figures whose biography has yet to appear. A telling example is the infamous first head of the Soviet political police, F.E. Dzerzhinskii.[11] True, biographies of scientists have featured in the growth since the 1970s of Western interest in the history of science, technology and engineering in revolutionary Russia. The subjects have included the geologist V.I. Vernadskii, the geographer P.P. Semenov-Tian-Shanskii, the mathematician S.V. Kovalevskaia, the chemist D.I. Mendeleev, the physiologist I.P. Pavlov and

---

[8]   Westwood, *Soviet Locomotive Technology*, especially pp. 10–21, 27–8, 39–47, 53–6.

[9]   Useful contributions include: E.A. Norman, 'Teplovoz professora Lomonosova – pervenets sovetskogo i mirovogo teplovozostroeniia', *Voprosy istorii estestvoznaniia i tekhniki*, 4 (1985): 116–25; L.L. Makarov, *Parovozy serii E: fotoal'bom* (Moscow, 2004), especially pp. 30–34, 40–42, 59–79; and especially V.L. Genis, *Nevernye slugi rezhima: Pervye sovetskie nevozvrashchentsy (1920–1933): kniga 1* (Moscow, 2009), pp. 440–479. Examples of the newspaper and magazine coverage are V.D. Kuz'mich, 'Pervye sovetskie', *Gudok*, 6 October 1987, p. 4; N.A. Zenzinov, 'Vydaiushchiisia rossiiskii zheleznodorozhnyi inzhener', *Zheleznodorozhnyi transport*, 4 (1993): 57–63; O.G. Kuprienko, 'Ostorozhno, Shelest! Pora postavit' tochku v istorii o tom, kto sozdal pervyi otechestvennyi teplovoz', *Gudok*, 25 December 1999, p. 2, and his 'Zabytye polpredy Rossii: 80 let nazad nachala rabotu za granitsei nasha zheleznodorozhnaia missiia', *Gudok*, 27 February 2001, p. 5. Note that Lomonosov does appear in V.A. Rakov, *Lokomotivy otechestvennykh zheleznykh dorog, 1845–1955* (Moscow, 1995), a much revised edition of the 1955 book mentioned in note 3 above. An example of the most strident criticism is A. Igolkin, 'Leninskii narkom: U istokov sovetskoi korruptsii', *Grazhdanin*, 3 (2004), with seemingly unfounded allegations of corruption with the Aleksandrov Gai–Emba Railway construction project (1920); see also Chapter 8, note 72.

[10]   H.A. Aplin, 'Iurii Vladimirovich Lomonosov (1876–1952)', in his *Catalogue of the Lomonossoff Collections* (Leeds, 1988), pp. vii–xx; and, in particular, A.J. Heywood, *Modernising Lenin's Russia: Economic Reconstruction, Foreign Trade and the Railways* (Cambridge, 1999).

[11]   Iain Lauchlan is now preparing a biography of Dzerzhinskii.

the physicist P.L. Kapitsa.[12] We also have valuable studies of the engineering profession as a whole in this period, together with the phenomenon of technocracy and the development of specific industries.[13] Yet there are very few studies of individual engineers. Loren R. Graham uses the career of the mining engineer P.A. Pal'chinskii (1875–1929) to explore the relationship between technology and Soviet society, contending that the misuse of technology disrupted Soviet industrialization and helped bring the USSR's demise. By contrast, Timothy O'Connor's biography of the electrical engineer L.B. Krasin (1870–1926) focuses on his political importance as a leading Bolshevik who became the first People's Commissar for Foreign Trade and, in effect, Soviet ambassador to Great Britain and France.[14]

A detailed study of Lomonosov is now feasible because a vast amount of relevant documentation has become accessible. Former Soviet archives contain numerous official papers about most aspects of his career. Admittedly, certain important documents may have been destroyed or lost. In particular, the archive

---

[12] On these examples see: K.E. Bailes, *Science and Russian Culture in an Age of Revolutions: V.I. Vernadsky and His Scientific School, 1863–1945* (Bloomington, 1990); D.P. Todes, *Pavlov's Physiology Factory: Experiment, Interpretation, Laboratory Enterprise* (Baltimore, 2001); W.B. Lincoln, *Petr Petrovich Semenov-Tian-Shanskii: The Life of a Russian Geographer* (Newtonville, 1980); A.H. Koblitz, *A Convergence of Lives: Sofia Kovalevskaia: Scientist, Writer, Revolutionary* (Cambridge, MA, 1983); M.D. Gordin, *A Well-Ordered Thing: Dmitrii Mendeleev and the Shadow of the Periodic Table* (New York, 2004); and, among other works, L. Badash, *Kapitza, Rutherford, and the Kremlin* (New Haven, 1985). More generally, for introductions to the history of Russian and Soviet science see A. Vucinich, *Science in Russian Culture, 1861–1917* (Stanford, 1970); L.R. Graham, *Science in Russia and the Soviet Union: A Short History* (Cambridge, 1994); and M.D. Gordin et al. (eds), *Intelligentsia Science: The Russian Century, 1860–1960* (*Osiris*, vol. 23) (Chicago, 2008).

[13] For example: H.D. Balzer, 'Educating Engineers: Economic Politics and Technical Training in Tsarist Russia' (unpublished PhD dissertation, University of Pennsylvania, 1980); 'The Engineering Profession in Tsarist Russia', in his (ed.), *Russia's Missing Middle Class: The Professions in Russian History* (Armonk, 1996), pp. 55–88; A.J. Rieber, 'The Rise of Engineers in Russia', *Cahiers du Monde Russe et Soviétique*, 31/4 (1990): 539–68; K.E. Bailes, *Technology and Society under Lenin and Stalin: Origins of the Soviet Technical Intelligentsia, 1917–1941* (Princeton, 1978); N. Lampert, *The Technical Intelligentsia and the Soviet State: A Study of Soviet Managers and Technicians, 1928–1935* (London, 1979); D.K. Rowney, *Transition to Technocracy: The Structural Origins of the Soviet Administrative State* (Ithaca, 1989). An excellent example of an industry case-study is J. Coopersmith, *The Electrification of Russia, 1880–1926* (Ithaca, 1992).

[14] L.R. Graham, *The Ghost of the Executed Engineer: Technology and the Fall of the Soviet Union* (Cambridge, MA, 1996); T.E. O'Connor, *The Engineer of Revolution: L.B. Krasin and the Bolsheviks, 1870–1926* (Boulder, 1992). Russian engineers, however, are not alone in this regard. For example, similar neglect of British engineers was highlighted in R.A. Buchanan, *The Engineers: A History of the Engineering Profession in Britain, 1750–1914* (London, 1989), pp. 11–29.

of the tsarist Ministry of Ways of Communication (MPS) no longer has his staff file (where one would expect to find correspondence by and about him with the minister and deputy ministers), and the archive of the Russian security service claims not to have any relevant papers.[15] Crucially, however, Lomonosov's private papers are now available in England at the Leeds Russian Archive, together with the papers of his second wife Raisa Nikolaevna and their son Iurii Iur'evich (George). These collections include Lomonosov's unpublished diary for 1918–1952 (32 handwritten volumes with about 6,000 folios), much official and private correspondence, copies of his technical books, approximately 7,000 photographs – he was a keen amateur photographer – and his unpublished memoirs. This last manuscript is extraordinarily large with about 15,500 folios grouped in 10 predominantly handwritten volumes: it contains a vast amount of information and has the rare advantage of incorporating hundreds of original typed documents such as letters, telegrams, reports and speeches. Unfortunately, his several hundred technical notebooks have probably perished.[16] Also missing, probably since the 1950s, are his schoolboy diaries, one volume of his post-revolutionary diary and much of his personal correspondence.[17] It is known that his son George 'threw a good deal away' in 1952 and that later Raisa may have lost or destroyed a box of her husband's papers.[18] Nonetheless, when the archival sources in Leeds and the former USSR are combined with other archival documents and contemporary publications, the result is an exceptionally large amount of material, most of which has never been used by historians.[19]

This book aims, as the first full biography of Lomonosov in any language, to provide an overall account and interpretation of his life that encompasses not just his professional activity, but also his childhood, education, politics and, at least to

---

[15]   D.P. Mikheev (Russian Consulate, London)–author, 13 August 2002.

[16]   In 1917 Lomonosov took all his technical notebooks to the USA. In May 1919, when he departed for Soviet Russia, he left them at the Soviet bureau of L.K. Martens in New York. Several weeks later the bureau's papers were seized by the US authorities in the shape of the so-called Lusk Committee (see Chapter 7, this volume), and Lomonosov never saw his books again. They are not among the papers of the Lusk Committee: see New York State Archives, Albany NY, Joint Legislative Committee to Investigate Seditious Activities (henceforth Lusk Committee papers). Nor are they with the papers of the Martens bureau at the Russian Foreign Ministry archive: N.P. Mozzhukhina–author, 6 November 2002. The fate of his technical notebooks from 1919–1952 is completely unknown.

[17]   In a diary entry Lomonosov mentioned his use of his schoolboy diaries for writing his memoirs and recalled having burnt his student diaries at the insistence of his first wife (in about 1898): 'Dnevnik', 2 November 1934: Leeds Russian Archive (LRA), MS 716 (G.V. Lomonossoff Collection), 716.3.21 (henceforth diary entries are referenced by date alone for simplicity and clarity). It is unlikely that he kept a diary during 1898–1917.

[18]   George Lomonossoff–R. Lomonosova, 15 June 1953: LRA, MS 718 (G.G. Lomonossoff Collection), 718.1.1.785; interview with G.M. Browning, 25–26 September 1995.

[19]   The Bibliography lists all the cited archival sources. The three Leeds collections are listed in detail in Aplin, *Catalogue*.

some extent, his private life. Hitherto the factual record has been patchy and often confused, and this attempt to clarify that picture will, one hopes, facilitate future research into the key aspects of his career.

In any case, it is impossible fully to understand many of his most important professional decisions – his choice of career, his reactions to the 1917 Russian revolutions and his departure from Soviet service, to give just a few examples – without reference to his private life and/or politics. Responding to criticism that his private life was too prominent in his memoirs, he himself explained that his text was 'not the history of my time but the history of my life, in which issues of sex and railway equipment always had dominant significance'.[20] Moreover, much of the contemporary criticism of him concerned his personality, behaviour and politics. Was he a pompous, cynical, scheming, ruthless and vindictive character, as many people alleged both before and after 1917? How accurate were the allegations of corruption? Were his politics genuinely radical? To use early Soviet political slang, was he a 'radish' after 1917 – Red on the outside, but White at heart, an anti-Soviet careerist and opportunist?

Given the wealth of available information in this case, this approach also allows us to try to reconstruct the experience of life as a Russian engineer in the first half of the twentieth century – an endeavour that can offer new perspectives on even very familiar issues in the history of revolutionary Russia.[21] True, the feasibility of this more general aim hinges on Lomonosov's typicality as a Russian engineer and the reliability of his private papers – especially his diary and memoirs – as an historical source, whereas the exceptionality of so much of his life helps make him such a strong candidate for a biography. Yet there is a critical nucleus of attitudes and experiences that can be considered typical for his profession in Russia at the time. He was first and foremost a practising engineer, whose career began with the usual challenge of finding a niche. His decision to specialize in railway engineering meant that he joined an industry that was commonly regarded as a driving force of the Russian economy and was among the largest employers of engineers. Further, it was common for Russian and indeed Soviet railway engineers also to have teaching responsibilities and do research. At least on the railways, many engineers with senior managerial roles before 1917 remained influential for some years thereafter.[22] Not least, as technical and political issues became more closely entangled after 1917 and Bolshevik activists started to regard Lomonosov as a politically unreliable 'bourgeois specialist', the question of how to survive in the Soviet milieu became increasingly fraught for him as for most other pre-revolutionary engineers.

---

[20]   'Vospominaniia', vol. 4, preface, dated 6 January 1939. The criticism came from friends who read his first draft.

[21]   A similar concern with life experience, though pursued in a different way, is evident in J. Hellbeck, *Revolution on My Mind: Writing a Diary under Stalin* (Cambridge, MA, 2006).

[22]   See Rowney, *Transition to Technocracy*, pp. 109–12.

As for whether Lomonosov's private papers offer any particular difficulties for historians, we should first address the allegation by a former Russian railway official that these papers were probably 'forged'.[23] All things considered, this possibility seems extremely unlikely, to say the least. Apart from the immense practical difficulty of perpetrating such a huge fraud, the provenance of all three collections is well documented.[24] Furthermore, the texts of the many unpublished official documents in the Leeds collections match copies of the same items in other archives. Likewise many of the photographic negatives in the collections are identical to images published in Lomonosov's books.

Lomonosov's diary, which he kept punctiliously from June 1918 until his death, is a valuable contemporary source for corroborating his memoirs for the period 1918–1930 and the largest single source for his life thereafter. Most entries were about 1–2 folios in length until the late 1930s, and somewhat shorter thereafter. The vast majority were written at the time, but some were transcribed from notes soon afterwards.[25] Intervals of more than a few days between entries are rare. The entries cover both professional and personal issues, including much detail about health matters and his relationships. Many of the entries are remarkably frank, but there are signs of circumspection due to security concerns during his seven years of Soviet service. Unfortunately, the entries tend not to be reflective.

Fundamental for our assessment of the memoirs is the matter of where and when they were written. Memoirs written in Soviet Russia could be published there only with approval from the official censor, and although many of the memoir articles and books published in the early years of Soviet power appear quite candid, censorship brought a marked and lasting deterioration in quality by the mid-1920s; Krylov's memoir, noted above, was an unusual exception. Conversely, memoirs that were published by Russians in the West during the inter-war years and Cold War – especially memoirs of erstwhile Soviet officials – often had a distinct and sometimes sensationalist anti-Soviet tone. Even the best ones, notably S.I. Liberman's *Building Lenin's Russia*, tend to have this unsurprising but potentially misleading characteristic.[26] In this context Lomonosov's memoirs compare favourably, be they his main, effectively complete text in Russian or his shorter, incomplete version in English.[27] He could ignore Soviet censorship because (with the exception of his short account of the February revolution, which he wrote and published in the United States during 1918–1919) he wrote them

[23]   Remark by O.G. Kuprienko in informal conversation with the author, Moscow, 31 March 2003.

[24]   Aplin, *Catalogue*, p. xxxi.

[25]   This point mainly concerns his extended visits to Nazi Germany in the mid- and late 1930s.

[26]   S.I. Liberman, *Building Lenin's Russia* (Chicago, 1945).

[27]   Iu.V. Lomonosov, 'Vospominaniia', vols 1–10: LRA, MS 716.2.1.1–10; the surviving English-language synopses and drafts, covering 1876–1901, are at LRA, MS 716.2.2.1–8. Henceforth the memoirs are referenced by volume and page.

from 1934 while living abroad.[28] True, he sent a publication proposal to Moscow in 1935, but he was neither surprised nor upset to receive a rejection.[29] In any case, he preferred to keep his text unpublished rather than compromise its integrity. In 1938, for example, he rejected an offer from the London publisher Martin Secker & Warburg because it stipulated a limit of just 120,000 words and the removal of lots of 'gossip', including remarks about the Russian royal family's lovers that the publisher feared would cause resentment in 'high quarters'.[30] And while he did try to produce a shorter (though still extremely long) English-language version, he retained much of the 'intimate' material and refused to incorporate an anti-Soviet tone.[31] Only at the end of his life did he begin to have doubts about the considerable amount of what he then termed 'pornography', and especially, given Raisa's vehement objections, about giving so much information about their marriage. These concerns were probably the main reason for his efforts from 1949 to shorten his Russian text and for his abortive bid to rework his story as an historical novel.[32]

If he drafted a discrete statement of motives for writing his memoirs, it has not survived. One can surmise, however, that he was contemplating a detailed memoir by the mid-1920s, to judge by his unsuccessful effort then to move a trunk of his pre-revolutionary papers from Moscow to Germany.[33] He began writing in 1934 due to encouragement from relatives and friends, especially Raisa, and because he had nothing else to do for the moment. Having started, he quickly became engrossed.[34] His original intention was to describe his life until his departure from Soviet service in January 1927; but having reached that point in 1943, and again finding himself at a loose end, he extended his account as far as 1930. He did

---

[28]   'Dnevnik', 2 September 1934; G. Lomonossoff, *Memoirs of the Russian Revolution* (New York, 1919).

[29]   R.N. Lomonosova–Lomonosov, 25 June, 22 July 1935: LRA, MS 717 (R.N. Lomonossoff Collection), 717.2.1.191, 206; 'Dnevnik', 30 October 1935. The proposal was forwarded to *Krasnyi arkhiv*, a journal of Russian revolutionary history, which deemed it as 'devoid of interest'. See also George Lomonossoff–Lomonosova, 29 December 1952: LRA, MS 718.1.1.783.

[30]   Martin Secker & Warburg–Lomonosova, 11 April 1938; Lomonosov–Martin Secker & Warburg, 16 May 1938; Martin Secker & Warburg–Lomonosov, 20 June 1938: LRA, MS 716.4.35.1–3.

[31]   For example: 'Dnevnik', 8 February 1944, 25 May 1949, 20 January 1950, 15 July 1951; George Lomonossoff–Lomonosova, 29 December 1952: LRA, MS 718.1.1.783.

[32]   For example: 'Dnevnik', 16 February 1949, 16 March 1949, 31 March 1951, 14 April 1951; Lomonosov–V. Yarros, 5 March 1951: LRA, MS 716.4.37.1. The novel is mentioned in 'Dnevnik', 17 April 1949, and may be the large unidentified text at LRA, MS 716.5.2.3.

[33]   'Vospominaniia', vol. 9, p. 1634. He could not find a vehicle with which to move the trunk to the station.

[34]   For example: 'Dnevnik', 2, 25 September 1934, 17 April 1949.

not go further because he regarded his life thereafter as uninteresting.[35] His main intended audience was Russian: his English-language version contains additional explanatory material for non-Russian readers.[36] Money was not his motive for producing his main Russian text: he did realize that it could not be published in Russia for many years and that no émigré publisher would accept his pro-Soviet stance.[37] Accordingly, he wrote as much, if not more, for posterity as for his contemporaries. Self-justification was surely a motive because of the controversies surrounding him before and after 1917. Fortunately for historians, he thought that his best tactic was to give extensive factual detail. Editing work was begun at an early stage and continued for the rest of his life. Indeed, during his final two years, partly because he saw this manuscript as his main legacy for his son George, he prioritized its completion over the task of finishing a long-planned textbook.[38]

Lomonosov admitted having 'a love for the art of deception', and Westwood cautions us to treat his remarks with a 'pinch of salt' given his 'tendency to make a good story even better'.[39] But although salt is certainly useful on occasion, one gains the strong impression that systematic deception was not Lomonosov's aim and that generally he wrote honestly. His habit of reporting conversations verbatim, even after several intervening decades, can be explained as a stylistic device that was popular at the time, and these sections should not, of course, be taken literally. Particular care is needed concerning his opinions: always highly competitive, he was deeply opinionated, and his remarks about former colleagues can be extremely harsh. Likewise he displays a tendency to direct responsibility and blame elsewhere when his decisions and conduct may seem particularly reprehensible. That said, an unusual feature of his memoirs is the inclusion of candid, detailed treatment of sensitive topics. For instance, he confesses that he embezzled a large sum of state money in 1922.[40] Similarly, he gives much information about his turbulent private life, including his extra-marital affairs – an account that can be verified to a degree with his diaries, letters and photographs. Whatever this detail suggests about his psychology, it reinforces the impression of his integrity as a memoirist. He does leave important gaps, but overall these seem to be relatively few in number. For example, he says little about his feelings for his first wife, Sof'ia Aleksandrovna (Sonia). Finally, extensive efforts to check assertions, events, dates and names have confirmed that while he did make mistakes, especially with dates and the spelling of names, the level of his factual accuracy is generally high. In short, his

---

[35] For example: 'Dnevnik', 7 December 1934, 24 October 1943, 23 January 1944.

[36] 'Dnevnik', 8 February 1944.

[37] See George Lomonossoff–Lomonosova, 29 December 1952: LRA, MS 718.1.1.783; Lomonosov's attitude towards the Russian émigré community is discussed in Chapter 10.

[38] For example: 'Dnevnik', 3 November 1935, 3 February 1951, 7 March 1951, 3 November 1951.

[39] 'Vospominaniia', vol. 1, p. 276; Westwood, *Soviet Locomotive Technology*, p. 12.

[40] See Chapter 8, this volume.

papers do need to be treated with caution, but they are certainly very valuable as an historical source.

The sequence of the following chapters is broadly chronological. Within these chapters, however, the arrangement of sub-sections is often thematic, and the analytical themes normally prevail where the interests of chronology and analysis diverge. This approach risks some repetition, but seems appropriate for the extremely varied nature of his professional activities and the way that he tried – with ever less success – to keep his professional and private lives separate, not least for reasons of his personal safety. In particular, Chapter 3 discusses what one might call his public life as a professor during 1901–1907, whereas Chapter 4 addresses the hidden world of his revolutionary politics during the same period. No attempt has been made to group the chapters. A case could be made that, for example, the revolutions of 1917 divided his life into two halves, for which a two-part structure could be appropriate. But Lomonosov did not see his story in that light, and as we shall see, an underlying theme of his life is continuity across the revolutionary watershed.

Chapter 1

The Making of a Russian Engineer

Iurii Vladimirovich Lomonosov was born on 24 April 1876 in a country where the rapid expansion of the engineering profession was soon to become a national priority.[1] Imperial Russia's defeat in the Crimean War of 1853–1856 had reinforced suspicions that the empire was economically and technologically backward relative to the other major European powers. This apparent weakness confronted Tsar Alexander II with a political dilemma. The well-being of his empire and dynasty seemed to require a process of socio-economic and technological modernization, yet that process might bring irresistible revolutionary pressure against the autocratic polity. The Tsar accepted this risk, authorizing important initiatives that included the abolition of serfdom, military, legal, financial, local-government and educational reforms, and measures to promote a 'boom' of railway construction. This process stuttered in the 1870s and 1880s under the pressures of recession, war and terrorism, including the murder of the Tsar by revolutionaries in 1881. However, the 1890s brought a state-led drive for rapid industrialization. In this campaign, which coincided with Lomonosov's transition to adulthood and was overseen by Count S.Iu. Vitte, Minister of Finances during 1892–1903, railway-building again served as a principal means to spur and support economic growth.[2] But success required, among other things, the availability of enough competent engineers to design and operate the industrial economy; and because the regime resolved to rely on home-produced talent, the need became urgent to recruit many more of the empire's young men (but not women, as yet) for the engineering profession.

In this context the intensely personal questions of why this youth chose to become an engineer, how he pursued his studies, why he specialized in railway engineering and how he sought to launch his career were at the same time broad matters of pivotal concern for the tsarist regime. Three of these broader themes are highlighted in this chapter. His choice of career pertains to the state's ability to define and achieve national economic priorities: was his decision primarily a response to state policy? Similarly, his experiences as a student illuminate the state's attempt to develop higher technical education instead of employing foreign experts. Not least, his decisions can inform the debate among historians about

---

[1] Certificate, 3 August 1884: Derzhavnii arkhiv m. Kyieva (DAK), f. 18, op. 2, spr. 159, ark. 232.

[2] See, for example, W.E. Mosse, *Alexander II and the Modernization of Russia* (London, 1992); T.H. von Laue, *Sergei Witte and the Industrialization of Russia* (New York, 1963); P. Gatrell, *The Tsarist Economy, 1850–1917* (London, 1986).

whether his social class – the minor rural hereditary nobility – lost their traditional position of influence in the society in the late nineteenth century or maintained it by adapting to the key social, economic and other changes.[3]

Childhood

The young Lomonosov's decision to study transport engineering broke a family tradition. Like so many of their ilk in pre-industrial Russia, the men of this 'ancient noble family' had normally served the state as army officers and then in some civilian capacity in their home area, in this case the north-east corner of the rural Smolensk province in western Russia.[4] This path was followed by the boy's father, Vladimir Grigor'evich Lomonosov (1841–1905): after schooling at the army's elite First Moscow Cadet Corps, he joined the Guards, retired in 1870 as a Major and made a career in law enforcement in and around the small district town of Sychevka, mainly as a Justice of the Peace.[5] In time Vladimir Grigor'evich steered his only son towards a similar life, sending him to the First Moscow Cadet Corps in 1887. But the youngster was desperate to avoid an army career by the time of his graduation in 1893 and entered the Institute of Ways of Communication in St Petersburg to study transport engineering.

This rebellion arose in part from a profound change in family circumstances that reflects to a degree the argument that the late nineteenth-century Russian nobility fell into decline. Lomonosov's childhood echoed that of his forefathers in most respects, not just in the prospect of an army life. Born in the town of Gzhatsk (now Gagarin), he spent his first decade in quiet environments in his family's heartland: he lived initially on isolated estates in the Sychevka district, before moving in 1883 to Sychevka itself with his parents and younger sister Ol'ga.[6]

---

[3]   On these broader issues see especially Balzer, 'Engineering Profession', pp. 55–88; S. Becker, *Nobility and Privilege in Late Imperial Russia* (DeKalb, 1985).

[4]   Quoted from N. Khomiakov, 'Pamiati Vladimira Grigor'evicha Lomonosova', *Dneprovskii vestnik*, 197 (1905); I am very grateful to V.V. Bersenev for this reference. Lomonosov's noble status is recorded in Certificate, 25 October 1884: DAK, f. 18, op. 2, zv. 159, ark. 233. The family was not related to the revered eighteenth-century polymath Mikhail Vasil'evich Lomonosov; on this latter character, who has been called the 'Russian Leonardo', see for example Graham, *Science in Russia and the Soviet Union*, pp. 20–24.

[5]   'Vospominaniia', vol. 1, pp. 16–18, 20–25; Khomiakov, 'Pamiati Vladimira Grigor'evicha Lomonosova'. According to his son, Vladimir Grigor'evich became seriously ill in 1905 and committed suicide in July, possibly to avoid becoming a burden for his family: 'Vospominaniia', vol. 3, pp. 1009–10, 1013, 1015–16, 1072–8. On Justices of the Peace see T.S. Pearson, 'Russian Law and Rural Justice: Activity and Problems of the Russian Justices of the Peace, 1865–1889', *Jahrbücher für Geschichte Osteuropas*, 32/1 (1984): 52–71.

[6]   Certificate, 3 August 1884: DAK, f. 18, op. 2, spr. 159, ark. 232; 'Vospominaniia', vol. 1, pp. 102–4, 109–14.

These surroundings, moreover, had barely changed in decades, as was typical for the Smolensk province. Most roads remained unpaved and the few factories were primarily small creameries and brickworks; not until the construction of the Rzhev–Sychevka–Viaz'ma railway began in the mid-1880s did the industrial world make any significant incursion in the Sychevka district.[7] However, this atmosphere of continuity and stability was contradicted for his family by a financial crisis that unfolded over just two generations. At least in terms of vanished wealth, decline was no myth here by the 1870s. Hence, unlike even his father, the young Lomonosov always faced the prospect of having to earn his living.

Prodigality – often assumed to cause such problems – did play a part, but so too did entailments and the emancipation of the serfs in 1861. Lomonosov's paternal great-grandfather, General G.G. Lomonosov, enjoyed substantial wealth, including many settlements with serfs in the Sychevka district and an estate in the Moscow province. At his death, probably in the 1810s or 1820s, his property passed to his five sons.[8] These inheritances may well have been debt-free, given that the two youngest beneficiaries built themselves estate houses at the settlements of Tatarka and Ekaterinovka.[9] But one brother, Mikhail, was often sued for non-payment of debts from 1832, if not earlier. Moreover, for unknown reasons two of his siblings needed mortgages by 1849, including the future engineer's grandfather Grigorii.[10]

---

[7]   On the Smolensk province in the nineteenth century see D.I. Budaev, *Smolenskaia guberniia v 1861–1917 godakh: Uchebnoe posobie k spetskursu* (Smolensk, 1990) and his *Smolenskaia derevnia v kontse XIX–nachale XX vv.: K voprosu o tempakh, urovne i stepeni razvitiia agrarnogo kapitalizma* (Smolensk, 1972). Contemporary socio-economic data about Sychevka district is in *Spisok naselennykh mest po svedeniiam 1859 goda, [vypusk] XL: Smolenskaia guberniia* (St Petersburg, 1868), especially pp. 3, 370–397; *Sbornik statisticheskikh svedenii po Smolenskoi gubernii, tom 2: Sychevskii uezd* (Smolensk, 1885); and *Materialy dlia otsenki zemel' Smolenskoi gubernii, tom 2: Sychevskii uezd: vypusk I: Estestvenno-istoricheskaia chast'* (Smolensk, 1904). The railway opened in 1888: G.M. Afonina (comp.), *Kratkie svedeniia o razvitii otechestvennykh zheleznykh dorog s 1838 po 1990 g.* (Moscow, 1996), p. 42.

[8]   'Vospominaniia', vol. 1, pp. 17, 56. The surviving Lomonosov files of the Ministry of Finances Main Redemption Administration contain information about several estates and the family: Rossiiskii gosudarstvennyi istoricheskii arkhiv (RGIA), f. 577, op. 37, dd. 3211–3215.

[9]   'Vospominaniia', vol. 1, p. 56. Tatarka and Ekaterinovka, roughly 6 km east of Nikolo-niz and 18 km south-east of Sychevka, were mapped on the *Voenno-tipograficheskaia karta Smolenskoi gubernii* (St Petersburg, 1858–1871), row 11.

[10]   Incomplete loan and repayment details are in RGIA, f. 577, op. 37, d. 3211, ll. 3ob., 10–11ob., 20–ob.; d. 3212, l. 32–ob.; d. 3213, l.27–8; d. 3214, l. 10–11, 16–17ob.; d. 3215, ll. 1, 3–ob., 13–16. See also Genealogy with account of documents about the nobility of the Lomonosovs, May 1868: RGIA, f. 1343, op. 24, d. 2777, l. 3–ob.; Service record for the service and qualities of former Sychevka district marshal of nobility Collegiate Assessor Mikhail Grigor'evich Lomonosov, 25 January 1868: RGIA, f. 1343, op. 24, d. 2777, ll. 6–15.

More financial trouble followed. Wealth dispersal threatened because at least 15 children were born to the five brothers, including three boys and four girls to Grigorii. Unhappily for the family, this storm broke soon with Grigorii's early death in 1851. Not all his seven children could inherit land: the future engineer's father Vladimir was one of two sons who received cash alone. The problem was eased by the generous inclusion of Grigorii's children in the wills of two uncles, which eventually made them part-owners of the Tatarka estate and other property.[11] However, when the Emancipation Act of 1861 forced the nobility to sell land to former serfs, the Lomonosovs were hit by the requirement to use the proceeds to repay any land-backed debts, which left the family with small or non-existent cash receipts.[12] Having moved into debt and an entailment crisis, they suddenly lost most of their regular income sources and potential collateral for loans, and had little capacity to invest in new income-generating ventures.

Could the Lomonosovs rebuild their fortunes? This question was now common for Russia's landed gentry, not least in the Smolensk province, where over half had large debts in 1865. The government hoped that credits and railway construction would revive this province.[13] But this hope proved illusory, certainly for Grigorii's children. Marriage into wealthy families produced better circumstances for three of the girls; the other lived modestly as a teacher and married a Justice of the Peace.[14] Meanwhile all three sons floundered. The eldest, Grigorii, apparently led a spendthrift, scandalous existence and was arrested for embezzlement in the 1890s.[15] Ivan netted a 300,000-ruble dowry in 1881, bought out his co-owners of Tatarka and a 'profitable' family estate at Nikolo-niz, and opened brick- and cheese-making businesses. But 'debauchery' consumed his fortune within six years; by 1903 Tatarka and Nikolo-niz were double-mortgaged, after which much of the land was sold.[16] As for the future engineer's father, Vladimir, he reportedly

―――――――

[11]   'Vospominaniia', vol. 1, p. 56; [Copy of statement] by widow A.A. Lomonosova and P.G. Lomonosov, 14 December 1855: RGIA, f. 577, op. 37, d. 3213, l. 7; Smolensk Criminal and Civil Court–Chair, Smolensk Province Office for Peasant Affairs, 12 December 1868: RGIA, f. 577, op. 37, d. 3215, ll. 13–16.

[12]   See the Lomonosov files of the Main Redemption Administration, RGIA, f. 577, op. 37, especially d. 3211, l. 24–ob.; d. 3212, l. 32–ob.; d. 3213, l. 31–ob.; and d. 3214, l. 20–ob. On the Emancipation see D. Moon, *The Abolition of Serfdom in Russia* (Harlow, 2001); the reform's impact on the Smolensk province is the focus of D.I. Budaev, *Krest'ianskaia reforma 1861 goda v Smolenskoi gubernii (K voprosu o realizatsii 'Polozhenii 19 fevralia')* (Smolensk, 1967).

[13]   *Spisok naselennykh mest po svedeniiam 1859 goda*, p. LIII.

[14]   'Vospominaniia', vol. 1, pp. 56, 60.

[15]   'Vospominaniia', vol. 1, pp. 56, 771.

[16]   'Vospominaniia', vol. 1, pp. 54–5; documents in RGIA, f. 593, op. 19: Loan pro forma, [November 1887]: d. 1831, l. 1; Resolution No.1, Smolensk division, State Noble Land Bank, 8 January 1904: d. 1832, ll. 3–4; Smolensk division, State Noble Land Bank– M.G. Lomonosov, 4 December 1891: d. 1833, l. 10; Loan pro forma, [March 1902]: d. 1833, l. 1; Smolensk division, State Noble Land Bank–State Noble Land Bank, 10 May

squandered his money while an officer, together with the dowry from his marriage in 1875 to Mariia Fedorovna Pegelau (1856–1921), a local noblewoman. Indeed, his decision to mortgage his two small estates at Borova and Podseevka in 1876 suggests that his marriage did not restore his finances. By the mid-1880s, moreover, he had no serious prospect of repaying his debts. His monthly salary was just 165 rubles, augmented only by his army pension and perhaps some rent from his two estates; significant aid from relatives was unlikely; modest household economies could not yield much money; and his wife planned to pass her own expected small inheritance to their daughter, trusting that their son could fend for himself. Vladimir Grigor'evich rejected corruption, the bureaucracy's notorious remedy for such personal difficulties, to judge by an obituary that praised him as 'a moral social force' and stated: 'in the modest environment where he worked an atmosphere of consciousness of duty and searching for the truth was created around him ... he earned himself a deserved reputation as not just a guardian of the law but also, more importantly, a searcher for truth in justice'.[17] By 1890, therefore, his estates were double-mortgaged, and his son needed a career.[18]

The related family debate in 1886 about whether to give the boy a civilian or a military education is striking for the enduring influence of old social sentiments. As yet the Lomonosovs' financial woes had scarcely affected their public standing and social attitudes. For instance, Lomonosov claims in his memoirs that his mother was regarded as Sychevka's de facto first lady.[19] Equally, many Lomonosovs scorned her humbler roots. Her father, Fedor Aleksandrovich Pegelau (1814–1892), was born a serf and sent to the army, where he earned ennoblement through promotion to the rank of Major. Her mother's family, the Loveikos, were nobles, but owned just two small estates; and their military tradition was tainted, for some Lomonosovs, by the friendship of Mariia Fedorovna's grandfather Loveiko with officer-participants in the failed Decembrist revolt against Tsar Nicholas I in December 1825.[20] These deep-rooted attitudes help to explain the prevailing assumption among Lomonosovs that the boy would enter state service, preferably

---

1902: d. 1834, l. 1. Sales information is in the latter file, such as: I.G. Lomonosov–Smolensk division, State Noble Land Bank, 30 September 1911: d. 1834, l. 48.

[17] Khomiakov, 'Pamiati Vladimira Grigor'evicha Lomonosova'.

[18] See 'Vospominaniia', vol. 1, p. 31, 56, 121, 209, 500, 578 and documents in RGIA, f. 593, op. 19, d. 1836: Loan 9231: l. 1–ob.; Loan 9230: l. 2–ob.; Request to Smolensk division, State Noble Land Bank, 26 May 1890: l. 4; List of peasant cases for Sychevka district, No.59: Estate at Podseevka wasteland: ll. 25–8; V.G. Lomonosov–Council, State Noble Land Bank, 2 September 1890. In 1893 Vladimir Grigor'evich reportedly affirmed that his salary was his only income: 'Vospominaniia', vol. 1, p. 372. However, since he was pleading impecunity to persuade his son to remain at the Cadet Corps, this may be an exaggeration.

[19] 'Vospominaniia', vol. 1, p. 126.

[20] 'Vospominaniia', vol. 1, pp. 1, 19, 30–31, 71–2, 252–7, 433, 601. On the Decembrists see, for example, A.G. Mazour, *The First Russian Revolution 1825: The Decembrist Movement* (Stanford, 1937).

in the army. Interestingly, Mariia Fedorovna and some of her relatives wanted him to attend a local civilian school – a sign, presumably, that they hoped for a civilian career. But Vladimir Grigor'evich successfully advocated his military Alma Mater in Moscow with the backing of most Lomonosovs and at least one Pegelau. He highlighted the financial advantage that the state would fund a cadet education because his army commission had lasted longer than ten years. Meanwhile his supporters stressed tradition and values: the Lomonosovs had always served in the military, civilian service was inappropriate, and a cadet corps would instil good discipline and self-discipline. A family friend, the district police officer, added some solidly traditional reasoning: a military appearance was imposing, it showed a well-born servant of the tsar, and the government disliked university students.[21]

The decision for a cadet education represented an emotive reassertion of army service as a sacred duty. It did not prioritize salary – officers were poorly paid – but the patriotic passion with which Vladimir Grigor'evich endorsed the army life. His attitude can also be sensed from his pride in his uniform, his habit of wearing his military coat for making legal judgements, and especially from this posthumous tribute:

> [Vladimir Grigor'evich was] a modest peace-loving person who was first and foremost a soldier to the root of his bones, in the best sense of the phrase … Military traditions, and a sense of honour, duty and discipline were inherent in him right up to the last days of his life. Only those close to him knew how he suffered as a powerless witness of the terrible indications of the decline of those principles on which he was brought up and which he professed … His feeling of love for his country and consciousness of a national duty of service to it were the guiding feeling of his whole life.[22]

Indeed, he was dissatisfied for his son to be offered merely a place at the nearby Polotsk Cadet Corps, where the boy's maternal uncles had studied. He secured permission for his boy to take the entry exams for the First Moscow Cadet Corps a year later than usual, and arranged coaching. Aged 11, Iurii Vladimirovich passed these tests in August 1887 and entered the school's second year.[23]

Founded in the eighteenth century with the military-political mission of preparing sons of the nobility for serving 'throne and Fatherland' as army officers, the Cadet Corps system remained steeped in exclusive tradition.[24] It was still permeated with the old aim of inculcating patriotic devotion to the throne as the

---

[21] 'Vospominaniia', vol. 1, pp. 21, 225–7.

[22] Khomiakov, 'Pamiati Vladimira Grigor'evicha Lomonosova'.

[23] 'Vospominaniia', vol. 1, pp. 227–8, 284–90.

[24] For a history of these schools see J.W. Steinberg, 'Cadet Corps Academies', paper presented to the 35th National Convention of the American Association for the Advancement of Slavic Studies, Toronto, November 2003. For a contemporary comparison see A.A. Ignat'ev's recollection of Kiev's corps: *Piat'desiat' let v stroiu* (Stalingrad, 1941),

political ethos of a first-rate officer corps. Because the admission criteria were very selective, most pupils in the late 1880s were either sons of wealthy hereditary nobles, or of men who had served as officers for over 10 years, been decorated for bravery, or been killed or wounded as senior officers. They were treated as soldiers, with parade-ground drill, applied military training and occasional army exercises. They had about 30 hours of academic tuition per week, with special emphasis on mathematics and languages.

This design appears to have been successful in terms of Lomonosov's academic achievements and general conduct. His graduation transcript lists 18 subjects and an average mark of 10.56 (out of 12). Also, in his final year he was appointed as an acting deputy non-commissioned officer, and won a prize 'for excellent conduct and very good success in sciences'. Interestingly, perhaps to help reinforce the notion that his politics were radical, he later recalled his character and behaviour in somewhat different terms: he described himself in his memoirs as 'obstinate by nature' and very 'wilful' except in personal matters, and claimed that his behaviour at school was initially poor, matching the standard view of younger cadets that parents and the school authorities were 'the enemy'.[25] Clearly, his conduct either improved by his final year or was better than he cared to remember.

The school failed with him, however, in its military-political mission. If he ever had any enthusiasm for an army career, his school experiences doused it. As for politics, his initial thinking had been shaped by relatives. By the mid-1880s he was aware of two political extremes among them: his late grandfather Loveiko's friendship with Decembrist rebels, and the traditionalism of his ex-serf grandfather Pegelau, who reportedly objected to the Emancipation by throwing the Tsar's portrait in a cesspit. But the boy was evidently more impressed by what he called the liberal reformist values of his parents. Also relevant, perhaps, was the experience of copywriting for his father, which gave him an intimate knowledge of rural life, including peasant poverty. Once at the Cadet Corps, however, he acquired conservative opinions from a prevailing 'cult of the tsar'. By 1891, when Alexander III visited the school, Lomonosov 'would have obeyed an order from the tsar to throw himself into a fire'.[26] Yet by his account he turned to radical republicanism during 1891–1892. The stimulus was an incident during the royal visit. Apparently the Tsar refused to visit the neighbouring Fourth Moscow Cadet Corps because it had mutinied in 1884, and in the ensuing demonstration his carriage hit and killed a cadet. Lomonosov's reactions included anger, religious

---

pp. 30–40. The First Moscow Cadet Corps was located in part of the Catherine Palace in Moscow's Lefortovo district.

[25] Certificate, 21 June 1893: DAK, f. 18, op. 2, spr. 159, ark. 235; Director, First Moscow Cadet Corps–Director, IPS, 30 November 1893: Tsentral'nyi gosudarstvennyi istoricheskii archiv Sankt-Peterburga (TsGIA SPb), f. 381, op. 1, d. 580, l. 600; 'Vospominaniia', vol. 1, especially pp. 44, 104, 130, 225–6, 228, 276, 294, 340, 352, 357, 369, 404, 417–18, 536, 557, 574, 592–600.

[26] 'Vospominaniia', vol. 1, pp. 23–4, 214, 256, 267, 438–9, 469.

fervour, greater antipathy to army service, and friendships with two radically minded cadets. By 1892:

> I was unconditionally opposed to autocracy, and dreamed of a democratic republic, but I considered freedom of conscience and of the press as the main things. Any restrictions on rights linked with origins, religion, gender or nationality seemed outrageous to me. I considered it as axiomatic that God created all people equal and free.[27]

Interpreting his fellow cadet's death as political murder, he began to reason that only violence would bring radical change. On the anniversary of the Decembrist revolt in 1892 he reportedly wrote in his diary: 'I believe that this day will always remain a memorial day for the Russian people: it was the first armed protest against autocracy'; and, he asserted, if the people rebelled in an organized manner, 'I will be the first to join that sacred army, which with sword in hand will defend the inalienable rights of the people'.[28]

Recalling these developments in retirement, Lomonosov was trying not simply to describe his youth but also to address accusations of political opportunism after 1917. So can we believe him here? It seems probable that he did witness a tragedy. Press coverage of this royal visit to Moscow was extensive, but the descriptions of the school excursion were cursory, no protests were noted, and most newspapers mentioned only the first and second corps. More informative was the liberal *Russkie vedomosti*, which reported that the Tsar headed for the fourth corps, where a choir was singing a hymn, but that he did not stop.[29] In short, a protest and accident probably did occur, but the reports were censored. As for Lomonosov's reaction, it is plausible that this event began the process of political radicalization that would lead him to risk his life for the revolution in 1905. At the same time, it is possible that his terminology reflects later experiences and that his political outlook as of 1891–1892 was less strident than he later claimed.

There may be some support for the latter contention in the uncritical nature of his attitude towards the sexual mores associated with class differences during the 1890s. In his relations with women at that time he seems to have been very conscious of his social standing as a member of the nobility. Like perhaps most Russian men of his class in those years, he carefully differentiated between noblewomen and non-noblewomen concerning heterosexual physical contacts. Essentially, he regarded casual intercourse as prohibited with noblewomen, but permissible with, for example, servants. By his own account he lost his virginity with a servant at the age of 14, yet did not experience his 'first kiss' for another two years (at a

---

[27] 'Vospominaniia', vol. 1, p. 563.

[28] 'Vospominaniia', vol. 1, pp. 466–73, 493, 563–4.

[29] See, in particular, *Pravitel'stvennyi vestnik*, 21 May 1891, p. 2 and 22 May 1891, p. 1; *Novoe vremia*, 21 May 1891, pp. 1–2; 23 May 1981, p. 1; *Russkie vedomosti*, 20 May 1891, p. 3.

Corps ball); his 'first love' (for a landowner's daughter) followed in 1892–1893. During 1893 he reportedly felt 'love' for four more women, including two of his cousins, with apparently no thought of a pre-marital physical relationship. Similar dualism characterizes his assertion that he experienced a moral 'resurrection' in the early 1890s, for he had sexual encounters with several servants during the same period.[30]

Given that sexual desire would shape several of his more important professional decisions as well as his personal life, it is worth adding that two experiences contributed to these attitudes. One was his father's conduct. Vladimir Grigor'evich had extra-marital affairs with women of his own class as well as servants; the most important case was a long-term relationship with a family servant called Nast'ia, which was known to his wife and son. The boy was aware of his mother's consequent unhappiness, yet was himself strongly attracted to Nast'ia and had his own affair with her. At school, peer pressure was relevant. For instance, he was teased for a (short-lived) resolution in 1892 to refrain from pre-marital sex; why he even mentioned this decision to his comrades is not explained.[31]

It was through his unrequited pursuit of a 24-year-old noblewoman, E.N. Kletnova, in the winter of 1892–1893 that the 16-year-old Lomonosov discovered engineering. With his school graduation impending, his future was on the family agenda, and three options were discussed: the army (with only a Guards commission or artillery school considered as socially acceptable); the priesthood in the Russian Orthodox Church; and a university education. He resisted the Guards option by arguing that he did not want the social stigma of being an officer who could not afford the extravagant lifestyle, and he pleaded that his grasp of mathematics was inadequate for the artillery.[32] His parents were horrified about the priesthood, but he adds – strangely, given his strong sexual appetite – that he rejected this idea because of reports of clerical debauchery and scandal; indeed, for reasons not specified but which presumably included his interest in Marxism, he had little time for religion thereafter.[33] By early 1893 a university course beckoned both as a family compromise and because he admired his relative N.I. Kareev, who was a professor of history at St Petersburg University.[34] At this juncture, however, Kletnova's father intervened. He sketched a patriotic, practical vision of the role of engineers: their work was an indisputable 'blessing' for Russia, and with the start of electrification the future clearly belonged to electricity. Lomonosov ignored his

---

[30] See 'Vospominaniia', vol. 1, pp. 404–5, 451, 465, 478, 553, 566, 580, 608–9, 613.

[31] 'Vospominaniia', vol. 1, pp. 30–31, 96–7, 519. More generally on sexual mores at this time see L. Engelstein, *The Keys to Happiness: Sex and the Search for Modernity in Fin-de-Siècle Russia* (Ithaca, 1992).

[32] 'Vospominaniia', vol. 1, p. 542.

[33] 'Vospominaniia', vol. 1, pp. 189, 441, 458, 472–3, 493–4, 564; interview with G.M. Browning, 25–6 September 1995. On his gradual abandonment of religion see, for instance, 'Vospominaniia', vol. 2, pp. 457, 493–4, 777, 817.

[34] 'Vospominaniia', vol. 1, pp. 226, 458, 493.

own doubts about his mathematical ability and accepted Kletnov's offer of help
to assess the army's electro-technical course and the civilian Electro-Technical
Institute in St Petersburg. For comparison two other civilian technical institutes
in the capital were also investigated. These were the Forestry Institute, which
produced engineers for Russia's vital timber industry, and the Institute of Ways of
Communication (IPS), which educated transport engineers for state service under
the Ministry of Ways of Communication (MPS).[35]

If, then, an idealized vision of engineering captured his imagination, his specific
preference to join the IPS was conditioned not by love, visionary patriotism, state
policy or even cost, but by social prejudices once again. The IPS was an excellent
compromise for all concerned. The parents rejoiced because, although its tuition
costs were high, the institute was very prestigious and exclusive. Thwarted in his
desire for his son to have a military career, Vladimir Grigor'evich was consoled
by the institute's roots in military engineering, its military-style uniform and the
fact that its graduates normally entered state service, albeit with the civilian MPS
rather than the army.[36] For the youngster, it enabled avoidance of the army without
alienating his family, whose support was needed for the costs. It also, perhaps,
implied social ambition on his part, for he noted the view of his old home tutor that
the other institutes were insufficiently grand for the Lomonosov name, position
and upbringing. And finally, having applied to the IPS and the Forestry Institute,
he was influenced by appearances. One was clean and well ordered, like his school,
and had admitted cadets; the other appeared dirty and unfamiliar with cadets.
Successful in the entrance examinations for both places, he chose the IPS.[37]

This decision marked a considerable personal achievement for him. He held
his own against family tradition, his father and, not least, the intense competition
for entry to the IPS. His success meant that he would gain an elite engineering
education with the prospect of earning the exclusive title of 'engineer of ways of
communication' (*inzhener putei soobshcheniia*, or colloquially *puteets*). In due
course he could expect to join the small community of IPS alumni who dominated
the management of Russia's transport sector, including privately owned as well as
state-owned railways. Kletnov did have a point about the importance of electricity,
but Lomonosov was joining the industry that drove the nineteenth-century's so-
called second industrial revolution and that lay at the heart of Russia's state-led
industrialization drive in the 1890s.

---

    [35] 'Vospominaniia', vol. 1, pp. 564–71, 574. No reference is made to any other
technical institutes. On Russian electrification in the 1880s and 1890s see Coopersmith,
Electrification of Russia, pp. 8–98.

    [36] An IPS-educated engineer who wanted to work on a private railway had to obtain a
temporary release from state service.

    [37] 'Vospominaniia', vol. 1, pp. 571–6, 602–4, 616–27; he requested permission to
take the IPS exams in a letter of 5 June 1893: TsGIA SPb, f. 381, op. 1, d. 580, l. 598.

An exclusive education

Conceived to address modern Russia's perennial aim of 'catching up' with the West, Count Sergei Vitte's strategy for rapid industrialization had profound implications for civilian higher technical education in the 1890s.[38] Hitherto this sector had comprised only a handful of institutes. Some were specialized, like the IPS: they operated under the auspices of the relevant government ministry, and their graduates normally worked in that ministry's domain. The others had a general character, notably the St Petersburg Technological Institute (founded 1828) and Moscow Higher Technical School (1868), and came under the Education Department of the Ministry of Finances. This system produced about 400 graduates in 1890, but Vitte's strategy needed more. In line with its aim of economic self-sufficiency the government decided not to import foreign expertise. Existing institutes were therefore pressured to increase admissions, and new institutes were established, including one in Moscow for transport engineers and the three polytechnic institutes that opened at Warsaw, Kiev and St Petersburg during 1898–1902 to teach a wide variety of disciplines. How, then, was the IPS affected by these developments? How far did IPS staff appreciate the need to look beyond quantitative targets at the capacity of the graduates to adapt to technological change? And how did Lomonosov fare there?

The 1890s were difficult years for the IPS.[39] The government's pressure for it to expand to support the programme of railway-building was at odds with the institute's unashamedly exclusive intellectual and social traditions. Founded in 1809 as the Institute of the Corps of Engineers of Ways of Communication, the IPS supported research as well as teaching, and its curriculum had a theoretical rather than practical bias, although railway officials did serve as tutors. It had kept student numbers low and, apart from an interlude from the mid-1860s to 1882, had largely restricted admission to sons of the nobility and military personnel. Accordingly there were only about 70 graduates per year in the late 1880s, or

---

[38] On technical education see Balzer, 'Educating Engineers'.

[39] The main institutional histories are: S.M. Zhitkov (comp.), *Institut inzhenerov putei soobshcheniia Imperatora Aleksandra I: Istoricheskii ocherk* (St Petersburg, 1899); A.M. Larionov (comp.), *Istoriia Instituta inzhenerov putei soobshcheniia Imperatora Aleksandra I-go za pervoe stoletie ego sushchestvovaniia* (St Petersburg, 1910); S.M. Grishukov, *Leningradskii ordena Lenina institut inzhenerov zheleznodorozhnogo transporta: kratkii ocherk* (Leningrad, 1959); and I.V. Veviorovskii, M.I. Voronin, P.L. Klauz et al. (eds), *Leningradskii ordena Lenina Institut inzhenerov zheleznodorozhnogo transporta imeni akademika V.N. Obraztsova, 1809–1959* (Moscow, 1960). See also *Kratkii istoricheskii ocherk uchebnykh zavedenii vedomstva putei soobshcheniia* (St Petersburg, 1900); and E.Ia. Kraskovskii and M.M. Uzdin (eds), *Istoriia zheleznodorozhnogo transporta Rossii, tom 1: 1836–1917 gg.* (St Petersburg, 1994), pp. 120–127.

some 17 per cent of Russia's annual cohort of engineers.[40] Perhaps unsurprisingly, the IPS attempted to appease Vitte without compromising its commitment to high standards. Its student population leapt from 182 in the 1889/1890 academic session to nearly 900 in 1897/1898. The entrance exams survived, but some restrictions from the 1880s were revoked, notably a requirement for two years' prior study elsewhere; thus, students would enter directly to a five-year course. Larger numbers of non-nobles were admitted, and as the intake grew, so did the number of applications – a sign of the empire's capacity to heed the regime's demand for more engineers. However, the cost for the IPS was serious. With the entrance competition remaining intense, the admission process gained a reputation for corruption. The commitment to preserving high academic standards caused such tension that the institute's ruling Council tried to restrain the expansion in the mid-1890s. Plans were confounded repeatedly: for instance, a building programme of 1892 for 410 students was outdated by its completion in 1895. The rates of student withdrawals, exclusions and deferrals increased.[41] Also, the political nature of the student body became more varied. Lomonosov records that adherents to the elitist traditions comprised only about one-third of his cohort, being known as 'ultra students' (*leib-studenty*), and that there were roughly 30 students, himself included, who considered themselves anti-elitist and radical.[42] This account is echoed by S.P. Timoshenko, who also studied at the IPS at about this time and was later called the 'father of engineering mechanics' in the United States, and it matches the assumption of contemporaries and historians that most of Russia's engineers were politically centrist by 1900.[43]

Ironically, there was less change in the curriculum than perhaps was needed. No great pressure came from the MPS for radical upheaval: presumably the ministry merely wanted more graduates of the same type and quality. The IPS could therefore continue to provide a firmly theoretical specialist education, devoting the first two years to general subjects and concentrating on transport matters thereafter. Following slight changes in 1896, the first two years covered physics, chemistry, advanced mathematics, geometry, mechanics (basic, theoretical and construction), geology, surveying, statics, construction arts, architecture and – a staple of tsarist

---

[40]   *Otchet o sostoianii Instituta inzhenerov putei soobshcheniia Imperatora Aleksandra I, s 1-go iiunia 1893 goda po 1-e iiunia 1894 goda* (St Petersburg, 1895), p. 5.

[41]   See Zhitkov, *Institut inzhenerov putei soobshcheniia*, pp. 300, 309–23, 381–8; Larionov, *Istoriia Instituta inzhenerov putei soobshcheniia*, pp. 238, 259, 270–271, 275, 302; *Otchet o sostoianii Instituta s 1-go iiunia 1893 goda po 1-e iiunia 1894 goda*, p. 4; *Otchet o sostoianii Instituta inzhenerov putei soobshcheniia Imperatora Aleksandra I, s 1-go iiunia 1897 goda po 1-e iiunia 1898 goda* (St Petersburg, 1899), pp. 23–4.

[42]   For instance, 'Vospominaniia', vol. 1, pp. 650, 654, 656–60; vol. 2, p. 86. See also Balzer, 'Educating Engineers', pp. 409–10.

[43]   S.P. Timoshenko, *As I Remember: The Autobiography of Stephen P. Timoshenko* (Princeton, 1968), pp. 39–41, 46–51; for the quotation see the foreword by J.M. Gere and D.H. Young, pp. v–vi.

higher education – theology. A modern foreign language – French, German or English – was compulsory in the first year and optional in the second, but this did not quieten the staff's lament that the students lacked the linguistic knowledge to follow the foreign technical press.[44] Thereafter some of these subjects were continued, such as theoretical and construction mechanics, geology and architecture. The transport courses included steam machines, railways, railway operations, locomotives, roads, port installations, waterways, bridges and hydraulic machinery. There were also courses on political economy, accounting procedures and jurisprudence.[45] Yet there was probably a need at least to review the balance of the theoretical and practical. Certainly Lomonosov had criticisms. In particular, he thought that his first-year tutor for technical drawing was more interested in artistic appearance than 'content' (by which presumably Lomonosov meant accuracy) and that the professor of railway design, Ia.N. Gordeenko, wanted just the regurgitation of his book. Courses were not coordinated, in Lomonosov's opinion, with the overarching aim of creating an engineer, and the practical courses were poorly organized and supervised. Such allegations may be tinted by Lomonosov's later interest in curricular reform for railway engineering, but again Timoshenko echoes them, and a debate about these matters did develop by 1909.[46]

The changes to the admission process had the most significance for the student Lomonosov. Under the previous system he would have needed two years of prior study at, for example, a university – a prospect that would surely have provoked insuperable family opposition. Entry remained difficult, but was unquestionably easier for him than hitherto: the application success rate of 24.8 per cent in 1893 compared with 12.8 per cent as recently as 1891.[47] Also, former cadets like him became more common in the institute.[48] By contrast, the teaching during his time at the institute differed little from earlier years, it would seem. Nor did the building work disrupt him unduly, to judge by the virtual absence of comment in

[44] For example: M. Gersevanov, *Institut Inzhenerov putei soobshcheniia Imperatora Aleksandra I, v period 1890–1896 g.* (St Petersburg, 1896), p. 5. The languages are specified at, for example, Larionov, *Istoriia Instituta*, p. 252.

[45] *Otchet o sostoianii Instituta s 1-go iiunia 1897 goda po 1-e iiunia 1898 goda*, pp. 25–6; Zhitkov, *Institut inzhenerov putei soobshcheniia*, pp. 304–9, 321–3.

[46] 'Vospominaniia', vol. 1, pp. 638, 640, 682, 711–21, 747–8; vol. 2, pp. 73, 124–5, 255, 276; Timoshenko, *As I Remember*, pp. 28–38. On Gordeenko see Veviorovskii et al., *Leningradskii ordena Lenina Institut*, p. 89. Among important contributions to the later debate were V. Nagrodskii, *Chego Institut Inzhenerov putei soobshcheniia ne daet svoim pitomtsam* (St Petersburg, 1909); Ne professor, 'Institutskii krizis', *Izvestiia Sobraniia inzhenerov putei soobshcheniia*, 5 (1912): 2–3; and A. Lavrov, 'Po voprosu o vvedenii v Institute Inzhenerov Putei Soobshcheniia spetsializatsii', *Izvestiia Sobraniia inzhenerov putei soobshcheniia*, 35 (1914): 570–575.

[47] See Zhitkov, *Institut inzhenerov putei soobshcheniia*, p. 309.

[48] *Otchet o sostoianii Instituta s 1-go iiunia 1893 goda po 1-e iiunia 1894 goda*, p. 3; *Otchet o sostoianii Instituta s 1-go iiunia 1897 goda po 1-e iiunia 1898 goda*, p. 24.

the memoirs, although this lifelong gastronome did display keen interest in the new canteen: his account of a student campaign for larger portions fills four pages of his memoirs.[49]

His academic performance was respectable. He certainly had his share of missed classes and inadequate preparation.[50] Nonetheless he was in the top third of his cohort at graduation, ranked forty-first. His result was surpassed by his friend G.D. Dubelir (twenty-ninth) and by a student with whom he would cross swords in 1917–1918, Count S.I. Shulenburg (thirty-fourth). On the other hand, he was awarded one of the two prizes, the L.A. Erakov prize, for his final-year design project in applied mechanics; his design for a locomotive is noteworthy for incorporating the latest thinking of the French engineer Alfred de Glehn, whose work on enhancing the efficiency of steam locomotives was becoming popular in many countries, including Russia.[51]

An intangible but significant and enduring product of his IPS studies was his sense of identity as an 'engineer of ways of communication', or colloquially a *puteets*. This identity was partly a function of his automatic transfer to MPS service at graduation and corresponding entry to the coterie of IPS graduates who dominated the management of Russia's transport sector. Strangely, for reasons not stated, he evidently did not cement this connection by joining the Assembly of Engineers of Ways of Communication (*Sobranie inzhenerov putei soobshcheniia*) – the professional body that doubled as an IPS alumni association, published a respected journal on transport engineering and economics, and also ran seminars and social events. But he did use the identity of *puteets* for career networking during, for instance, research visits to individual railways and factories. Many of these contacts lasted for years and often involved practical help. Antipathies, however, were equally important for his career. In particular, there was mutual resentment between MPS *puteitsy* and engineers who came to the railways from institutes outside the MPS hierarchy, such as the St Petersburg Technological Institute, and who were termed 'production engineers' (*tekhnologi*). Unfortunately for Lomonosov, *puteitsy* usually left questions of locomotive design to *tekhnologi*;

---

[49]   'Vospominaniia', vol. 2, pp. 81–5.

[50]   For instance: 'Vospominaniia', vol. 1, pp. 659–60, 698, 758, 808–9; vol. 2, pp. 108, 174, 190–191, 287, 291.

[51]   *Otchet o sostoianii Instituta s 1-go iiunia 1897 goda po 1-e iiunia 1898 goda*, pp. 30, 33–4; 'Vospominaniia', vol. 2, pp. 258–9. He misremembered the prize as the Borodin prize: ibid., p. 282. His actual Borodin prize, won in 1911, is discussed in Chapter 5, this volume. Dubelir (1874–1942), who specialized in electric railways, had a varied career that included teaching at the IPS from 1916. He died during the siege of Leningrad: see his file at DAK, f. 18, op. 2, spr. 81 and A.I. Melua, *Inzhenery Sankt-Peterburga: Entsiklopediia* (St Petersburg and Moscow, 1996), p. 231.

the latter would duly ridicule him as a theory-obsessed dilettante, and this demarcation quarrel would blight his career.[52]

It was during his student years that he acquired his abiding interest in improving the operating efficiency of railways. He first contemplated this matter during his 1896 summer work placement as an assistant land surveyor, which took him to a construction project in the distant Samara province. There he realized that a railway's operating costs would be reduced if decisions about the location and design of the route and structures were based on information about the likely operating requirements – a connection that, so far as he could see, was not being made. Apparently his insight flowed from his need to choose between two options for bridging a ravine. Such decisions were normally resolved according to the engineer's judgement, based on experience accumulated over years. But Lomonosov lacked such experience, and wanted theory-based guidance about whether to prioritize the minimum cost of construction or the minimum cost of operation (probable fuel expenditure, wear-and-tear on locomotives and rolling stock, etc.). Neither his books nor supervisor could help, he felt:

> This conference in a Samara hut had decisive meaning for my engineering career.
> I understood that my superiors themselves had little grasp of theory, and that in
> order to become a genuinely good surveyor, I still had to do a lot of studying.
> I dimly began to guess that all these cursed questions about radii and gradients
> were somehow linked with the work of steam locomotives.[53]

In other words, as he continued his studies he began to believe not simply that surveyors required theoretical criteria, but that such criteria had to be based on, among other things, scientific knowledge about how locomotives functioned.[54]

This reasoning may well have been shaped by the work of the American engineer A.M. Wellington (1847–1895). The author of a much used textbook on railway surveying, Wellington was a leading figure in the growing community of European and American railway engineers who were keen to develop systematic, cost-efficient approaches to the design and operation of railways. Famously

---

[52] He encountered this tension on his first day of employment after graduation, at Khar'kov: 'Vospominaniia', vol. 2, pp. 302–3. More generally, see Balzer, 'Engineering Profession', pp. 71–5.

[53] See 'Vospominaniia', vol. 2, pp. 164–5, 189, 191, 276. He had in mind books by such professors as Ia.N. Gordeenko and A.D. Romanov: 'Vospominaniia', vol. 2, pp. 164–5, 255; slightly later editions of their textbooks are, respectively, *Kurs zheleznykh dorog po programme utverzhdennoi G. Ministrom putei soobshcheniia 4-go aprelia 1888 goda dlia ekzamena na zvanie tekhnika putei soobshcheniia* (St Petersburg, 1898) (with a supplement called *Dopolnitel'nyi kurs zh.d. dlia studentov–inzhenerov putei soobshcheniia*) and *Parovozy: Kurs, prepodavaemyi v Institute Inzhenerov Putei Soobshcheniia Imperatora Aleksandra I*, 2nd edn (St Petersburg, 1900).

[54] 'Vospominaniia', vol. 1, p. 818; vol. 2, pp. 189, 254–5.

defining engineering as 'the art of doing that well with one dollar, which any bungler can do with two after a fashion', he aimed to cut waste in railway construction and operation by replacing traditional trial-and-error design methods with new techniques founded on scientific theory.[55] Furthermore, he regarded the fixed assets (infrastructure), locomotives and rolling stock of a railway as a single system that encompassed mechanics and economics. In other words, to use a metaphor that Lomonosov would employ later, he treated the railway like a factory, the output of which was the movement of freight and people. In fact, engineers like Wellington were developing the concept of systems or network analysis, which now lies at the heart of engineering theory and practice. Other engineers, however, remained reluctant to base decisions on what they saw as abstract theory: they valued practical experience and techniques that were at least tried-and-tested, if not ideal in theory. These differences could be especially troublesome when, as in continental Europe, the advocates of change relied heavily on mathematical techniques. The same conflict would hinder Lomonosov's effort to link management, economics, mechanics and design on Russia's railways.[56]

Against this background two developments steered Lomonosov to his lifelong fascination with what became known as 'traction theory' – scientific theory about the many technical processes involved in the operation of a railway locomotive. He grew very uneasy about his initial idea to become a railway surveyor. Although he insisted even late in life that surveying was his 'vocation', he claims that he was deterred by the corruption prevalent among private contractors – an issue which seems to have hardened his aversion to private enterprise. Also, he wished to avoid any form of dependence on his prospective father-in-law: one of Moscow's leading railway-building contractors was A.I. Antonovich, for whom Lomonosov worked as a surveyor in summer 1895, and to whose second daughter Sof'ia Aleksandrovna (Sonia) he became engaged in the autumn of the same year.[57] Meanwhile, Lomonosov became aware that Russia had two schools of thought about locomotive design, which reflected the *puteets–tekhnolog* dichotomy noted above. One was associated with IPS *puteitsy* and used theoretical formulae developed from Prussian locomotive tests. Initially promoted in Russia by L.A. Erakov (1839–1885), this approach was championed by such authorities as the IPS professor of locomotives, A.D. Romanov (1853–1920?), and the Moscow–Kazan' Railway's superintendent for locomotives and rolling stock, E.E. Nol'tein (1854–

[55]   A.M. Wellington, *The Economic Theory of the Location of Railways: An Analysis of the Conditions Controlling the Laying Out of Railways to Effect the Most Judicious Expenditure of Capital*, 6th edn (New York, 1911), p. 1; 'Vospominaniia', vol. 3, p. 1342.

[56]   On Wellington and systems analysis in railway design and management see R. Whittaker, 'A.M. Wellington and the Idea of Network Effects', paper presented to the Institute of Railway Studies, University of York, 9 February 2000.

[57]   See, for instance, 'Vospominaniia', vol. 2, pp. 28–30, 34–6, 39–44, 66, 69–70, 108–9, 189. Lomonosov's summary of Antonovich's career is at ibid., pp. 2–5. For his talk of 'vocation' see, for example, 'Vospominaniia', vol. 4, pp. 1289–90.

1934). The rival camp, which was dominated by *tekhnologi* from the St Petersburg Technological Institute, preferred to use formulae from, among others, N.P. Petrov (1836–1920), a deputy minister at the MPS. In this group was the designer of the latest standard 2-6-0 passenger locomotive for the state railways, Professor N.L. Shchukin (1848–1924), a future deputy minister with whom Lomonosov would always be at odds.[58]

Yet it is important to stress that Lomonosov did not wish to design locomotives. He was attracted to the steam locomotive partly as an object of applied mechanics but mainly as 'an instrument of railway operating'. Specifically, he 'wanted to study the haulage of trains in a scientific way, as one of the factors that shape the route and operation of railways'. In time, as his thinking developed, he found a phrase that expressed his aim: to 'arrive at theory from life'. In other words, he wanted to study how a steam locomotive functioned in practice in the context of its working environment, acquire an intimate knowledge of the machine, its constituent parts and its employment, and develop mathematical theories that described how the locomotive operated as a machine; he wished then to apply those theories to improve the efficiency of railway construction and operation.[59]

To fulfil this ambition he hoped to make a career in 'science' (*nauka*), preferably as a professor. Perhaps thanks to an IPS tradition of encouraging the ablest students to consider research, he had begun to contemplate this possibility as early as his second year. As he neared graduation in 1898 the real attraction of academe was the scope for doing research without the distractions of full-time employment on the railways. That said, he intimates that his fiancée Sonia helped to crystallize his thinking during 1896. She steered him towards 'science' by arguing that, firstly, he was a good teacher and that, secondly, scientific knowledge was the only enduring thing in the world. As for how to proceed, he relied on advice from two IPS professors whom he saw as mentors – a good illustration

---

[58] 'Vospominaniia', vol. 2, pp. 188–90, 255, 668–9. On Erakov and Nol'tein see Veviorovskii et al., *Leningradskii ordena Lenina Institut*, p. 92; E.B. Kriger-Voinovskii, *Zapiski inzhenera* (Moscow, 1999), p. 61; J.N. Westwood, *Locomotive Designers in the Age of Steam* (London, 1977), p. 241. For Romanov's career see *Materialy k biobibliografii deiatelei Instituta*, vol. 5 (Leningrad, 1959), pp. 152–3. For Petrov and Shchukin see D.N. Shilov, *Gosudarstvennye deiateli Rossiiskoi Imperii, 1802–1917: Biobibliograficheskii spravochnik* (St Petersburg, 2001), pp. 507–13; Kriger-Voinovskii, *Zapiski inzhenera*, pp. 59–60; *Materialy k biobibliografii deiatelei Instituta*, vol. 6 (Leningrad, 1959), pp. 73–5; *Pamiati Nikolaia Leonidovicha Shchukina, zasluzhennogo professora, chlena Soveta Nauchno-Tekhnicheskogo Komiteta* (Moscow, 1925); Westwood, *Locomotive Designers*, pp. 248–9; N.A. Zenzinov and S.A. Ryzhak, *Vydaiushchiesia inzhenery i uchenye zheleznodorozhnogo transporta*, 2nd edn (Moscow, 1990), pp. 64–78, 108–19. Shchukin's MPS service record is at RGIA, f. 229, op. 19, d. 4833.

[59] 'Vospominaniia', vol. 2, pp. 252, 276–7, 293. This thinking echoes the approach of Macquorn Rankine in constructing the science of thermodynamics: see C. Smith, *The Science of Energy: A Cultural History of Energy Physcis in Victorian Britain* (London, 1998), pp. 150–169. Ben Marsden is completing a biography of Rankine.

of how such contacts could shape his career. Professor A.A. Brandt proposed a two-year study tour in Europe – a formula beloved by generations of Russian students – to deepen his knowledge of thermodynamics, with a view to becoming an assistant to the professor of locomotives, Romanov. But the latter proposed a period of work at a Russian locomotive-building factory, which would provide practical familiarity with locomotives. Lomonosov chose the latter path, and with Romanov's help found a job in the Technical Department of the new Khar'kov Locomotive-building Works. He states in his memoirs that he made this choice to gain some practical experience.[60] But this explanation requires some caution because of his later concern to refute criticism that his theorizing was divorced from reality. One can speculate that other reasons for choosing Khar'kov were his close friendship with Romanov and especially his urgent need for an income with which to support his family, following his marriage to Sonia in May 1897 and the birth of their daughter Mar'ia in April 1898.

On balance it does seem likely that he married Sonia for career reasons rather than love. Their relationship needs to be seen in the context of three long-term affairs that he had during 1893–1896, not to mention physical relations with at least five peasant women and prostitutes in those years.[61] The first affair was with his family's servant, Nast'ia; the next was with Ekaterina L'vovna fon Bakh, the step-daughter of his Petersburg landlord; and the third was with Liudmila Sherer, the wife of his manager for his 1895 summer work placement. He claims defensively that at times he was in love with Nast'ia and Ekaterina L'vovna, and that he contemplated marriage to each of these three women; he even calls Ekaterina L'vovna his 'first fiancée'. But he also accepts that his behaviour and attitude towards women was cynical and exploitative.[62] In this light and given his readiness to record professions of even fleeting attraction and love for many women, the almost complete absence in his memoirs of expressions of attraction and love for Sonia is striking. Furthermore, his decision to marry someone who, by his account, supported the political status quo fits uncomfortably with his purported moral opposition to autocracy.[63] Also telling is his failure to accept responsibility for the engagement, which he attributes to her mother, and for the marriage's shaky start. Apparently his doubts began within weeks of the wedding, when Sonia became furious about his earlier affair with Ekaterina L'vovna and forced him to burn his IPS diaries; he claims that Sonia was diagnosed as an

---

[60]   'Vospominaniia', vol. 2, pp. 276–7, 290–294. Brandt was an authority on steam technology: see his *Kurs parovykh mashin: Lektsii, chitannye v Institute inzhenerov putei soobshcheniia Imperatora Aleksandra I*, 2nd edn (St Petersburg, 1896). He also published memoirs in emigration: *List'ia pozheltelye: Peredumannoe i perezhitoe* (Belgrade, 1930).

[61]   'Vospominaniia', vol. 1, pp. 695–8, 718, 784–9, 816–17; vol. 2, pp. 31, 56.

[62]   'Vospominaniia', vol. 1, pp. 629–31, 675, 694–5, 701, 703–4, 707–8, 721–2, 733–42, 750–751, 760–768, 782–4, 796–7, 800, 804–10, 812–19, 854; vol. 2, pp. 46, 49–51, 61–3, 65, 71–2, 96–7, 102–3, 107–9, 113–18, 122–4, 193.

[63]   'Vospominaniia', vol. 2, p. 129.

'hysteric' in July 1897 and then with 'hereditary syphilis'; that the first pregnancy was difficult; and that Sonia's refusal to support him in a row with her father in 1898 caused a 'crack' in their relationship.[64] Finally, his claim that he wanted professional and financial independence from Antonovich is unconvincing. True, he avoided direct contact with his father-in-law after the row in 1898. But he clearly liked being labelled a 'rising star' by Antonovich. He even assisted Antonovich's ultimately unsuccessful bid in 1897 for the contract to build the Moscow ring railway by deceitfully using an acquaintance to obtain the plans of a rival bid.[65] He was irritated that Sonia's sister received a larger dowry. And his decision to sever contact with the family may have been simply an attempt to save face concerning his extra-marital liaisons: he still 'permitted' Sonia to accept money from her family.[66]

If, then, there was opportunism in his decision to marry, can the same be said about the evolution of his politics during the 1890s? He claims that he arrived at Marxism by 1900, with several main influences during the intervening years. One was his relative Professor Kareev, whose political sympathies in the mid-1890s reportedly lay with agrarian socialism. Also influential were several IPS friends, especially Dubelir, whom he called his 'teacher of Marxism'; and illegal Russian revolutionary literature, which he encountered during his first trips to Western Europe in 1897 and 1900.[67] Possibly also relevant at this time were several conversations with industrial workers, although he complained that the workers' political horizons did not extend beyond the factory or even the workshop, and that 'the interests of the working class as a whole were completely foreign to them'.[68]

The reliability of this description hinges on his explanation for why he eschewed revolutionary activism at this time. His memoirs point to a critical moment in the student politics of late Imperial Russia – the suicide in 1897 of an imprisoned student, M.F. Vetrova – and overall his account does seem credible.[69] The student Council of United Mutual-Aid Societies in St Petersburg responded

---

[64] 'Vospominaniia', vol. 2, pp. 9, 66, 68–70, 108, 128–9, 227–9, 233–4, 264–5, 270–272.

[65] 'Vospominaniia', vol. 2, pp. 5, 16, 66, 71, 124, 173, 189, 192–3, 202–3, 250–251, 268–73, 299. See also A.I. Antonovich, *Kak stroit' Moskovskuiu Okruzhnuiu Dorogu s shirokim rel'sovym kol'tsom ili kol'tsom priblizhennym k gorodu* (Moscow, 1897); S.V. Kalmykov et al., *Moskovskoi okruzhnoi zheleznoi doroge 100 let* (Moscow, 2008), pp. 12–17.

[66] 'Vospominaniia', vol. 2, pp. 150, 206, 229–31, 273, 299. Lomonosov claims that Antonovich's wife boasted of the dowry being 300,000 rubles, but that actually the couple received 400 rubles per month; he did acknowledge that this sum exceeded his salary of 125 rubles: ibid., p. 299.

[67] 'Vospominaniia', vol. 1, pp. 647, 650, 656, 658, 673, 676–7, 687–8, 753, 773–4, 778; 'Vospominaniia', vol. 2, pp. 86, 537, 575. By 1905 Kareev was a prominent liberal.

[68] 'Vospominaniia', vol. 2, pp. 222–3, 615–17.

[69] On Vetrova's suicide see S.K. Morrissey, *Heralds of Revolution: Russian Students and the Mythologies of Radicalism* (Oxford, 1998), pp. 178–82; and N. Rostov,

to Vetrova's death by summoning the capital's students to a memorial service at the city's Kazan' Cathedral. Lomonosov and Dubelir walked there together, but returned home when the authorities prevented the service from starting. However, other IPS students joined a march to the Winter Palace, and 17 were among the hundreds arrested when Cossacks dispersed the crowd. The IPS student body was divided about how to respond. Some thought that the arrests brought the institute into disrepute, whereas others demanded a statement of gratitude. For himself Lomonosov claims that he adopted a 'compromise formula':

> To struggle with autocracy is the duty of every honest person. But I never had any intention of being a professional revolutionary. I set myself a different task: to become an educated railwayman. For this I needed to study, not march in demonstrations. But if a real revolution began, I would have to drop everything and if necessary lay down my life.[70]

This caution may have been a conscious recipe for inaction. Yet his credibility seems stronger for his candour about rejecting political activism as a student, and his claim is consistent with his decision in 1905, just eight years later, to risk his life in much this way.

Nonetheless it may be inappropriate to accept his claim that he was a Marxist by the turn of the century. To judge by his memoirs, he shared the Marxist preoccupation with promoting social justice, ending the material poverty of the labouring masses and, unsurprisingly, reorganizing the economy on a scientific basis. Evidently, too, he agreed with the Marxist analysis of class consciousness, class war and revolution, and about the need to fight with the incumbent regime for control of the state. But if Marx stressed the subsequent importance of destroying the old state, Lomonosov seems, without discussing the issue explicitly, simply to differentiate between the tsarist regime, which he saw as intrinsically corrupt like private enterprise, and the Russian state, which he saw as the main instrument for improving national prosperity. Also, despite accepting the centrality of class consciousness and class war, he endorsed the anti-determinist position that an individual could play significant a role – a conclusion that stemmed in part from seeing how the manager of the Poltava railway workshops successfully placated his workforce in 1900. Similarly, significant differences can be identified between Lomonosov's views and those of Marxist groups that were forming in the Russian empire at about this time. For him the national priority was to eliminate peasant poverty, and the type of state structure was only a secondary matter. Accordingly, he abhorred what he saw as the willingness of the Marxists to await

---

'Samoubiistvo M.F. Vetrovoi i studencheskie besporiadki 1897 g.', *Katorga i ssylka*, 2 (1926): 50–66.

[70] 'Vospominaniia', vol. 1, pp. 774, 778; 'Vospominaniia', vol. 2, pp. 194–200. However, Lomonosov describes a similar episode of contemplation and compromise in 1900: ibid., pp. 537–9.

the proletarianization of the peasantry and world revolution, since that would condemn the peasants to perhaps hundreds more years of poverty. Equally, he was scathing about the Socialist Revolutionary (SR) Party, which emerged as such from the Populist tradition in 1902–1903 and at least in theory focused on the peasantry. Guided by his intimate knowledge of the countryside from childhood and by his experience of two long-term affairs with Ukrainian peasant women during 1899–1907, he felt that the Socialist Revolutionary Party's concept of the peasantry as 'the god-bearing people' (*narod bogonosets*) had little basis in reality.[71]

Yet the typically liberal overtones that recur in his political remarks were there largely by default. He would support a constitutional monarchy if that would help the masses quickly. Similarly, he wanted the creation of a parliament together with freedom of speech and assembly – preferences that were strengthened by witnessing peaceful labour demonstrations during a brief stay in Britain in 1897. But such liberal aims were important for him mainly because he believed that the worldwide proletarian revolution might not come for perhaps 500 years due to the proletariat's low level of class consciousness.[72] Indeed, he became convinced that the moderate liberalism advocated by his parents could not solve the fundamental problems of the countryside, especially the widespread poverty. In short, then, his recollections of his politics at the turn of the century may be sincere, but, as with his description of his schoolboy politics, they probably overstate his radical commitment at the time. His politics were actually a somewhat confused combination of socialism (including Marxism) and liberalism.

*   *   *

Lomonosov became a railway engineer mainly for personal reasons that had little to do with the tsarist government's rapid industrialization drive in the 1890s. To be sure, his attention was first attracted by the notion of engineering as a patriotic vocation, and he would probably not have been able to join the IPS had the regime not been pressuring the institute to increase admissions. But the state's employment of railway construction to drive economic growth had no bearing on his preference for the IPS: given his need to earn a living and his family's tradition of state service, the key considerations were his aversion to an army career, his reliance on his family to fund his higher education, and social snobbery. Beyond that point his future depended on his ability to pass the IPS's entrance exam. At a more general level, then, his experiences challenge the dichotomy of decline or adaptation for the rural nobility in late nineteenth-century Russia; instead they reflect both of these models. The youth was mindful of traditional social values,

---

[71]   See 'Vospominaniia', vol. 2, pp. 537–9, 617, 672. For an introduction to the SR Party see M. Melancon, 'Neo-Populism in Early Twentieth-Century Russia: The Socialist Revolutionary Party from 1900 to 1917', in A. Geifman (ed.), *Russia under the Last Tsar: Opposition and Subversion, 1894–1917* (Oxford, 1999), pp. 73–90.

[72]   'Vospominaniia', vol. 2, pp. 538–9, 617.

yet acutely conscious of the current financial pressures and open to new ideas, and in the generational conflict that ensued he played his cards more adeptly than did his elders.

Once enrolled at the IPS he was always likely to specialize in a railway-related discipline. Railway construction and operation were, after all, the dominant concerns of the institute's supervising ministry, the MPS, and were given far more attention at the IPS than the road and waterway sectors. In that environment intellectual curiosity – not any specific sense of patriotism or state policy – led Lomonosov to identify a railway-related 'vocation' – railway surveying – that happened to match one of the state's immediate and future transport priorities. On moving to Khar'kov after graduation in 1898 Lomonosov could expect to spend several years gaining practical railway experience, in the hope of becoming a teaching assistant to a professor like Romanov and then finding a professorial position. But since the number of academic posts in his field was miniscule despite the ongoing expansion of Russian engineering education, the chances were that he would make a career on the railways, at least in the short and medium terms.

Chapter 2

First Steps in Railway Engineering

Lomonosov's first three years of full-time employment set the direction of his entire research career. His start was inauspicious: he was effectively dismissed from the Khar'kov locomotive-building factory for misconduct after just three months. But having quickly found a job with the nearby Khar'kov–Nikolaev Railway, he developed his first assignment there into a large-scale research project in locomotive engineering that even attracted interest from the distant MPS in St Petersburg. This project not only generated his lifelong fascination with locomotive testing, but also provided practical experience that helped him earn junior faculty appointments at two of the new polytechnic institutes, in Warsaw and Kiev, without having even started the necessary postgraduate studies.

The first section of this chapter sets the technical and intellectual context for these developments by sketching the general aims and evolution of locomotive testing from its beginnings in Britain in the early nineteenth century to the 1890s. The second section explains why and how Lomonosov conceived and developed his own distinctive method of locomotive testing, while the final section focuses on his career ambition to become a professor of railway engineering, showing how he contrived to make such rapid progress towards this aim. More generally, this chapter also considers the question of whether his early experiences confirm the established view that a lack of interest from the tsarist state was the main reason why Russia's industrial research activity lagged behind the efforts of her major European rivals and the United States in both investment and achievement during the late nineteenth and early twentieth centuries.

The science of locomotive testing

Locomotive testing has always been important for the world's railway industry. The term does not refer to the many routine quality-control and maintenance tests conducted on the constituent materials and components of each locomotive – like any heavy industrial machine – throughout its manufacture and life. Rather, it is a process whereby a selected locomotive is monitored under power to obtain data about its operating performance. The specific objectives of this exercise have evolved in line with the growing complexity of the railway locomotive and its working environment, but the fundamental concerns have remained unchanged over nearly two centuries of railway operation: safety, reliability, haulage capacity, efficiency and publicity.[1]

All these concerns framed the world's first formal railway locomotive tests, which were held at Rainhill, England in 1829 for the Liverpool and Manchester Railway ahead of its planned opening in 1830. The key questions here for the company's directors were whether the steam locomotive was a safe, reliable and, not least, publicly acceptable machine for pulling the trains and whether it was more effective than the perceived main alternative – rope haulage by a series of stationary winding engines. The advocates of locomotives combined serious technical questions with clever publicity to stimulate enormous popular interest, which would help to distinguish this railway as the first in the world to provide a regular public passenger and freight railway service between two major cities powered exclusively by steam locomotives.[2] Famously, the victor at Rainhill was George and Robert Stephenson's *Rocket*, thanks especially to its new type of boiler. Notwithstanding some dubious manipulation of the competition rules, the *Rocket* alone safely met all the core technical requirements for reliability, fuel economy and haulage power, and its unprecedented maximum speed of 29 miles per hour amply justified its canny name, dazzling the crowds of spectators. This success

---

[1]   Arguably the classic technical account of the evolution of the steam locomotive is by the distinguished French engineer A. Chapelon: *La Locomotive à Vapeur*, 2nd edn, English translation (Rode, 2000). For accessible surveys of nineteenth-century locomotive development see, in particular, J. Simmons, *The Victorian Railway* (London, 1995), pp. 69–101 and D. Ross, *The Steam Locomotive: A History* (Brimscombe Port, 2006), especially pp. 13–149. Valuable introductions to the principles and history of locomotive testing include: E.L. Diamond, 'The Horse-power of Locomotives: Its Calculation and Measurement', *Monthly Bulletin of the International Railway Congress Association (English Edition)*, 18/2 (1936): 150–88; S.O. Ell, 'The Testing of Locomotives', in P. Ransome-Wallis (ed.), *The Concise Encyclopaedia of World Railway Locomotives* (London, 1959), pp. 386–410; D.R. Carling, 'Locomotive Testing Stations (Parts I and II)', *Transactions of the Newcomen Society*, 45 (1972–1973): 105–82; and A. Tester, 'An Introduction to Steam Locomotive Testing', parts 1, 2, 2A, 3A, 3B, *Backtrack*, 23/4 (April 2009): 199–203; 23/5 (May 2009): 308–15; 23/9 (September 2009): 564–9; 24/2 (February 2010): 100–107; and 24/4 (April 2010): 242–50. In 1928 one of Lomonosov's long-time colleagues published a useful survey of how engineers had been seeking to improve locomotive efficiency since the dawn of the steam railway: A.I. Lipetz, 'Attempts to Increase Steam Locomotive Efficiency: A Review of the Progress Attained up to the Present Time, and of Recent Developments', *Railway Mechanical Engineer*, 102/7–8 (1928): 384–91 and 437–42.

[2]   Ironically the image was not quite accurate: due to a parliamentary ban on using steam locomotives in a long tunnel between Edge Hill and Liverpool docks, the company initially used rope-haulage by winding engines for this section: F. Ferneyhough, *Liverpool and Manchester Railway, 1830–1980* (London, 1980), p. 37. During the first half of the nineteenth century various attempts were made to move trains using alternatives to the steam locomotive, horse and winding engine, such as the 'atmospheric railway', but these were not commercially successful.

was a major step towards defining the basic technical layout of the railway steam locomotive, and it also marked the steam railway's coming of age – a defining moment of the nineteenth century.[3]

The very nature of the steam locomotive as a form of heat engine ensured that locomotive designers, operators and testers were always very concerned with improving its efficiency. Simplicity and ruggedness were great advantages of this machine: it could usually get itself home even when ailing. Conversely, low thermal efficiency was its Achilles heel. In optimum conditions only about 8–12 per cent of the heat energy generated through burning the fuel in the firebox of a steam locomotive could be made available for the job of moving trains; some of the rest of the energy had to be used for initial steam-raising and moving the locomotive itself, but most was simply ejected through the chimney as waste gases. Moreover, in practice the thermal efficiency of any steam locomotive at work was usually well below 10 per cent for all sorts of reasons that could include low-grade fuel, inadequate maintenance, driving techniques, poor design, changes in the surrounding landscape and even the climate and weather. Thus, with the fuel bill being one of the largest operating costs of every railway, the Holy Grail for steam locomotive designers was a substantial increase in the machine's overall thermal efficiency. Many ideas were tested for this purpose throughout the steam age, but as of 1900 no dramatic breakthrough had been made. Indeed, the question often arose of whether a given change was technically and economically worthwhile, especially if it involved complex equipment that had higher build and running costs and compromised the machine's simplicity. It followed that designers and researchers keen to test a new idea could face a struggle to justify the cost of the necessary modifications and tests.

Sophisticated measuring equipment was used in locomotive testing from the 1830s. The main specialized instruments were an indicator and a dynamometer. The indicator monitored pressure variations inside a locomotive's cylinder to show the work done in one cycle of the engine, and its information was recorded in a diagrammatic form that was called an indicator diagram. The dynamometer showed the amount of 'useful work' done by the locomotive, which was defined as the pulling power (tractive force) exerted at the point (called the drawbar) where the trailing load (the train) was attached. By 1900 it was becoming the norm to locate the dynamometer in a special research vehicle known as a dynamometer car; several such vehicles existed in Russia by this time, including one on the railway that Lomonosov joined at Khar'kov in 1898. The dynamometer car was marshalled immediately behind the locomotive (i.e. at the drawbar) and also had equipment for measuring the distance travelled and time taken. Together with measurements of fuel and water consumption, this information enabled engineers to calculate how many horsepower were generated by the locomotive for hauling

---

[3]   On the Rainhill trials see, for example, Ferneyhough, *Liverpool and Manchester Railway*, pp. 44–59; B. Marsden and C. Smith, *Engineering Empires: A Cultural History of Technology in Nineteenth-Century Britain* (Basingstoke, 2005), pp. 136–45.

its train over a given section of track, and to calculate the mean efficiency of the work done at the drawbar.[4]

How far, then, could such 'road tests' be considered 'scientific'? Certainly it was feasible to run a series of tests systematically: the researchers could use, in particular, the same locomotive, schedule and route, the same type of coal, the same set of vehicles and the same locomotive crew for their set of tests. However, many engineers believed that it was impossible to create constant experimental conditions out on the line that could allow any given test to be fully replicated. One obstacle was that many of the significant variables could not easily be controlled. Important examples included the weather conditions, the traffic situation and, not least, the fact that the locomotive was consuming fuel and water constantly, yet the amounts of fuel and water usage could not be measured at will out on the line because of traffic and other considerations. In any case, compromise was fundamental in locomotive design. A locomotive being designed for heavy freight traffic, for instance, would need maximum adhesion on the rails and maximum capacity to make steam, yet these demands could be technically incompatible.[5] Similarly, standardization of equipment became attractive to many railways for its simplicity and economies of scale, but often meant that a given locomotive had to do jobs for which it was not designed. In short, the sceptical engineers believed that neither the machine itself nor its working environment were conducive to the development and implementation of ideal design solutions or to running very precise, repeatable scientific tests.

A serious challenge to this mindset was made in the 1880s by the Russian engineer A.P. Borodin (1848–1898).[6] He was scathing about the research done previously. To the Institution of Mechanical Engineers in London he declared that 'the whole of the experiments and investigations hitherto made, on the work of locomotive engines as well as on their consumption of fuel and water, appeared ... to be deficient in completeness and accuracy, and based on data

---

[4]   On this equipment see, for instance, R.P. Johnson, 'Dynamometer Cars', *Baldwin Locomotives*, 5/3 (January 1927): 35–43; M. Rutherford, 'Measurement not Mystification: The British Dynamometer Car', *Backtrack*, 9/8 (August 1995): 436–44; and A. Tester, 'An Introduction to Steam Locomotive Testing', parts 3A, 3B, *Backtrack*, 24/2 (February 2010): 100–7; and 24/4 (April 2010): 242–50.

[5]   Steam-making capacity depends partly on the size of the grate in the firebox, whereas adhesion is a function of the load that bears on the powered axles, with maximum adhesion obtained when all the axles are powered. The problem in this example is that a large grate to maximize steam-making would be heavy and might need a non-powered axle underneath as support.

[6]   On Borodin see S.M. Zhitkov, *Biografii inzhenerov putei soobshcheniia*, vypusk 3 (St Petersburg, 1902), pp. 24–37; Zenzinov and Ryzhak, *Vydaiushchiesia inzhenery*, pp. 92–108; and Westwood, *Locomotive Designers*, pp. 109–11, 185–6. This Borodin should not be confused with the chemist and composer A.P. Borodin (1833–1887).

not sufficiently scientific'.[7] His response brought major changes to the practice of locomotive testing. He devised an apparatus that allowed a locomotive to be tested under power in controlled constant conditions inside a building. His own device was a temporary affair that soon had to be dismantled, but he took the trouble to publicize it internationally, mainly by presenting a paper in London to the Institution of Mechanical Engineers. His concept, which became known as a locomotive test or testing plant, testing station or laboratory, was taken up and refined by an American university professor, W.F.M. Goss, on whose initiative the world's first permanent locomotive test plant was built at Purdue University, Lafayette, Indiana during 1890–1891. By 1905 there were three more test plants in the United States and two in Europe – one at the Swindon workshops of Britain's Great Western Railway (opened in 1903) and the other at the Putilov engineering factory in St Petersburg (opened in 1905). That said, the immense cost of building and operating such plants ensured that few were built. Also, they could be criticized as an artificial environment that lacked such significant phenomena as wind and train resistance.[8]

Russia's prominence in these developments may seem surprising given that, compared with her European and American rivals, the country was not regarded as a major player in industrial and scientific research in the late nineteenth century.[9] Her research activity was located mainly in universities, whereas the emphasis in the major Western powers was shifting to research institutes and industrial laboratories. Moreover, to judge by the tiny proportion of the state budget assigned for this purpose, the Russian government was uninterested in such research – a reflection in part of its perception of universities as hotbeds of political opposition and unrest. Nor were private companies very active. One reason was that a sizeable proportion of Russian industry was foreign-owned, relying on imported technologies; another reason was a lack of interest among Russian entrepreneurs. That scientists like the physiologist I.P. Pavlov and chemist D.I. Mendeleev made profound contributions to their disciplines was despite, not because of, government policy.

---

[7]   A.P. Borodin, 'Experiments on the Steam-Jacketing and Compounding of Locomotives in Russia', *Proceedings of the Institution of Mechanical Engineers* (1886): 297.

[8]   In 1925 L.H. Fry described the locomotive laboratory as a 'purely American' idea, but Lomonosov and another Russian engineer, A.I. Lipets, pointed to Borodin's work: see Fry, 'The Locomotive Testing Plant and its Influence on Steam Locomotive Design', *Transactions of the American Society of Mechanical Engineers*, 47 (1925): 1267–85 and discussion 1285–93. Borodin's role is emphasized in Carling, 'Locomotive Testing Stations', pp. 105–7. Goss is described in Westwood, *Locomotive Designers*, pp. 214–15.

[9]   R. Lewis, *Science and Industrialisation in the USSR: Industrial Research and Development, 1917–1940* (London, 1979), pp. 1–5. As in other countries, the testing of materials for production purposes was concentrated in factories and workshops, including dedicated laboratories.

But the situation with Russia's railway-related research was more akin to the position in major Western countries. The academic input came largely from a specialist institute: the IPS. Furthermore, although the MPS did not promote a national railway research programme as such, it did allow individual railways to conduct research, which effectively meant that the state funded any research conducted by state-owned railways. There was some collaboration, too, between the railways and Russian private industry, especially the few locomotive-builders like the Putilov company. However, Russia's railways were generally not international leaders in research: the running was made in Germany, France, Belgium, Austria–Hungary and the United States. Indeed, the Russian railways adopted various foreign technical developments, such as the locomotive valve gear devised by Egide Walschaerts in the 1840s; the Westinghouse air brake (1868); and compounding, whereby steam was expanded in cylinders twice – a technology that was developed for railway locomotives from the 1870s by French and German engineers such as Anatole Mallet, Alfred de Glehn and August von Borries, and then by the likes of Samuel Vauclain in the United States from the 1890s.[10] Similarly, French and German railways were regarded as the leaders in the theory and practice of locomotive testing. It should be added, however, that whereas at least some Russian engineers did follow leading Western engineering periodicals, the level of foreign awareness about Russian railway research was low. Western engineers did not see the Russian railway technical journals because they were not translated for foreign audiences, and Russian railway engineers rarely published their findings abroad. The international fame of Borodin was due not simply to the merits of his work but also to his uncommon initiative to speak and publish abroad.[11]

In the late nineteenth century there was a marked growth of European and American railway interest in scientific research. It was driven by heightened concern to improve safety and efficiency, including the notion of optimal economy. The science of materials, for example, was always relevant for railways, but now it acquired even more importance as scientists deepened their understanding of issues like metal fatigue and stresses. The analysis of materials was further encouraged by the railways' growing need to cut costs due to such pressures as intense competition and tighter state regulation, and by the desire to avoid the

---

[10] For technical details of these examples see, for instance, Ransome-Wallis, *Concise Encyclopaedia*, pp. 296–304 (valve gears), 305–6 (Westinghouse brake), 312–14 (compounding); and Ross, *The Steam Locomotive*, pp. 69–74 (valve gear), 127–36 (compounding).

[11] Aside from Borodin's work the best-known Russian research on locomotives as of 1900 concerned experiments on the Transcaucasus Railway on using oil waste products as fuel. The researcher was the line's British locomotive engineer, Thomas Urquhart, who gave several papers about it in London. See, for example, his 'Supplementary Paper on the Use of Petroleum Refuse as Fuel in Locomotive Engines', *Proceedings of the Institution of Mechanical Engineers* (1889): 36–84.

expense of over-engineering. The railway industry was thus heavily involved in improving and developing the inspection and testing techniques for materials, including the now routine concept of non-destructive testing.

Locomotive engineering, which encompassed the major part of research into railway equipment, was likewise influenced by this increased concern for safety and efficiency. Here an important example of safety concerns helping to drive scientific enquiry was the question of how a locomotive's wheels interacted dynamically with the rail: a better understanding of this issue could enable engineers to identify, for instance, the maximum safe speed for a given class of locomotive on a particular section of track. Among the economic motives for research was, naturally, the steam locomotive's low thermal efficiency, with results that included the widespread application of technologies like compounding from the 1880s and 1890s to help improve the use of the steam. Such concerns inspired Goss's research at Purdue and help to explain why there was at least some investment in locomotive test plants in Europe. Even more radical was a response that was beginning to gain serious momentum by 1914: to abandon steam for electric power or internal combustion engines, where an increase in thermal efficiency to over 20 per cent was anticipated and with which the problems caused by locomotive smoke in tunnels and cities could largely be eradicated.[12]

By the early 1900s there was a general consensus about the main directions of research for improving the steam locomotive, but disagreement about the relative merits of road tests and test plants. Five topics were identified: fuel type and combustion; heat transfer to the heating surfaces in the boiler; boiler design and performance; cylinder design and performance; and the process, known as draughting, whereby the exhaust gases and ash were ejected into the atmosphere. Put simply, researchers sought to discover the scientific nature of these processes – how precisely these processes occurred – in the hope of identifying scope for design changes to improve performance. Thus, for example, Goss discovered in 1902 that the jet of exhaust gases from the cylinders did not have a pump-like single action in the blastpipe beneath the chimney, but a two-fold action – a finding that led eventually to the widespread use of double exhaust pipes and chimneys.[13] As for how to study locomotives under power, Borodin had put an important question mark over road tests, yet most engineers had to rely on them by default, happily or not: few test plants could be commissioned due to the great expense of building and operating them. In any case, there was also the argument that the conditions of a test plant were artificial.[14]

---

[12] On the thermal efficiency of steam traction see Ransome-Wallis, *Concise Encyclopaedia*, p. 288. The use of the internal combustion engine in railway traction is the focus of Chapter 9, this volume.

[13] Ransome-Wallis, *Concise Encyclopaedia*, pp. 394–5.

[14] For instance: E.A. Phillipson, 'Notes on Locomotive Running Trials', *Journal of the Institution of Locomotive Engineers*, 14/65 (1924): 364.

In this light one can say that Lomonosov's work on locomotive testing was both an affirmation and a rejection of Borodin's intellectual legacy. He agreed with Borodin's premise that previous testing had been insufficiently scientific, that rigorously scientific testing was actually feasible and also that it was desirable for the purpose of improving railway efficiency in many ways from locomotive design to the compilation of traffic schedules. He even dedicated his first book to Borodin in 1907.[15] However, Lomonosov did not accept that laboratory-type constant experimental conditions could be created only in a purpose-built testing plant: he believed that they could be created in road tests too. This claim, which can be seen as the main long-term product of his research in 1898–1901, potentially offered a cheaper alternative to a test plant. But was it credible?

'Scientific road experiments'

The immediate background to Lomonosov's involvement in locomotive testing was the loss of his job at the Khar'kov locomotive-building factory in October 1898. His troubles there included the hostility of the factory manager, P.P. Ritstsoni, and of colleagues who were graduates of the Khar'kov Polytechnic Institute and disliked his status as a *puteets*. Also, Lomonosov had no opportunities to design locomotive components because the factory was merely fulfilling large contracts to assemble locomotives of the MPS's latest standard freight design, known as the '1897 type'. Fortunately, the superintendent of the locomotive shops was sympathetic. E.M. Gorokhov spent hours discussing mechanics, components, tooling and materials with Lomonosov, and found him a niche as supervisor of the pre-delivery fine-tuning of new locomotives. This job entailed responsibility for about 20 staff and two or three locomotives at any given time. Each engine was operated gently at the factory for two days before making a short journey with a train. Lomonosov thus had an agreeable chance to wield managerial responsibility while learning without close supervision. Indeed, with typical gusto he seized the chance of learning to drive a locomotive. Yet his quick temper and pride soon brought trouble. An official certificate states that he resigned, but Lomonosov's memoirs paint a livelier picture: he disobeyed an order from Ritstsoni and capped the ensuing row by assaulting him.[16]

This episode betrays a dark side of Lomonosov's character that often impinged on his work. He was an ambitious, competitive, energetic and determined person who liked to wield power far more than to account for his actions. He was also stubborn, mercurial and impetuous, with a tendency to lose his sense of proportion

---

[15]   Iu.V. Lomonosov, *Opytnoe issledovanie tovarnykh vos'mikolesnykh parovozov kompaund normal'nogo tipa, proizvedennoe v 1898–1900 gg. na Khar'kovo–Nikolaevskoi zh.d.* (Kiev, 1907), p. III.

[16]   'Vospominaniia', vol. 2, pp. 303–8, 314, 318–20, 322, 326–8, 335–7; Certificate, 7 October 1898: DAK, f. 18, op. 2, spr. 159, ark. 231.

and, at worst, to indulge a certain cantankerous pugnacity. His memoirs offer many indications of violent conduct during his student years and early career, including some that are corroborated elsewhere. With regard just to his three years in Khar'kov they report that he challenged a neighbour to a duel over the use of a garden; participated in a drunken brawl; held a doctor at gunpoint for a night to monitor his year-old daughter's reaction to an incorrect prescription; threw a teapot at a nanny; assaulted a traffic official of the Khar'kov–Nikolaev Railway because of an operating problem; struck his landlord with a bottle; and hit a power-station official for attempted bribery. Why he was never prosecuted is unknown, but he was evidently fortunate that Ritstsoni merely recorded his resignation.[17]

On this occasion, Lomonosov landed firmly on both feet. On Romanov's advice he approached N.D. Baidak, who was superintendent of the Traction Department of the Khar'kov–Nikolaev Railway – the person responsible for all the railway's locomotives, carriages, wagons and associated depots and workshops. Baidak arranged Lomonosov's appointment as an assistant superintendent without portfolio at the Romny locomotive depot some 200 km north-west of Khar'kov. But this was merely an accounting device: he sent Lomonosov to the Traction Department's Technical Office at the line's Khar'kov headquarters.[18] There Baidak set Lomonosov a task that was inspired by a troublesome operating problem. Crews were upset that they could not earn bonuses for meeting fuel economy targets because their freight engines known as the '1893 standard type' were using more, not less, coal than the older locomotives that they had replaced. In consultation with the MPS Baidak had organized road tests and made modifications during 1895–1896, but to no avail.[19] Lomonosov's challenge was to explain and solve the problem. Thus, instead of accumulating practical knowledge and experience gradually in a factory or depot, he already had a chance to conduct advanced design-related research. Moreover, he had greater freedom of action than would typically have been the case at a factory. And of course the question of fuel efficiency returned him to the core issue that had seized his student imagination, namely the relationship between technical equipment, mechanics and economics. Baidak doubtless anticipated a report within a few months; in fact, this project occupied Lomonosov for nearly three years, and yielded his concept of 'scientific road experiments' (*nauchnye opyty v puti*).

Although this task concerned one railway, it must be seen in the broader context of the government's railway policy. By 1898 Russia's public railway network had some 42,700 km of track and was growing by over 2,000 km per

---

[17] 'Vospominaniia', vol. 2, pp. 332–4, 373–4, 437–8, 470–471, 525–9, 614, 640; although the memoirs describe a prosecution for striking his landlord, his service record does not show a conviction: Service record, 1 January 1908: DAK, f. 18, op. 2, spr. 159, ark. 188.

[18] 'Vospominaniia', vol. 2, p. 337. The depot appointment was announced on 3 April 1899: Service record, 1 January 1908: DAK, f. 18, op. 2, spr. 159, ark. 187zv.

[19] Lomonosov, *Opytnoe issledovanie*, pp. 10–25.

year. This system was divided among 36 railway companies, of which most were state-owned like the Khar'kov–Nikolaev line. The private companies were either being taken into state ownership or consolidated into a few large private entities such as the South Eastern Railways, Moscow–Kazan' Railway and Vladikavkaz Railway; these private companies operated on a long-term concession from the state and were scheduled for eventual sale into state ownership.[20] Characteristic of this situation was a high degree of equipment standardization compared to other large railway systems: logistical problems experienced during the 1877–1878 Russo–Turkish war had persuaded the ministry of the virtues of standardization. Thus, for example, the MPS was trying to replace a plethora of locomotive designs with a few standard designs that would each be built in enormous numbers for the state sector and would also be wanted by the private railways. In so doing the MPS sought to improve efficiency by introducing technologies such as compounding. It reasoned that the new standard compound designs – notably the 1893 freight type with which Lomonosov was now confronted – would be cheaper to run than their 'simple' predecessors for the same output of useful work. Any rise in maintenance costs due to the higher boiler pressure and compound system would be outweighed by fuel savings – a vital consideration for any railway. If, then, Russia's railways were buying several thousand locomotives that were actually using more fuel, the implication was a technical deficiency that would waste many millions of rubles over the next two or three decades.[21]

Lomonosov's response was, for his humble circumstances, decidedly bold. He persuaded Baidak that the existing data was contradictory, chaotic and insufficient to confirm and quantify the problem. The coal and water consumption of the old and new locomotives needed to be measured carefully and correlated with new dynamometer data about the work performed by each design.[22] Road tests were the only way to obtain these data, notwithstanding their shortcomings, because Russia had no test plant as yet: lobbying for a national plant by a committee of railway engineers that included Borodin and Baidak had achieved merely an agreement to locate it in Khar'kov.[23] Sadly, Borodin could not advise Lomonosov due to his untimely death earlier that year; almost certainly they never met or corresponded.

The fact that Lomonosov immediately looked beyond his remit shows the force of his ambition. In a further report that Baidak ratified on 3 November he defined his aims far more broadly than was warranted: 'to clarify the relative strengths and weaknesses of the compound steam locomotives and to find ways

[20] For the list and mileage see Afonina, *Kratkie svedeniia*, pp. 56–7.

[21] Rakov, *Lokomotivy*, pp. 127–9.

[22] Iu.V. Lomonosov, 'Ob opytakh nad parovozami normal'nogo tipa', *Zheleznodorozhnoe delo*, 3–4 (1911): 2Д–3Д; 'Vospominaniia', vol. 2, pp. 293, 335–7, 343; Rakov, *Lokomotivy*, p. 144.

[23] On the committee see *Protokoly zasedanii XXII Soveshchatel'nogo s"ezda inzhenerov sluzhby podvizhnogo sostava i tiagi russkikh zheleznykh dorog, sozvannogo v Kieve na 25 noiabria 1900 g.* (St Petersburg, 1901), p. 92.

for possible improvement in their economy'.[24] Specific tasks would include determining 'as accurately as possible the hourly coal and water consumption per unit of work for both designs of locomotive in the most varied traffic conditions', and using an indicator to record the steam pressure in the cylinders.[25] In other words, he was looking to compile a full thermodynamic picture of each locomotive design – a project that, he later admitted, would take several years. In due course, moreover, the momentum of his work took him still further.[26] And, crucially, he became fascinated by the challenge of minimizing, and ideally overcoming, the flaws of road tests.

A key to his progress with such a complex project was his ability to get support from colleagues other than Baidak. At first, apparently, many people displayed hostility. In particular, an objection arrived from G.N. Teodorovich, the head of the locomotive depot at Nikolaev, who managed the dynamometer car and had overseen the 1895 tests. Also hostile was I.I. Vaterlei, superintendent of the Liubotin locomotive depot where Lomonosov wished to base his research: apparently Vaterlei detested Teodorovich and everything associated with him, including the dynamometer carriage. More reasoned and reasonable was the Traffic Department's insistence that regular freight trains should have priority over test trains. Generally Lomonosov handled these politics adeptly, usually by persuading Baidak to intervene or cultivating other colleagues.[27]

Nor did Lomonosov intend to reinvent wheels. Conscious of his inexperience, he sought guidance from foreign as well as Russian sources when devising his programme of tests for November–December 1898. He relied heavily on technical journals, having his own subscriptions to French, German and British periodicals. Thus, for example, the two selected locomotives were prepared according to French norms, not least to ensure the comparability of his findings with those.[28] As was customary both in Russia and abroad he used regular fast freight trains with the rostered driver, who was allowed to drive as he desired. The usual procedure to monitor fuel and water consumption was employed: the coal was weighed and the water plumbed in the tender tank at depots or yards before and after trips. Similarly, the dynamometer carriage was used to measure the locomotive's useful work; indicator diagrams were taken at the cylinders; and the train weights, distances,

---

[24] Lomonosov, *Opytnoe issledovanie*, p. 26.

[25] See A.J. Heywood, 'Iu.V. Lomonosov and the Science of Locomotive Testing in Russia: First Steps, 1895–1901', *Transactions of the Newcomen Society*, 72/1 (2000–2001): 9 (Document 1).

[26] Lomonosov, *Opytnoe issledovanie*, prilozhenie II, p. 8; 'Vospominaniia', vol. 2, pp. 320, 350, 357.

[27] 'Vospominaniia', vol. 2, pp. 343, 346, 350–353. On Teodorovich and Vaterlei see Lomonosov, *Opytnoe issledovanie*, pp. 13–17, 20–24.

[28] Lomonosov, *Opytnoe issledovanie*, pp. 58–9.

times and speeds were recorded.[29] It was evidently Lomonosov's idea, however, to insist on priority for his test trains over all other freight traffic. This measure could be justified as a way to avoid the expense of repeating interrupted test trips, but the eventual approval was probably more a product of Baidak's influence, especially given the railway's liability for fines for late delivery of freight consignments.[30]

Reliance on foreign expertise notwithstanding, and despite attempts to improve the methodology as the work proceeded, the results were meagre. At first Lomonosov did not try to ensure the repeatability of his tests, but once he realized the importance of maximum consistency and repeatability for making comparisons, he chose one driver for all tests and made him use prearranged speeds and rates of steam-expenditure (known as 'cut-off') for such variables as distance travelled and train weight. Nonetheless the coal and water data contained substantial variations, which Lomonosov interpreted as confirmation that a crew's working habits were relevant. Further, most indicator diagrams and the dynamometer data were found to be useless.[31] These problems were caused partly by technical faults and partly by the human factor, including the general lack of experience. Having had to rely on published information, Lomonosov concluded that not much guidance was available, and that Borodin's complaint about the scientific weakness of road testing remained valid.[32]

Lomonosov's response to this setback set the direction for his future work. He proposed a three-stage programme of tests that would greatly extend his parameters. The first task, logically, was to check his results from 1898. The second stage involved the standard '1897 type' locomotive, which he knew from the Khar'kov factory and which was now entering service on his railway; in other words, he wanted to study the latest equipment in advance of any local or national disquiet about it. The final tests were to study modifications that he was starting to formulate for improving the locomotives' thermodynamic characteristics. Baidak was again supportive, and the line superintendent, F.I. Shmidt, confirmed the plan on 9 June 1899. Doubt was cast over it shortly afterwards when Baidak moved to another railway: for reasons now unknown his replacement, I.A. Buteskul,

---

[29]   Lomonosov, *Opytnoe issledovanie*, prilozhenie II, p. 12; Lomonosov, 'Ob opytakh nad parovozami normal'nogo tipa', p. 3Д; 'Vospominaniia', vol. 2, pp. 350–351, 353, 355, 363, 423.

[30]   In his memoirs the question of traffic priority is first mentioned at vol. 2, pp. 350–351; it was formalized in line circular No.110 of 24 April 1900: translated in Heywood, 'Lomonosov: First Steps', pp. 12–13.

[31]   Lomonosov, *Opytnoe issledovanie*, prilozhenie II, p. 12; on p. 30 he reports: 178 water measurements, 1098 pairs of diagrams, and total distance travelled 2,029 km.

[32]   Lomonosov, *Opytnoe issledovanie*, prilozhenie II, p. 12; 'Vospominaniia', vol. 2, pp. 353–5, 361–3.

demanded Lomonosov's resignation. But Shmidt upheld Lomonosov's appeal, and the programme was implemented during 1899–1900.[33]

The stage one tests during June–August 1899 were important both for their methodological innovations and for their results, which had national ramifications. The driver was ordered to move the regulator and reversing lever as rarely as possible, and to use specified settings for the cut-off. Most trips were operated with special trains on just one quiet route with a fairly even gradient profile (Liubotin–Poltava, 115 km). Hence it was practicable to agree a special schedule with the Traffic Department that needed only two or three different combinations of regulator and reversing lever setting. Just one type of coal was used for all trips, with an identical trailing load. The result, in Lomonosov's opinion, was reliable data that allowed rigorous comparison of the two locomotives. In this way he ascertained that the compound locomotives did use more coal than their predecessors in certain conditions and that they had seven significant shortcomings – a result sufficiently worthwhile for Shmidt to request an interim report for dispatch to the MPS.[34]

The results of the next tests in September 1899 were more far-reaching. First, they seemed to show that the 1897 type had four of the same shortcomings. In other words, the MPS was bulk-buying engines that were hardly better than their predecessors.[35] Second, they spurred Lomonosov to a radical thought. Because he could not identify uniform rates of steam consumption despite having what he called 'ideal' indicator diagrams and good dynamometer data, he concluded that constant experimental conditions were essential:

> I just could not obtain [graph] curves for steam consumption at given combinations of regulator setting, cut-off and speed. I didn't get anything from studying German and French authors: their [graph] curves were based on hypotheses which were dubious to a greater or lesser extent. At Easter 1900 I concluded that I had either to keep the reversing lever and regulator positions unchanged between water measurements or to abandon scientific road experiments. In other words, I set myself the task of creating, at least with

---

[33] 'Vospominaniia', vol. 2, pp. 415–19. On Shmidt see Kriger-Voinovskii, *Zapiski inzhenera*, pp. 61–2.

[34] See Lomonosov, *Opytnoe issledovanie*, pp. 32–3 and prilozhenie II, p. 14; and 'Vospominaniia', vol. 2, pp. 423–5, 460. For an English translation of the 'prilozhenie' in which the perceived shortcomings were detailed, see Heywood, 'Lomonosov: First Steps', pp. 9–11. The memoirs date Shmidt's request as November 1899, but a corresponding report has not been found; however, Lomonosov did submit a report exactly a year later: Explanatory note about rebuilding the freight 8-axle compound steam locomotives of the 1893–1897 government orders, 17 November 1900: RGIA, f. 273, op. 6, d. 2606, ll. 28–39.

[35] Lomonosov, *Opytnoe issledovanie*, pp. 33–6 and prilozhenie II, pp. 14–15; 'Vospominaniia', vol. 2, pp. 448–50, 460.

regard to experiment trains, the same conditions as Borodin and Goss created in stationary test-plants.[36]

Again, his thinking was controversial. Essentially he was condemning the theories of the engineer and deputy minister N.P. Petrov that were the main tool of the Russian railways for calculating the parameters of locomotive performance for such important purposes as traffic scheduling. Moreover, like Borodin, Lomonosov rejected most earlier research, which usually used regular trains on changing gradients and took water measurements only at stations and yards. He believed that it was essential to have a long constant gradient for sustaining hard work with a special train for at least 35 minutes, and to stop immediately before and after the incline to measure the water usage.[37] Yet this conclusion implied disruption to regular traffic, especially as most Russian routes were still only single-track with passing places.

Lomonosov's surprising, unprecedented achievement was to persuade the line's management to support his new thinking. Specifically, his ideas formed the basis of an official circular in April 1900. One of his two purposes was bureaucratic:

> I wanted to strengthen the legal position of the experiments and protect myself from [being given] unrelated assignments. To this end I drafted a circular for distribution throughout the railway, called 'On the procedure for conducting experiments on locomotives', which apart from anything else established the job of 'experiments manager'.[38]

The other was methodological. For the first time anywhere his document specified the aim of creating constant, repeatable 'scientific' conditions during locomotive road tests. To this end it institutionalized a wide-ranging series of principles and procedures. For instance, paragraph 8 entitled these trains to have special schedules, and paragraph 13 gave the researchers priority over most regular freight trains. Crucially, test trains could be halted wherever needed for the measurement of water consumption.[39]

That Lomonosov secured such extensive rights was largely due to an unusual, fortunate combination of factors. The document related only to his railway, where some of the principles and procedures had already been agreed incrementally. Doubtless also helpful was the support of formerly antagonistic colleagues like the chief of the Liubotin depot, Vaterlei. The availability of several long, fairly even gradients on the railway minimized the first, often insuperable hurdle in arranging steady sustained hard work for the locomotive. Essential for access to

---

[36]   'Vospominaniia', vol. 2, pp. 473–4.

[37]   'Vospominaniia', vol. 2, p. 474.

[38]   'Vospominaniia', vol. 2, p. 475.

[39]   Lomonosov, *Opytnoe issledovanie*, prilozhenie II, pp. 3–4; see also Heywood, 'Lomonosov: First Steps', pp. 12–13.

these locations, and for suitable schedules, was the management's support, and especially its rare willingness to waive operating restrictions and commercial priorities – decisions that doubtless would not have been agreed in the private sector. Not least, the railway's head, Shmidt, had some flexibility with his budget: he had to account to the MPS, but he could usually create some room for manoeuvre if he wished. Finally, for reasons that will be explained below, the MPS instructed Shmidt to give Lomonosov every assistance.

In the event, the launch of this new framework in the third batch of tests in June–August 1900 was not the conscious triumph that Lomonosov later claimed. His memoirs describe the inaugural trip on 4 June as the world's 'first road trip of a laboratory nature', which could justifiably be called an 'experiment' rather than a mere 'test'.[40] Certainly his work was innovative: such procedures had never been used anywhere before. But there is political exaggeration here too. Throughout his career he struggled to convince colleagues that such tests were genuinely scientific 'experiments' and that they produced data that could have valuable practical applications. Furthermore, it is significant that, as we shall see, his reports to the MPS at the time barely mentioned methodology.

Ironically, these tests were not a resounding success. Their purpose was to study modifications that were intended to reduce fuel consumption by improving steam distribution at low speeds. Operationally they went well. The only memorable glitch was a policeman's suspicion that the calorimeter's protective barrel contained an illicit vodka still; the misunderstanding was resolved amicably in a nearby brothel.[41] But the tests showed that Lomonosov's proposed modifications were less beneficial than hoped. The initial changes improved forward running, but operation in reverse (known as tender-first running) was worse. Subsequent changes only partially remedied this fault. Hence the railway's Traction Department endorsed only some of the proposals, on condition that tender-first operation was unimpaired. The first such modified engine was tested in February 1901, but disputes developed about how to interpret the results.[42] Fortunately, Buteskul was no longer antagonistic. Indeed, he mentioned the fuel data at the XXII congress of senior Russian traction engineers at Kiev in November 1900.[43]

This congress helped to bring Lomonosov into a national debate about improving the standard locomotives. It did so by prompting the establishment in February 1901 of an MPS commission to take evidence about problems with the '1897 type' locomotive and develop a replacement. This group evolved

[40]   'Vospominaniia', vol. 2, p. 503.

[41]   Lomonosov, *Opytnoe issledovanie*, pp. 33–6, 38; 'Vospominaniia', vol. 2, pp. 497–506.

[42]   'Vospominaniia', vol. 2, pp. 517, 561–2, 618.

[43]   *Protokoly zasedanii XXII Soveshchatel'nogo s "ezda*, p. 159; 'Vospominaniia', vol. 2, pp. 517, 563–4. First convened in the 1870s and loosely associated with the MPS, the traction congresses were held on more or less an annual basis as a forum for senior traction engineers from both the state and private sectors to discuss all aspects of the operation and development of the locomotive, carriage and wagon stocks.

into a permanent, influential committee for locomotive policy, which became known as the Shchukin Commission in honour of its chairman, Professor N.L. Shchukin. Its minutes for 1901 show that Lomonosov's research was easily the most thorough study of the standard locomotives.[44] Uniquely, a two-day meeting was organized in March 1901 to hear and discuss a report by him.[45] This event was his first *démarche* on the national scene, and as it coincided with the end of his project, it is an appropriate juncture to assess the initial impact of his methodology and findings.

The evidence suggests that as yet his impact was very limited in Russia and non-existent abroad. With regard first to his methodological ideas, the Shchukin Commission discussed them only briefly, and no one applied them elsewhere in Russia. Part of the explanation was limitations in the geographical, technical and administrative environment that had helped nurture his thinking. For example, long, constant gradients were scarce; a less supportive manager would have vetoed such unconventional ideas as stopping on the line to take measurements; and private railways were unlikely to risk delay to commercial traffic, or even to shoulder the direct research costs. Also, in Russia's centralized railway administrative environment the propagation of his methodology required a combination of MPS pressure and enthusiastic kindred spirits on the other lines. Yet the ministry gave no such lead, for the Shchukin Commission had reservations about his work. Some of the objections were justified, such as the point that the gradients were not perfectly constant. But others were contentious, notably that research using instruments like dynamometers (of which Lomonosov's project was just one example) merely gave 'useless academic' comparisons, as opposed to data that could have practical applications – a traditionalist criticism that Borodin had suffered back in the 1880s.[46]

Also relevant was the fact that Lomonosov made no effort to promote his road-test methodology. His reports of 1900–1901 downplayed the underlying philosophy and techniques. In contrast to Borodin, he did not publish any articles, and he gave only one lecture, to the Nikolaev branch of the Imperial Russian

---

[44]   The main report was: 'O sposobakh k ustraneniiu nedostatkov, zamechennykh v tovarnom parovoze normal'nogo tipa, zakaza 1897 goda', in *Protokoly zasedanii XXII Soveshchatel'nogo s"ezda*, pp. 157–201. The Shchukin Commission's minutes are in RGIA, f. 273, op. 6, d. 2606, and were later widely circulated in printed form as Inzhenernyi sovet MPS, *Zhurnaly Kommissii podvizhnogo sostava, tiagi i masterskikh* (from 1911: *Zhurnaly Kommissii podvizhnogo sostava i tiagi*). For a brief history of the commission see N.L. Shchukin, *Istoricheskii ocherk deiatel'nosti Kommissii podvizhnogo sostava i tiagi pri Inzhenernom sovete* ([Petrograd, 1916]). See also Westwood, *Soviet Locomotive Technology*, pp. 5–7.

[45]   Minutes of meeting no. 5, 12–13 March 1901: RGIA, f. 273, op. 6, d. 2606, ll. 24–7; Explanatory note, 17 November 1900: RGIA, f. 273, op. 6, d. 2606, ll. 28–39. Supplementary evidence dated 24 and 29 March 1901 is at ibid., ll. 59–61ob.

[46]   Minutes of meeting no. 1, 12 February 1901: RGIA, f. 273, op. 6, d. 2606, l. 3.

Technical Society in November 1899.[47] For this uncharacteristic reticence there was a simple explanation that was later forgotten or ignored: he did not wish to promote his methodology because he was far more interested in the proposed national testing plant. He ended his presentation to the Shchukin Commission in March 1901 by rehearsing the arguments for that project. Indeed, he had just made an unsuccessful attempt to have it built at the Warsaw Polytechnic Institute (WPI), where he was due to begin teaching in September 1901.[48] His move to the polytechnic system was also relevant, for it transferred him from the hierarchy of the MPS to that of the Ministry of Finances. Consequently his road-test methodology would not have a champion on the railways – a weakness fatal for any new thinking in this large bureaucracy.

The impact of his recommendations was similarly marginal. He was pleased that the Shchukin Commission approved some of them for the next upgrading of the standard freight locomotive (duly known as the '1901 type'), and that it told the MPS Directorate of Railways to circulate his reports throughout the network.[49] Also, a report about his work was discussed at the XXIII congress of traction engineers in November 1901.[50] Yet in practice his findings were virtually ignored. In 1908 he lamented that only about 50–70 locomotives had been modified to his specifications; by 1916 only 194 had been altered, which left 3,507 untouched.[51] For this neglect Lomonosov identified two reasons. One was the ministry's failure to make them compulsory, which was naturally interpreted by hard-pressed staff as a reason or hint not to spend time and money on them. The second was that

---

[47]   Explanatory note, 17 November 1900: RGIA, f. 273, op. 6, d. 2606, ll. 28–39; Minutes of meeting no. 5, 12–13 March 1901: RGIA, f. 273, op. 6, d. 2606, ll. 24–ob.; 'Vospominaniia', vol. 2, p. 460. The lecture cannot be corroborated because the society's Nikolaev branch did not submit an annual report for 1899: see *Zapiski Imperatorskogo Russkogo Tekhnicheskogo Obshchestva*, 9 (1900): 70.

[48]   Minutes of meeting no. 5, 12–13 March 1901: RGIA, f. 273, op. 6, d. 2606, l. 26.

[49]   For instance, it agreed with his recommendation for using Walschaerts instead of Joy valve gear. See Minutes of meeting no. 5, 12–13 March 1901: RGIA, f. 273, op. 6, d. 2606, l. 27; 'Vospominaniia', vol. 2, p. 626.

[50]   'O nedostatkakh, zamechennykh v tovarnom parovoze normal'nogo tipa, zakaza 1897 goda: rassmotrenie doklada inzh. Iu.V. Lomonosova po tomu zhe predmetu', in *Protokoly zasedanii XXIII Soveshchatel'nogo s"ezda inzhenerov sluzhby podvizhnogo sostava i tiagi russkikh zheleznykh dorog, sozvannogo v Moskve na 23 noiabria 1901 g.* (St Petersburg, 1902), pp. 439–62.

[51]   Lomonosov, 'Ob opytakh nad parovozami normal'nogo tipa', 14,Д; Rakov, *Lokomotivy*, pp. 144, 152. See also: 'O nedostatkakh, zamechennykh v parovozakh normal'nogo tipa, zakaza 1897 goda' and 'O rezul'tatakh sluzhby parovozov normal'nogo tipa, peredelannykh soglasno ukazaniiam, pomeshchannym v zakliuchenii XXIII soveshchatel'nogo s"ezda po voprosu 19, punkt b', in *Protokoly zasedanii XXIV soveshchatel'nogo s"ezda inzhenerov sluzhby podvizhnogo sostava i tiagi russkikh zheleznykh dorog, sozvannogo v Varshave na 23 noiabria 1902 g.* (St Petersburg, 1903), pp. 265–93 and 294–300.

railways which did experiment with his proposals were often discouraged by poor results. On this point Lomonosov was especially annoyed. He told the Technical Society in 1908 that the modified locomotives often belied expectations because railways had introduced additional changes without prior research, or had deviated from the specification, or even used incompatible components. As one might expect, he claimed that where the changes were made 'more or less correctly', railways were getting 'significant economies'.[52]

These explanations may well be valid, but they are incomplete. Lomonosov could hardly expect the MPS to advocate changes which, by his own admission, were not fully successful. Delicate fine-tuning was required, which could be awkward or impracticable in the rough-and-tumble of daily operation, and might imply higher maintenance costs. In other words, hard-pressed railway staff could easily see such precision engineering as more trouble than it was worth. Also, members of the Shchukin Commission rightly stressed the potential influence on efficiency of cultural, geographical and other factors. Older drivers, for example, tended to disagree that compound locomotives needed to be driven in a different way to the simples; hence, efforts might be needed to re-educate them. Similarly, the terrain appeared to have a significant influence on the fuel efficiency of locomotives.[53] Finally, as with the methodology, considerable scope existed at the local level for enthusiastic individuals like Baidak and Lomonosov to pursue new ideas, but in the absence of any central directive, the proliferation of these ideas would depend on informal contacts.[54] Yet Lomonosov's move to non-MPS academia meant that he could not play any such developmental role on the railways. And the fact that no one assumed his mantle on the Khar'kov–Nikolaev Railway was scarcely an encouraging endorsement of his work for the rest of the network.[55]

Towards a professorship

Lomonosov's early opportunity to join the education sector arose from Vitte's expansion of higher technical education. This policy included the creation of several polytechnic institutes that would offer four-year programmes with a more general curriculum than in specialist institutes like the IPS and that, at a time of rapid technological change, were to provide their students with a 'solid grounding in basic skills that would enable engineers to master additional specialities when their work required'.[56] Overseen by the Education Department at the Ministry of

---

[52] Lomonosov, 'Ob opytakh nad parovozami normal'nogo tipa', 13Д–14Д.

[53] For instance, Minutes of meeting no. 1, 12 February 1901: RGIA, f. 273, op. 6, d. 2606, l. 3.

[54] Precisely this problem occurred with some research on the Nicholas Railway: Minutes of meeting no. 1, 12 February 1901: RGIA, f. 273, op. 6, d. 2606, l. 3.

[55] Lomonosov, *Opytnoe issledovanie*, p. VI.

[56] Balzer, 'Educating Engineers', pp. 383–4.

Finances (relocated in 1905 to the Ministry of Trade and Industry), these new institutes were modelled on the Khar'kov Polytechnic Institute and were opened at Warsaw (1898), Kiev (1898), St Petersburg (1902) and Novocherkassk (1907). The Warsaw Polytechnic Institute focused on mechanics, chemistry and engineering construction, but ignored agriculture, which was covered elsewhere in the locality. The Kiev Polytechnic Institute (KPI) was 'the flywheel' of the policy: located in Vitte's home region, it had as founding director the renowned engineer V.L. Kirpichev (1845–1913), formerly head of the Khar'kov institute. Unlike Warsaw it had agriculture and, at Kirpichev's insistence, it emphasized experimental research; thus, for example, the institute had agricultural research premises near the city.[57] Crucially for Lomonosov, Warsaw and Kiev recruited their initial staff over several years to reflect the four-year unfolding of their teaching programme, and both required an expert in locomotives for September 1901.

Warsaw's vacancy was the first to attract Lomonosov's attention. The WPI began its search for a teacher (*prepodavatel'*) for locomotives in 1898 and invited Lomonosov to apply on Romanov's advice. Lomonosov was intrigued because the post was more senior than an assistantship and could lead to a regular professorship if he completed the postgraduate examinations and dissertation to qualify as a junior scientific researcher (*ad"iunkt*). Thus, although the salary was lower than his railway salary at Khar'kov, he sent an application to Warsaw and registered for the *ad"iunkt* programme at the IPS.[58] Late in 1899 the WPI marked him as its preferred candidate, and the Ministry of Finances accepted this recommendation. Accordingly, Lomonosov was ordered to spend two years preparing himself in Russia and abroad, with an annual bursary of 1,800 rubles; then he was to join the WPI with the title of teacher.[59]

The process of defining his preparatory programme epitomizes his consummate abilities to spot opportunities, play the bureaucratic system and alienate his immediate superior. The WPI's head, A.E. Lagorio, expected him to study abroad, as the ministry stipulated. But Lomonosov wanted to complete his Khar'kov project both for its intrinsic interest and for use in his dissertation; doubtless the advantage of keeping his railway salary was also relevant. The matter was complicated by the fact that he would lose the right to conduct locomotive experiments upon his

---

[57] See Balzer, 'Educating Engineers', pp. 371, 376–86. For an appreciation of Kirpichev see I. Ganitskii (ed.), *Illiustrirovannyi sbornik materialov k istorii vozniknoveniia Kievskogo Politekhnicheskogo Instituta: Pamiati Viktora L'vovicha Kirpicheva* (Kiev, 1914), pp. 13–24.

[58] The memoirs say that he registered in 'mechanical sciences', but his service record shows that he graduated in 'construction arts': 'Vospominaniia', vol. 2, p. 398; Service record, 1 January 1908: DAK, f. 18, op. 2, spr. 159, ark. 187. This change arose from the way that his appointment to the Kiev Polytechnic Institute was defined in 1901: see Chapter 3, this volume.

[59] 'Vospominaniia', vol. 2, pp. 370, 372, 374–8, 398, 460; Service record, 1 January 1908: DAK, f. 18, op. 2, spr. 159, ark. 188zv.

departure from MPS service in 1901. Referred to St Petersburg by a frustrated Lagorio, Lomonosov won over not just officials at the finances ministry but also V.K. Timiriazev, the head of personnel in the Minister's Chancellery at the MPS and henceforth a valuable contact for Lomonosov. A key result was apparently a letter from Count M.I. Khilkov, Minister of Ways of Communication, to Shmidt in Khar'kov, instructing him to help Lomonosov implement the instructions of the Minister of Finances. Also, Lomonosov received an 'open pass' (*otkrytyi list*), which gave him access to traction (i.e. locomotive- and rolling stock-related) facilities of all Russian state and private railways. Lagorio had been outmanoeuvred, though at the cost of his goodwill.[60]

Lomonosov's preparations thus encompassed three principal activities during 1900–1901. One was the experimental work: his project's expansion in 1899–1900 was undoubtedly motivated partly by the requirements for his dissertation, and Khilkov's letter to Shmidt helps explain the latter's exceptional support. Another activity was the reading for his first postgraduate examinations, which he passed in June 1901.[61] His third concern was a series of railway visits to collect research material and solicit advice about which subjects to cover in his courses. These trips were made during April and December 1900 and June–August 1901, and covered most railways of European Russia as well as the Putilov and Briansk locomotive-building factories.[62] Similarly, his attendance at the 1900 Paris International Exhibition was facilitated by the WPI and Ministry of Finances as part of his preparations. He was supposed to participate in expert commissions, but arrived too late for this work. Consequently he spent most of his time showing Russian visitors the railway exhibits. According to his memoirs, his free time was mostly devoted to illegal Russian political literature – useful preparation, perhaps, for working with students.[63]

Two of his consultations with Russian engineers would greatly influence his career. His discussions with E.E. Nol'tein, the traction superintendent of the Moscow–Kazan' Railway, concentrated on four topics: the dynamic action of wheels on rails; the ways of calculating speed limits; the mechanics of the train; and the resistance of trains. Lomonosov's later research and teaching would owe much to Nol'tein's thinking, and his recollections of Nol'tein were warm and generous.[64] By contrast, he would always have a poor relationship with Shchukin,

---

    60   'Vospominaniia', vol. 2, pp. 461–6, 570–571. Although these ministerial papers have not been traced at RGIA, there are similar exchanges from 1901–1907 in f. 25, op. 1, d. 2732.

    61   'Vospominaniia', vol. 2, p. 660.

    62   'Vospominaniia', vol. 2, pp. 478–81, 580–587, 668–77, 683–92, 697–700, 705–17.

    63   See 'Vospominaniia', vol. 2, pp. 524–5, 533–45, and Service record, 1 January 1908: DAK, f. 18, op. 2, spr. 159, ark. 188zv. On Russia's railway exhibits in Paris see 'Russkii zheleznodorozhnyi otdel na Vsemirnoi Vystavke v Parizhe', *Inzhener*, 8–9 (1900): 348–50.

    64   'Vospominaniia', vol. 2, pp. 707–13.

the chair of the MPS traction commission and Professor of Applied Mechanics at the St Petersburg Technological Institute. Technical disputes were always important here, but so too at the outset were Lomonosov's lack of deference, which bordered on insolence, and their differing educational backgrounds: Shchukin was a 'production engineer' (*inzhener–tekhnolog*), not a *puteets*.[65] Lomonosov had no time for Shchukin's view that lectures were secondary to practical experience in workshops, and that with regard to locomotives the students needed to learn only the theory of boilers and of machines, since a locomotive was merely 'a boiler on wheels'. In other words, Shchukin dismissed Lomonosov's view that the locomotive was unlike other machines and that special theories were needed to describe its operation. Furthermore, he rejected Lomonosov's contention that locomotive design needed to consider both purpose and environment, and he doubted the value of the Lomonosov-type road tests.[66] For the next few years these disputes had no great personal or practical significance because of Lomonosov's transfer to academia; but they would become extremely problematic once Lomonosov rejoined the railways in 1908.

Ironically, Lomonosov's careful preparation for the WPI was called into question at virtually the last minute. A trip to the Kiev Polytechnic Institute in May 1901 led to a discussion with the director, V.L. Kirpichev, about working there. That visit may well have been motivated by the fact that the proposed national locomotive test plant was now due to be built at the Kiev institute instead of Khar'kov – a move that was designed to ensure Kirpichev's continued involvement in the steering committee; as will be shown in Chapter 3, Lomonosov had tried and failed to get the plant reassigned to Warsaw. Also, Lomonosov doubtless knew that the KPI had a vacancy for a locomotive specialist due to the unexpected withdrawal of the appointed candidate. Unsurprisingly, the job was discussed, and Lomonosov was seriously tempted. His impressions of Kiev were positive, he liked Kirpichev, who seemed keen to acquire a *puteets*, and the terms of service were better than in Warsaw. He was especially attracted by the opportunity to teach railway design. However, the KPI Council was not expected to nominate a replacement until the autumn and Warsaw might not release him. Moreover, Sonia, who was half-Polish, preferred Warsaw.[67]

---

[65] Shchukin's education and appointments are detailed in his MPS service record: RGIA, f. 229, op. 19, d. 4833, ll. 1–14.

[66] 'Vospominaniia', vol. 2, pp. 661–3, 707–13.

[67] N.E. Putiat, 'Ob ustroistve ispytatel'noi stantsii dlia opredeleniia kachestva vody, topliva i parovykh kotlov', in *Protokoly zasedanii XXII Soveshchatel'nogo s"ezda*, p. 92; 'Vospominaniia', vol. 2, pp. 648–57, 666; vol. 3, p. 45. Awareness of the laboratory issues is evident from: Lomonosov–WPI Director, 28 November 1900: RGIA, f. 25, op. 1, d. 5814, ll. 349–56; Lomonosov–Minister of Finances, 11 December 1900: RGIA, f. 25, op. 1, d. 5814, l. 348–ob.; Ministry of Finances–KPI Director, 14 December 1900: DAK, f. 18, op. 1, spr. 155, ark. 1; and KPI Director–Ministry of Finances, 9 January 1901: RGIA, f. 25, op. 1, d. 5814, ll. 345–7. This issue is discussed in Chapter 3.

Ultimately, apart from the immense attraction of the locomotive test plant, four considerations brought Lomonosov to Kiev. One was irritation with the WPI: Lagorio was unwilling or unable to advance the date of his appointment from September to 1 July 1901, the date when his employment at Khar'kov ended; additionally, it was rumoured that Lagorio would not give him the locomotive department due to new circumstances. A related point was that Kiev could give him the dual status of teacher (*prepodavatel'*) and acting associate professor (*ekstraordinarnyi professor*), which Warsaw would not do. The third consideration was that Timiriazev at the MPS arranged for him to remain nominally on the books of the MPS for a further year on the basis of his IPS education. Lomonosov could thus have free rail travel for that year, which in turn meant that he could offer to teach at Warsaw for the 1901/1902 academic year while also beginning his career at Kiev. This arrangement, he thought, would discharge his immediate moral obligation to the WPI and give that institute enough time to find a long-term replacement. Additionally, he would thereby earn some much-needed extra income, and he would be better able to restructure his personal life.[68]

Those personal circumstances were the fourth consideration.[69] He recalled in his memoirs that his relationship with Sonia improved following their move to Khar'kov in 1898, which led to the birth of their son Vsevolod on 17 September 1899. However, the couple drifted apart again during the pregnancy, and thereafter he had liaisons with many women.[70] Adultery was illegal, but neither this nor his wife's feelings troubled him. In particular, he began two affairs that would last intermittently for many years, sometimes in parallel, with women who were briefly servants to the family. The first, with Mariia Ivanovna Shelkoplasova (Masha), began in April 1900 and produced two sons named Iurii (born February 1903) and Anatolii (born December 1906); the second, with Ksenia Andreevna Zabugina (Oksana), began in early 1901.[71] To judge by his memoirs, he told Sonia about these affairs in June 1901 to justify his rejection of her reconciliation overtures. The upshot was an agreement with Sonia on his terms. He would work in both Warsaw and Kiev for one year if possible, Sonia would spend the year in Warsaw as his wife, and he would keep Oksana as his mistress in Kiev. Indeed he began

---

[68]   'Vospominaniia', vol. 2, pp. 649, 680, 725–6.

[69]   'Vospominaniia', vol. 2, p. 726.

[70]   See especially 'Vospominaniia', vol. 2, pp. 365, 404–7, 416, 431–5, 439–40, 451–2; the birth date is confirmed at Service record, 1 January 1908: DAK, f. 18, op. 2, spr. 159, ark. 188. An obituary appeared in *Elektrichestvo*, 12 (1962): 88.

[71]   Memoir references for 1900–1901 include: 'Vospominaniia', vol. 2, pp. 476, 481–6, 508–9, 610–612, 619–20, 632–3, 641–4. The births are noted at 'Vospominaniia', vol. 3, pp. 733, 757–8, 1599, 1772, and the dates are confirmed in a family tree provided by N.S. Skachkova.

living with Oksana several days before the KPI Council conditionally approved his candidacy in August 1901.[72]

*   *   *

Lomonosov became involved in locomotive testing by chance, but he made the most of his opportunity. He devised a method of road testing that in his view was more rigorous than previous practice and promised greater accuracy, consistency and comparability of the performance data. This work was very much a product of his imagination, audacity and energy in his specific professional and geographical environment. Yet its impact was negligible at this time. His results – in terms of improvements to the performance of the 1893- and 1897-type standard freight locomotives – were insufficient to persuade the influential Shchukin Commission to adopt his modifications for the large stock of these locomotives. Furthermore, his transfer to the academic world left his research without a necessary champion on the railways. That said, Lomonosov did not press this matter very hard. By 1900 he wanted to work with a locomotive testing plant and for him the main value of his road tests was, literally, academic, as the subject of the dissertation needed for his educational career.

This tale qualifies the contention that a lack of interest and especially funding from the state largely explains the desultory nature of industrial research in Russia compared with Western Europe and North America. The state regulated and to a great extent owned the railway system, and it did not have a large central budget for research and development as such at this time. Yet research was certainly conducted under MPS auspices, not to mention work done by the privately owned railways and locomotive-building companies. Moreover, the organizational framework for that research was relatively flexible. Apart from the efforts of MPS-funded institutes, especially the IPS, the state-owned railways had the scope to initiate large projects without, apparently, any requirement for prior approval from the ministry provided that funds were found locally, at least initially. The annual congresses of traction engineers evidently had an important research role as a forum for senior specialists from the ministry, railways, institutes and industry to identify priorities, promote projects and disseminate results. The creation of the Shchukin Commission in 1901 was perhaps partly an attempt by the MPS to assert greater control over traction research. But its lukewarm attitude to Lomonosov's work was not merely a matter of discouraging local initiative: at issue fundamentally was the question of what constituted a rigorously scientific road test and whether such a test was practicable and worthwhile.

---

[72]   'Vospominaniia', vol. 2, pp. 493–4, 658–9, 718–20; vol. 3, pp. 8–9, 11–13; Minutes of KPI Council meeting, 28 August 1901: DAK, f. 18, op. 1, spr. 203, ark. 62–zv.

Chapter 3
Engineering Professor

Recession, war and revolution shook Russia during Lomonosov's six years in Kiev. Amid a European economic downturn there were slumps in the iron, steel, fuel and engineering industries and in foreign investment, with state expenditure on railway building and improvements falling sharply. These troubles overshadowed continuing growth in the net national product and net investment, and buoyant demand for consumer goods. Hopes and fears of revolution were fanned by rising social tension and violence, and by the emergence of oppositionist political parties and movements like the Russian Social Democratic Workers' Party, the Socialist Revolutionary Party and the liberal Union of Liberation. The shooting of many unarmed demonstrators in St Petersburg in January 1905 provoked public outrage, and the pressure on the tsarist regime was intensified by events such as strikes, the formation of trade and professional unions, and defeats in the Russo–Japanese War of 1904–1905. In October 1905 a railway strike became a general strike that threatened to topple the autocracy. Yet the tide of revolution turned in late 1905, and the regime survived to regain the political initiative during 1906–1907.[1]

This chapter and Chapter 4 together examine how Lomonosov fared during this period – years that to his mind included possibly the 'most interesting' and also the 'most shameful' episodes of his life.[2] The present chapter analyses his career as a polytechnic teacher and professor of engineering, initially at both Warsaw and Kiev for the 1901–1902 academic year and then wholly at Kiev. It explores his teaching and research activities, traces the impact of the 1905 revolution on his work and explains the ending of his polytechnic career in 1907. Through these experiences it also considers the broader question of how effectively the polytechnic institutes fulfilled the educational needs of the industrializing economy and of whether the tsarist regime was capable of making essential reforms in technical education.[3] The next chapter will investigate Lomonosov's political opinions and activities, showing how and why the revolution became a moment of truth for him.

---

[1]   See, for example, Gatrell, *Tsarist Economy*, pp. 29–47, 141–87; A. Ascher, *The Revolution of 1905*, 2 vols (Stanford, 1988–1992); J.D. Smele and A.J. Heywood (eds), *The Russian Revolution of 1905: Centenary Perspectives* (London, 2005); and J.W. Steinberg et al. (eds), *The Russo–Japanese War in Global Perspective: World War Zero*, 2 vols (Leiden, 2005–2006).

[2]   'Vospominaniia', vol. 3, preface. He does not explain 'shameful', but since he displays no regrets about his revolutionary activity, it presumably refers to his marriage, affairs and/or certain actions at work.

[3]   Balzer, 'Educating Engineers', p. 455.

The battle for curricular reform

The Kiev Polytechnic Institute's status as the 'flywheel' of progress in Russian higher technical education was embodied in its imposing, modern, purpose-built premises three kilometres west of the city centre. But in all other major respects the KPI was similar to existing Russian technical institutes. In command, as usual, was the Director (until 1902, Professor V.L. Kirpichev), who answered to the Education Department at the Ministry of Finances (from late 1905, at the Ministry of Trade and Industry). The Director chaired the KPI's Council, which comprised all the full and associate professors and (in 1902) one teacher. The Council was responsible for policy-making (including curricular matters), although its decisions were subject to approval by the ministry. For routine administrative issues there was a Board consisting of the Director, Inspector and deans of faculties. The KPI had four faculties (*otdeleniia*) for the science of mechanics (statics, dynamics, etc.), engineering (the design, construction and use of machines, buildings, etc.), chemistry and agriculture; as normal these faculties were sub-divided into departments (*kafedry*) and sections (*otdely*). The permanent staff, all of whom were state employees, consisted of professors, associate professors, teachers and support personnel, and additionally there was a contingent of temporary teaching staff. Teachers were paid per class-contact hour, whereas professors got a monthly salary for an agreed number of class contact hours. Lomonosov found that the pay was low compared with the state railways, but professors could earn additional money through hourly-paid overtime teaching, project supervision, overseeing a laboratory and so forth. Most of the student population, which had reached nearly 1,150 by September 1901, still came from the gentry and, to a lesser extent, the middle classes.[4]

Political reliability remained a key criterion for academic staff appointments, as the paperwork for Lomonosov confirms. His nomination depended on his delivery of a formal lecture to the Council's satisfaction; this process was completed in mid-September 1901. Only then did Kirpichev seek the ministry's approval, and only in December, upon receiving political clearance from the police, did the Education Department recommend the responsible deputy minister to agree. Lomonosov meanwhile fulfilled his duties with no option but to trust in confirmation. His final hurdle would be his *ad"iunkt* dissertation, which he had to submit within two years of his appointment; success would secure his job, whereas failure implied dismissal.[5]

---

[4] *Kievskii Politekhnicheskii Institut (Kratkii istoricheskii ocherk), 1898–1973* (Kiev, 1973), pp. 4–5; E. Kushch and M. Glovatskii, 'Kievskii Politekhnicheskii Institut nakanune i v period pervoi russkoi revolutsii', in P.N. Troitskii (ed.), *Kievskii Politekhnicheskii Institut: uchenye zapiski: Trudy kafedry Marksizma–Leninizma* (Kiev, 1956), pp. 124–6. The main published memoir by a KPI academic in the late tsarist period is E.O. Paton, *Vospominaniia: Literaturnaia zapis' Iuriia Buriakovskogo* (Kiev, 1962).

[5] Minutes of KPI Council, 28 August and 14 September 1901: DAK, f. 18, op. 1, spr. 203, ark. 62zv., 70; 'Vospominaniia', vol. 3, pp. 15, 26–7; Ministry of Finances

Unfortunately for Lomonosov the process of his appointment raised problems that would overshadow his entire career at the institute. Kirpichev had approval and funds to appoint a teacher in the Faculty of Mechanics for the subject of railway locomotives. But the ambitious Lomonosov wanted the rank of associate professor for the better status, pay and research opportunities. Despite his inexperience his wish was not unreasonable in the sense that such appointments before the dissertation defence did occur. But with neither approval nor funds for such a post in the Faculty of Mechanics, Kirpichev proposed to appoint him as teacher and acting associate professor in the Railways Section of the Department for Construction Arts in the Faculty of Engineering, which had a vacancy for an associate professor of railway design and operation. This idea was either a clever ruse or an endorsement of Lomonosov's inter-disciplinary thinking, or perhaps both. It seemed feasible because students from both faculties took the courses in locomotives and railways, and because Lomonosov could (and did) couch his locomotive research in terms of railway design and operation as well as mechanics. But the Council's 16:4 vote in favour upset the Faculty of Mechanics, presumably because it lost the post for locomotives. The decision also caused trouble between Lomonosov and the Dean of Engineering, V.V. Perminov, who taught surveying and railway design, and who perhaps felt threatened. Also, some colleagues like A.A. Radtsig, an expert on steam machines and mechanics, would never accept the notion of inter-disciplinary analysis in railway studies.[6]

That financial constraints may have compromised the KPI's academic standards, especially during the 1905 revolution, is evident from the arrangement of Lomonosov's initial salary and his dissertation defence. His dual status as teacher and acting professor raised the question of which salary regime should apply. Kirpichev selected the cheaper option. That meant a basic salary for the 1901/1902 academic year of just 1,200 rubles (six contact hours per week at 200 rubles per hour per academic year) – a 20 per cent reduction from Lomonosov's salary as a junior railway engineer. A concession was that, like a professor, he could be paid for supervising the students who elected to do design projects on locomotives.[7] In these circumstances Lomonosov's arrangement to teach at

---

Education Section–Department of Police, 16 October 1901: RGIA, f. 25, op. 1, d. 2732, l. 3; Department of Police–Education Section, 23 November 1901: ibid., l. 10; Ministry of Finances–KPI Director, 22 December 1901: DAK, f. 18, op. 2, spr. 159, ark. 7.

[6]   Minutes of KPI Council, 28 August 1901: DAK, f. 18, op. 1, spr. 203, ark. 62–zv.; KPI Director–Deputy Minister of Finances, 22 September 1901: RGIA, f. 25, op. 1, d. 2732, l. 1–2; KPI Director–Deputy Minister of Finances, 26 August 1904: DAK, f. 18, op. 1, spr. 454, ark. 22–zv.; Minutes of KPI Council, 18 January 1903: DAK, f. 18, op. 1, spr. 338, ark. 3–zv.; Minutes of Engineering Faculty, 12 December 1907: DAK, f. 18, op. 1, spr. 648, ark. 140; 'Vospominaniia', vol. 3, pp. 1–4. He remembered the vote incorrectly as 26:2: ibid., pp. 12–13.

[7]   Minutes of KPI Council, 3 and 17 November 1901: DAK, f. 18, op. 1, spr. 203, ark. 79–zv., 84–91zv.; Resolution, 5 November 1901: DAK, f. 18, op. 2, spr. 159, ark. 18; Traction

Warsaw as well as Kiev for the 1901/1902 academic year provided a useful fixed-term source of supplementary income, and may indicate generally that hourly-paid teachers felt pressured to work a second job. As for the dissertation, his plan was to offer a book-length analysis of his Khar'kov experiments. However, he fell behind schedule, and in 1903 he obtained a one-year extension.[8] A year later, having been denied a second extension, he sought and obtained permission to substitute an article about locomotive dynamics.[9] Surprisingly, he was examined at Kiev instead of at the IPS, but more importantly, the institute did not appoint an external examiner. This irregularity, the purpose of which was to save money during a budgetary crisis, was compounded by the unavoidable absence of one of the two internal examiners, Perminov, from the public defence on 24 April 1905. The institute was saved only by a third colleague's willingness to act as an unofficial examiner. The Faculty of Engineering and the Council ratified the defence, and hence the ministry confirmed Lomonosov's appointment as an associate professor.[10]

To judge by his memoirs, Lomonosov's initial preoccupations in relation to his teaching duties were typical for a novice tutor. He worried about his lecture preparation and inexperience, and endured some excruciating moments. For instance, some of his Warsaw students began disputing statements in his lectures.[11] That said, his situation was unusual in that he did much of the lecture-writing during the overnight train journeys that were required by his presence in Warsaw from Sunday to Wednesday, and Kiev from Thursday to Saturday – a demanding schedule, but one that was mitigated in his view by his 'bigamy' with his wife Sonia in Warsaw and mistress Oksana in Kiev.[12] Also, he was soon sufficiently confident to use his own research in his lectures and supervision. For example, when discussing how to calculate the performance parameters of locomotives, he

---

Department, Khar'kov–Nikolaev Railway–KPI Director, 12/15 November 1901: ibid., ark. 22.

[8]   Extensions were sought for five other colleagues in May 1902: Minutes of KPI Council, 4 May 1902: DAK, f. 18, op. 1, spr. 271, ark. 34zv. The beneficiaries included M.M. Tikhvinskii: Service record of M.M. Tikhvinskii, 15 March 1911: DAK, f. 18, op. 2, spr. 256, ark. 104zv.–105; and V.V. Perminov, whose case is discussed below. See also 'Vospominaniia', vol. 3, pp. 1001–2.

[9]   Service record, 1 January 1908: DAK, f. 18, op. 2, spr. 159, ark. 190zv.–91; Minutes of KPI Council, 28 August 1904: DAK, f. 18, op. 1, spr. 445, ark. 65zv.; Iu.V. Lomonosov, 'Sovremennye zadachi passazhirskogo dvizheniia na russkoi seti s tochki zreniia parovoznoi sluzhby', *Inzhener*, 10 (1903): 332–41; 12 (1903): 404–11; 1 (1904): 15–21; 'Vospominaniia', vol. 3, pp. 807–8, 872–5, 902, 1001–2.

[10]   Minutes of KPI Council, 1 June 1905: DAK, f. 18, op. 1, spr. 557, ark. 73zv.–74; Service record, 1 January 1908: DAK, f. 18, op. 2, spr. 159, ark. 191zv.–93; 'Vospominaniia', vol. 3, pp. 1002–6.

[11]   'Vospominaniia', vol. 3, pp. 42–3.

[12]   'Vospominaniia', vol. 3, pp. 27, 79–95.

quoted his Khar'kov findings and foreign information in preference to the Petrov formulae used by Russian railways.[13]

By the start of his second year in academia, when he was working solely at Kiev, his apprehensions were beginning to give way to irritation and frustration. A fundamental cause was the government's failure to ensure that the teaching of railway studies at the polytechnic institutes could really benefit the railways. At the heart of this problem was the KPI's subordination to the Ministry of Finances in the context of a very hierarchical, compartmentalized state bureaucracy. The bureaucratic boundaries meant, for example, that unlike the MPS institutes, the KPI did not automatically receive railway documents, locomotive components and suchlike for teaching purposes. Given that neither the MPS nor the Ministry of Finances seemed interested in changing this situation, Lomonosov tried to circumvent it by approaching his own contacts – a tactic that did yield useful donations, but which could not compare with a regular supply under MPS auspices.[14] Similarly, his contacts were helpful for organizing railway work placements and study visits for students, but he had no guarantee of continuity for these informal arrangements, especially as a state railway normally required an MPS directive to permit collaboration with a non-railway body like the KPI.[15]

Yet it was the opposition of a few colleagues to curricular reform that became, for Lomonosov, the most frustrating aspect of teaching at the Kiev institute. He may well have instigated the review of the engineering curriculum that began in early 1902, and he was certainly a central figure in those discussions.[16] His passion for his subject explains some of his vigour, but so too does his long-term feud with the Dean of Engineering, Perminov. Lomonosov underplays this tension in his memoirs, which define Professor D.P. Ruzskii as his most serious 'enemy' at the institute.[17] But other evidence suggests that the trouble with Perminov became a leitmotif of Lomonosov's KPI career. Undoubtedly Lomonosov considered Perminov incompetent. Also, Perminov was vulnerable: he had yet to defend his dissertation; colleagues were indignant at his disorganized handling of

---

[13] For example: 'Vospominaniia', vol. 3, pp. 15–16, 21–5. A published example of work that he supervised is: A.N. Turchaninov, 'Poiasnitel'naia zapiska k proektu dachnogo parovoza dlia Kh.–N. zh.d.', *Izvestiia Kievskogo politekhnicheskogo instituta*, Otdel mekhanicheskii i inzhenernyi, 7/4 (1907): I–IV, 175–252.

[14] For example: Report about the state of the Kiev Polytechnic Institute named after Emperor Alexander II for 1902: DAK, f. 18, op. 1, spr. 294, ark. 72; Minutes of KPI Council, 1 December 1904: DAK, f. 18, op. 1, spr. 446, ark. 225; 'Vospominaniia', vol. 3, pp. 30, 36.

[15] For example: Kiev Governor–KPI Director, 10 April 1902: DAK, f. 18, op. 1, spr. 289, ark. 2–zv.; 'Vospominaniia', vol. 3, pp. 29–33, 834, 841.

[16] Iu.V. Lomonosov, Note about the teaching of applied mechanics in the Engineering Faculty of the Institute, 20 January 1902: DAK, f. 18, op. 1, spr. 282, ark. 30–3zv. It is possible that his report provoked the review.

[17] 'Vospominaniia', vol. 3, p. 14.

faculty administration; and with four children to support, he feared the loss of his overtime income.[18]

Lomonosov's main proposals, the essentials of which were accepted by the KPI Council during the spring of 1902, were for the engineering faculty to provide a broader curriculum in applied mechanics and for his preferred holistic approach to railway design to be embedded in the railway-related curriculum. His thinking was heavily influenced by the railway-oriented IPS curriculum, but was also relevant for much of the industrializing economy. For instance, he complained about the lack of coverage of earth-movers, lifting machines, pumps and steam boilers, and he noted large gaps in the syllabi for metals and tooling. As for his proposals, he wanted the existing six courses reorganized as five. He thought that the theory of machines should be expanded into a broad course on machine-building, and two courses on thermodynamics, steam machines and locomotives should be merged into a single course on steam mechanics, with new material added. Finally, he wanted parts of the locomotive syllabus to be moved into a course on railway operations – a change that would match his research interests. He did not, however, broach the more fundamental question of the skills and knowledge that engineering students at a polytechnic institute were expected to acquire, or discuss the relationships between the disciplines and subjects.[19]

The feud between Lomonosov and Perminov helped to ensure that the process of developing these principles into detailed proposals became very fraught. Perhaps because Lomonosov had not consulted Perminov, the latter did not convene a meeting of the engineering faculty for over two months. At length the faculty formally instructed the Dean to chair a curriculum review, yet nothing happened before the autumn of 1902. A later investigation cited these delays as typical of Perminov's incompetent management, though ironically this charge was unfair: from April until August 1902 Lomonosov was away in the Far East supervising placement students, and then Perminov was ill during September.[20] In October, however, Perminov tried to seize the initiative by blaming personnel problems – including Lomonosov's 'periodic' lecturing – for much of the faculty's disarray. The way forward, he argued, was to make some new appointments and restructure the curriculum so that compulsory courses were supported by advanced final-year option courses. As for railway studies, he reasserted the status quo: he should

---

[18]   See, for example, Report of the commission for reviewing the organization of teaching in the institute's engineering faculty, 6 May 1904: DAK, f. 18, op. 1, spr. 282, spr. 454, ark. 9–zv.; Service record of V.V. Perminov, 1 January 1908: DAK, f. 18, op. 2, spr. 195, ark. 1–12; and 'Vospominaniia', vol. 3, pp. 5–6. Perminov's dissertation is discussed below.

[19]   Note about the teaching of applied mechanics, 20 January 1902: DAK, f. 18, op. 1, spr. 282, ark. 30–3zv.

[20]   [V. Perminov], Short note about the organisation of teaching in the engineering faculty, 5 October 1902: DAK, f. 18, op. 1, spr. 282, ark. 21; Report of review commission, 6 May 1904: DAK, f. 18, op. 1, spr. 282, spr. 454, ark. 9–zv.; and 'Vospominaniia', vol. 3, p. 501. The Far East trip is discussed briefly below: see note 43.

supervise all railway design and special projects, while Lomonosov took projects on traction and operations.[21]

If this retort gave the engineering faculty an opportunity to rethink its mission, that chance was squandered. Again no attempt was made to reassess the requirements of graduates and employers, or the relationships between the engineering disciplines and subjects. Instead, the review focused narrowly on disciplines and subjects, mainly in terms of identifying topics for inclusion. The first five meetings covered, respectively, applied mechanics, construction arts, hydro-technical sciences, architecture and railways (construction; operation). The sixth meeting agreed personnel recommendations for these proposals, and the last meeting finalized a report. References to context and aims were inconsequential; of more concern were the curricula of other institutes like the IPS and WPI and perceived gaps in the KPI's expertise.[22]

For Lomonosov the engineering faculty's unanimous endorsement of this report in May 1903 was a pyrrhic victory. His proposals for applied mechanics were accepted, and he could use the changes to expand his teaching in construction arts and especially railways, partly at Perminov's expense.[23] Perminov's reputation was damaged, and after more criticism he was ousted as Dean in 1904.[24] Overall, Lomonosov was becoming the more influential player in shaping the provision of railway-related subjects. Yet he was badly shaken. During 1903 he requested a transfer to the mechanics faculty, arguing disingenuously that his engineering faculty commitments were requiring a lot of work outside his speciality to the detriment of his research; nothing came of this *démarche*, possibly because his faculty objected to losing his post.[25] Meanwhile, he investigated potential jobs elsewhere, including some on the railways and one at the Moscow Higher Technical School.[26]

During the same period Lomonosov's morale and reputation were undermined by several unrelated scandals and disputes. For example, in July 1903 he took photographs of a river in which allegedly a lady could be seen bathing naked with her family. The lady complained to the Minister of Education. Lomonosov denied that he had done anything untoward, but he was summoned to St Petersburg,

---

[21] [V. Perminov], Short note, 5 October 1902: DAK, f. 18, op. 1, spr. 282, ark. 17, 20–21.

[22] Minutes of meetings of the commission concerning the organisation of teaching in the engineering faculty, 1, 8, 27 November 1902, 4 December 1902, undated, 3 May 1903: DAK, f. 18, op. 1, spr. 283, ark. 1–14.

[23] See Minutes of first, second and fifth meetings of the commission concerning the organisation of teaching in the engineering faculty, 1, 8 November 1902 and undated: DAK, f. 18, op. 1, spr. 283, ark. 1–4, 9–12.

[24] See Report of review commission, 6 May 1904: DAK, f. 18, op. 1, spr. 454, ark. 9–zv.; Service record of D.P. Ruzskii, 15 March 1911: DAK, f. 18, op. 2, spr. 227, ark. 78zv–79zv.; 'Vospominaniia', vol. 3, pp. 793–5.

[25] Minutes of KPI Council, 18 January 1903: DAK, f. 18, op. 1, spr. 338, ark. 3–zv.

[26] 'Vospominaniia', vol. 3, pp. 547–9, 730–33, 737–8, 753–4, 862, 868.

relieved of the negatives and told to destroy any prints.[27] Also, following Kirpichev's resignation in autumn 1902 Lomonosov quarrelled frequently with successive directors of the institute.[28] All in all, it would seem, he was not a favourite at the KPI. He received just one vote in a ballot for Dean in May 1906, and only three votes in a later ballot.[29] Certainly he always felt unsettled after 1902.

Not just personnel problems but also Russia's revolutionary crisis hampered the considerable efforts that Lomonosov made after 1903 to develop the railway curriculum. The KPI Council restarted the debate about curricular reform in May 1904, but only two meetings could be held before the summer vacation.[30] During the next academic year the discussions were suspended owing to the political crisis. In August 1906, with the general tension receding, the engineering faculty appointed Lomonosov to chair a commission to prepare proposals urgently. The outline recommendations won Council approval as early as 8 September, but the details took two more months of talks.[31] Ironically, Lomonosov's resignation a year later merely provoked more discussion that lasted until at least 1910. Radtsig, now Director, endeavoured to separate the teaching of railway mechanics from operations, which he saw as discrete topics linked only 'by chance' through Lomonosov, and eventually he got his way.[32] For the KPI, then, the long-term impact of Lomonosov's work on the curriculum was negligible.

---

[27]   M.V. Antonovskaia–Minister of Education, 23 August 1903; Minister of Education–Acting Minister of Finances, 16 September 1903; Deputy Minister of Finances–KPI Director, 4 October 1903; Lomonosov–KPI Director, 9 October 1903; KPI Director–Deputy Minister of Finances, 9 October 1903; Ministry of Finances–M.V. Antonovskaia, 9 November 1903: all in RGIA, f. 25, op. 1, d. 2732, ll. 41–6; 'Vospominaniia', vol. 3, pp. 798–809. For another case see KPI Director–Deputy Minister of Finances, 21 June 1904: RGIA, f. 25, op. 1, d. 2732, ll. 51–2; 'Vospominaniia', vol. 3, pp. 884–8.

[28]   For instance, see 'Vospominaniia', vol. 3, pp. 601, 604, 606.

[29]   Minutes of KPI Council, 27 May and 9 September 1906: DAK, f. 18, op. 1, spr. 639, ark. 54–zv., 71zv.

[30]   Report of review commission, 6 May 1904: DAK, f. 18, op. 1, spr. 454, ark. 9–zv.; Minutes of KPI Council, 6 May 1904: DAK, f. 18, op. 1, spr. 454, ark. 4–12; Minutes of eighth and ninth meetings of the commission concerning the organisation of teaching in the engineering faculty, [early/mid-May 1904], 15 May 1904: DAK, f. 18, op. 1, spr. 283, ark. 15–16zv., 28–37; 'Vospominaniia', vol. 3, pp. 793–5 (misdates the start of the special review as autumn 1903).

[31]   Minutes of Engineering Faculty, 24 August 1906: DAK, f. 18, op. 1, spr. 648, ark. 17zv.–18; Minutes of Engineering Faculty, 8 September 1906: DAK, f. 18, op. 1, spr. 648, ark. 21–3zv.; Minutes of Engineering Faculty Commission, 9 September 1906: DAK, f. 18, op. 1, spr. 283, ark. 44–zv.; Minutes of Engineering Faculty, 29 November 1906: DAK, f. 18, op. 1, spr. 648, ark. 53zv., 50.

[32]   See Minutes of Engineering Faculty, 12 December 1907: DAK, f. 18, op. 1, spr. 648, ark. 140; and the papers of DAK, f. 18, op. 1, spr. 714, which relate to 1907–1910.

Research: vision and realities

Lomonosov's research at the KPI was intimately connected with the proposed national state-funded locomotive test plant. His desire to join the institute in 1901 arose partly from the decision to situate the plant at the KPI instead of Khar'kov, and his involvement continued for some years; the project's failure epitomizes the bureaucratic difficulties inherent in research collaboration between state organizations located within differing ministerial hierarchies.

Lomonosov put the proposed test plant at the heart of his research plan during 1900 when still doing his road tests on the Khar'kov–Nikolaev Railway and preparing to move to Warsaw. Above all, the plant would enable him to fulfil his dream to develop theories of how the steam locomotive functioned. But for this to happen he would need easy, constant access to the equipment, which in turn meant that the plant had to be built in Warsaw. By November 1900, therefore, he was lobbying for the project to be transferred there. He deployed four main arguments. The first was the necessity of a test plant for high-quality teaching of the subject of locomotives, as was demonstrated by Goss's plant at Purdue. Second, with a canny eye on the bureaucracy's penchant for penny-pinching, he noted that if certain existing buildings were employed, the plant could be completed for 95,000 rubles at Warsaw, whereas Kiev's basic budget was 300,000 rubles. Third, he suggested that whereas Purdue's plant was used solely for teaching with one locomotive, a Warsaw plant could do contract research for railways and locomotive-builders: this would not only save the treasury from having to pay the operating costs but also offer great scope for systematic theoretical and applied research. Finally, he stressed Warsaw's greater importance as a railway centre compared to Kiev, and its ability as a border city to host a dual Russian- and European-gauge plant where foreign locomotives could also be tested on a commercial basis for the state's benefit.[33]

This vision had profound implications. One was Warsaw's potential to become the first international centre for locomotive teaching and research. In technical terms his proposal was an ideal way for the polytechnic sector to support this branch of the Russian economy: it encompassed fundamental research, the refinement of domestic equipment designs and the possibility for faster inward technology transfer in a field – locomotive design – where German and Austrian influence had been strong in Russia since the 1860s. Had his strategy been followed, it would have given Russia a unique facility with real potential to demand the attention of foreign traction engineers. To be sure, Purdue's laboratory had already been imitated by the Chicago and North Western Railroad (1895) and Columbia University in New York (1899); but those plants did not have scientific and commercial ambitions to

33 Lomonosov–WPI Director, 28 November 1900: RGIA, f. 25, op. 1, d. 5814, ll. 349–51; Lomonosov–Head of Education Section, Ministry of Finances, 11 December 1900: ibid., ll. 348–ob. The memoirs credit Lagorio with the idea of testing foreign locomotives: 'Vospominaniia', vol. 2, p. 568.

match Lomonosov's proposal. No other plants existed as yet, and the other plans were more limited in scope.[34]

One might expect this ambitious proposal to have captured the imagination of a strategic thinker like Vitte. Yet Vitte ignored the opportunity. Kirpichev responded to the Lomonosov report by defending the KPI's interest. He argued that because both institutes wanted such facilities and neither could cope alone with such a large project, the issue needed to be pressed more broadly in conjunction with the Russian railways and one of the new technical associations. As for the location, he asserted that Kiev had a larger concentration of people wanting the plant and that the KPI had easy rail access.[35] But Kirpichev had no answer to Lomonosov's intriguing international idea and financial argument, which he duly ignored. Nor was there any obvious substance to his claim that Warsaw could not cope. With the possible exception of rail access, Kirpichev's other points were hardly unique to Kiev. Nonetheless, in early 1901 Vitte backed Kirpichev. His reasons are unknown, but they presumably included Vitte's pet interest in the KPI as standard-bearer of the polytechnic system in his home region.[36]

The ramifications of this decision were far-reaching. One result was Lomonosov's desire to move to Kiev; indeed, Kirpichev stressed the proposed plant to the Minister of Finances as a key justification for Lomonosov's appointment.[37] Another result was that the railways began to expect that the plant would be mainly MPS-funded; and on that basis they demanded control and a more conveniently central location than Kiev, by which they meant St Petersburg or Moscow. Accordingly, a location debate became a major distraction yet again during 1901. That November, representing the KPI at the XXIII Congress of Russian Railway Traction Engineers, Lomonosov defended Kiev as the plant's home, but Moscow won the vote.[38] Lomonosov responded, it would seem, by persuading Vitte to insist that a plant be built at Kiev; whether this would be additional to, or instead of, a railway-controlled plant is unclear. A condition, however, was that construction should not start until after the completion of a plant at the privately owned Putilov

---

[34]   Carling, 'Locomotive Testing Stations', pp. 106–10.

[35]   Ministry of Finances–KPI Director, 14 December 1900: DAK, f. 18, op. 1, spr. 155, ark. 1; KPI Director–Head of Education Section, Ministry of Finances, 9 January 1901: RGIA, f. 25, op. 1, d. 5814, ll. 345–7.

[36]   See the minute of 16 January 1901 on KPI Director–Head of Education Section, Ministry of Finances, 9 January 1901: RGIA, f. 25, op. 1, d. 5814, l. 345.

[37]   KPI Director–Deputy Minister of Finances, 22 September 1901: RGIA, f. 25, op. 1, d. 2732, l. 1–2.

[38]   See S.F. Stempkovskii, 'Ob organizatsii zheleznodorozhnoi ispytatel'noi stantsii', in *Protokoly zasedanii XXIII Soveshchatel'nogo s"ezda inzhenerov sluzhby podvizhnogo sostava i tiagi russkikh zheleznykh dorog, sozvannogo v Moskve na 23 noiabria 1901 g.* (St Petersburg, 1902), pp. 899–940 (unusually, the debate was not minuted); 'Vospominaniia', vol. 3, pp. 124–7. Lomonosov's participation in the debate is confirmed at: Minutes of KPI Board, 12 February 1902: DAK, f. 18, op. 1, spr. 272, ark. 22zv.

engineering works in St Petersburg – a proviso that suited the finance ministry for postponing the expenditure, and Lomonosov for the chance to learn from Putilov's experience.[39] However, the delay proved fatal. By the time that Lomonosov got local planning permission to begin surveying work in 1903, the KPI Council was slashing expenditure, having grossly exceeded its construction budget, and it refused to give him any funds.[40] Soon afterwards, when the Japanese war and 1905 revolution pitched the empire into financial crisis, the impetus for building a state-owned national test plant was lost, and was never regained.[41] Lomonosov found scant consolation in visits to the Putilov test plant and cagey friendship with its chief engineer, M.V. Gololobov, because the latter denied him access to its data.[42] Road tests would thus remain Lomonosov's sole recourse when he returned to locomotive testing in 1908.

The receding prospect of a Kiev test plant, and Lomonosov's consequent need to rethink his research priorities, significantly altered the value of his academic affiliation for his research. The main advantages were his right to attend the congresses of senior traction engineers and undertake occasional study visits abroad as well as in Russia, plus the opportunity to conduct research during trips organized for student learning, such as his four-month visit to the Far East in

---

[39] 'Vospominaniia', vol. 3, pp. 136, 571, 574, 576, 581–5.

[40] Lomonosov–KPI Director, 28 April 1903: DAK, f. 18, op. 1, spr. 155, ark. 9–zv.; KPI Director–Superintendent of South Western Railways, 22 May 1903: ibid., l. 10; Superintendent of South Western Railways–KPI Director, 3/5 June 1903: ibid., l. 11; KPI Director–Kiev Governor, 22 May 1903: ibid., l. 13; Kiev Governor–KPI Director, 11 June 1903: ibid., ll. 15–16; Minutes of KPI Council, 25 October 1903: ibid., d. 338, ll. 46–ob.

[41] This national project was extensively discussed at the traction congresses and in the technical press during 1907–1914. Examples include: M.V. Gololobov, 'Proekt parovoznoi laboratorii dlia russkikh zheleznykh dorog', in *Protokoly zasedanii XXVII soveshchatel'nogo s"ezda inzhenerov sluzhby podvizhnogo sostava i tiagi russkikh zheleznykh dorog, v Varshave, v avguste 1909 g.* (St Petersburg, 1910), pp. 251–90; L.M. Levi, 'Eshche po povodu zheleznodorozhnoi opytnoi stantsii imeni A.P. Borodina', *Inzhener*, 8 (1911): 227–39; and 'Protokol No.64 zasedaniia Obshchego biuro tekhnicheskikh soveshchatel'nykh s"ezdov', *Izvestiia Obshchego biuro soveshchatel'nykh s"ezdov*, 9 (1914): 800–801. Apparently money for preparatory work was released in early 1917: B.B. Sushinskii, 'Opytnaia parovoznaia stantsiia imeni A.P. Borodina russkikh zheleznykh dorog', *Izvestiia Obshchego biuro soveshchatel'nykh s"ezdov*, 3 (1917): 176–7. However, the project was lost in the chaos of the 1917 revolutions and civil war.

[42] 'Vospominaniia', vol. 3, pp. 581–5, 892–4, 941, 944, 1018–19. On the Putilov works see J.A. Grant, *Big Business in Russia: The Putilov Company in Late Imperial Russia, 1868–1917* (Pittsburgh, 1999). So far as one can tell, only a few papers of this test plant survive, and are in TsGIA SPb, f. 1418, op. 7, d. 91. This collection and f. 1309 (papers of the company's Board) were inaccessible at the time of my visit to this archive. Gololobov published a full technical description: *Opytnaia parovoznaia stantsiia Putilovskogo zavoda* (St Petersburg, 1907). For brief biographical notes about Gololobov (1870–1919) see Melua, *Inzhenery Sankt-Peterburga*, p. 189.

1902.[43] It was through these activities, for example, that he got to know many of the network's traction engineers. One such acquaintance was Borodin's former assistant, L.M. Levi, who by 1900 was influential in policy-making for locomotive design and testing; in the event Levi was always dubious about Lomonosov's work. Among other notable acquaintances were P.I. Krasovskii, who was deputy head of the Moscow works of the Moscow–Kazan' Railway in 1904, and would be one of the key technical policy-makers on the Soviet railways in the 1920s; M.E. Pravosudovich, a long-standing critic of Lomonosov who became traction superintendent of the South Eastern Railways in 1906, and would become a deputy to Krasovskii in the 1920s; and B.I. Rozenfel'd, the traction superintendent of the Transcaucasus Railway, an early advocate of internal combustion engines for railway use, and father of the future Bolshevik leader L.B. Kamenev.[44] Other benefits of his travels included information about the latest practice in traction design, operation and maintenance, and ideas for two projects on fundamental aspects of railway operation and mechanics. One was to devise a method for identifying safe maximum speeds for a given class of locomotive, which he contemplated for over a decade. The other project was to devise a formula to determine the costs of railway operation; it was prompted by discussions during a visit to France in 1903, and still intrigued him 30 years later.[45]

Yet the organization of research activity at the KPI posed serious difficulties for him. The financial and administrative support was inadequate: decisions about research leave and funding were evidently made on an ad hoc basis, and the institute's ongoing financial crisis hardly helped. Also, the KPI's subordination to the Ministry of Finances isolated Lomonosov from the railways in important bureaucratic and intellectual ways. As a polytechnic professor he was not entitled to MPS support in the form of open access to railway installations, help with publication costs and so forth. Lomonosov therefore had to cultivate contacts in his own ministry's Education Department and at the MPS. But he had no answer

---

[43]   His Far East recollections are at 'Vospominaniia', vol. 2, pp. 197–501, 519, 544; and prints of some of his many photographs are in two albums: LRA, MS 716.6.1.1 and 716.6.1.2. An unexpected by-product of that trip was his appointment as a Mandarin of the Blue Opaque Button; he later recalled that his powers 'were limited to two executions a day': Reed, 'Lomonossoff', p. 47. For other examples of student trips see ibid., vol. 3, pp. 830–44, 1489–505.

[44]   See 'Vospominaniia', vol. 3, pp. 117–18, 124, 176–80 (Levi); 890 (Krasovskii) 759–62 (Voskresenskii, Sheglovitov); 730 (Pravosudovich) and 116, 772–3 (Rozenfel'd). Lomonosov learnt of the connection with Kamenev in 1919: 'Vospominaniia', vol. 7, p. 732.

[45]   For example: KPI Director–Kiev Governor, 12 November 1902, and reply, 15 November 1902: DAK, f. 18, op. 2, spr. 159, ark. 37–8; Report about the state of the Kiev Polytechnic Institute named after Emperor Alexander II for 1903: DAK, f. 18, op. 1, spr. 340, ark. 41, 49zv.; *Protokoly zasedanii XXIV Soveshchatel'nogo s"ezda*, p. 602; Report by the Chair of the Rail Commission, Member of the Engineering Council S.K. Kunitskii, 15 October 1915: RGIA, f. 240, op. 1, d. 1175, ll. 1–13ob.; 'Vospominaniia', vol. 3, pp. 66–78, 526–9, 563–70, 608–51.

to a change of MPS policy in 1903 whereby his system-wide travel permit was replaced with just four specific tickets per year.[46] As for intellectual barriers, he sensed scepticism among railway officials that a non-railwayman like himself could have expertise about locomotives. He also found, like Borodin two decades earlier, that they often dismissed his emphasis on science and theory as irrelevant to 'life'– an infuriating impasse for one so passionate about the practical value of his research. Whereas he was proud to represent 'science' (*nauka*) at traction congresses, some delegates criticized him precisely for that connection. Similarly, Shchukin and Pravosudovich dismissed his views about a proposed locomotive design for the South Eastern Railways in 1906 as 'academic rubbish'.[47]

Such complaints might appear to be just routine professorial grumbling. But the lack of financial support nearly brought him personal ruin. His difficulties stemmed from the enthusiasm with which he fulfilled his duty to publish. First to appear, for the benefit of his students, was a lithograph edition of his locomotive lectures.[48] For the same audience he edited and annotated a translation by Sonia of an influential textbook on steam machines by J.A. Ewing, Professor of Engineering at the University of Cambridge; Lomonosov wanted this book translated because it was close to the syllabus for steam mechanics used in the KPI's Faculty of Engineering and because 'the originality of its method and elegance of its exposition' would have wider appeal for specialists in steam machines.[49] His inaugural scientific article appeared in the winter of 1903–1904 in *Inzhener* (*Engineer*), a significant Kiev-based journal; it presented his 'formula

---

[46]   For example: KPI Director–[Ministry of Finances Education Section], 5 November 1901: DAK, f. 18, op. 2, spr. 159, ark. 24; Deputy Minister of Finances–Minister of Ways of Communication, 2 January 1903, and reply 20 January 1903: RGIA, f. 25, op. 1, d. 2732, l. 31–2; Lomonosov–KPI Director, 14 August 1903: DAK, f. 18, op. 2, spr. 159, ark. 41; Report, 12 March 1905: ibid., l. 56–ob.; Acting Minister of Ways of Communication–Deputy Minister of Finances, 30 April 1905: ibid., l. 64; KPI letters of introduction at ibid., ll. 51–2, 112, 116–21; 'Vospominaniia', vol. 3, pp. 28–33, 575–7, 589, 593, 892.

[47]   For example, 'Vospominaniia', vol. 3, pp. 123–4, 126, 553. The SER-related criticism is at ibid., p. 1411.

[48]   *Konspekt kursa parovoza, chitannogo v Kievskom i Varshavskom politekhnicheskikh institutakh i. ob. ekstraordinarnogo professora Iu.V. Lomonosovym* (Kiev–Warsaw, 1901–1903); see also: Report about the KPI for 1902: DAK, f. 18, op. 1, spr. 294, ark. 45.

[49]   Iu.V. Lomonosov (ed. with annotations) *Parovaia mashina i drugie teplovye dvigateli*, by J.A. Ewing (Kiev, 1904); 'Vospominaniia', vol. 3, pp. 103, 845–55. The English source was J.A. Ewing, *The Steam-Engine and Other Heat-Engines*, 2nd edn (Cambridge, 1897). Ewing remarked 'jestingly' in 1904 that the translation 'might become a best seller' (because Russia was at war with Britain's ally, Japan): A.W. Ewing, *The Man of Room 40: The Life of Sir Alfred Ewing* (London, 1939), p. 147. On Ewing at Cambridge, including his reorganization of teaching in the Department of Engineering, see T.J.N. Hilken, *Engineering at Cambridge University, 1783–1965* (Cambridge, 1967), pp. 107–26, and E.I. Carlyle, rev. W.H. Brock, 'Ewing, Sir (James) Alfred (1855–1935)', in *Oxford Dictionary of National Biography*, vol. 18 (Oxford, 2004), pp. 822–3.

of the 4Z', which mathematically described the dynamic loading on a locomotive wheel, and was also the article that he used as his dissertation.[50] His book about his Khar'kov experiments was eventually published in 1907.[51] His other writings included articles about the mechanics of the train, which formed part of his attempt to supersede Petrov's formulae, and would lead to his textbook *Tiagovye raschety* (*Traction Calculations*) by 1912.[52] However, most of these publications required many drawings, which were expensive to prepare and print. Lomonosov always requested financial assistance, but KPI grants, if given, were much smaller than the costs, and the Ministry of Finances was uninterested. The MPS often assisted authors employed in its realm when their topic was unattractive for commercial publishers, but it did not normally support writers who were not on its payroll.[53] The result for Lomonosov was the question of whether to use his own money.

This situation produced a personal crisis because of bad luck and his cavalier attitude towards savings. Lomonosov was perfectly capable of living economically: he saved over 30 per cent of his expense allowance for his trip to the Far East in 1902. But generally he lived for the moment. When Sonia and their two children moved to Kiev in late 1902, he reportedly gave Oksana 1,000 rubles – virtually all his Far East savings – to buy a plot of land in the country so that their affair could continue. And he decided to use his own money to fund his publications if necessary. Thus, for instance, he funded the Ewing translation out of a monthly income of nearly 600 rubles (nearly half of which came from Sonia's parents), from which about 350 rubles went on family expenses (including 30 rubles per month to Masha) and about 100–200 rubles could be saved; the point here is his readiness to commit all that reserve to the Ewing book.[54] But whereas the sales of the translation covered his costs, disaster ensued with the Khar'kov book. This enterprise began well when, in 1904, he obtained an exceptional MPS promise to contribute 1,500 rubles upon publication. Encouraged, he decided to print each chapter as he finished it and pay the bills himself ahead of reimbursement in about 1906. But after the printing had started the political crisis intervened. In September

---

[50] 'Vospominaniia', vol. 3, pp. 538–9, 808. See also note 9 above.

[51] Lomonosov, *Opytnoe issledovanie.*

[52] For example: Iu.V. Lomonosov, 'Naivygodneishii sostav tovarnykh poezdov', *Inzhener*, 2 (1904): 52–5; 'Nabliudeniia nad stepen'iu sukhosti para v reguliatornoi trube parovoza normal'nogo tipa', *Inzhener*, 5 (1905): 135–40; and 6 (1905): 181–6; 'Tochnyi vyvod uravneniia dvizheniia poezda', *Izvestiia Kievskogo politekhnicheskogo instituta, Otdel mekhanicheskii i inzhenernyi*, 3 (1905): 1–6. Also: Iu.V. Lomonosov, *Tiagovye raschety i prilozhenie k nim graficheskikh metodov* (St Petersburg, 1912).

[53] For examples of his requests for publication subsidies see: Minutes of KPI Board, 22 April 1903, 28 August 1903, 29 October 1903: DAK, f. 18, op. 1, spr. 339, ark. 112zv., 200, 249; Minutes of KPI Board, 27 January 1904: DAK, f. 18, op. 1, spr. 446, ark. 17–zv.; Minutes of KPI Council, 4 May 1904, 18 June 1904: DAK, f. 18, op. 1, spr. 445, ark. 27zv., 51zv.; notes in his personnel file: DAK, f. 18, op. 2, spr. 159, ark. 49–50, 57–9.

[54] 'Vospominaniia', vol. 3, pp. 512–13, 532, 846, 848, 855.

1904 students refused to pay their tuition fees, and the cash-strapped KPI withheld all supplementary payments to staff for such duties as project supervision. The consequent loss of income was catastrophic for Lomonosov and so, presumably like many other professors, he sought temporary work elsewhere to cover the shortfall. Unfortunately his search took months and prevented him from working on the book. Meanwhile, to pay his printing bills and eat, he had to negotiate loans and pawn the family silver. Nonetheless by 1906 he faced a printing debt of perhaps 900 rubles, which the Ministry of Trade and Industry and the MPS refused to cover, and which neither he nor the KPI could pay. The printer got his money several years later via a lawsuit.[55]

Lomonosov's search for other work ended in May 1905. In normal circumstances the landmarks of that year would have been his successful dissertation defence and his confirmation as an associate professor. But these achievements were overshadowed by his financial crisis, not to mention the revolution. It was an enormous relief to be offered a short-term position as a consultant 'engineer for technical work' with the private South Eastern Railways company (SER).[56] At about the same time he became a consultant to the Briansk locomotive-building works. Nothing more is known of the latter job, which conceivably was a sinecure arranged by E.M. Gorokhov, his former colleague from the Khar'kov factory.[57] But the SER would dominate his professional life for over a year.[58]

The values of consulting

Except for the locomotive test plant, the polytechnic sector was never likely to become an important source of consulting expertise for the Russian railways. For any state-owned railway the easiest recourse – and cheapest too, if payments were ever needed – was the expertise abundantly available within MPS institutes like the IPS, on other railways or at the ministry. In theory the small private railway sector offered a more likely clientele for paid consulting by polytechnic academics, but in practice the scope was minimal: those railways also had the option of obtaining assistance from MPS sources, and they were mindful of their dependence on the usually grudging goodwill of the MPS as regulator of their activities. Moreover, polytechnic academics could not normally be absent from their institute for the large amount of time that substantial work for a distant company could entail.

---

[55] See KPI Director–Minister of Trade and Industry, 23 December 1906, and Minister of Trade and Industry–Ministry of Ways of Communication, 9 January 1907: RGIA, f. 25, op. 1, d. 2732, ll. 90–91ob.; 'Vospominaniia', vol. 3, pp. 846, 852–3, 952–4, 968–9, 1000, 1718, 1781–2; vol. 5, pp. 25–6, 30–31, 36.

[56] 'Vospominaniia', vol. 3, pp. 1022–3.

[57] 'Vospominaniia', vol. 3, pp. 834–6.

[58] For contemporary information about the company see Obshchestvo Iugo-Vostochnykh Zheleznykh Dorog, *Otchet po ekspluatatsii dorog za 1905 g.* (St Petersburg, 1906).

Lomonosov would undoubtedly have had opportunities for consulting work and close cooperation with the Russian railways if a locomotive test plant had been built at Kiev or Warsaw. In practice he gave little if any thought to consulting during his first three years at Kiev, and he began searching for such work only because of the sudden problems with his income in 1905. Once he had found a position with the SER, he was able to continue for so long only because teaching remained suspended and because he could still devote sufficient time to his KPI administrative responsibilities, including numerous Council meetings about the political situation.[59] He was lucky too that, with its company headquarters in St Petersburg and operating headquarters in Voronezh (about 750 km east of Kiev via Kursk), the SER was relatively accessible.

This relationship offered him several professional benefits. One was his first opportunity to scrutinize a large, important private railway: the SER covered a huge area of southern Russia, including major cities like Khar'kov, Voronezh, Rostov, Tsaritsyn (now Volgograd) and Novocherkassk, and carried huge amounts of locally mined coal and transit freight from the south. Second, he encountered a wide variety of issues. For example, in the realm of traction policy he had to assess proposals for improvements to the Voronezh workshops, new designs of locomotive and a new type of high-pressure locomotive boiler that had been developed in Austria by Johann Brotan; operating and commercial issues included yard design, depot location and traffic costs.[60] Finally, he experienced the workings of private ownership, including many practices that were new to him. He was struck especially by the considerable power of the Chairman; the Board's preoccupation with short-term financial considerations; the strong influence of private banks; and the obstructiveness of MPS officials. A notable irritation was the prevalence of corruption in the SER's dealings with the ministry. He reports, for instance, that

---

[59] The educational context of these meetings was the reform of higher education in August 1905, on which see, for instance, S.D. Kassow, *Students, Professors and the State in Tsarist Russia* (Berkeley and Los Angeles, 1989), especially pp. 228–36 and D. Wartenweiler, *Civil Society and Academic Debate in Russia, 1905–1914* (Oxford, 1999), pp. 43–81. During 1906 Lomonosov was especially involved with reorganizing the institute's laboratories: see, in particular: Minutes of Engineering Faculty, 26 January, 23 March, 4 May 1906: DAK, f. 18, op. 1, spr. 648, ark. 1–zv., 8, 13zv.–14zv.; Minutes of Commission for Reorganising and Coordinating the Laboratories and Study Rooms, 16 February 1906: DAK, f. 18, op. 1, spr. 648, ark. 4; Minutes of Commission for Reorganising and Coordinating the Laboratories and Study Rooms, 22 February, 6 March 1906: DAK, f. 18, op. 1, spr. 283, ark. 45–8zv.

[60] 'Vospominaniia', vol. 3, numerous references in the page ranges 1022–188 and 1301–477. His first tour is covered in pp. 1099–172. His photographs of the Voronezh workshops survive at LRA, MS 716.6.1.6, p. 13. On the Brotan boiler see Iu.V Lomonosov and A.I. Lipets, *Poiasnitel'naia zapiska k proektu kotla Brotana s peregrevatelem Piloka–Slutskogo dlia tovarnogo parovoza kompaund normal'nogo tipa* (Kiev, 1906); 'Vospominaniia', vol. 3, pp. 1046–8; and G. Szontagh, 'Brotan and Brotan–Dreffner type Fireboxes and Boilers applied to Steam Locomotives', *Transactions of the Newcomen Society*, 62 (1990–1991): 21–51.

the railway spent 50,000 rubles per year on bribes, and that the head-office payroll included several women whose responsibility was to give sexual favours to MPS officials who could not be bribed.[61]

One of the traction debates was a portent of future controversy, revealing the interplay of personalities and technical criteria in design decision-making. It concerned a proposed ten-wheel freight locomotive, for which the main options were a 0-10-0 wheel arrangement where all the wheels were powered and a 2-8-0 arrangement with eight powered wheels and two non-powered wheels at the front. The first option would have better wheel/rail adhesion and hence be more powerful when climbing gradients because its 'coupled weight' would be higher – in other words, because the driving wheels carried the engine's total weight (excluding the tender). By contrast, the 2-8-0 could have a higher maximum speed on level track because its non-powered leading axle would cause lower stresses in the rails than a 0-10-0. For various reasons, including the prevalence of hills on the SER, Lomonosov prioritized the speed on gradients, and therefore the 0-10-0; he thereby reflected, probably consciously, an influential school of thought in Germany and Austria, where the same question was being asked. But the 2-8-0 was advocated by Professor Shchukin at the MPS and the SER's traction superintendent, Pravosudovich. Each case had merits, but personalities and prejudices predominated in the absence of agreed technical and economic criteria for comparing costs and benefits. Lomonosov being the junior player, the SER supported Shchukin. Significantly, Shchukin was also winning the same debate within the MPS about the specification for the next generation of standard freight locomotive for the state railways; but Lomonosov would reopen this issue from 1908, much to Shchukin's annoyance.[62]

If Lomonosov gained so much from his consulting work, and had been unsettled at the KPI for so long, why did he not join the SER on a permanent basis? He did have opportunities, and he was certainly tempted, for all his dislike of private enterprise. For instance, in May 1906 he welcomed the possibility of a 12,000-ruble salary for managing the technical and operations departments. However, this prospect evaporated in June with the failure of his company patron, R.G. Salome, to be elected as chairman. And the issue came to a head in September 1906, just as teaching was about to resume at the KPI. The new chairman, N.L. Markov, appointed S.N. Kul'zhinskii, the son of the former owner, to head the operations department and decreed that Lomonosov could remain a consultant only if based in St Petersburg. Lomonosov preferred to stay at the KPI. One reason, he claims, was his renewed interest in academic work: he had just been promoted to full professor, and was pleased at the prospect of teaching again. Other considerations were the security of a regular state salary, his dislike

[61] On the MPS, the company's short-termism and corruption see 'Vospominaniia', vol. 3, especially pp. 1029, 1060–71, 1351–64, 1422–3.

[62] 'Vospominaniia', vol. 3, pp. 1410–1411, 1463. On the origins of the 2-8-0 arrangement in Russia during 1896–1905 see Rakov, *Lokomotivy*, pp. 160–176.

of Markov and Kul'zhinskii, and his fear of being unable to do research in the private environment.[63]

Markov's ultimatum raises the question of whether the SER gained much benefit from Lomonosov. A definitive response is impossible because virtually all the relevant papers have not survived. But the available information strongly implies that his work had only marginal importance for the railway. His main recommendations were evidently ignored, such as his ideas for traffic reorganization, his case for a 0-10-0 freight engine and his proposals for experiments with the Brotan boiler. And in the case of a proposed 4-6-0 passenger locomotive, where he did have some influence, Lomonosov himself was disappointed at his failure radically to improve the design: the class Z, as it became, won only a modest reputation in service.[64]

After this somewhat hectic episode the direct longer-term effects of the 1905 revolution on Lomonosov's professorial career were minimal: the main changes in his routines were unrelated to that crisis. In his research he concentrated on completing the Khar'kov book, but new interests continued to emerge. From the autumn of 1906 he participated in an informal KPI discussion group about electric and diesel traction, which arose from the interests of his old IPS friend Dubelir, who had just been appointed as the professor of local transport; these seminars marked the start of Lomonosov's important association with diesel traction. Also, in 1907 he began writing about the mechanics of a train – what he called his theory of traction; this too would become one of his best-known projects.[65] As for teaching, his promotions enabled him to have a former student, A.I. Lipets, appointed as an assistant, and to resist pressure to introduce new courses.[66] His administrative duties were again dominated by curricular change, as discussed above, and his conflict with Perminov resumed.

---

[63]   'Vospominaniia', vol. 3, pp. 1301–2, 1306–7, 1385, 1387, 1422, 1435–7, 1457, 1477–9.

[64]   On the 4-6-0, which became class Z in 1912, see Rakov, *Lokomotivy*, pp. 226–7; 'Vospominaniia', vol. 3, pp. 1033–5, 1059–60, 1455–7. The Brotan boilers were tried on these locomotives, but were soon removed: Rakov, *Lokomotivy*, p. 226; 'Vospominaniia', vol. 4, pp. 249–50.

[65]   'Vospominaniia', vol. 3, pp. 1482–3, 1681–2, 1698. Note that his request for expenses associated with preparing his congress report was granted: Minutes of KPI Board, 6 February 1907: DAK, f. 18, op. 1, spr. 707, ark. 27–zv.

[66]   Minutes of Engineering Faculty, 26 January 1906: DAK, f. 18, op. 1, spr. 648, ark. 2; Minutes of Engineering Faculty, 24 August 1906: DAK, f. 18, op. 1, spr. 648, ark. 18; Minutes of Engineering Faculty, [September 1906]: DAK, f. 18, op. 1, spr. 648, ark. 19; Minutes of KPI Council, 9 September 1906: DAK, f. 18, op. 1, spr. 639, ark. 74; 'Vospominaniia', vol. 3, pp. 1200–1201, 1331–2, 1482, 1720–1721. After an eventful career Lipets died in the United States in 1950: see his KPI file at DAK, f. 18, op. 2, spr. 158; '[A. I. Lipetz]', *Railway Mechanical Engineer* (June 1936): 276; and 'Lipetz, 68, Retired Alco Engineer, Dies', *Schenectady Gazette*, 18 April 1950, p. 18. I am grateful to Cliff Foust for the obituary reference.

Yet Lomonosov's polytechnic career was now approaching its conclusion. The overture was Perminov's controversial dismissal, in which Lomonosov played a leading role. Ostensibly the sacking was due to Perminov's failure to defend his dissertation. He had submitted it in 1904, and the external examiner from the IPS had condemned it as basically an unoriginal digest of the literature; now the same criticisms were levelled at the revised version of 1906. A controversy arose because one of those critical examiners was Lomonosov, probably Perminov's staunchest enemy. Perminov sought permission to defend elsewhere on the grounds that his KPI examiners would be biased, but his request was denied. The upshot in October 1907 was the institute's refusal to accept the revised text for formal public defence, and thus Perminov had to leave. Whether Lomonosov's involvement became unsavoury is impossible to ascertain, but one does sense a hint of malevolent satisfaction in his remark 40 years later that Perminov was 'chased away'.[67]

Lomonosov's own resignation a few weeks later was partly a product of shabby treatment by colleagues. Following criticism of Perminov for not organizing any summer practical placements in land surveying for several years – hardly a fair complaint given the revolution – the KPI Board appointed Lomonosov to do this task for 1907. The plan was to assist the Kiev provincial authority by checking calculations for a land drainage scheme on the River Ros'. All concerned, including the provincial Governor-General and future Minister of War, V.A. Sukhomlinov, were nervous that radical students would try to provoke revolutionary unrest among the peasantry. However, the project passed off peacefully, not least, according to the memoirs, because Lomonosov secured the cooperation of the revolutionary parties. The scandal came later. When Lomonosov submitted his report and accounts in September 1907, Ruzskii and other colleagues complained about financial irregularities. These allegations were checked and rejected, and Lomonosov's report was accepted, but the furious professor decided to resign at the earliest opportunity.[68] The denouement followed quickly. His preference, apparently,

---

[67] This paragraph is based on Perminov's personnel files (DAK, f. 18, op. 2, spr. 195, 196) and a file about his dissertation (DAK, f. 18, op. 1, spr. 452). On the final decision see: Minutes of Engineering Faculty, 29 November 1906: DAK, f. 18, op. 1, spr. 648, ark. 51zv., 48. Lomonosov misdates the episode as about 1905, instead of 1906–1907: 'Vospominaniia', vol. 3, p. 1201.

[68] For example: Minutes of Engineering Faculty, 8 November and 13 December 1906: DAK, f. 18, op. 1, spr. 648, ark. 41, 44zv., 58–9zv.; Minutes of KPI Council, 14 December 1906: DAK, f. 18, op. 1, spr. 639, ark. 131; Minutes of Engineering Faculty, 19 March 1907: DAK, f. 18, op. 1, spr. 648, ark. 75–6zv.; KPI Director–Governor-General, 3 April 1907: Derzhavnii arkhiv Kyivskoi oblasti (DAKO), f. 2, op. 44, spr. 171, ark. 1–zv.; Head of Okhrannoe Otdelenie–Governor-General, 16 May 1907: DAKO, f. 2, op. 44, spr. 171, ark. 4; Governor-General–Cherkassy District Police Officer, 30 May 1907: DAKO, f. 2, op. 44, spr. 171, ark. 6–7; Minutes of Engineering Faculty and attachments, 21 September 1907: DAK, f. 18, op. 1, spr. 648, ark. 104, 108–10; Minutes of Review Commission for Reports on the Engineering Faculty's Summer and Winter Practicals in Land Surveying in the 1906/7 Academic Year, 26 September 1907: DAK, f. 18, op. 1, spr. 648, ark. 111–15zv.; Minutes of

was to move to the St Petersburg Polytechnic Institute, where Kirpichev was director; but no vacancy existed. He also approached the SER, again in vain. In late October he decided to try to rejoin the state railways. Here his motive was research. An acquaintance, the Polish-born engineer A.O. Chechott (Czeczott), mentioned some locomotive tests being done by an MPS commission chaired by B.B. Sushinskii. Ambitious as ever, Lomonosov approached his MPS contacts about getting himself appointed as a traction superintendent – a senior managerial position. He had audiences with the acting Head of the Directorate of Railways, P.N. Dumitrashko, and the Minister, N.K. Shaufus. He did not get his wish as such, but he did find himself nominated as second deputy traction superintendent of the Catherine Railway – a top-rank state railway with its headquarters in the southern Ukrainian city of Ekaterinoslav (now Dnipropetrovsk).[69]

Lomonosov seized this opportunity for both career and personal reasons. He was, he claims, unsure now that teaching was his 'realm', and he certainly wanted to make a point at the KPI: he submitted a written statement about his reasons for leaving.[70] Further, he accepted the argument that he could not become a traction superintendent without any managerial experience. He was pleased, too, at being appointed to a major state railway. But above all, he was delighted at having negotiated permission to conduct comparative experiments, like Sushinskii, on the standard freight locomotives and other designs.[71] The moment was propitious for a fresh start in his private life. He and Sonia had separated more or less formally in January 1906, and Sonia was now studying medicine at Berne University in Switzerland.[72] His intermittent relationships with Oksana and Masha had recently ended. He was now committed to a young Jewish girl named Raisa Nikolaevna (Revekka Izrailovna) Rozen. She had been working as his secretary as well as

---

Engineering Faculty, 28 November 1907: DAK, f. 18, op. 1, spr. 648, ark. 132–zv.; Minutes of KPI Board, 4 December 1907: DAK, f. 18, op. 1, spr. 707, ark. 302zv.; 'Vospominaniia', vol. 3, pp. 1546–53, 1562–9, 1571–9, 1581, 1605–36, 1653–83, 1694–8, 1725–9.

69　MPS Directorate of Railways–KPI Director, 10–11 December 1907: DAK, f. 18, op. 2, spr. 159, ark. 159; 'Vospominaniia', vol. 3, pp. 1730–1732, 1737, 1752–5, 1757–8, 1761–4. On Sushinskii's tests see 'Sravnitel'nye ispytaniia tovarnykh parovozov', *Zhurnal Ministerstva Putei Soobshcheniia*, 10 (1908): 57–151.

70　'Vospominaniia', vol. 3, pp. 1521, 1770; Minutes of Engineering Faculty, 28 November 1907: DAK, f. 18, op. 1, spr. 648, ark. 132–zv.; Minutes of KPI Council, 14 December 1907: ibid., spr. 706, ark. 162zv.–4. The statement was attached to the minutes but is now missing.

71　'Vospominaniia', vol. 3, pp. 1761–3.

72　'Vospominaniia', vol. 3, especially pp. 1204–5, 1313–18, 1332. Sonia's registration at Berne University's medical faculty in early 1906 is confirmed in Universität Berne, *Behörden, Lehrer und Studierende*, WS 1906/7, Medizinische Fakultät (Berne, [n.d.]), p. 32. She completed her final dissertation in 1911: S. Lomonossoff, 'Ueber die Beeinflussung der Wirkung narkotischer Medikamente durch Antipyretica', Diss. med. (Berne, 1910/1911). I am grateful to Franziska Rogger of Berne University Library for these references and associated references in the university's book collection in Berne State Archives.

looking after Mar'iia and Vsevolod, and her first pregnancy was confirmed in late 1907; the birth of their son Iurii would follow on 22 August 1908.[73] Finally, as Chapter 4 will show, he may have wanted to leave Kiev as a way to terminate his involvement in revolutionary activism.

*   *   *

Lomonosov does not clarify whether he regarded any of the main developments in his work and family life during 1901–1907 as especially 'shameful'. His dealings with his colleague Perminov and treatment of his wife Sonia are just two salient examples of conduct for which perhaps he did feel shame when writing his memoirs. Be that as it may, it seems safe to conclude that disappointment predominated in his career during these years. He managed to put his stamp on the KPI's curriculum for railway-related engineering, launched a number of important long-term research projects and developed a strong publication record. However, he was deeply unsettled at work as early as 1903 and remained so for most of the next four years. His quarrels with colleagues took much of his time, his teaching was disrupted almost every year by political troubles, his work was complicated by his isolation from the MPS and, above all, nothing came of his plan to put the national locomotive testing plant at the heart of his research. Consulting work for the SER had its advantages, but basically it remained for him an uncongenial necessity. To some extent his troubles were self-inflicted, but circumstances beyond his control were more important. Had the testing plant been built at the KPI as planned, he would surely not have been so keen to leave.

These experiences tend to endorse the view that the tsarist system could not make essential reforms in higher technical education. Notwithstanding its mission to support the industrial economy, the KPI was in a poor position to assist the railways. There was a serious problem in the form of the administrative barriers that existed between the MPS and the ministries responsible for overseeing the polytechnic sector, and those barriers hampered railway-related teaching, research and consulting. The unhappy consequences included the shortages of teaching materials and research funds, a psychological chasm between railway staff and polytechnic academics and, of course, the collective failure to build the planned national locomotive testing plant. Nor was there any real prospect of a marked improvement in this situation in 1907, not least for want of any ministerial disquiet about it. Meanwhile, there were also problems within the institute – not just the minefield of academic politics, which one would expect, but especially the structural obstacles that hindered the efforts to reform the railway-related curriculum. In short, the polytechnic sector was not an especially encouraging

---

[73] Minutes of KPI Board, 30 January 1907: DAK, f. 18, op. 1, spr. 707, ark. 24; 'Vospominaniia', vol. 3, especially pp. 1531, 1578–9, 1581, 1642–4, 1709, 1769. See also H.A. Aplin, 'Raisa Nikolaevna Lomonosova (1888–1973)' and 'George Lomonossoff (1908–1954)', in his *Catalogue of the Lomonossoff Collections*, pp. xxi–xxvi and xxvii–xxix.

environment for a young railway engineer keen to make a difference through teaching and especially research.

Chapter 4

The Russian Revolution of 1905

The outbreak of revolution in 1905 was a moment of truth for Lomonosov, as for many of his compatriots. After months of rising political tension the regime's 'Bloody Sunday' massacre of unarmed petitioners in St Petersburg on 9 January 1905 sparked a wave of protest demonstrations and strikes in Russia. For the first time, liberals and radicals made common cause against the autocracy – a crucial change for fostering a popular sense of revolutionary purpose and momentum. In this unprecedented situation Lomonosov had to decide whether actively to support the revolutionary cause in line with his student vow. In the event, he played a leading role in efforts to build liberal–socialist cooperation in Kiev, and then participated in the revolutionary activity of the Russian Social Democratic Workers' Party (SD Party), risking his life by helping this Marxist organization to make bombs and guns.

This chapter explores these personal experiences in two sections that reflect the two main phases of Lomonosov's political activities. So doing, it also attempts to cast new light on wider issues in the history of the revolution. For example, the course of events in the provinces has had less attention from Western historians than events in St Petersburg and Moscow: what do his activities reveal about the political situation in the major city of Kiev?[1] Also, it is generally thought that the regime managed to regain the political initiative in late 1905 largely because the Tsar's Manifesto of 17 October 1905 drove a wedge between the liberal and radical camps, effectively halting their cooperation: what light does Lomonosov's first-hand experience cast on this analysis? A third issue is the nature of political violence in late Imperial Russia: does Lomonosov's experience help us to understand why at least some revolutionaries opted for terrorism, and in particular does it support the contention that psychological factors, not ideology, were the main reason?[2]

---

[1]   See B. Williams, '1905: The View from the Provinces', in Smele and Heywood, *Russian Revolution*, pp. 34–54. Many books and articles about Ukraine and Kiev in the revolution appeared from Soviet publishers during the 1920s and early 1930s, some of which are cited below. Later works by Soviet historians include: P.M. Shmorgun, *Bol'shevistskie organizatsii Ukrainy v gody pervoi russkoi revoliutsii 1905–1907* (Moscow, 1955); *Do istorii bil'shovits'kikh organizatsii na ukraini v period pershoi rosiis'koi revoliutsii (1905–1907 rr.)* (Kiev, 1955); I.N. Bondarenko, *Bol'sheviki Kieva v pervoi russkoi revoliutsii (1905–1907 gg.)* (Kiev, 1960); T.V. Glavak et al. (eds), *Ocherki istorii kievskikh gorodskoi i oblastnoi partiinykh organizatsii* (Kiev, 1981).

[2]   See in particular A. Geifman, 'Psychohistorical Approaches to 1905 Radicalism', in Smele and Heywood, *Russian Revolution*, pp. 13–33; her chapter 'The Russian

Liberalism and socialism

As a professor in Kiev Lomonosov was always likely to find himself at a major battleground if a revolution occurred. Kiev was one of the largest cities of the Russian empire and had gained a reputation for revolutionary and socialist activity.[3] By 1904 its panoply of illegal political organizations included the liberal Union of Liberation and, on the Left, the Bolshevik and Menshevik factions of the Russian Social Democratic Party, the Socialist Revolutionary Party, the Jewish Bund and the main Ukrainian party, the Spilka Ukrainian Social Democratic Union.[4] Revolutionary literature was relatively common, much of it being smuggled into the area from Austria–Hungary. Kiev's revolutionaries were particularly mindful of the large contingent of industrial workers, most of whom lived in poor, overcrowded areas like the Podol'sk district just north of the city centre and the Shuliavka district near the KPI; soldiers of the garrison were another important target for revolutionary agitators. Furthermore, the city was home during much of the year for several thousand university, polytechnic and other students – an element that the regime traditionally regarded as disruptive. That suspicion perhaps owed more to stereotype than numbers at the KPI: Lomonosov perceived strong nationalist sentiment among the small band of Polish students, but estimated that only about 10 per cent of the mostly Russian and Ukrainian majority were members of revolutionary parties in 1902, and that student political unrest was less problematic at the KPI than at Kiev University during 1901–1904. Nonetheless, during his time in Kiev only the 1906/1907 academic year was completed without serious political disruption.[5]

Intelligentsia, Terrorism, and Revolution', in V. Brovkin (ed.), *The Bolsheviks in Russian Society: The Revolution and the Civil Wars* (New Haven, 1997), pp. 25–42; P. Holquist, 'Violent Russia, Deadly Marxism? Russia in the Epoch of Violence, 1905–21', *Kritika*, 4/3 (2003): 627–52; L. Engelstein, 'Weapon of the Weak (Apologies to James Scott): Violence in Russian History', *Kritika*, 4/3 (2003): 679–94; and M. Geyer, 'Some Hesitant Observations Concerning "Political Violence"', *Kritika*, 4/3 (2003): 695–708.

[3]   On the city's revolutionary tradition see M.F. Hamm, *Kiev: A Portrait, 1800–1917* (Princeton, 1993), pp. 173–5.

[4]   On the Union of Liberation see S. Galai, *The Liberation Movement in Russia, 1900–1905* (Cambridge, 1973); the Bund is surveyed in A. Lokshin, 'The Bund in the Russian–Jewish Historical Landscape', in Geifman, *Russia under the Last Tsar*, pp. 57–72. The complex history of the social democrats has been discussed in numerous works; a biography of Lenin can be a helpful means to approach it, such as R. Service, *Lenin: A Political Life*, 3 vols (Basingstoke, 1985–1995).

[5]   For example, 'Vospominaniia', vol. 3, pp. 182–97. By contrast, the Provincial Governor-General claimed credit for helping to dissuade KPI students from following the rebelliousness of Kiev University students, and stressed that the KPI authorities displayed a united front with him: Sukhomlinov, *Vospominaniia* (Berlin, 1924), p. 132. Archival evidence disproves the latter assertion, such as: Kiev Governor–Kiev Governor-General, 18 April 1906: Tsentral'nyi derzhavnii istoricheskii arkhiv Ukraini (TsDIAU), f. 442, op.

The actual disturbances in Kiev during 1905, although ultimately violent, were less extensive than the authorities feared and the revolutionaries hoped. Like many other places, the city witnessed protest demonstrations after Bloody Sunday, which prompted among other things the indefinite closure of the KPI and university. Local branches were formed of the new trades and professional unions, including those for engineers, doctors, teachers, academics and railway workers. Many strikes were declared, including one on the railways in the spring. Yet no great street violence occurred during the first half of the year, and the summer was fairly quiet. In the autumn, however, the mood hardened. The revolutionary upsurge throughout the empire was strongly reflected in Kiev. Strikes and demonstrations became commonplace, and student activism burgeoned, the university and KPI having just reopened. Leaders of revolutionary parties like M. Ratner (Socialist Revolutionary) and A.G. Shlikhter (Bolshevik) addressed many public meetings, workers elected a council of representatives that was called the Kiev Soviet of Workers' Deputies, and the general strike of October 1905 was well supported. When news came of the Tsar's Manifesto of 17 October, which offered an elected legislative assembly to be called the State Duma, a large crowd gathered outside the Kiev city hall for speeches by opposition leaders. That day, however, extremist supporters of the regime launched a three-day 'pogrom' in the city, killing nearly 50 people while the city authorities remained passive. Then the police began arresting revolutionaries. In mid-November troops loyal to the regime fired on mutinous army sappers in the city centre; several dozen people died. In December, a workers' republic was declared in the Shuliavka district; but again the authorities deployed loyal troops and many rebels were killed or arrested, including most of the Soviet's leaders. Thereafter, the city remained subdued.[6]

What, then, were Lomonosov's politics on the eve of the revolution? It is often thought that liberal-minded people dominated the professions in late tsarist Russia, yet as Balzer has noted, that assumption may be inaccurate.[7] That caution seems appropriate for Kiev's professoriate. The police believed that many KPI academics supported the Socialist Revolutionary Party.[8] Lomonosov's calculation

---

855, spr. 322, ark. 192–204. Lomonosov dismisses Sukhomlinov as a 'naive general': 'Vospominaniia', vol. 3, pp. 1783. On Kiev University during these years see: V.A. Zamlinskii (comp.), *Kievskii universitet: dokumenty i materialy, 1834–1984* (Kiev, 1984), pp. 68–71 and G.I. Marakhov, *Kievskii universitet v revoliutsionno-demokraticheskom dvizhenii* (Kiev, 1984).

[6]   For an overview of the 1905 revolution in Kiev see Hamm, *Kiev*, pp. 173–207.

[7]   H.D. Balzer, 'The Problem of Professions in Imperial Russia', in E.W. Clowes et al. (eds), *Educated Society and the Quest for Public Identity in Late Imperial Russia* (Princeton, 1991), p. 193.

[8]   The police tried to identify the politics of individual academics: see, for example Kiev Polytechnic Institute, [March 1908?]: TsDIAU, f. 275, op. 1, spr. 1250, ark. 75a, where most of the 15 people listed are identified as Socialist Revolutionary or Social Democrat. The idea that KPI staff had 'extreme political convictions' was noted in police

was that only about half of his colleagues backed the liberal Union of Liberation in 1904 and that as many as 20 per cent had right-wing views, as did most professors at the university.[9] By 1906 he felt that the contingent of liberal KPI professors had grown slightly: now 20 Council members supported the Constitutional Democrats (Kadets), seven were 'rightist' and seven were 'leftist'. That said, his idea of 'rightist' may have included Octobrists – liberals who were satisfied by the concessions in the Tsar's October manifesto, in contrast to the Kadet position that the manifesto was merely a step towards further changes.[10]

Lomonosov's public behaviour probably suggested that his politics were non-aligned with a hint of centrism. His main KPI friendships were with colleagues whose politics were thought to be radical: a geology professor, A.V. Nechaev, who became an informal mentor; the mechanics professor A.A. Radtsig; a fellow engineer named N.A. Artem'ev; and the chemistry professor M.M. Tikhvinskii, who assisted the Bolsheviks' central Military–Technical Group in St Petersburg under the codename 'Ellips'.[11] But friendship did not necessarily signify agreement with their politics. He reportedly voted in KPI Council meetings with 'leftist' professors on 'political' matters such as student demonstrations but with 'rightists' on academic issues, thinking that academic discipline should be strict.[12] Perhaps more demonstrative was his choice of newspaper: he preferred the liberal *Russkie vedomosti* to the left-wing *Kievskaia mysl'* and right-wing *Kievlianin*.[13] His most overt political statement was effectively his attendance at a banquet in Kiev on 20 November 1904. This event ostensibly marked the fortieth anniversary of

---

correspondence with the Ministry of Finances: see F.E. Los' et al. (eds), *Iz istorii Kievskogo Politekhnicheskogo Instituta, tom 1 (1898–1917 gg.)* (Kiev, 1961), pp. 237–8. If accurate, this picture suggests that KPI staff were more radical than their counterparts in Khar'kov, the other Ukrainian city with a university and polytechnic institute. See M.F. Hamm, 'On the Perimeter of Revolution: Kharkiv's Academic Community, 1905', *Revolutionary Russia*, 15/1 (June 2002): 58–60.

[9]   'Vospominaniia', vol. 3, pp. 153–68, 190, 829.

[10]   'Vospominaniia', vol. 3, p. 1476. For introductions to the Kadets and Octobrists during 1905–1917 see M. Stockdale, 'Liberalism and Democracy: The Constitutional Democratic Party', and D.B. Pavlov, 'The Union of October 17', in Geifman, *Russia under the Last Tsar*, pp. 153–78 and 179–98 respectively.

[11]   In Russian *ellips* can mean both ellipse and ellipsis; Tikhvinskii's thoughts about it are unknown. Lomonosov's friendship with Radtsig did not survive the latter's appointment as Director of the KPI in 1907. On Artem'ev and Nechaev see their KPI files at DAK, f. 18, op. 2, spr. 8 and spr. 180 respectively.

[12]   'Vospominaniia', vol. 3, pp. 168–9. Unfortunately, this claim cannot be verified because the minutes lack such information. To judge by the minutes, Lomonosov rarely spoke in KPI council debates.

[13]   'Vospominaniia', vol. 3, pp. 157–8, 160–62, 164, 653, 919, 1192–3. Tikhvinskii's underground activity and his execution by the Soviet regime in 1921 are confirmed in S.M. Pozner (comp.), *1905: Boevaia gruppa pri TsK RSDRP (b) (1905–1907 gg.): Stat'i i vospominaniia* (Moscow and Leningrad, 1927), pp. 44, 46, 48–50, 52, 54, 151–3.

Tsar Alexander II's legal reform, but was part of a 'banquet campaign' organized by the Union of Liberation to promote liberal reforms; inspired by the banquet campaign of 1847–1848 in France, these dinners likewise became part of the overture to a revolution.[14]

A measure of the secretiveness that could surround an individual's politics at this time in Russia is the fact that one can identify very diverse assessments of Lomonosov's outlook. He acquired a certain reputation as a right-winger, to judge by his report that a servant asked his advice about joining the secret police department (*Okhrannoe otdelenie*, popularly known as the Okhrana). Similarly, concern that he might be a police agent may explain why, despite having radical friends, he was rebuffed when he tried to contact the revolutionary movement in 1903 via Tikhvinskii's partner, V.A. Solomon.[15] But one can assume that at least some acquaintances interpreted his presence at the 1904 banquet as a mark of sympathy for liberalism. Meanwhile the Kiev office of the secret police saw him as a likely troublemaker: it noted in 1904 that he had caused a KPI employee to be sacked as 'a gendarme spy'.[16]

It seems reasonable to accept his contention in his memoirs that his stance was oppositional. No evidence has been found to show that he had any political sympathy for the regime at this time. That he swore the oath of allegiance while employed with the Khar'kov–Nikolaev Railway can be discounted because every state official had to do so. There is a ring of truth in his report that initially he shared the regime's belief in victory over Japan in the war of 1904–1905, in contrast to most of his colleagues, but this was most likely a product of his military background and patriotism, rather than of support for the autocracy. Hindsight may partly explain his claim that by mid-1904 he hoped for military disasters to discredit the autocracy, but it is conceivable he did take this view at the time, which of course followed the logic popular among revolutionaries of 'the worse [for the regime], the better [for the revolution]'.[17]

To define his politics more precisely is difficult both for want of corroboration and because his own description is unclear. He states that he was vacillating between the positions of the future Kadet liberals and the Marxist SDs, and that ideally he

[14]   On the banquet campaign see T. Emmons, 'Russia's Banquet Campaign', *California Slavic Studies*, 10 (1977): 45–86. Lomonosov states that Kiev had 10 banquets: 'Vospominaniia', vol. 3, p. 965; but Emmons indicates only the banquet of 20 November.

[15]   'Vospominaniia', vol. 3, pp. 169, 661–2, 869–70. Apparently Lomonosov was spurred to contact Solomon by news of a horrific pogrom in April 1903 in the south-western city of Kishinev (now Chişinău, capital of Moldova). On the tsarist secret police see, in particular, J.W. Daly, *Autocracy under Siege: Security Police and Opposition in Russia, 1866–1905* (DeKalb, 1998); his *The Watchful State: Security Police and Opposition in Russia, 1906–1917* (DeKalb, 2004); and F.S. Zuckerman, *The Tsarist Secret Police in Russian Society, 1880–1917* (Basingstoke, 1996).

[16]   Kiev Polytechnic Institute, 16 February 1904: TsDIAU, f. 274, op. 1, spr. 888, ark. 34.

[17]   'Vospominaniia', vol. 2, pp. 365, 390; vol. 3, pp. 821–4, 919.

wanted a radical party akin to the Radical Party in France.[18] In Russian terms this description implies that he was located in what is known as the 'interstitial left', basically at the point on the political spectrum just to the left of the Kadet liberals. Yet he never mentions the groups that emerged there in 1906: the Popular Socialist Party and the Titleless (*Bez zaglaviia*) group of Marxists. Moreover, other evidence shows that he rejected the peasant-oriented populist tradition of the Popular Socialists, and the Titleless group was averse to the violent 'Jacobin tactics' that Lomonosov did support.[19] Presumably, too, given his distaste for private enterprise, he rejected the French Radical Party's support for laissez-faire economics. By way of explanation, he wrote that his heart favoured revolution to secure a constitution, whereas his mind accepted that reformism had more practical advantages, provided that the Tsar adopted it in good faith.[20] One conclusion might be that as a memoirist he was trying to rationalize a certain lust for violent confrontation; a more generous and perhaps fairer view might be that he was confused.

Finally on this point, it is worth considering his attitude towards Freemasonry given the enduring popularity in Russia of theories that Masonic conspiracies explain the revolutions of 1905 and 1917. The writer Nina Berberova has asserted that Freemasonry was important to him given that he was closely acquainted with M.V. Alekseev, one of the Russian army's most senior generals.[21] But this deduction seems unwarranted. Although his memoirs do mention Freemasonry, these remarks are few, cursory and if anything derogatory. One of them suggests that he may have attended a Kiev lodge regularly, but he also explains dismissively that the KPI's professors comprised a lodge merely for discussing Council affairs separately from junior staff. Generally his papers contain no indication that he was a Freemason by conviction. No evidence has been found to indicate any close contact with Alekseev, let alone friendship, and one should reiterate that Lomonosov is exceptionally frank about his private life.[22]

---

[18]   'Vospominaniia', vol. 3, pp. 997. The French Radical Party was formed in 1901 as a diverse alliance of radical groups and became the largest party in the parliament, with mainly peasant and bourgeois support. By 1905 it was a centre-left party. Popular democracy, private property and secularism were key concerns, but a programme as such was not devised for some years. Unfortunately Lomonosov does not describe his understanding of the party's platform. See S. Bernstein, *Histoire du Parti Radical: La recherche de l'âge d'or, 1919–1926* (Paris, 1980), especially pp. 23–86.

[19]   On the term and groups see T. Emmons, *The Formation of Political Parties and the First National Elections in Russia* (Cambridge, MA, 1983), pp. 78–88. Lomonosov's support for violence is shown by his conspiratorial activity, discussed below; his rejection of populism is discussed in Chapter 1, this volume.

[20]   'Vospominaniia', vol. 3, p. 965.

[21]   N.N. Berberova, *Liudi i lozhi: Russkie masony XX stoletiia* (New York, 1986), pp. 38, 137, 170.

[22]   For example, 'Vospominaniia', vol. 3, pp. 974, 997, 1188–9. Neither Lomonosov nor any reference to a KPI lodge appear in A.I. Serkov, *Russkoe masonstvo, 1731–2000:*

Much like the unfolding of the whole crisis, Lomonosov's political journey was fitful during the year that followed the Kiev banquet. In late 1904 he attended a secret meeting of 'leftist' KPI professors, who agreed to demand a constitution and parliament at a meeting of the institute's Council, and to resist any attempt by the local authorities to force KPI staff to oppose a political strike by the students.[23] He was in St Petersburg at the time of Bloody Sunday, heard the commotion and saw bodies in the street afterwards; in reaction, apparently, he added his name to the demand for higher education reform that would become known as the Declaration of 342 Professors.[24] Back in Kiev he suggested that fire hoses be turned against troops who were surrounding the KPI – the incident in mid-January that prompted the Provincial Governor to close the institute indefinitely.[25] By the spring Lomonosov was active in the Kiev branches of two liberal-dominated unions, the Academic Union and the Union of Engineers, and he described the situation in Kiev to a meeting of the Academic Union in the capital in April.[26] But then came a lull in his political activity. One reason was probably the summer softening in Kiev's political mood; another was the urgency of his initial assignments for the South Eastern Railway.

Absent from his account is any indication that his identity and role as an engineer had any notable effect on the evolution of his political outlook at this juncture. As noted in Chapter 1, he felt an affinity as an engineer with the general idea of organizing the society and especially the economy on what could loosely be termed a scientific basis. But to judge by his memoirs, he did not differentiate

---

Entsiklopedicheskii slovar' (Moscow, 2001). Alekseev may have intervened in Lomonosov's career in January 1917: see Chapter 6, note 47.

[23] 'Vospominaniia', vol. 3, pp. 966–8. This demand, if made, was not minuted.

[24] 'Vospominaniia', vol. 3, pp. 972–6, 979–81. The declaration is discussed in Kassow, *Students*, pp. 219–22. The original declaration was published in January 1905 and reprinted with a further 1,420 names in April 1905 in *Vsemirnyi vestnik*: Brandt, *List'ia pozheltelye*, pp. 28–9. I have yet to trace a copy to verify Lomonosov's claim.

[25] 'Vospominaniia', vol. 3, pp. 985–8. The incident certainly occurred: Protocol of search of KPI, 15 January 1905: TsDIAU, f. 275, op. 1, spr. 734, ark. 4–6; Kiev Police Chief–Kiev Governor, 16 January 1905: DAKO, f. 2, op. 221, spr. 55, ark. 4–5 (covers a report about the events at ll. 6–10); M.I. Mebel' (ed.), *1905 god na Ukraine: Khronika i materialy, tom 1: ianvar'–sentiabr'* ([Khar'kov], 1926), pp. 296–7.

[26] 'Vospominaniia', vol. 3, pp. 997–9. His membership of one union was suspected by the police: List of members of the bureau, [1905]: TsDIAU, f. 274, op. 1, spr. 1113, ark. 142ob. But the Kiev Okhrana's enquiries did not identify him as a member of these unions in March 1906: TsDIAU, f. 275, op. 1, spr. 1122, ark. 22, 23–zv. Nor does he feature in the surviving papers of the two unions in the archive of the Union of Unions: Gosudarstvennyi arkhiv Rossiiskoi Federatsii (GARF), f. 518, op. 1, d. 16, 18, 29, 36. On the two unions at this time see Kassow, *Students*, pp. 219–44, 265–71, 293–8; and J.E. Sanders, 'The Union of Unions: Economic, Political, and Human Rights Organizations in the 1905 Russian Revolution' (unpublished PhD dissertation, Columbia University, 1985), pp. 222–43, 434–59, 565–87.

between liberals and radicals over this issue, and he defined his ideal society mainly in terms of social justice. Indeed, only in one specific respect does he discuss the Union of Engineers as an overtly political organization in 1905: he became disillusioned with this and similar professional unions for what he saw as their unspoken preference not to unite with workers, peasants and soldiers in the battle against autocracy. He acknowledged that such an alliance would be risky, but he suspected by the autumn that these unions were overestimating their moral victory over the autocracy and that a bloody showdown was becoming inevitable. That being the case, he was increasingly anxious for a solution to the land question: if denied land, the peasants would 'happily impale us all on a stake'. Aside, then, from this anger with the professional unions and from any prior Marxist beliefs on his part, he indicates three reasons why he began helping the Social Democrats in the autumn of 1905. One was his perception that the mood was becoming radical: for instance, the mutiny on the battleship *Potemkin* in June encouraged him to believe that the military might revolt. In other words, the autocracy's collapse seemed a distinct possibility. Another reason was his friendship with Tikhvinskii and Solomon: on about 1 October Solomon invited him to help the radical cause by collecting money for political prisoners. Finally, his character was relevant: abrasive, somewhat reckless and with a mischievous sense of adventure, he had been fishing for this invitation.[27]

Politically Kiev's Social Democratic Party organization was troubled. Some of its problems were typical for the party throughout the empire. For example, its membership was small, at only about 500, not least because most workers preferred not to join illegal political parties.[28] Police surveillance was hindering movement, meetings and agitation work. Crucially, the party was internally divided, having recently split into a Bolshevik faction that supported Lenin's notion of a small, highly organized revolutionary party, and a larger Menshevik faction that wanted a mass-membership party. The seriousness of this rift varied across the empire, but the two factions were certainly operating separately in Kiev by early 1905, and cooperated uneasily thereafter. Other problems for the SD Party in Kiev included its lack of a printing press, and strong competition from the many other leftist groups in the city.[29]

---

[27]   'Vospominaniia', vol. 3, pp. 974, 998, 1053–4, 1206–7. On the revolutionary situation in Kiev see Hamm, *Kiev*, pp. 185–7.

[28]   Ascher, *The Revolution of 1905, vol. 1: Russia in Disarray*, p. 184; Hamm, *Kiev*, pp. 179, 184. Ascher's estimates for SD Party membership and 'organized workers' in other cities include: St Petersburg – 1,200 Mensheviks and several hundred Bolsheviks; Ivanovo-Voznesensk – 600 Bolsheviks; Khar'kov – 300; Ekaterinoslav – 1,000.

[29]   See, for instance, T. Stepanenko, 'Kievskaia sotsial-demokraticheskaia organizatsiia nakanune pervoi russkoi revoliutsii 1905–1907 gg.', in Troitskii, *Kievskii Politekhnicheskii Institut*, p. 109; M. Maiorov, *Bol'sheviki i revoliutsiia 1905 g. na Ukraine* (Tashkent, 1934), p. 67; Hamm, *Kiev*, pp. 178–80. The police were always trying to find and confiscate illegal printing presses.

That said, in Shlikhter the Bolshevik faction had an experienced, energetic and persistent leader. Described by the historian Michael Hamm as the 'dominant personality and most notorious symbol of the "heroic days" of Kiev's labour movement', Shlikhter was the son of a skilled worker from the Ukrainian town of Poltava. Having participated in Populist groups during the 1880s, he joined the Social Democratic movement at the end of that decade. After several years in Switzerland he returned to Ukraine in 1892, but was soon arrested for conducting propaganda, imprisoned, and then sent into internal exile, latterly at Samara on the river Volga. Released in 1902, he moved to Kiev, joined its Social Democratic committee and held a job as an office worker at the headquarters of the South Western Railways until his dismissal in spring 1905 for involvement in the All-Russian Union of Railway Workers and Employees. He was also a founder member of the discrete Bolshevik group in Kiev, where his associates included the engineer G.M. Krzhizhanovskii – a fellow veteran of exile in Samara and, from 1921, the founding chairman of the Soviet regime's State Planning Commission (Gosplan).[30]

The portrayal of Shlikhter as a courageous, uncompromising Leninist is central to the Soviet accounts of the 1905 revolution in Kiev.[31] Among other things, this image was used for criticizing the Mensheviks, whose inclination to compromise and cooperate with the liberals was 'class treachery' towards the proletariat. For instance, Shlikhter was said to differ from 'opportunist' Mensheviks in deriding the 1904 banquets and generally the liberal bourgeoisie as agents of political change.[32] Unsurprisingly Shlikhter himself cultivated this image in his writings after 1917 and highlighted his work with Lenin in the Petersburg area and Finland in the winter of 1905–1906.[33] However, a later Soviet account claimed that Shlikhter spoke at liberal banquets to applause, and Lomonosov confirms that he used the Kiev banquet of 20 November to demand a Constituent Assembly.[34] Admittedly

---

[30] Hamm, *Kiev*, p. 178; J.K. Libbey, 'Shlikhter, Aleksandr Grigor'evich', in A.T. Lane (ed.), *Biographical Dictionary of European Labour Leaders*, vol. 2 (Westport, 1995), pp. 888–9; O.S. Rovner and R.Ia. Korolik, *Oleksandr Grigorovich Shlikhter, 1868–1940: Bibliografichnyi pokazhchik* (Kyiv, 1958), pp. 6–11; D.F. Virnyk, *Aleksandr Grigorevich Shlikhter* (Kiev, 1979), pp. 7–9, 26; H. Reichman, *Railwaymen and Revolution: Russia, 1905* (Berkeley, 1987), p. 176. Some police sources describe Shlikhter as petty bourgeois (*meshchanin*), though this may be due to his occupation as an office manager: see Head of Kiev Gendarmerie–Manager of Kiev Governor-General's Chancellery, 7 November 1905: TsDIAU, f. 442, op. 855, spr. 342, ark. 5–zv.; 'Oktiabrskie sobytiia 1905 goda v Kieve v izobrazhenii okhrannogo otdeleniia', *Letopis revoliutsii*, 2 (1925): 114.

[31] For example, Iu.Iu. Kondufor et al. (eds), *Istoriia Kieva, tom 2* (Kiev, 1983), p. 337.

[32] See S. Kokoshko, *Bilshovyky u Kyievi naperedodni i za revoliutsii 1905–1907 rr.* (Kharkiv and Kiev, 1930), pp. 57–8.

[33] See, in particular, A.G. Shlikhter, *Ilich, kakim ia ego znal: Koe-chto iz vstrech i vospominaniia* (Moscow, 1970).

[34] Stepanenko, 'Kievskaia sotsial-demokraticheskaia organizatsiia', pp. 105–6; 'Vospominaniia', vol. 3, pp. 964–5.

Lomonosov's account needs caution because the police surveillance reports fail to mention Shlikhter's presence.[35] Lomonosov's memory may be inaccurate, but it seems more likely that, for example, the agents did not recognize Shlikhter, especially if he spoke from the large crowd of onlookers, as seems probable. In Lomonosov's favour is the fact that the banquet's resolution was one of only six in the empire to call for a Constituent Assembly elected by four-tail suffrage (in other words, by a universal, direct, equal, secret ballot), which the SD Party had been demanding since 1903.[36] Also, Lomonosov did become acquainted with Shlikhter in 1905. On balance, then, Shlikhter almost certainly did seek liberal support in 1904.

This impression is strengthened by Lomonosov's description of his first meetings with Shlikhter in October 1905, which reveals that the Bolshevik leader regarded him as a left-liberal and that Shlikhter wanted cooperation with liberals as well as Mensheviks. The professor had just begun collecting money for political prisoners, working in tandem with the actress N.A. Smirnova. He had praised the growing national railway strike at meetings of the Union of Engineers and Academic Union, and was conveying local strike news to a railway friend, A.M. Arkhangel'skii, who was a strike leader as a delegate to the All-Russian Railway Congress in St Petersburg.[37] On about 12 October Lomonosov joined Krzhizhanovskii and others in a delegation of engineers that lobbied the head of the South Western Railways about the arrest of various South Western engineers. These delegates were then taken to meet Shlikhter at a public meeting, where the latter lauded them as members of the 'honest intelligentsia'.[38] That evening, at Shlikhter's invitation, Lomonosov attended a meeting at the apartment of E.M. Ianovskii – a Menshevik later described by a veteran revolutionary as a 'party worker of central significance'.[39] Present were Shlikhter, a woman identified as E.V. Ianovskaia and chair of Kiev's regional SD committee, a Menshevik called F.A.

[35]   Report about banquet on 20 November 1904: TsDIAU, f. 275, op. 1, spr. 552, ark. 12–zv.; [Report about banquet of 20 November 1904], by Spiridovich: ibid., l. 43–7.

[36]   Emmons, 'Russia's Banquet Campaign', pp. 53, 59, 61.

[37]   'Vospominaniia', vol. 3, pp. 1208–10, 1212–16; on Arkhangel'skii, the congress and the general strike see Reichman, *Railwaymen and Revolution*, pp. 186–223.

[38]   'Vospominaniia', vol. 3, pp. 1217–23.

[39]   S.L. Gel'zin, 'Iuzhnoe Voenno-Tekhnicheskoe Biuro pri TsK RS-DRP', *Katorga i ssylka*, 12 (1929): 39. The best known Menshevik leader in Kiev is A.A. Vannovskii: see V. Nevskii, *Materialy dlia biograficheskogo slovaria sotsial-demokratov, vstupivshikh v rossiiskoe rabochee dvizhenie za period ot 1880 do 1905 g, vypusk I: A–D* (Moscow and Petrograd, 1923), pp. 121–2, and *Revoliutsiia 1905–1907 gg v Rossii: Dokumenty i materialy: Vysshii pod"em revoliutsii 1905–1907 gg: vooruzhennye vosstaniia, noiabr'–dekabr' 1905 goda, chast' 3, kniga 1* (Moscow, 1956), p. 1028 (note 30). According to Nevskii, *Materialy dlia biograficheskogo slovariia*, p. 122, Vannovskii's wife was V.V. Vannovskaia, and she did not serve as committee leader; identifying the lady here as E.V. Ianovskaia, Lomonosov appears not to be confusing the two men.

Dudel' ('comrade Nina') and two other 'professional revolutionaries'. Shlikhter shared the Menshevik preference to bring liberals into Kiev's Revolutionary Coalition Council, which was a clandestine group formed of Bolsheviks, Mensheviks, Socialist Revolutionaries, Polish and Ukrainian Social Democrats, Jewish Bundists and Ukrainian nationalists. Specifically, he suggested that Kiev's professional unions should form a local branch of an organization called the Union of Unions, which had been founded in St Petersburg in the spring to facilitate cooperation between liberals and socialists. Shlikhter wanted this new entity to contact the Coalition Council and have two or three seats in it.[40]

Lomonosov's attempt to implement this proposal – his first significant activity as a revolutionary – was a failure. He began by organizing a founding meeting of a Kiev branch of the Union of Unions, which was held on about 15 October under the guise of an expanded meeting of the Literary Society. He made the opening speech, stressing the intelligentsia's obligations towards the Russian people. A suitable resolution was agreed, and a Presidium was chosen, which consisted of Lomonosov, the Socialist Revolutionary leader Ratner in his capacity as a lawyer, and a university academic named Zheleznov, with three alternates in case of arrests, of whom one was Lomonosov's KPI colleague Artem'ev. The first meeting of the Presidium followed: Lomonosov was chosen as secretary, and a decision was taken to contact the Coalition Council. However, the next day Ratner was arrested and Zheleznov withdrew. After consulting Shlikhter, who was now in hiding, Lomonosov and Artem'ev met representatives of the Coalition Council. The latter seemed cautious, but agreed that the new union should issue a proclamation about its existence and its solidarity with the Coalition Council. The printing could be done at a clandestine press in Kiev. Lomonosov drafted a combative statement that alarmed other members, who stressed peaceful protest, but he sent it for printing anyway. Meanwhile, however, the Coalition Council decided not to admit the Union of Unions to its ranks. Lomonosov therefore withdrew his text from the press late on 17 October.[41]

Ultimately, then, Lomonosov discovered that Kiev's liberals and radicals lacked sufficient common ground for joint action against tsarism. Notwithstanding

---

[40]   'Vospominaniia', vol. 3, pp. 1222–8. Further, Lomonosov reports that Shlikhter addressed a meeting of the Academic Union on 9 October seeking support for the railway strike: ibid., pp. 1213–14. See also V. Manilov, *Kievskii Sovet Rabochikh Deputatov v 1905 g.* (Kiev, 1926), pp. 8, 10–11, 59; Glavak et al., *Ocherki*, p. 67. On the Union of Unions see D. Sverchkov, 'Soiuz soiuzov', *Krasnaia letopis'*, 3 (1925): 149–62; L.K. Erman, *Intelligentsiia v pervoi russkoi revoliutsii* (Moscow, 1966), pp. 86–102; S. Galai, 'The Role of the Union of Unions in the Revolution of 1905', *Jahrbücher für Geschichte Osteuropas*, 24/4 (1976): 512–25; and Sanders, 'Union of Unions'. Lomonosov's project is the focus of A.J. Heywood, 'Socialists, Liberals and the Union of Unions in Kyiv during the 1905 Revolution: An Engineer's Perspective', in Smele and Heywood, *Russian Revolution*, pp. 177–95.

[41]   'Vospominaniia', vol. 3, pp. 1228–43.

the growing outspokenness of the central Union of Unions in St Petersburg, most professional union members in Kiev rejected violence, while the Coalition Council defaulted on Shlikhter's assurance of membership. This affair shows that even before the news about the Tsar's October Manifesto reached the city on 18 October there was a fundamental and probably irreconcilable disagreement between Kiev's liberals and radicals about the way forward. In Kiev, therefore, the manifesto cannot be said to have had the decisive role in breaking the liberal–radical rapport, and the Kiev Soviet's later rejections of an alliance with the Union of Unions on 30 October and 6 November merely formalized an established fact.[42] Lomonosov does not mention this project again, and he played little if any part in the union's subsequent activities, which soon ceased.

Socialism, terrorism and the future

Lomonosov's memoirs identify two main reasons for his involvement with revolutionary terrorism in 1905. One was the revolutionary mood of September–October, in which the autocracy's demise seemed a real possibility and in which, as we have seen, he sought to join the revolutionary movement. The other was the climax of this crisis in October. His first reaction to the promises of the October Manifesto was liberal: people should support the constitutional monarchy. Thus, standing in the large crowd that gathered outside the city hall on 18 October, he was alarmed to hear Shlikhter and Ratner calling the people to arms. But his mind was changed by the ensuing three-day pogrom perpetrated by right-wing extremists. The Tsar, it seemed to him, could not be trusted. Having witnessed several assaults and also police intervention to shield the perpetrators, he began helping victims and sheltering refugees at the KPI, and on about 21 October he told Ianovskii that he wanted to help the Coalition Council make bombs.[43]

How far, then, can we trust this explanation? The historian Anna Geifman has argued that the terrorists of late Imperial Russia were psychologically impaired, that self-loathing was a crucial determinant of their destructive activity and that ideology had little bearing on their behaviour. Hence she warns against relying on the predominantly ideological rationalizations that such terrorists usually offered in later publications such as memoirs. This concern about these sources, especially memoirs, is surely appropriate because their potential for distortion is so high. Yet we should not simply discount this evidence when, as in Lomonosov's case, it is the only available source for a particular person; indeed Geifman herself employs memoirs for her psychological analysis.

---

[42]   On the Soviet's stance see the report of 8 November 1905 in *Kievskaia gazeta*, reprinted in F.E. Los' et al. (eds), *Revoliutsiia 1905–1907 gg. na Ukraine: Sbornik dokumentov i materialov v dvykh tomakh*, vol 2/1 (Kiev, 1955), pp. 626–30; Manilov, *Kievskii sovet*, pp. 8, 13–20 and Hamm, *Kiev*, pp. 208–9.

[43]   'Vospominaniia', vol. 3, pp. 1249–61. On the pogrom see Hamm, *Kiev*, pp. 189–207.

Lomonosov's memoirs are typical in stressing idealistic and altruistic motives, and minimizing or ignoring issues such as psychological needs and criminality. They also offer a logic that could loosely be called 'ideological': armed struggle could be legitimate where effectively the regime was at war with the society.[44] Further, they reveal a certain propensity to aggression and violence, and a self-confessed love for deception, intrigue and conspiracy.[45] One might thus infer that Lomonosov got a certain thrill from the dangers of revolutionary activism. Yet so far as one can see, his character does not fit other parts of the psychological profile developed by Geifman. His self-confidence and ego were still strong: not for him the suggested feelings of self-hatred, inadequacy and frustration with the self, and the abandonment of the self in the cause. Also, feelings of deep anxiety did not cause him to become involved in revolutionary activism; on the contrary, his activism eventually caused chronic anxiety, such that he felt relieved when the bomb-making operation ended in 1906 – a reaction that, if accurately described, suggests that he was not addicted to this activism.[46]

In fact, paradoxically, there was still much indecision or confusion in his politics at this time. So far as one can see, he never formally joined any Russian political party. He felt uncomfortable with being designated a Social Democrat in 1905 even on a secret basis. In conversation with Shlikhter in November 1905 he described himself as a 'non-party Marxist', and he would repeat this assertion many times after the Bolshevik revolution. Pressed by comrades as to whether he was a Menshevik or Bolshevik, he reportedly explained that he was not a party member, that he opposed such labels, that he favoured unity in the battle against autocracy, but that reluctantly he could accept being called Bolshevik because the Bolshevik emphasis on organizational discipline was vital for the current situation of class war.[47] He later acknowledged, however, that his stance on some issues was far from Bolshevik policy: 'My tragedy was that in those days I found myself far to the right of my own party.' In particular, he opposed the Bolshevik boycott of the first Duma, for he saw this new institution as a 'school for free speech' and first step towards popular power. Likewise, he could see no justification for the Bolsheviks' disastrous armed uprising in Moscow in late 1905. Also, he claims that by April 1906 he shared the outlook of the peasant-oriented 'Trudovik' group,

---

[44] 'Vospominaniia', vol. 3, pp. 1244–61. See also Chapter 1, this volume.

[45] See, in particular, 'Vospominaniia', vol. 1, pp. 275–6.

[46] 'Vospominaniia', vol. 3, pp. 1365, 1378.

[47] In July 1921 he reported to Moscow that 'during 1903–1907 he was a member of (*sostoial v*) the RSDRP (Bolshevik fraction)': Rossiiskii gosudarstvennyi arkhiv sotsial'no-politicheskoi istorii (RGASPI), f. 76, op. 2, d. 76, l. 24. But apart from his exaggeration concerning the dates, this statement begs the question of how membership of illegal political parties was registered and recognized.

though understanding that it was poorly organized, had no leaders and was ill-prepared for working in the Duma.[48]

On balance, then, we can suggest several reasons for his radical activism. In the context of his perception of a state of war between regime and society, the key spur for his involvement was the prospect of the autocracy's imminent defeat. That he was less immediately concerned about the longer-term future is suggested by the fluidity of his political outlook and his disinclination to join a political party. Some psychological factors such as a propensity to aggression probably did contribute to his decision, but it seems unlikely that they constituted his main motivation or that they indicate an aberrant personality of the type suggested by Geifman.

Lomonosov did not join the bomb-making operation immediately: Ianovskii responded evasively to his offer, presumably wanting more evidence of his reliability. But there is no doubt that the professor did become involved in SD activities, not least because Krzhizhanovskii and others provided confirmation in the Soviet era.[49] Apparently his code name was 'Boris Ivanovich'.[50] He undertook more fundraising, now for pogrom victims and Jewish self-defence groups; but this activity was less successful than before, and probably ceased after he complained about the use of the money to evacuate party activists from Kiev.[51] More importantly, he facilitated Shlikhter's escape abroad on 1 November. He arranged for the night train to Kovel' (near the Austro-Hungarian border) to halt at a crossing outside Kiev where he would be waiting with Shlikhter in disguise; the pair then travelled to Kovel', where another comrade took charge.[52] This account confirms that Shlikhter was absent from Kiev throughout November and

---

[48]   'Vospominaniia', vol. 3, pp. 1198–9, 1279, 1289, 1310, 1365, 1367, 1369–72, 1378. He does not explain his leaning towards the Trudoviki.

[49]   For Krzhizhanovskii's remarks see Kushch and Glovatskii, 'Kievskii Politechnicheskii Institut', pp. 134–5; also noted in Erman, *Intelligentsiia v pervoi russkoi revoliutsii*, p. 230. These references are surprising given that Lomonosov was basically *non grata* in the USSR at the time of publication in the 1950s.

[50]   'Vospominaniia', vol. 3, p. 1289. Probably because Lomonosov did not formally join the party, Nevskii does not record this code name: the *Materialy dlia biograficheskogo slovariia*, p. 95, merely gives a cross reference to a certain Goloshchekin, for whom there is no entry (because only this first volume was published). The choice of Lomonosov's code name is not explained.

[51]   'Vospominaniia', vol. 3, pp. 1260–1263. Such activities may explain why the Okhrana logged him as leader of a Jewish self-defence group, and why the Kiev Governor wanted him sacked as one of the more politically active KPI professors (along with Chirvinskii, Vagner, Tikhvinskii, Konovalov, Radtsig, Izhevskii, Ivanov and Perminov): Head of Kiev Okhrannoe otdelenie–Kiev Governor, 16 May 1907: DAKO, f. 2, op. 44, spr. 171, ark. 4; Kiev Governor–Governor-General, 18 April 1906: TsDIAU, f. 442, op. 855, spr. 322, ark. 204.

[52]   'Vospominaniia', vol. 3, pp. 1269–81. For confirmation from Shlikhter, see M.K., 'Bol'sheviki v 1905 g. v Kieve (Po materialam vechera vospominanii)', *Letopis' revoliutsii*, 4 (1925): 53. Shlikhter was arrested in St Petersburg in 1908, and sentenced to permanent

December. Lacking another leader of similar calibre, Kiev's Bolsheviks were thus considerably weaker than hitherto – an important factor for explaining the bloody fates of the sappers' mutiny and the Shuliavka republic.[53]

It was through the crushing of the sappers' mutiny in November that Lomonosov became more deeply and riskily involved in revolutionary activity. When the mutineers were attacked he and Tikhvinskii's partner, Vera Solomon, joined other bystanders in assisting the victims. Specifically, they conveyed the badly wounded rebel leader B.P. Zhadanovskii to Tikhvinskii's KPI apartment, Tikhvinskii being out of town. During the next few days Lomonosov helped plan a successful raid on a hospital to free four injured sappers from arrest. He then assisted with moving Zhadanovskii to the KPI's agricultural research station outside the city. There the fugitive was cared for by Solomon and others, including a 17-year-old girl calling herself Raisa Rozen. The plan was to transfer Zhadanovskii to Zhitomir, a town some 150 km to the west, when he was well enough to be moved. But the police found him before this could happen. Years later, Lomonosov deduced from the memoirs of the then Provincial Governor-General, Sukhomlinov, that the police had probably followed the unsuspecting Raisa Rozen, his future second wife.[54]

Siberian exile, but was freed in 1917. He recalled Kiev in 1905 in his article 'Kiev v oktiabr'skie dni 1905 goda', *Puti revoliutsii*, 1 (1926): 37–58.

[53]   For example, Glavak et al., *Ocherki istorii kievskykh partiinykh organizatsii*, p. 70; Hamm, *Kiev*, pp. 193, 208. As of 7 December 1905 the police had no information about Shlikhter's activities and location after 18 October: Kiev Judicial Procurator–Governor-General, 7 December 1905: TsDIAU, f. 442, op. 855, spr. 342, ark. 18.

[54]   'Vospominaniia', vol. 3, pp. 1282–8, 1294–6, 1689–90. He incorrectly records Zhadanovskii as Zhadovskii. Descriptions of this soldier's time in hiding, which confirm details of Lomonosov's account, are in the police indictment of 10 March 1906: *Revoliutsiia 1905–1907 gg v Rossii: Dokumenty i materialy: Vysshii pod"em revoliutsii 1905–1907 gg: vooruzhennye vosstaniia, noiabr'–dekabr' 1905 goda, chast' 3, kniga 1*, pp. 261–2; a report of 13 April 1906 by the Kiev Okhrana chief A.M. Eremin to the Director of the Department of Police E.I. Vuich: ibid., pp. 1028–9 (note 31); Head of Army Staff–Military Investigator, Kiev Military District, 27 November 1905: TsDIAU, f. 275, op. 1, spr. 276, ark. 62–3; and Kiev Okhrannoe otdelenie–Military Investigator, Kiev Military District, 21 June 1906: TsDIAU, f. 275, op. 1, spr. 678, ark. 115–zv. A slightly different, and possibly inaccurate, description of Zhadanovskii's rescue is in I. Genkin, 'Revoliutsioner-podvizhnik – B.P. Zhadanovskii', *Katorga i ssylka*, 3 (1925): 203. According to a witness called A. Shapiro, the people tending Zhadanovskii expected him to die: 'Iz vospominanii o B.P. Zhadanovskom', *Katorga i ssylka*, 7 (1927): 190. Krzhizhanovskii later informed Lenin that Lomonosov had 'helped selflessly' in the mutiny: Typescript of notes by Lenin, circa 24 May 1920: RGASPI, f. 2, op. 1, d. 14079, l. 7 (also published in Iu.N. Amiantov et al. (eds), *V.I. Lenin: Neizvestnye dokumenty, 1891–1922* (Moscow, 1999), pp. 341–7). Concerning the mutiny see also Sukhomlinov, *Vospominaniia*, pp. 108–9; V. Manilov, 'Kievskaia voennaia organizatsiia RSDRP i vosstanie saper v noiabre 1905 g.', *Letopis' revoliutsii*, 5–6 (1925): 176–225; and M. Akhun and V. Petrov, 'Vosstanie inzhenernykh voisk v Kieve (iz istorii revoliutsionnogo dvizheniia v armii v 1905 g.)', *Krasnaia letopis'*, 3 (1925): 126–48.

Accounts of bomb production by the SD Party in the south of the empire focus on the cities of Rostov and Batum, whereas propaganda and agitation work are indicated as the main SD activities in Kiev. In reality, the scale of Kiev's SD bomb-making operation may have been considerable.[55] Like, for example, the Rostov operation, it was overseen by the Southern Military–Technical Bureau of the SD Central Committee. Lomonosov states that it made approximately 10,000 bombs – codenamed 'small balls' (*shariki*) – during 1905–1906.[56] This figure seems very unlikely because other activists recalled that the southern bureau produced between 200 and 600 bombs in total (it aimed to make 1,000), and no other source shows more than about 200 bombs in Kiev. However, Lomonosov's figure may indicate the number of bombs that could have been made using the amounts of raw materials that to his knowledge were acquired. Also, some of these materials were sent to Rostov and perhaps other cities.[57]

Lomonosov first contributed to this operation in December 1905 in a venture that has been virtually unknown: cooperation between SD bomb-makers in St Petersburg and Kiev. His memoirs reveal that his initial task exploited his engineering expertise and his ability, thanks to his consulting work, to travel quite freely despite severe government restrictions on travel. The assignment was to brief L.B. Krasin, head of the Bolshevik Central Committee's Military–Technical Group in St Petersburg, about a new type of small bomb developed in Kiev. He had to memorize the design and show Krasin how to make it. He met Krasin at

---

[55]   'Vospominaniia', vol. 3, pp. 1133–40. See also A. Trofimenko, 'K istorii Voenno-Tekhnicheskogo Biuro iuga Rossii v 1905–1906 godakh', *Letopis' revoliutsii*, 5–6 (1925): 99–105; V.I. Iurchuk et al., *Ocherki istorii kommunisticheskoi partii Ukrainy*, 4th edn (Kiev, 1977), pp. 86–109; Glavak et al., *Ocherki istorii kievskikh partiinykh organizatsii*, pp. 57–79; and L.T. Senchakova, *Boevaia rat' revoliutsii: ocherk o boevykh organizatsiiakh RSDRP i rabochikh druzhinakh 1905–1907 gg.* (Moscow, 1975). Lomonosov's involvement with the bomb-making operation is confirmed by: G. Mikhailov, 'Iz spogadiv pro pratsiu v Kiivs'kii bil'shovits'kii grupi', in V. Manilov and G. Marenko (eds), *1905 rik u Kiivi ta na Kiivshchini: zbirnik stattiv ta spogadiv* (Kiev, 1926), p. 310; Kushch and Glovatskii, 'Kievskii Politekhnicheskii Institut', pp. 135, 138.

[56]   'Vospominaniia', vol. 3, pp. 1441–2.

[57]   Gel'zin, 'Iuzhnoe Voenno-tekhnicheskoe biuro', p. 39; N. Rostov, 'Iuzhnoe Voenno-Tekhnicheskoe Biuro pri Tsentral'nom Komitete RS-DRP', *Katorga i ssylka*, 1 (1926): 98, 107–8. Circumstantial evidence includes: the serial numbers of bombs found by the police in Kiev in July 1907 ranged from 58 to 134: Kiev Police Chief–Kiev Governor, 16 July 1907: DAKO, f. 2, op. 223, spr. 252, ark. 1–2; the Kiev Social Democratic committee spent 1,522.95 rubles on bomb-making from November 1905 to January 1906, and had approximately 120 bombs in stock as of early/mid-April 1906: Bondarenko, *Bol'sheviki Kieva*, pp. 239, 241; 550 bombs were reported to have been made in Rostov, and 1,500 in Batum – evidently large operations, but far smaller than Lomonosov's figure: Trofimenko, 'K istorii Voenno-Tekhnicheskogo Biuro iuga Rossii', p. 104; Senchakova, *Boevaia rat' revoliutsii*, p. 86. See also *Tekhnika Bol'shevistskogo podpol'ia: sbornik statei i vospominanii*, vypusk 1 (Moscow and Leningrad, 1924).

least twice in December 1905, and again during visits to the capital in early 1906 under cover of his consulting, including one trip to Krasin's country *dacha* when Lenin was present. Apparently Krasin gave him a Browning pistol, which he kept for years, and the bomb was tested successfully in Finland. In March 1906 he also helped Krasin to design a heavy gun, but because of technical difficulties it was deemed easier to steal or import guns.[58]

The main value of Lomonosov to the Kiev bomb-makers lay in the unusual combination of his academic position and railway contacts. Contrary to the Okhrana's suspicion, the bombs were not made at the KPI but in other towns and villages at small 'laboratories' that were frequently relocated. Teams of KPI students smuggled nitric acid to them, and Lomonosov supervised this operation with Tikhvinskii. Meanwhile, his SER travel permit and occasional access to a railway managerial saloon carriage, courtesy of the unsuspecting SER Board of Directors, meant that he could help move 'party cargoes' such as bombs and bomb materials, or assist the relocation of a 'laboratory'. To maintain his security his regular contacts were limited to just Tikhvinskii, Ianovskii and possibly a lawyer called S.N. Gus'kovskii. He had to destroy all private correspondence, was not permitted to keep address notes and non-technical notebooks, and was encouraged to socialize with his female fundraising colleagues – Smirnova and another actress, V.N. Il'narskaia – as a cover. The authorities did harbour some suspicions about him, as will be shown in Chapter 6, but his standing as a state official remained sufficiently solid that, ironically, he was once asked to escort the commanding officer during a police search of the KPI.[59]

His memoirs indicate that this bomb-making activity ended in the early summer of 1906. One reason was that the renewed closure of the KPI hampered

[58] 'Vospominaniia', vol. 3, pp. 1296–8, 1302–4, 1331, 1333–40. According to S.M. Pozner, the designer was I.K. Mikhailov and this was Kiev's second type of bomb: 'Rabota boevykh bol'shevistskikh organizatsii 1905–1907 gg.', *Proletarskaia revoliutsiia*, 7 (1925): 87, 92–3. Krasin's home was at Kuokkola (now Repino), where Lenin also lived at times during February–April 1906: O'Connor, *Engineer of Revolution*, pp. 88, 102; G.N. Golikov et al. (eds), *Vladimir Il'ich Lenin: Biograficheskaia khronika, tom 2: 1905–1912* (Moscow, 1971), pp. 228–37. Lomonosov is not mentioned in Krasin's recollections, notably: 'Bol'shevistskaia partiinaia tekhnika', in *Tekhnika Bol'shevistskogo podpol'ia*, pp. 7–15; *Delo davno minuvshikh dnei* (Moscow, 1931); and *Bol'sheviki v podpol'e* (Moscow, 1932). See also M.N. Liadov and S.M. Pozner (eds), *Leonid Borisovitch Krasin ('Nikitich'): gody podpol'ia: sbornik vospominanii, statei i dokumentov* (Moscow and Leningrad, 1928).

[59] 'Vospominaniia', vol. 3, pp. 1333–40, 1407. KPI students also made bombs for the Socialist Revolutionary (SR) Party: A. Spiridovich, *Zapiski zhandarma*, 2nd edn (Khar'kov, 1928), pp. 201–5. Il'narskaia, from the Solovtsov drama theatre, was known to the police: Kiev City Governor–Kiev Governor-General, 1 March 1905: TsDIAU, f. 442, op. 855, spr. 77, ark. 1–zv.; later the partner of the writer L.G. Munshtein in emigration, she died in 1946: V. Alloy (ed.), *Diaspora: Novye materialy*, vol. 1 (Paris and St Petersburg, 2001), p. 365. Lomonosov's presence during a search of the KPI was officially recorded on at least one occasion: Protocol, 21 September 1906: TsDIAU, f. 275, op. 1, spr. 1027, ark. 4–zv., 8–zv.

the procurement of materials. Also important was the lack of any money from St Petersburg once Krasin departed for Stockholm to attend the Fourth SD Party Congress in April 1906. Shortly afterwards the Kiev SD organization was instructed to cease its bomb-making activity in accordance with the congress's Menshevik-inspired resolution to renounce terrorism. By now a related local concern was that the authorities knew that Tikhvinskii's flat had been used for hiding Zhadanovskii. Interestingly, the coordinator of these Kiev bomb-makers, a Menshevik, disagreed so strongly with the renunciation of terrorism that he passed the stock of bombs and supplies to activists of the Socialist Revolutionary Party.[60]

Lomonosov worked in the SD underground movement for one further year. For example, he couriered money to Shlikhter in Finland in July 1906, possibly in connection with the attempted mutiny at the Sveaborg fortress in Helsingfors harbour. He did more fundraising work, allowed party visitors to use his KPI apartment and hid some bombs there for a week. He helped to organize rooms and security for party meetings at the KPI, such as a regional conference in March 1907.[61] Above all, he participated in the election campaign for the Second Duma in the winter of 1906–1907, after the end of the Bolshevik boycott of this institution. He was co-opted to a committee that coordinated the electoral activity of parties to the left of the Kadets in the Kiev, Volynsk, Podol'sk, Chernigov and Poltava

---

[60] 'Vospominaniia', vol. 3, pp. 1365, 1378, 1471. Pozner confirms that SR activists acquired some of the Kiev bombs, and also reports that the police captured the bomb-making works: 'Rabota boevykh bol'shevistskikh organizatsii', p. 94. The debate about revolutionary terrorism and armed 'expropriations' at the 1906 congress of the social democrats is discussed in, for example, L. Schapiro, *The Communist Party of the Soviet Union*, 2nd edn (London, 1970), pp. 89–90; O'Connor, *Engineer of Revolution*, pp. 80–86. The SR Party's council suspended terror actions during May–July 1906, but certain local groups ignored this policy: M. Perrie, 'Political and Economic Terror in the Tactics of the Russian Socialist-Revolutionary Party before 1914', in W.J. Mommsen and G. Hirschfeld (eds), *Social Protest, Violence and Terror in Nineteenth and Twentieth-century Europe* (London, 1982), pp. 66, 69. The Kiev SR group was one such exception: N. Komarov, 'Ocherki po istorii mestnykh i oblastnykh boevykh organizatsii partii sots.-revoliutsionerov, 1905–1909 gg.', *Katorga i ssylka*, 4 (1926): 56–81. The militancy of the SR organization in the Ukraine is noted by M.I. Leonov: *Partiia sotsialistov-revoliutsionerov v 1905–1907 gg.* (Moscow, 1997), pp. 274–6, 309–34. See also M. Hildermeier, 'The Terrorist Strategies of the Socialist-Revolutionary Party in Russia, 1900–14', in Mommsen and Hirschfeld, *Social Protest*, pp. 80–87; and C. Rice, *Russian Workers and the Socialist Revolutionary Party through the Revolution of 1905–07* (Basingstoke, 1988), pp. 64–70. The secret police suspected that Tikhvinskii was an SR and kept a file about him from October 1903: TsDIAU, f. 275, op. 1, spr. 152. So too did the regular police: TsDIAU, f. 274, op. 1, spr. 1113. For their suspicions about his involvement with Zhadanovskii see, for instance, the report of 13 April 1906 to the Director of the Department of Police, in *Revoliutsiia 1905–1907 gg v Rossii: Dokumenty i materialy: Vysshii pod"em revoliutsii 1905–1907 gg: vooruzhennye vosstaniia, noiabr'–dekabr' 1905 goda, chast' 3, kniga 1*, pp. 1028–9 (note 31).

[61] 'Vospominaniia', vol. 3, pp. 1409, 1445–7, 1486, 1535. On the Sveaborg mutiny see A. Ascher, *The Revolution of 1905, vol. 2: Authority Restored*, pp. 227–36.

provinces. He was even discussed as a candidate for the Kiev province, but could not stand there because he was registered to vote in the city itself.[62] As in October 1905, he focused on bridging the divide between the radicals and liberals and again was unsuccessful. On behalf of the coordinating committee he agreed with a Kadet representative, the KPI's Professor Ivanov, that the Left would assist the Kadets in the city of Kiev and vice versa in the province. The committee criticized Lomonosov for being too pliable, but ratified the deal. However, the Kadets defaulted a week before the election, forcing the committee to produce new leaflets hastily.[63]

Lomonosov's retirement from this fray in mid-1907 is puzzling. His explanation – disillusionment with politics and his belief that the revolution had failed – fits with his old logic of holding himself in reserve for the right moment. But that does not explain the timing: he was disillusioned and pessimistic by the spring of 1906, so why did he not quit then?[64] To speculate, he may have feared reprisals because he knew dangerous secrets of the Kiev SD group and the Central Committee. By 1907, however, he may have begun to suspect police surveillance: at least one letter to him was intercepted.[65] An incident in May 1907 probably gave him and his co-conspirators pause for thought. It concerned his task to organize the summer project in land surveying for the KPI's engineering students, described in Chapter 3. He was disconcerted when party colleagues demanded the appointment of Ianovskii as a site supervisor, a job for which this Menshevik activist was technically qualified. Knowing that the Provincial Governor-General, Sukhomlinov, was nervous about the project, Lomonosov feared that Ianovskii's appointment might lead to disturbances that would endanger his own position, despite his belief that most students were not revolutionary. Hence he obtained assurances from the Socialist Revolutionary and SD parties that they would not exploit the exercise. Ianovskii was appointed, and caused no trouble. Given that Lomonosov does not report any further party work in Kiev, it is likely that this affair cooled relations. Be that as it may, the break was reinforced by the abortive rising in Kiev in June 1907 (which precipitated the collapse of the city's Bolshevik group) and then by his move to Ekaterinoslav.[66]

---

[62] A copy of the city electoral register is at TsDIAU, f. 275, op. 1, spr. 941, ark. 63–90zv.; Lomonosov features on ark. 76zv.

[63] 'Vospominaniia', vol. 3, pp. 1505–15, 1518–21.

[64] 'Vospominaniia', vol. 3, pp. 1365, 1371, 1521.

[65] Extract from unsigned letter, Kiev, 6 January 1907, to Iurii Vladimirovich Lomonosov: GARF, f. 102, op. 265, d. 154, l. 24.

[66] 'Vospominaniia', vol. 3, pp. 1546–53, 1558–67, 1605–9, 1662–4. The river Ros' project was sufficiently important to earn a mention in Sukhomlinov's *Vospominaniia*, p. 136. On the failed uprising, which was led by the Socialist Revolutionaries, see: *Revoliutsiia 1905–1907 gg v Rossii: Dokumenty i materialy:Vtoroi period revoliutsii, 1906–1907 gody: ianvar'–iiun' 1907 goda, kniga 2* (Moscow, 1965), pp. 388–9; and A. Troianovskii, 'Vosstanie v Kieve v 1907 g.', *Katorga i ssylka*, 3 (1922): 181–8.

It is worth adding that the priorities of his personal life do not appear to have significantly affected, or been affected by, his revolutionary activism during 1905–1907. Clearly the fact that he was supporting three children – two with Sonia and one with Masha – did not deter him from risking his life in the underground. His activism was not encouraged by, or an attempt to impress, either his estranged wife or any of the six other women with whom he was romantically and sexually involved during these years. Only one of these women had similar politics: Raisa Rozen. The daughter of a Jewish merchant in the small south-west Ukrainian town of Letichev, she was associated with the Menshevik-oriented Spilka Ukrainian Social Democratic Union. Detained briefly in Odessa in April 1905 as a suspected terrorist, she moved to Kiev about two months later. Lomonosov first met her in August 1905 when Sonia gave her shelter at Vera Solomon's request, and their paths crossed again in the Zhadanovskii fiasco.[67] But no love affair developed until May 1907, by which time Lomonosov's political work had largely ceased. After he and Sonia went their separate ways in January 1906, albeit without a divorce as yet, his thoughts about marriage concerned other women in typically earnest yet feckless fashion.[68] These were Oksana, whom he had hardly seen since 1904, but to whom he was still strongly drawn; Masha, for whom too his feelings were strong, and who would give birth to a second son in December 1906; a distant relative; a servant named Marfusha, whose pregnancy from autumn 1906 may also have been his doing; and, less ardently, two other acquaintances.[69] Raisa began working as his secretary in January 1907, assisted with his election work and tutored his children Vova and Marusia in Russian and mathematics. He writes that he sensed interest on her part in February, but that he did not reciprocate. Only in mid-May did an affair suddenly blaze.[70]

This event seems to have bemused Lomonosov himself. He recalled that a crucial stimulus was Raisa's barefoot participation in a family gardening session, for he had a 'fetish' for bare female feet, and this was the first time that he had seen

---

[67] 'Vospominaniia', vol. 3, pp. 1175–6, 1296. Background details were also recorded during a police search of her apartment: Information about person searched at 01.30: R.I. Rozen, 16 May 1906: TsDIAU, f. 275, op. 1, spr. 1141, ark. 12–zv. This file also has reports about her correspondence, which was intercepted from about January 1906. On her arrest see Head of Odessa City Police Directorate–Police Department VII d-p [St Petersburg], 31 May 1905: GARF, f. 102, D-VII, 1905, d. 2974, l. 14–15ob.; and 9 July 1905: GARF, f. 102, D-VII, 1905, d. 2057, l. 49.

[68] Although Lomonosov and Sof'ia Aleksandrovna were basically separated from January 1906, they were not formally divorced until February 1916: 'Vospominaniia', vol. 5, pp. 1145–9.

[69] See 'Vospominaniia', vol. 3, thus: Oksana: pp. 1318–22, 1536, 1553–7, 1558–61, 1578–9, 1596, 1599, 1603–5; Masha: pp. 957, 1132–3, 1336–7, 1530, 1536, 1599, 1772; Aunt Asia: pp. 1438–9, 1517, 1530, 1536, 1542, 1557; Marfusha: pp. 1483–4, 1530, 1536, 1599; Variia: pp. 1415–21, 1461–2; and Dina: pp. 1485–6, 1536, 1553–7, 1561.

[70] 'Vospominaniia', vol. 3, pp. 1519, 1523–4, 1531, 1578–9, 1581, 1591–6, 1603–5.

such behaviour from an intelligentsia girl.[71] He liked her attitude to the children, and, more cynically, he thought that marriage to a Jewish girl would demonstrate his scorn for social prejudices. Also, he could be very impulsive despite – or perhaps in conjunction with – his chronic indecision in affairs of the heart and politics. Yet he states that he was not attracted to her physically, and was even afraid of her – an odd confession given his strong libido.[72] Be that as it may, this relationship was undoubtedly unique for him. His philandering ceased for several years, and although he had occasional liaisons with Masha around 1912–1916 and then similarly intermittent relationships with her and several other women during 1919–1925, he displayed a far deeper and more durable emotional commitment to Raisa than to any other woman. As he made a fresh start in Ekaterinoslav, all thoughts of other possible long-term partners had disappeared.

*   *   *

Lomonosov made a noteworthy contribution to the revolutionary activities of the SD Party during the 1905 revolution, at considerable personal risk. By October 1905, if not before, he even endorsed terrorism for overthrowing the autocracy. So far as one can sense from the available evidence, feelings of self-loathing and inadequacy had no place in his motivation; rather, he was generally a self-confident character with a genuine aversion to the institution of autocracy and perhaps a certain relish for battle, at least in the heat of the moment in late 1905. Yet his politics remain difficult to judge. He did not formally join any political party, and although he called himself a 'non-party Marxist', the most accurate description would seem to be a rather confused mixture of liberalism and Marxism, as though he selected ideas from a long and varied liberal-radical menu.

His experiences point to several broader conclusions about the 1905 revolution. In Kiev the liberal–socialist relationship remained tentative, and the decisive rift between these two camps occurred before the publication of the October Manifesto – a fact at odds with the usual assumption that the regime caused that rift by announcing major political concessions in the manifesto. Likewise, the Bolshevik leader in Kiev, Shlikhter, appears in a new light as a pragmatic, flexible character who – sadly for the Bolsheviks – left the city for good on 1 November. The SD bomb-making operation in Kiev was bigger, and more closely connected to the party leadership in St Petersburg, than has been

---

[71]   'Vospominaniia', vol. 3, pp. 1592–3. Earlier he reports that during the winter of 1900–1901 he consulted a 'doctor' about his passion for female bare feet. He learnt that this 'common abnormality' was called 'fetishism', that this doctor had not previously encountered it in a member of the gentry, and that the cause was possibly genetic; the advice was to take a barefooted wife or lover: 'Vospominaniia', vol. 2, pp. 602–3. If this account is accurate, the doctor was ignoring the usual Russian medical prescription for such sexual desire, which was restraint: Engelstein, *Keys to Happiness*, pp. 221–5.

[72]   'Vospominaniia', vol. 3, pp. 1593–4, 1640.

thought. Finally, Lomonosov's tale provides more support for the view that the boundaries between the radical parties and factions were rather more fluid than party leaders cared to think, to the extent that a Menshevik could supply arms to the rival Socialist Revolutionary Party.

Chapter 5

Applications of Science on the Russian Railways, 1908–1914

Lomonosov's retirement from revolutionary activism and his return to railway employment coincided with the end of the first Russian revolution and the start of a seven-year interval until the outbreak of the world war in 1914. That period of Russian history, and especially the empire's trajectory of political, economic and social development by 1913, has since become the focus of much historical debate. Put simply, two interpretations have predominated. For some historians, Russia was evolving more or less peacefully into some form of capitalist liberal democracy. For example, the economy was booming from about 1909, the new parliamentary institutions were becoming more firmly established, and there was increasing evidence of a civic society, including the creation of scientific and technical associations and societies. By contrast, other scholars have argued that the empire was heading for crisis. The economic boom was patchy and could not last, the constitutional experiment was effectively dead, revolutionaries were killing thousands of people, and after troops massacred striking workers at the Lena goldfields in Siberia in 1912, a wave of strikes developed to the point that a general strike was brewing in July 1914. Respectively for these two perspectives, then, the impact of the world war was either to turn Russia towards revolution or – perhaps after a delaying display of patriotic unity – confirm her on that course.[1]

In this context the development of Lomonosov's railway career to 1914 raises the fundamental question of whether, at least in the important realm of railway transport, the tsarist state was capable of delivering economic reform and modernization in the early twentieth century. This chapter shows how he quickly earned national prominence in both management and research. After just one year as a deputy traction superintendent on the Catherine Railway he became a traction superintendent in his own right, first on the Tashkent Railway and then on the premier line, the Nicholas Railway. As early as 1912 he joined the upper

---

[1]   Good introductions to these issues include H. Rogger, *Russia in the Age of Modernisation and Revolution, 1881–1917* (London, 1983), pp. 229–50; and P. Gatrell, *Russia's First World War: A Social and Economic History* (Harlow, 2005), pp. 1–16. Particularly important have been works by L.H. Haimson, such as '"The Problem of Political and Social Stability in Urban Russia on the Eve of War and Revolution" Revisited', *Slavic Review*, 59/4 (Winter 2000): 848–75. On the scientific societies see J. T. Andrews, *Science for the Masses: The Bolshevik State, Public Science, and the Popular Imagination, 1917–1934* (College Station, 2003), pp. 26–35.

ranks of the MPS itself as a deputy superintendent of the Directorate of Railways, the main institution for overseeing the common-carrier railway network.[2] A year later he was appointed to the ministry's Engineering Council – the committee of influential engineers that determined technical policy. Meanwhile, still making time for research, he established a reputation as an expert on locomotive testing, and in 1914 he became the founding director of an MPS research unit known as the Experiments Office (*Kontora opytov*). He also served part-time from 1911 as a professor at his Alma Mater, the Institute of Ways of Communication, where he led a reform of the railway-related curriculum. How, then, is this record of rapid promotion and varied achievement to be explained, and how might it enhance our understanding of the late tsarist economy?

Management

Disagreement about the health of the economy has been central to the historical debate about the condition of the Russian empire on the eve of the Great War. Some analyses stress the signs of stability and success, such as the growth of the net national product by over 5 per cent per year in 1909–1913, the relatively high level of net capital investment, the growing diversity in the sources of wealth creation, the large increases in industrial and agricultural production, and the boom in railway-building. Other accounts emphasize the signs of trouble. For example, industrial and agricultural output per capita remained low compared with leading industrialized countries, as did productivity; Russia lagged behind her main rivals in her overall level of industrial development and had to import vital industrial products such as many types of chemical; and the national debt was enormous, with nearly half of this borrowing held abroad.[3]

This ambiguity is evident in the railway sector, which contemporary experts still generally saw as crucial for driving economic growth (albeit with less justification than hitherto thanks to growing diversification in the economy). On the one hand, traffic and revenue recovered strongly from about 1908 following the years of recession, war and revolution.[4] Still regarded as underdeveloped by politicians,

---

[2]   A common-carrier railway is required by law to accept all traffic that is presented for shipment. Russia's public railway network – as opposed to non-public railways such as internal factory systems – had this status throughout the imperial and Soviet periods. The mixed system of state and private railway ownership in the imperial period had no bearing on the common-carrier status.

[3]   Helpful summaries of these economic issues include: R.W. Davies (ed.), *From Tsarism to the New Economic Policy: Continuity and Change in the Economy of the USSR* (Basingstoke, 1990), pp. 1–26; A. Nove, *An Economic History of the USSR* (Harmondsworth, 1982), pp. 11–45.

[4]   Concerning the railways in the Russo–Japanese war and 1905 revolution see F. Patrikeeff and H. Shukman, *Railways and the Russo–Japanese War: Transporting War*

the MPS and public alike, the railway network continued to expand: almost 1,500 km of new routes were opened in 1913, a figure close to the average for the boom of the 1890s.[5] On the other hand, there was a crisis of public confidence in the railways. The treasury found that the growth in railway revenue was outpaced by the increase in the huge cost of operating the state-owned bulk of the network and servicing the vast railway-related portion of the public debt. Meanwhile deputies in the new legislative assembly, the State Duma, complained that the railways were not delivering value for money. Railway troubles had apparently disrupted the war effort against Japan, notably on the trans-Siberian route; press reports were legion about corruption, delays, accidents and shortages of freight wagons; and still money had to be found for building the many lines that the MPS considered essential. So by January 1908, when Lomonosov rejoined the railways, the tsarist regime was starting to define greater railway efficiency and investment as national priorities. It commissioned an inquiry that lasted for five years and was led by N.P. Petrov, the former deputy head of the MPS whose formulae for locomotive performance so upset Lomonosov; it appointed the tough-minded legal, trade and shipping expert S.V. Rukhlov as minister in 1909; and most radically, the Ministry of Finances prioritized private funding for new railways for the first time in decades despite the MPS preference for using state funds.[6]

Although this situation would seem propitious for engineers like Lomonosov with a passion for improving efficiency and the ambition to make a difference, its impact on his career was modest. Much more important for the trajectory of his career was the way in which he rejoined the railways. Highlighting his railway-based IPS education and practical experience on two railways, and using his considerable self-confidence and energy, he managed to talk his way into the managerial hierarchy at a fairly senior level. In other words, he did not follow the usual sequence of promotions through the lower ranks. This situation had potential benefits both for him and for the railways: he could soon seek promotion into the highest ranks of the railway bureaucracy, and the fact that he was not steeped in traditional railway thinking conceivably implied that he was well placed to develop and introduce fresh ideas. On the other hand, his lack of day-to-day experience meant that he faced not just a steep learning curve, but also the

(Abingdon, 2007); Reichman, *Railwaymen and Revolution*.

   [5]   See Afonina, *Kratkie svedeniia*, pp. 44–87.

   [6]   Westwood, *History of Russian Railways*, pp. 129–66. For Rukhlov's career see Shilov, *Gosudarstvennye deiateli*, pp. 575–7. The issue of state/private ownership of railways was much debated during this period. Examples of publications are: L. El'kin, 'Narekannia na kazennoe khoziaistvo zheleznykh dorog', *Izvestiia Sobraniia inzhenerov putei soobshcheniia*, 9 (1911): 1–5; Zheleznodorozhnik, 'Kazna i zheleznodorozhnoe stroitel'stvo', *Izvestiia Sobraniia inzhenerov putei soobshcheniia*, 11 (1911): 5–7; and A. Frolov, 'K voprosu o kazennom i chastnom khoziaistvakh na zheleznykh dorogakh', *Izvestiia Sobraniia inzhenerov putei soobshcheniia*, 16 (1913): 253–6. On Lomonosov's involvement see also notes 61 and 63 below.

distrust of colleagues and subordinates, the more so because his appointment to the Catherine Railway thwarted a proposed internal promotion.[7]

The importance of practical experience is very apparent in the light of the structure of Russia's common-carrier railway bureaucracy and the typical career path of a junior traction engineer.[8] The main organizational entity below the MPS was the individual railway. There were 47 such railways in the empire (excluding Finland) in 1908, and 55 by 1914 thanks to construction and reorganizations. The biggest lines had over 4,000 route kilometres and 20,000 staff with annual budgets of tens of millions of rubles, which ranked them among Russia's largest industrial enterprises. Most were state-owned and the remainder were private companies, but all had a departmental structure that reflected the primary concerns of day-to-day operations. In the state sector, where Lomonosov would remain, each railway was headed by a superintendent. His main tasks included supervision of the general office, legal department and chief accountant's office. He had several deputies to oversee departments for commercial performance, income, material supplies, traffic (concerned with operating the train service), traction (mainly locomotives, rolling stock and associated infrastructure) and track (including bridges). These departments maintained contact with the MPS (primarily the Directorate of Railways), coordinated their work with each other and their counterparts on neighbouring lines, and supervised a variety of subordinate entities out on the line. For example, a traction department had sub-departments or sections for locomotives, rolling stock, workshops, contracts and accounts, and a general technical sub-department; it negotiated with the railway's traffic, track and other departments about train schedules, speed limits, fuel and spare-part procurement and suchlike, and it supervised the line's workshops and traction sections. The latter were based at the main locomotive depots to handle the day-to-day traction affairs of a given district.[9] Correspondingly, traction engineers often began their career as an assistant superintendent of a locomotive or carriage depot, as Lomonosov had begun nominally in 1898. All being well, they would seek experience as a depot superintendent, either on the same line or elsewhere. In due course they might work as a deputy superintendent and superintendent of a traction section, and earn promotion to a line headquarters as deputy head and head of a sub-department. Beyond that, they might aim to become a deputy head of department (this was the relatively senior point at which Lomonosov joined the Catherine Railway). Then they might hope to head a department, before becoming a line superintendent and/ or moving to the MPS.

No concession was made to Lomonosov for his inexperience: the MPS assigned him to one of the largest and busiest railways. Named after Catherine the Great,

---

[7]   'Vospominaniia', vol. 4, pp. 1–3.

[8]   For a detailed description of the railway bureaucracy in English see Reichman, *Railwaymen and Revolution*, pp. 15–40.

[9]   Private railways had a broadly similar structure and procedures, with a chairman and board of directors to provide overall leadership.

the inimitable eighteenth-century tsarina, the Catherine Railway was opened in 1884 with a roughly east–west route across southern Ukraine and headquarters at Ekaterinoslav, a city of 200,000 inhabitants on the River Dnepr. Soon profitable thanks to bulk freight traffic, it expanded eastwards in the 1890s, encompassing the new heavy industry around the city of Iuzovka (now Donets'k). When congestion became a problem, a parallel route further south was developed via Nikolpol', Aleksandrovsk and Pologi, which was known locally as the Second Catherine Railway. Eastbound traffic was dominated by mineral ores and westbound by Donbass coal. By 1908 some 3,000 freight wagons were loaded daily, which was about 20 per cent of the empire's total. As befitted its status the line had excellent workshops, which were located at Ekaterinoslav, Nizhnedneprovsk, Lugansk, Aleksandrovsk and Taganrog. Similarly, it was enjoying priority for deliveries of the latest standard freight engine, a 2-8-0 championed by Shchukin and known as the 'modified Chinese' type.[10]

Finally, Lomonosov's holistic outlook was likely to cause tension because of the compartmentalized way in which the world's major railways were organized. As the industry expanded and became more complex in the 1830s and 1840s it joined the vanguard of technical specialization. Although the basic distinction between civil and mechanical engineering formed part of that change, the distinction soon went further. By the 1850s a budding railway engineer usually specialized in one significant aspect of the enterprise, the largest areas being (in the Russian terminology) traffic, traction and track; and increasingly expertise was defined even more narrowly, such as signalling or bridge design. Unsurprisingly, the traffic, traction and track specialists usually developed strong departmental identities, and even though close cooperation between them was obviously essential, their inter-departmental relations were often adversarial. This phenomenon had already hampered Lomonosov in his research on the Khar'kov–Nikolaev Railway during 1898–1901. But although he duly became a staunch advocate of traction interests, he did often relate those interests to their broader context – an outlook that challenged traditional professional and organizational boundaries based on specialist knowledge.

How, then, did he fare in the rough-and-tumble of everyday railway life, and how successful was his attempt to apply scientific principles to it? As a deputy departmental superintendent he was unlikely to have a major impact at Ekaterinoslav, yet he probably failed to exploit all the scope that he did have. The extent to which a deputy was expected to show initiative depended on his immediate superior. In this

---

[10]   N.S. Konarev et al. (eds), *Zheleznodorozhnyi transport: Entsiklopediia* (Moscow, 1994), p. 130; Afonina, *Kratkie svedeniia*, pp. 35–6, 48, 69; 'Vospominaniia', vol. 4, pp. 16–22. For the regional railway context see A.M. Solovyova, 'The Railway System in the Mining Area of Southern Russia in the Late Nineteenth and Early Twentieth Centuries', *Journal of Transport History*, Third Series, 5/1 (March 1984): 66–81. The few surviving records of the Catherine Railway for 1897–1919 (in Derzhavnii arkhiv Dnipropetrovs'koi oblasti, f. 566) have no information about Lomonosov: V.A. Iurkova–author, 22 March 2000.

case Lomonosov felt that the traction superintendent, A.S. Gutovskii, did not want such independence of mind.[11] Be that as it may, Lomonosov's inexperience meant that he first had to study his jurisdiction – the sub-departments and sections for carriages and wagons, water supply and accounting. Yet he was quick to generate new ideas. For example, he told the water-supply sub-department to devise a plan for infrastructure improvements to reduce the damage caused to locomotive boilers by poor-quality water. Similarly, he reportedly reduced workshop costs by changing employment arrangements – ideas that would help inform a new MPS labour policy some years later.[12] But because he lacked influence over budgetary planning, he always needed Gutovskii's support. Here, as occurred repeatedly throughout his career, Lomonosov became his own worst enemy. Taking to excess the idea of battling for his beliefs, he gained a reputation for insubordination and arrogance as well as conscientiousness and imagination. For example, believing passionately in the value of inspection tours, he disobeyed an order not to make such tours when deputizing for Gutovskii in March 1908.[13]

Probably the most important experience that he gained at Ekaterinoslav concerned financial administration and management. As we have seen, he believed that considerable scope existed to cut the costs of railway construction and operation by basing decisions on scientifically defined criteria – an analysis that in principle could help the state to curb its escalating railway-related deficit. Yet his earlier experience of railway finances had been minimal. Now he acquired a deeper knowledge of the system, its shortcomings and the likely obstacles to remedying them. A telling example concerned minor papers. Each day he had to countersign as many as 300 documents that were mostly small invoices or staff travel warrants. This duty was devised to minimize corruption and fraud – a problem and remedy found throughout the Russian bureaucracy and economy, not least because salaries were low. But because he could not scrutinize these papers systematically he felt that his involvement was meaningless and merely slowed the paper flow at the cost of his time. Yet it was impossible for him to change this system.[14]

Unsurprisingly, the changes that he did initiate caused discontent wherever staff interests seemed threatened. He was courting trouble, for instance, by making reforms that – ironically for a Marxist – reduced the wages of some staff. Even more contentious was his campaign against labour indiscipline, theft and corruption, for it involved many reprimands, dismissals and recommendations for prosecution. As yet, however, trouble was not caused by the resumption of his effort to devise a formula for costing railway operations, despite its purpose to reveal inefficient working practices: this research was still at a preliminary stage,

---

[11]   For example, 'Vospominaniia', vol. 4, pp. 9–10, 39–40, 45.

[12]   'Vospominaniia', vol. 4, pp. 34–6, 61–7, 78–89. See also note 56 below.

[13]   'Vospominaniia', vol. 4, pp. 39–40, 45, 58–61.

[14]   'Vospominaniia', vol. 4, pp. 9–10, 41–5, 73–92.

and several years would pass before it appeared in his book *Nauchnye problemy ekspluatatsii zheleznykh dorog* (*Scientific Problems of Railway Operation*).[15]

Some of the opposition to his reforms was extreme, and indicates a continuity between the pre- and post-revolutionary epochs that can help explain the tolerance of corruption, theft and inefficiency in both the tsarist and Soviet economies. Specifically, anonymous letters of complaint about him were sent to his superiors.[16] This sinister phenomenon tends to be associated with the Soviet era, especially Stalin's purges in the 1930s, yet it was also characteristic of late Imperial Russia. Indeed, such denunciations became so common and disruptive that in 1898 the Directorate of Railways officially decided to ignore them, although it abandoned this principled stance after the shock of the 1905 revolution.[17] Lomonosov would become the object of such unpleasant letters on at least one other railway before 1914.[18] Moreover, he had to take seriously any threats in them: as a prominent state official he was a potential target for murder by revolutionaries – a danger that was highlighted for him as early as February 1908 by the killing of the superintendent of the Lugansk workshops.[19] That he nonetheless risked provoking staff hostility reflects his commitment to organizational rationality and especially his character – his drive, ambition, pugnacity and occasional recklessness. His broader dilemma, of course, was that however much he tried to treat staff fairly, his toughness would appear arbitrary unless his fellow managers applied the same approach consistently throughout the railway, which never happened.

Given his emphasis on competence and fairness, it is ironic that his promotion to traction superintendent in 1909 probably owed more to personal connections than to his achievements. Apparently in 1907 he had requested a guarantee of promotion after a year and Dumitrashko, the then head of the Directorate of Railways, had acquiesced. When the time came, however, Dumitrashko's successor, D.P. Kozyrev, demurred. Lomonosov then secured permission to exchange jobs with the head of the Catherine Railway's Aleksandrovsk workshops. Yet just days later he was appointed as the acting traction superintendent of the Tashkent Railway. His memoirs attribute this change to recognition from the new minister, Rukhlov, for his energy and strict enforcement of the law. They also note, however, a change in the attitude of Kozyrev, who, it transpired, had known the cadet Lomonosov during 1888–1891 when serving with the Sychevka railway's parent company.[20] Very probably Lomonosov ingratiated himself with

[15] 'Vospominaniia', vol. 4, pp. 83–90, 263–5; Iu.V. Lomonosov, *Nauchnye problemy ekspluatatsii zheleznykh dorog* (Odessa, 1912).

[16] 'Vospominaniia', vol. 4, pp. 90–92.

[17] See *Izvestiia Sobraniia inzhenerov putei soobshcheniia*, 17 (1911): 10.

[18] See note 43 below.

[19] 'Vospominaniia', vol. 4, pp. 72–3, 90–92; V.F., 'Nekrolog', *Vestnik Ekaterininskoi zheleznoi dorogi*, 56 (1908): 222.

[20] 'Vospominaniia', vol. 4, pp. 248–50, 262–3, 271–83. For Kozyrev's career see: Service record, 1917: RGIA, f. 229, op. 18, d. 4046, ll. 3–4.

Kozyrev, who persuaded Rukhlov; certainly Lomonosov's association with Kozyrev would become sufficiently close to attract comment from the anonymous letter-writers, as we shall see.

Lomonosov's two-year stint on the Tashkent Railway merits attention both as an important stage in his career and as a well documented example of the challenges, opportunities and constraints for reform-minded managers on the railways during the last pre-war years. This line was a distinctive, difficult component of the railway network. A single-track railway rendition of an ancient silk road, it stretched some 2,200 km from Kinel' (near Samara) eastward to Orenburg and thence south-east through steppe, desert and river valleys to Tashkent, where it joined the Central Asian Railway (Krasnovodsk–Samarkand–Tashkent).[21] Its headquarters were at Orenburg, a city of nearly 100,000 inhabitants on the River Ural that was a major Russo–Chinese trading centre. The railway had been created in 1905 through a merger of the Kinel'–Orenburg section (opened 1877) of the Samara–Zlatoust Railway with the Orenburg–Tashkent line, which opened in two stages during 1905–1906. It was partly intended to enhance Russian political control and influence over Central Asia and – so feared London – Afghanistan, Tibet and China. But its economic potential was also a major consideration, especially its capacity to facilitate the shipment of cotton to European Russia and grain to Central Asia. Naturally the line also became an important employer, the Orenburg workshops being that city's biggest factory-type enterprise. As yet, however, the line's facilities were inadequate: for example, the Tashkent workshops were incomplete.[22] For Lomonosov, then, a transfer to this distant desert line was no sinecure; indeed, now responsible for some 5,900 staff, he relished the tough environment as a chance to make his managerial name. It seems to have given him a long-term psychological boost, with the self-confident outlook and bearing of, to use a Russian slang term, a *bol'shoi nachal'nik* (a big boss). Not least,

---

[21]   The route is described in K. Baedeker, *Russia, with Teheran, Port Arthur and Peking: Handbook for Travellers* (London, 1914, facsimile reprint London and Newton Abbot, 1971), p. 522. For the official descriptions of its operations during Lomonosov's time see the two annual reports *Otchet po ekspluatatsii Tashkentskoi kazennoi zheleznoi dorogi za 1909* (Orenburg, 1910) and *Otchet po ekspluatatsii Tashkentskoi kazennoi zheleznoi dorogi za 1910* (Orenburg, 1911).

[22]   Konarev, *Zheleznodorozhnyi transport*, p. 426; Afonina, *Kratkie svedeniia*, pp. 70–72; 'Vospominaniia', vol. 4, pp. 274–6. Railway development in this region and its economic impact are analysed in D. Spring, 'Railways and Economic Development in Turkestan before 1917', in L. Symons and C. White (eds), *Russian Transport: An Historical and Geographical Survey* (London: Bell, 1975), pp. 46–74. On the strategic context and British fears see A. Marshall, *The Russian General Staff and Asia, 1800–1917* (Abingdon, 2006), especially p. 132; D.N. Collins, 'The Franco–Russian Alliance and Russian Railways, 1891–1914', *The Historical Journal*, 16/4 (1973): 777–88 (779–80); K. Neilson, *Britain and the Last Tsar: British Policy and Russia, 1894–1917* (Oxford, 1995), pp. 124, 133, 225–6, 234, 252; and J. Siegel, *Endgame, Britain, Russia and the Final Struggle for Central Asia* (London, 2002), pp. 4–5, 81.

the increase in his salary from 4,700 to 7,600 rubles transformed his straitened domestic circumstances. In due course his work there would include some notable initiatives, but it would also raise serious problems. And although some of these difficulties derived from the bureaucratic system, others were inherent in his ideas, and as ever his activities inspired strong reactions.[23]

Indeed, his 20 months on this railway were clouded by his dreadful relationship with the line's superintendent, engineer V.A. Shtukenberg. Similarity of character would perhaps have sparked their conflict eventually: an obituary described Shtukenberg much as Lomonosov saw himself – energetic, intolerant of disorder and abuses, principled, and hence a target of numerous enemies and vicious slander.[24] But Lomonosov started the trouble at the outset by acting, perhaps innocently, in a way that Shtukenberg considered insubordinate; and for reasons unknown Lomonosov preferred conflict to conciliation.[25] There was evidently fault on both sides, but the crucial point is the harm to the interests of both the railway and Lomonosov. For instance, he made a solid case for reforming, or even abolishing, the Materials Department, whereas Shtukenberg perceived only a personal attack.[26] Eventually their feud would reach Rukhlov and overshadow Lomonosov's move from the railway in 1910.

As a departmental superintendent Lomonosov had greater responsibilities than hitherto and little time for research. Much of his attention was devoted to five priorities: accounting (broadly defined); personnel; technical improvements; freight traffic delays; and reorganizing the departmental structure and procedures. Only the final item was perhaps unusual: it would become characteristic of his management style, but here it mainly reflected his frustration at having his central offices located in two non-adjacent buildings. Much time disappeared on routine meetings and paperwork. There were also exceptional matters, notably the

---

[23]   See his service record, 21 February 1908: Gosudarstvennyi arkhiv Orenburgskoi oblasti (GAOO), f.r-316, op. 1, d. 10034, l. 9ob.; and Vremennyi Prikaz po Tashkentskoi zh.d. No. 44, 16 March 1909: in GAOO, f. 142, op. 1, d. 433. With bonuses Lomonosov hoped to earn 10,000 rubles: 'Vospominaniia', vol. 4, p. 283. The staff statistic is in a draft letter by Lomonosov, circa May 1910, at GAOO, f.r-316, op. 1, d. 10034, l. 49. See also 'Vospominaniia', vol. 4, pp. 276, 279–82.

[24]   'Vladimir Antonovich Shtukenberg (Nekrolog)', *Izvestiia Sobraniia inzhenerov putei soobshcheniia*, 29 (1913): 506–8. His Tashkent personnel file is at GAOO, f.r-316, op. 1, d. 2205.

[25]   See, for example, 'Vospominaniia', vol. 4, pp. 298, 343, 346–8, 352–3, 355–7, 419–21.

[26]   For example, Minutes of General Office of Tashkent Railway, 14 April 1910: GAOO, f. 142, op. 1, d. 360, ll. 76ob–8ob; Iu.V. Lomonosov, Memorandum, circa May 1910: GAOO, f.r-316, op. 1, d. 10034, ll. 73–82; Vysochaishe uchrezhdennaia Osobaia Vysshaia Komissiia dlia vsestoronnego issledovaniia zheleznodorozhnogo dela v Rossii, *Doklady po obsledovaniiu zheleznykh dorog: Doklad No. 17: Tashkentskaia zheleznaia doroga* (St Petersburg, 1913), pp. 35–40; 'Vospominaniia', vol. 4, pp. 335, 557–60, 570–575, 690–692.

preparations for a (cancelled) royal visit, two inspections by Kozyrev and deputy minister Shchukin, and interviews with Petrov's national review commission. His research was delegated to subordinates or pursued during evenings, although he did have some research leave in summer 1910.[27]

Underpinning his managerial activity was his determination to improve efficiency and minimize costs, and to do so by using, as far as possible, science-based principles and criteria. What this meant in practice can be illustrated by three examples, the first two of which also indicate how he could combine the defence of departmental interests with a broader railway perspective. First, he responded to the high cost of fuel by, among other things, initiating experimental work on the development of a locomotive powered by a diesel engine, and commissioning local geological surveys for coal and oil deposits; this diesel research became part of the foundation of the important Soviet work in this field in the 1920s (see Chapter 9, this volume), while the oil surveys anticipated major Soviet investment in oil extraction in this region.[28] The second example is his complaint that flawed accounting and procurement procedures inflated his department's costs. For instance, he criticized the methods for identifying the work done by locomotives and dividing the cost between the Traction and Traffic budgets. Similarly, he attacked the Materials Department for continuing, despite requests to desist, to buy components for locomotive classes no longer present on the railway. For these particular problems he won agreement to monitor 'locomotive-hours' (the time that an engine was away from the depot) instead of mileage – an idea that he took from Germany for a more accurate estimation of operating costs, especially for the shunting engines that might stand for hours awaiting their next task; he introduced new norms for fuel consumption that were derived from his road tests; and as noted above he sought the abolition of the Materials Department, taking the provocative step of approaching Kozyrev without Shtukenberg's consent.[29]

The third example was among his principal concerns: the Orenburg workshops. Their equipment was inadequate, the boiler shop was tiny, materials were in short supply, productivity was low and production quality was poor. The pooling of large components like boilers, which was becoming popular abroad to maximize workshop productivity, was impossible owing to MPS accounting rules: they valued a boiler awaiting overhaul as scrap, but a repaired boiler as new. He

---

[27]   He recalls his time on this railway in 'Vospominaniia', vol. 4, pp. 284–933.

[28]   For example: Minutes of General Office of Tashkent Railway, 21 May 1911: GAOO, f. 142, op. 1, d. 363, ll. 15–21; 'Vospominaniia', vol. 4, pp. 536–7, 571, 590, 595–600, 664–5, 681. This diesel research is discussed in Chapter 9, this volume.

[29]   For instance: 'Vospominaniia', vol. 4, pp. 557–60, 570–575, 594–5, 647–60. He reported about locomotive-hours to the Eastern Regional Committee for regulating traffic: Zhurnal No. 40, 20–21 February 1910: RGIA, f. 273, op. 10, d. 2076, l. 25ob. Two months later, it would seem, he gave a further report that formed the core of his book *Nauchnye problemy ekspluatatsii zheleznykh dorog*: ibid., ll. 25ob., 35ob. Unfortunately a copy of the report itself has not been traced.

responded with some investment, reorganization and pressure for reform. Though small, the investment funded the staff for an extra shift, some new tooling and a temporary extension to the boiler shop. He introduced a production line whére the small groups of workers would each concentrate on one designated task instead of completely overhauling a given locomotive. He also introduced bonuses for exceeding work norms, at the same time tightening the quality-control system. He tried to persuade the auditing organization, State Control, to facilitate the pooling of major parts, he devised and submitted a workshop development plan for Orenburg to the MPS, and he lobbied Kozyrev, Shchukin and the Petrov commission, again bypassing Shtukenberg.[30]

This project may well lie behind assertions in the 1920s that Lomonosov was a proponent of Taylorism.[31] At issue were the theories of the American manager and writer F.W. Taylor about the 'scientific organization of work', which included a differential piece-rate system and a model for functional workshop management; these ideas attracted much attention in North America and Europe in the last few years before the First World War and were tried at a number of enterprises in tsarist Russia, usually with changes to suit local conditions. That Lomonosov was later called a Taylorist was important both because Lenin strongly endorsed Taylorist principles (as he understood them) for the Soviet economy and because the term was also often associated with using piece-rates to raise productivity – a tactic with obvious connotations of exploiting workers. One might expect Lomonosov to have taken an interest given Taylor's emphasis on 'scientific' principles, but unfortunately no record of his attitude survives. Overall, it seems most likely that his changes at the Orenburg workshops were motivated simply by the latest railway workshop practices in western Europe, which he certainly monitored in the engineering periodicals.[32]

His persistent concern with efficiency and costs accorded with the government's need to reduce the railway burden on the treasury and restore public confidence.

---

[30]   For instance: Explanatory note about the proposal to expand the Main Orenburg Workshops, 19 February 1910: RGIA, f. 273, op. 6, d. 2528, ll. 46–98; Vysochaishe uchrezhdennaia Osobaia Vysshaia Komissiia dlia vsestoronnego issledovaniia zheleznodorozhnogo dela v Rossii, *Vypusk 96: Materialy po obsledovaniiu zheleznykh dorog:Tashkentskaia zheleznaia doroga* (St Petersburg, 1913), pp. 22–3, 82–3, 182–8, prilozhenie 9; 'Vospominaniia', vol. 4, pp. 335–7, 365–6, 458–71, 568–9, 699–701, 864–5, 870–871, 874–6, 881–2.

[31]   For example: Typescript of notes by Lenin, circa 24 May 1920: RGASPI, f. 2, op. 1, d. 14079, l. 2.

[32]   F.W. Taylor, *The Principles of Scientific Management* (New York, 1911). On Lenin's interest and Taylorism in Soviet Russia see Service, *Lenin*, vol. 3, pp. 4–5; K. Bailes, 'Alexei Gastov and the Soviet Controversy over Taylorism, 1918–1924', *Soviet Studies*, 29/3 (1977): 373–94; Z. Sochor, 'Soviet Taylorism Revisited', *Soviet Studies*, 33/2 (1981): 246–64; and R.H. Jones, 'Taylorism and the Scientific Organisation of Work in Russia, 1910–1925' (unpublished PhD dissertation, University of Sussex, 1986), especially pp. 219–27.

But how effective was his activity, and how far does it show that the railways acknowledged these priorities? Unsurprisingly, Lomonosov assessed his work positively. He was especially proud that the Orenburg workshops raised their annual productivity from 40 locomotive overhauls in 1908 to 100 by 1910, and that he achieved a substantial departmental saving of some 1.62 million rubles in 1909–1910.[33] And one might suspect from his transfer to the Nicholas Railway in December 1910 that the MPS was pleased with him, for that move gave him a measure of national prominence. It brought him close to the centre of power in St Petersburg with the most prestigious position in railway-level traction management, included an associate professorship in locomotive engineering at the IPS, and enhanced his standing *vis-à-vis* M.V. Gololobov, now Russia's foremost authority on locomotive test plants and a rival for the IPS professorship.[34]

Yet Lomonosov's long-term impact on the Tashkent line was probably negligible. Even a department manager could achieve little in isolation, for major reforms required regional or system-wide impetus; and in reality the MPS was not exerting strong pressure for radical gains in productivity.[35] Thus, for instance, his ongoing struggle against theft, corruption and incompetence could not bring much benefit without a vigorous railway-wide campaign. His successor largely abandoned his system for recording 'locomotive-hours'. And although the Petrov commission expressed interest in, for example, his criticisms and reforms of the accounting system, generally the MPS was unsympathetic. Indeed it was even embarrassed by the Orenburg workshops, for by exceeding national norms they undermined Shchukin's campaign for a large new regional repair facility at Samara, and could assist the ministry's critics: accordingly, the ministry was pleased that productivity fell back to target levels by 1912.[36]

In any case, merit was not the main reason for Lomonosov's departure from Orenburg. His desire to move was becoming firm by spring 1910, just one year after his appointment, and yielded a resignation letter on 13 May. The reasons included his poor relationship with Shtukenberg and 'illness'; he was especially irritated by

---

[33]   'Vospominaniia', vol. 4, pp. 699–701, 929.

[34]   [Curriculum vitae], 20 November 1910: TsGIA SPb, f. 381, op. 1, d. 569a, ll. 706–7; IPS Council minutes, 2 and 9 December 1910: *Zhurnaly Soveta Instituta putei soobshcheniia Imperatora Aleksandra I za 1910 god*, vyp. 2 (St Petersburg, 1911), pp. 106–8, 118–19.

[35]   See V. N-skii, 'Kazennyi dukh', *Izvestiia Sobraniia inzhenerov putei soobshcheniia*, 26 (1910): 95–6.

[36]   Vysochaishe uchrezhdennaia Osobaia Vysshaia Komissiia dlia vsestoronnego issledovaniia zheleznodorozhnogo dela v Rossii, 'Zhurnal No. 43' (17 December 1912), p. 2; Vysochaishe uchrezhdennaia Osobaia Vysshaia Komissiia, *Doklad No. 17*, p. 45; 'Vospominaniia', vol. 4, pp. 874–77, 880–882, 1191–2; Mikul'skii–Traction Chief, Samara–Zlatoust Railway, 30 September 1912: RGIA, f. 273, op. 6, d. 3099, ll. 102–26.

Shtukenberg's scepticism about his request for sick leave.[37] Also important was the fact that Professor Brandt, now director of the IPS, wanted Lomonosov to succeed his other former mentor, Romanov, who was retiring. A full-time IPS professorship would have involved a large drop in salary for Lomonosov, but Brandt envisaged a parallel appointment on a nearby railway – a traditional means whereby the IPS employed experienced railwaymen for teaching, and for Lomonosov an attractive way to combine theory, practice and a good income.[38]

Further impetus for Lomonosov's malaise and resignation probably came from tension with subordinate staff. Echoing his personnel troubles on the Catherine Railway, this matter provoked more anonymous denunciations together with press criticism and conceivably an attempted murder. Lomonosov charted a confrontational course from the outset by making sweeping changes of personnel. For instance, most traction sections gained a new superintendent, such as V.B. Baron fon Tizengauzen at Aktiubinsk. To a degree Lomonosov was indulging an inclination typical of newly appointed managers to install familiar, trusted subordinates: he enticed various acquaintances from Ukraine, including his former student and colleague Lipets, who took charge of some locomotive testing and the diesel locomotive research. Moreover, some staff liked his management style: a Soviet secret police agent reported in about 1921 that 'ordinary honest workers' of the Tashkent Railway spoke well of him, recounting stories about how he dragged specialists in their best white suits under locomotive boilers and his attention to detail when inspecting the work of subordinates.[39] Yet the turnover of traction staff on the Tashkent line was unusually high throughout his period of office.[40] Unsurprisingly, this situation, in conjunction with some of his other measures, did cause deep anger. In March 1910 the far-right nationalist newspaper *Russkoe znamia* accused him and Tizengauzen of victimizing workers who belonged to the Aktiubinsk branch of the Union of the Russian People – an extreme anti-Semitic pro-government organization.[41] An attempt to murder

---

[37]   Tashkent Railway Chancellery–Lomonosov, 18 March 1910; medical certificate, 14 April 1910; Lomonosov–Shtukenberg, 30 April 1910; Report, 13 May 1910: GAOO, f.r-316, op. 1, d. 10034, ll. 41, 43–5; 'Vospominaniia', vol. 4, pp. 694, 710, 728–32, 743–5. The gossip column of the unofficial journal *Puti soobshcheniia* quickly reported his impending departure from MPS service: 22 (27 May 1910): 336. The medical certificate, which was obtained at Shtukenberg's insistence, specified the trouble as general obesity and rushes of blood to the head, and prescribed a two-month mineral-water cure in Karlsbad or Piatigorsk.

[38]   'Vospominaniia', vol. 4, pp. 283, 742, 744, 868, 912–13.

[39]   OKTChK (Moscow–Kazan' Railway)–Blagonravov, [circa 1921]: RGASPI, f. 76, op. 2, d. 76, l. 27.

[40]   This situation is evident from the railway's Orders for 1909–1910, which are at: GAOO, f. 142, op. 1, dd. 433/a/b/v, 434/a/b/v.

[41]   *Russkoe znamia*, 16 March 1910, p. 4. Lomonosov had to explain himself to Kozyrev: Shtukenberg/Lomonosov–Directorate of Railways, 1–8 May 1910: RGIA, f. 273, op. 12, d.

Tizengauzen may have been related to this confrontation.[42] Also, venomous letters were sent anonymously to minister Rukhlov. These combined elements of truth and wild exaggeration. For instance, one condemned Lomonosov as 'an amoral, dissolute charlatan, libertine and bribe-taker': he had abandoned his wife and children to live illegally with a Jewess, he had sacked most subordinates upon his arrival, he shielded Jews, outperformed all typical bribe-takers, and intimidated Shtukenberg. The letter continued:

> [Lomonosov is] established on such a lofty pedastal that he is the tsar himself, God himself. Who can give him orders? Perish the thought! People joke that he's a protégé of General Kozyrev, a friend, a comrade on first-name terms, even a close relative; what will be next? Who will dare to say anything to him, or complain about him?[43]

And similar sentiments were expressed in an article in a newspaper called *Moment*.[44]

These developments, much more than disputed achievements, were the reasons for Lomonosov's move to the Nicholas Railway. True, he retained Rukhlov's support, despite the latter's own membership of the Union of the Russian People: the minister ignored the complaints (merely intimating that Lomonosov should reorganize his family life if he wanted to become a line chief), recommended him for the Order of St Anna (II class) for outstanding service and rejected his resignation.[45] Yet Lomonosov's authority on the line had been damaged. Moreover, he had lost his enthusiasm for his job; even two months of research leave could not banish his malaise, which included heavy drinking. During the autumn of 1910 his relations with Shtukenberg virtually collapsed. Meanwhile, Brandt continued

---

1025, l. 16. On right-wing radicalism during 1900–1917 see, for example, A. Bokhanov, 'Hopeless Symbiosis: Power and Right-Wing Radicalism at the Beginning of the Twentieth Century', in Geifman (ed.), *Russia under the Last Tsar*, pp. 199–213; and W. Laqueur, *Black Hundred: The Rise of the Extreme Right in Russia* (New York, 1994), pp. 3–57.

[42]   See the MPS case file: RGIA, f. 273, op. 12, d. 1089; and 'Vospominaniia', vol. 4, pp. 771–4, 818, 833–4, 838–9. A private dispute was thought to have motivated the (unsolved) murder of another strict manager in early 1910: 'Vospominaniia', vol. 4, pp. 626–46.

[43]   'Suppliers'–Rukhlov, received 24 March 1910: RGIA, f. 229, op. 2, d. 1581, ll. 242–5. The original Russian is semi-literate, and I am grateful to Richard Davies and Galya Bradley for assistance with this translation. See also: Anonymous–MPS, received 8 March 1910: RGIA, f. 273, op. 12, d. 996, l. 4–ob.

[44]   See Lomonosov–Editor, *Moment*, circa May 1910: GAOO, f.r-316, op. 1, d. 10034, ll. 48–50. This paper was apparently short-lived, and has not been seen.

[45]   Notification of award, circa April 1910: GAOO, f.r-316, op. 1, d. 10034, l. 84; Directorate of Railways–Shtukenberg, 22 June 1910: GAOO, f.r-316, op. 1, d. 10034, l. 51; 'Vospominaniia', vol. 4, pp. 705, 757–8, 760–761, 908–9.

to press the ministry for his appointment to the IPS.[46] Certainly Lomonosov would not have been transferred to the Nicholas Railway if Rukhlov had lost confidence in him. But ultimately that move was an attempt more to solve a personnel problem than to reward success.

Nonetheless Lomonosov's stint as the Nicholas Railway's traction superintendent did boost his career.[47] Its significance did not emanate from his work as such: his aims, activities and lack of long-term impact were reminiscent of Orenburg. More important for his career was the fact that the Nicholas Railway was the empire's premier line, so called because its core was the 652-km route between the two greatest cities, St Petersburg and Moscow. Indeed this busy route and a long, strategically important line from Bologoe through Velikie Luki towards Poland comprised the bulk of the railway's 1,635 route kilometres. Traffic was heavy. Most freight was long-distance, moving to and from St Petersburg and other railways. The inter-city passenger service enjoyed the latest equipment and was supplemented by intensive local passenger traffic out of St Petersburg and Moscow. The Alexander workshops in the capital dated from the mid-nineteenth century, but were well equipped and were complemented by good facilities at Bologoe and Velikie Luki. The Traction Department also maintained the St Petersburg-based imperial trains and dozens of special saloons for dignitaries.[48]

Furthermore, as noted above, the location of the line's headquarters in the capital gave Lomonosov the opportunity to teach at the IPS, which he continued to do until 1917. Being part-time, this academic work was much less substantial than his work at the KPI, yet it was still important.[49] Although, typically for him, there were chronic conflicts with certain colleagues, these problems were generally containable because he worked for the IPS for only one day per week. He aimed to change the railway-related curriculum in accordance with his principle of linking equipment design to the nuances of operation, as at Kiev, and he put his textbook *Tiagovye raschety* (*Traction Calculations*) at the heart of the programme. That

---

[46]   For example, Directorate of Railways Chancellery–Shtukenberg, 30 June 1910: GAOO, f.r-316, op. 1, d. 10034, l. 87; Lomonosov–Shtukenberg, 1 November 1910: GAOO, f.r-316, op. 1, d. 10034, l. 63; 'Vospominaniia', vol. 4, pp. 778, 800–802, 808–18, 885–6, 911–12.

[47]   The records of the railway's headquarters are at TsGIA SPb, f. 1480 and RGIA f. 343, but no significant papers concerning his work survive.

[48]   Lomonosov recalls his time on this railway in 'Vospominaniia', vol. 4, pp. 910–912, 938–1361. For brief historical notes about the line see Konarev, *Zheleznodorozhnyi transport*, pp. 259, 271–3.

[49]   The main parts of Lomonosov's memoirs that concern the IPS in 1911–1916 are: 'Vospominaniia', vol. 4, pp. 1001–14, 1172–6, 1257–8; 'Vospominaniia', vol. 5, pp. 213–32, 308–11, 510–517, 560–62, 691–4, 1170–1171, 1357–62. Other sources for his IPS work include the minutes of the institute's Council, which were published bi-annually as *Zhurnaly Soveta Instituta putei soobshcheniia Imperatora Aleksandra I*, and the IPS collection in TsGIA SPb (f. 381).

said, his colleagues abandoned the essence of his curricular reform after he stopped teaching temporarily in 1915 to focus on war work. Probably his most enduring (though fortuitous) contribution was his idea of a locomotive 'passport'. This term originated as student slang for the graphs of locomotive performance that he displayed in the drawing hall. He adopted it for his pamphlets that summarized the performance data for each locomotive class, and which he published from about 1912. Subsequently 'technical passport' became the official MPS term for the logbook of each locomotive – a usage that continues on railways of the former USSR.[50]

His move to the capital in 1911 also had two other important implications for his career. First, he could play a direct role in national railway policy-making. Significantly, he did not participate regularly in the Shchukin Commission, of which he was now an *ex officio* member: initially he chose not to attend, presumably because of his dislike of Shchukin, and later other commitments prevented his attendance. However, he did attend meetings of the Engineering Council's Rails Commission: his appointment to this body in 1911 suggested top-level recognition of the potential practical value of his theoretical work on defining safe maximum speeds.[51] Second, he became responsible for the many royal and other VIP carriages based on the Nicholas Railway, and had to accompany the line superintendent, I.K. Ivanovskii, and other senior colleagues in numerous ceremonies for travelling dignitaries. He even served as the MPS chaperone for the visit of a British Parliamentary Delegation in January 1912, earning a commendation from his guests in the London *Times*.[52] He was thus one of the few railway officials regularly entrusted with detailed information about the

---

[50]   Lomonosov, *Tiagovye raschety*; 'Vospominaniia', vol. 5, pp. 219–20, 746–7. This usage may explain why, in modern Russian, 'passport' can denote 'the registration document of a machine, building, piece of apparatus, item of household equipment and so forth': S.I. Ozhegov, *Slovar' russkogo iazyka* (Moscow, 1975), p. 452.

[51]   'Vospominaniia', vol. 4, pp. 1216–24; vol. 5, pp. 260–268, 357–8, 558, 699–720. The work of the rails commission, and related discussions about specifying maximum speeds, are detailed in, for example, RGIA, f. 240, op. 1, dd. 921, 1016, 1074, 1175; f. 273, op. 6, d. 2751, 3115; *Trudy Komissii pod predsedatel'stvom chlena Inzhenernogo soveta d.s.s. Kunitskogo za vremia s 31-go maia po 23-'e oktiabria 1912 goda, po voprosam: 'O napriazheniiakh v rel'sakh i v ostal'nykh sostavnykh chastiakh verkhnego stroeniia zheleznodorozhnogo puti'* (St Petersburg, 1913). See also Iu.V. Lomonosov, 'Opredelenie predel'nykh skorostei dvizheniia v zavisimosti ot konstruktsii puti i parovoza', in *XXIX Soveshchatel'nyi S"ezd inzhenerov podvizhnogo sostava i tiagi v Rostove 24 maia–10 iiunia 1912 g.* (St Petersburg, 1913), pp. 289–305 (the last of his five reports and other congress contributions on this topic since 1902).

[52]   'Vospominaniia', vol. 4, pp. 944, 946–7; he describes the British visit on pp. 1296–135. See also 'The British Visit to Russia', *The Times*, 13 February 1912, p. 5 and the MPS file about this visit at RGIA, f. 273, op. 10, d. 1269. For the diplomatic context see Neilson, *Britain and the Last Tsar*, pp. 322–3.

movements of royalty and other VIPs. Whether he forwarded this information to the revolutionary movement is addressed below in Chapter 6.

After just one year on the Nicholas Railway Lomonosov was reassigned to the MPS itself from February 1912. Enormous, complex and popularly considered haughty and inefficient, this ministry was created in 1865 to oversee the transport sector and was situated by the Fontanka canal in central St Petersburg. Its principal railway functions were to regulate the public network, including privately owned railways, and to manage the state-owned railways.[53] Needless to say, these responsibilities produced some serious conflicts of interest. However, all the railways were subjected equally to a quasi-military code of discipline by the MPS, albeit perhaps more because of Russia's authoritarian bureaucratic culture than because of any concern for public safety. The minister, who was normally a railwayman, was assisted by two deputy ministers, and his papers (including all MPS staff files) were administered by his general office, where Lomonosov still had friendly contacts. Most of the ministry paperwork, however, was handled by directorates. Six directorates existed as of 1912 to cover the railways; the construction of railways; waterways and roads; buildings; the pension committee; and invalid care homes. The important one for Lomonosov at this time was the Directorate of Railways, because it was his workplace during 1912–1913 and oversaw the network. Headed by a superintendent with three deputies, it had a general office, four departments, three smaller sections, a pensions office and a group of inspectors; of most relevance for Lomonosov were the Operations Department, Technical Department and medical section. The directorate also had an inter-ministry committee for broad policy questions. This committee was chaired by the superintendent, and had about 10 MPS members together with representatives from the Ministries of Finances, State Control, Internal Affairs, Agriculture, Trade and Industry, War and the Navy. Finally, the MPS had numerous permanent and temporary committees and commissions; most important for Lomonosov's interests were the Engineering Council, which handled technical policy, and its standing commissions for traction and workshop affairs, which were chaired by Professor Shchukin.[54]

Contrary to appearances, Lomonosov's move to the ministry was not simply a spectacular promotion. In fact, it reconfirmed the centrality of conflict in his career. Formally he joined the Committee of the Directorate of Railways, where a vacancy had arisen through Kozyrev's appointment to chair the Engineering Council. More precisely, Lomonosov became one of the three assistants to the new head of the

---

[53] On the ministry's origins see A.S. Turgaev (ed.), *Vysshie organy gosudarstvennoi vlasti i upravleniia Rossii, IX–XX vv.: Spravochnik* (St Petersburg, 2000), pp. 175–6.

[54] For a list of principal MPS institutions and staff as of 1914 see *Spravochnaia knizhka po lichnomu sostavu sluzhashchikh Tashkentskoi zh.d. 1914 goda* (Orenburg, 1914), pp. 1–15. See also D.I. Raskin and O.P. Sukhanova (comps), *Fondy Rossiiskogo Gosudarstvennogo Istoricheskogo Arkhiva: Kratkii spravochnik* (St Petersburg, 1994), pp. 39–42, 45–7.

directorate, G.N. Viktorov. His remit was to oversee the departments concerning traction and medical services, and to serve in the inter-ministry committee that assessed proposals for new railways. Thus he was well placed to shape policy on two of his favourite issues. Yet this transfer endangered his prospects. Experience as a line superintendent was considered essential for promotion to the ministry's highest ranks. Accordingly Rukhlov, apparently convinced of his potential, wanted him to stay on the Nicholas Railway before serving somewhere as a line chief. But again the obstacle was Lomonosov's inability to coexist with his immediate superior, Ivanovskii. This time, bluffing, he threatened to accept an offer from the private Moscow–Kazan' Railway.[55] His stratagem worked, but at a cost. Rukhlov endorsed the transfer to the MPS, but remarked angrily that Lomonosov would probably be stuck in that post for good.[56]

Although this authoritative prediction proved hopelessly inaccurate – true to form he stayed for only a year – Lomonosov did, for perhaps the first time, have a significant long-term impact as a manager. The main matter was his project to replace the network's confusing plethora of locomotive classifications with a standard uniform alpha-numerical system – a reform with important military as well as economic benefits. This change was so effective that it survives in much of the former USSR at the time of writing, albeit in modified forms, despite attempts to replace it with numerical systems in 1925 and 1984.[57] Other traction-related examples of his power and influence include his intervention to end years of debate about whether to increase the power of standard freight locomotives by raising their boiler pressure limit; his leadership roles in the preparation of a new statute for traction accounting and a new 'labour book' for railway workshops; and his success at promoting research into diesel traction.[58]

---

[55]   See *Izvestiia Sobraniia inzhenerov putei soobshcheniia*, 42 (1911): 7. Ivanovskii (1857–1917) is described at 'Vospominaniia', vol. 4, pp. 985–92, and in L. Davidova, 'Gosudar' oschastlivil ego svoim rukopozhatiem: Ivanovskii, Ippolit Konstantinovich', *Oktiabr'skaia magistral'*, 13 January 1993; L. Davidova, letter to the author, 15 August 2002.

[56]   *Izvestiia Sobraniia inzhenerov putei soobshcheniia*, 42 (1911): 7; 'Vospominaniia', vol. 4, pp. 1292–5. The move took effect from 26 January: *Vestnik putei soobshcheniia: Ukazatel'pravitel'stvennykh rasporiazhenii po Ministerstvu putei soobshcheniia* (henceforth *Vestnik putei soobshcheniia: Ukazatel'*), 11 (1912): 124. Fortunately for Lomonosov, the rumour that Ivanovskii was to head the directorate proved unfounded: *Izvestiia Sobraniia inzhenerov putei soobshcheniia*, 39 (1911): 9.

[57]   *Vestnik putei soobshcheniia: Ukazatel'*, 20 (1912): 250–51; 13 (1914): 199–200; 'Vospominaniia', vol. 5, pp. 56–7; A.J. Heywood and I.D.C. Button, *Soviet Locomotive Types: The Union Legacy* (Malmo, 1995), pp. 149–51.

[58]   On the question of boiler pressure: *Vestnik putei soobshcheniia: Ukazatel'*, 48 (1914): 771; 'Vospominaniia', vol. 5, pp. 37–8, 46–8. On traction accounting: Iu.V. Lomonosov, 'O novoi organizatsii schetovodstva', in *XXIX Soveshchatel'nyi S"ezd inzhenerov podvizhnogo sostava*, pp. 402–22; 'Vospominaniia', vol. 4, pp. 1185–90 and vol. 5, pp. 170–178, 297–305, 550–553. On the labour book: RGIA, f. 273, op. 6, d. 2533; 'Vospominaniia', vol. 5, pp. 531–5.

However, his experiences show that the constraints on even such a senior official were considerable. Much of his work needed smooth cooperation with other departments, especially the Operations Department, which was far from guaranteed. Also, his projects could be thwarted by subordinate institutions and personnel. For instance, he instructed railways to use a new method for calculating the loaded weight of freight trains, which he believed would increase the use of wagon capacity and thereby help address an alleged shortage of wagons that was generating intense public anger; but various line officials opposed his directive as over-complicated and time-consuming.[59] Another major problem was Shchukin. Formally the latter's commissions on traction and workshops were sub-commissions of the MPS Engineering Council, but they were virtually autonomous in practice thanks to his influence as deputy minister. Moreover, these commissions were closely associated with the private locomotive-building industry, and were not clearly demarcated from the traction department of the MPS Directorate of Railways. Indeed, Lomonosov felt unable to reform that department for fear of antagonizing Shchukin.[60] His resultant frustration helps explain why he soon wanted another move.

Lomonosov paid particular attention to his duties concerning new railways. As one would expect, he promoted his own views about the scientific-technical design of railways when assessing proposals for new lines. He consistently defended his ministry's positions in meetings of the inter-ministry Commission for New Railways, which was the main state planning committee for analysing railway construction proposals and was based at the Ministry of Finances. Like the MPS as a whole he strongly opposed the regime's new policy of using private capital to expand the railway system – a policy that, for the treasury, was now the sole means to find the necessary money, but which contradicted the statist tradition of Russian railway management. Lomonosov's opposition stemmed partly from his own long-standing disgust with private enterprise, but it was also a direct function of his MPS remit: one of his tasks until late 1916 was to coordinate the work of all the MPS members of the Commission for New Railways, and in this capacity he frequently had to express his ministry's opposition to the new financing strategy – sometimes successfully, sometimes not.[61]

[59]   *Vestnik putei soobshcheniia: Ukazatel'*, 17 (1912): 210–212; 21 (1912): 282; 23 (1912): 299; Iu.V. Lomonosov, 'Opredelenie sostavov tovarnykh poezdov po vesu', in *Protokoly zasedanii XXVIII soveshchatel'nogo s"ezda inzhenerov sluzhby podvizhnogo sostava i tiagi russkikh zheleznykh dorog v Rige, 27 maia–8 iiuniia 1911 g.* (St Petersburg, 1913), pp. 349–85; 'Vospominaniia', vol. 5, pp. 37–8, 57–9, 200, 203–4. That there were problems of implementation is also evident from *Vestnik putei soobshcheniia: Ukazatel'*, 49 (1913): 844–7; 32 (1914): 491.

[60]   'Vospominaniia', vol. 5, pp. 4–7, 37–46.

[61]   'Vospominaniia', vol. 5, pp. 62–81, 1354–5. A short history of the commission, produced by the Ministry of Finances for the Tsar following its transfer to the MPS in October 1916, is at RGIA, f. 268, op. 3, d. 1372, ll. 1–15.

These bureaucratic details have general importance because railway planning and investment have been fundamental to analyses of tsarist economic strategy, the Russian war effort in 1914–1917 and the outbreak of the 1917 revolutions. The prevailing view has echoed the claim, already popular by 1918, that pre-war investment in the railways had been inadequate for peacetime needs, let alone the huge wartime demand for transport; the railways therefore buckled under that strain, which spurred the military and economic collapse, food shortages and revolution.[62] Be that as it may – historians have yet to study this issue in detail – the important issue for present purposes is that although Lomonosov basically endorsed this analysis, his own evidence may suggest a different verdict for the pre-war role of the state. He offers a practical justification for the ministry's abhorrence of private capital, namely the fear that new private railways would harm the profitability of existing state-owned lines, except when these new lines were located far from those state railways.[63] This analysis made sense for the ministry as state-railway operator, but it constituted an abdication of responsibility for system-wide planning. In other words, the MPS's dual role as regulator and operator had the unintended and unwelcome effect of limiting the network's expansion.

Lomonosov had been at the Directorate of Railways for just one year when, in February 1913, he moved to the MPS Engineering Council.[64] This appointment has been seen as a mark of exceptional esteem for him.[65] To be sure, from its inception in 1892 the council had been the preserve of distinguished servants of the railways – a fact reflected in the very high average age of its members, which was about 70 in 1913.[66] Conceivably Rukhlov and Kozyrev wanted to rejuvenate it: the 35-year-old Lomonosov became the youngest member by about 20 years, and the minister was anxious to promote relatively young people to high office.[67] Yet the move was another setback for his administrative ambitions, in that it isolated him from day-to-day railway management with no obvious way back, and it could be interpreted as a form of early retirement.

That said, this change proved exceptionally productive for all concerned during the 18 months to July 1914. Not only did Lomonosov find the work interesting, but for the first time in years he found his environment generally congenial. His responsibility was to produce and deliver reports about the technical aspects of the detailed proposals that were submitted by or through the MPS for council approval. For the most part he studied plans for new railways or infrastructure improvements,

---

[62]   See Chapter 6, this volume.

[63]   'Vospominaniia', vol. 5, pp. 71–2.

[64]   'Vospominaniia', vol. 5, pp. 306–7, 311–13.

[65]   Aplin, *Catalogue*, pp. xi–xii.

[66]   'Vospominaniia', vol. 5, p. 339. All current (in 1917) and past members of the Council are listed in S.K. Kunitskii (ed.), *Kratkii istoricheskii ocherk deiatel'nosti Inzhenernogo soveta za XXV let s 1892 po 1917 g.* (Petrograd, 1917), pp. 116–27.

[67]   'Vospominaniia', vol. 5, p. 307; Inzhener, 'Omolozhenie', *Izvestiia Sobraniia inzhenerov putei soobshcheniia*, 3 (1912): 10–11.

such as a project to extend the water-supply facilities on part of the trans-Siberian route. He also became involved in debates about designs for new locomotives. He thus had many opportunities to promote his vision for railway design, and he did begin to make an impression by summer 1914. Interestingly, he rarely attended Shchukin's traction commission, apparently because its meetings clashed with those of the Commission for New Railways. But he enjoyed the compensation of being able to give more time to his research and continue working at the IPS. Tellingly, aside from his friction with Shchukin, he had no major confrontations at the council during this period. He even remained on good terms with Kozyrev. It is thus conceivable that his life would have continued in this manner for at least a few years had the war not intervened.[68]

Research

Of the various research interests that Lomonosov pursued in 1908–1914 locomotive road testing was the one that occupied him most and for which he became best known to his contemporaries; it also underpinned his research into traffic costs and other work. It thus provides the most appropriate focus for analysing the significance of his research during this period. The prevailing assumption, at least among railway engineers outside Russia, has been that his type of road test became the standard method of locomotive road testing in late tsarist Russia.[69] This view does seem apposite if we observe that by 1914 all new locomotive designs were required to undergo road experiments with his Experiments Office. Yet actually it exaggerates the status of his research for the Russian railways. Like much of his managerial work and despite some notable achievements, his research career at this time is ultimately a tale of unfulfilled potential.

Conducting his road tests he usually had several related aims. Information about the performance characteristics of the given locomotive class could be used, he argued, to inform decisions about the operation and improvement of these machines, about the design of future locomotives, and also about operating and improving the network, including the design and operation of new railways. Additionally, he hoped to develop theories that described the workings of the steam locomotive as a machine. To improve the quality of his data he refined the methodology by incorporating variables such as wheel/rail friction and wind

---

[68]   On the trans-Siberian route's water supply: RGIA, f. 240, op. 1, d. 1068 (Lomonosov's report of 21 April 1914 is at ll. 18–44); 'Vospominaniia', vol. 5, p. 697 (misdated as October 1914). For examples of Lomonosov's involvement in debates about new locomotive designs in 1911–1914 see Inzhenernyi sovet, *Zhurnal Kommissii podvizhnogo sostava i tiagi*, 25 (3 November 1912): 370–373; 30 (8 December 1912): 438–42; 22 (30 September 1914): 504, 513–14; 'Vospominaniia', vol. 5, pp. 232–42, 566–7, 669–73.

[69]   For example, Chapelon, *La Locomotive à Vapeur*, p. 358; Ransome-Wallis, *Concise Encyclopaedia*, p. 404.

resistance, introducing a stricter operating regime, and so forth. In time, too, he studied passenger as well as freight classes, and extended the scale of his research. For example, in 1908 he compared the two most modern standard freight classes of the state railways: the 0-8-0 of 1901 and its successor, the 'modified Chinese' 2-8-0. During 1913, by contrast, Lomonosov-type experiments were conducted on 10 locomotive classes and sub-classes, not to mention experiments to compare different types of coal and cancelled experiments.[70]

This work was undertaken in the context of significant changes in American and European locomotive practice that were reflected in Russia. A generational change was occurring in locomotive size to handle heavier trains: in continental Europe, for instance, the long-dominant 0-8-0 pattern for freight traffic was surrendering to 2-8-0, 0-10-0 and 2-10-0 designs, while the old 4-4-0 and 4-4-2 passenger types were yielding to 4-6-0 and 4-6-2 designs. Meanwhile the use of high-temperature superheating was starting to oust the complex and expensive practice known as compounding; superheating technology offered worthwhile economies in fuel and water consumption by adding further heat to the steam after it left the boiler space, which expanded the steam and hence enabled it to be used more efficiently in the cylinders.[71] Furthermore, the first substantial steps were being taken to supersede steam in railway traction. There were large-scale electrification projects in areas like the Alps and New York City where there were very long tunnels in which smoke was problematic; and attempts were being made to use internal combustion engines in railway traction, such as the building of an experimental main-line 'diesel' locomotive for the Prussian State Railways.[72] Also, the advantages of standardization for locomotives and their components were becoming more widely recognized – a process that was remarkably protracted on the world's railways, and in which Russia's MPS became an early participant in the 1880s.[73] Finally, as noted in Chapter 2, change was occurring in the process of locomotive development, with growing interest in locomotive testing. The relative merits of road tests and testing plants were much debated, although in practice most investigators had no access to a testing plant.

So how significant was Lomonosov's work on road testing at this time? Without question it was an original development for both Russia and the wider world. Most road testing was concerned with either a given locomotive class or

---

[70] For a list of his experiments see A.J. Heywood, 'Iu.V. Lomonosov and the Science of Locomotive Testing in Russia: Consolidation, Methodology and Impact, 1908–1917', *Transactions of the Newcomen Society*, 72/2 (2000–2001): 285–7.

[71] On superheating see, for example, Ransome-Wallis, *Concise Encyclopaedia*, pp. 284, 286, 316–17, 394; Ross, *The Steam Locomotive*, pp. 145–9.

[72] The main history of railway electrification is M.C. Duffy, *Electric Railways, 1880–1990* (London, 2003). On the Sulzer locomotive see, for example, 'The Sulzer–Diesel Locomotive', *Engineering* (5 September 1913): 317–21 and associated editorial 'The Internal-Combustion Locomotive', p. 326.

[73] Westwood, *History of Russian Railways*, pp. 95–8.

the development of a new design or the introduction of a particular service; no other engineer, it would seem, was using it like Lomonosov for a wide variety of practical and theoretical aims. Further, his notion of a road test as a laboratory-type experiment, with elaborate procedures to ensure constant experimental conditions, appears to have been unique; most road tests were still being conducted in the ways that Borodin had condemned in the 1880s. Finally, no other national railway system had a department analogous to the MPS Experiments Office.[74]

Nonetheless there was little if any contemporary foreign interest in Lomonosov's road tests or indeed his other research. In fact, his work remained virtually unknown outside Russia. Although he published extensively at home in these years, he made little effort to promote his ideas and findings abroad, presumably because he saw no particular benefit for himself in doing so. He published just two articles in foreign journals, both of which were co-authored and discussed marginal aspects of his research.[75] Also, he made no research trips abroad at this time. Meanwhile, foreign railway engineers still did not normally see Russian periodicals because, it would seem, these were unavailable in English, German or French.[76] In any case, one can speculate that his road tests would not quickly have achieved any great international impact had they become widely known because, apart from anything else, they remained very controversial within Russia, where they were certainly not regarded as 'the Russian method'.

This controversy flowed partly from the institutional environment. As noted above, the rigidly hierarchical character of the tsarist bureaucracy meant that young would-be innovators in railway engineering needed support from their superiors; and at the same time the power of the MPS to veto any appointment throughout the state and private railway sectors encouraged junior engineers to avoid controversy. This situation was especially difficult for locomotive specialists, for the only other jobs in locomotive engineering were at the few non-MPS technical institutes and locomotive-building firms. Further, as we have noted, relations between the major railway departments were often poor. To advocate closer collaboration, as did Lomonosov and a few of his colleagues, was to challenge the bureaucratic grain.[77] Compounding this situation was the

---

[74]   For instance: Phillipson, 'Notes on Locomotive Running Trials'; A. Mascini and G. Corbellini, 'Methods Employed by the Italian State Railways for Testing Locomotive Performances', *Monthly Bulletin of the International Railway Congress Association (English edition)*, 7/13 (1925): 2523–55; Diamond, 'Horse-power of Locomotives', pp. 150–189.

[75]   Iu.V. Lomonosov and A. Krukovskii, 'Die Temperaturmessungen im Feurraum der dampflokomotive während der Fahrt', *Zeitschrift des Vereines Deutscher Ingenieure*, 53/9 (1909): 345–6; Iu.V. Lomonosov and A.O. Czeczott, 'Zur Erforschung der Lokomotivüberhitzer', *Zeitschrift des Vereines Deutscher Ingenieure*, 56/5 (1912): 184–5.

[76]   I am very grateful to Katherine Dike at the library of the Institution of Mechanical Engineers for her efforts to check whether any translations were made.

[77]   On the Russian bureaucracy see, in particular, W.M. Pintner and D.K. Rowney (eds), *Russian Officialdom: The Bureaucratization of Russian Society from the Seventeenth*

fact that the private locomotive-building firms had considerable influence over locomotive design, not least by operating a cartel and sponsoring the Shchukin Commission. Anyone else who sought a role here was automatically questioning the prerogatives of the companies and the commission.[78]

Additionally, the experts continued to argue about Lomonosov's claim that road tests could have constant experimental conditions and about the purposes of road tests. They tended to adopt extreme positions, usually ignoring or rejecting the possibility that test plants and road tests could complement each other. Personal enmities were doubtless partly to blame, but also important was the fact that these engineers were competing for large amounts of money amid pressure on the MPS to reduce costs. Thus, for instance, the proponents of test plants claimed that even the most rigorously executed road tests could not fully eliminate variations in driving and firing techniques, weather conditions and gradient changes. Lomonosov replied by trying to reduce these variations to a point where, in his view, they were insignificant. For example, he aimed to use only routes where the gradient variations, if any, were tiny; to demonstrate that weather variations would normally have minimal effects; and to perfect a method to calculate the resistance of the train (the possibility of including this phenomenon was a potential advantage of road research over a test plant). He sought maximum rigour by developing measures to improve discipline and communication among the research team, data recording and the data analysis. For example, a bell code was used to signal key moments such as the start and end of the experiment; personnel had to remain silent during an experiment except for commands and reports; forms were devised for recording the data; and the research team did the analysis promptly so as to minimize the risk of having incomplete or unclear data. As for the purposes of road tests, Shchukin held that they were useful to ascertain how a given locomotive class would perform in regular service. For that purpose, he argued, it was logical to use ordinary service trains in routine non-constant conditions. He denied that Lomonosov-type tests had any theoretical or practical value, and contended that Lomonosov was creating 'abnormal' conditions to achieve spurious precision. For his part Lomonosov derided Shchukin's tests as uselessly unscientific. He

---

to the Twentieth Century (London, 1980), especially the analyses of the Ministry of Internal Affairs in 1855–1914, pp. 250–315. To some extent Lomonosov's difficulties may reflect the tension between autocracy and the scientific spirit with its critical attitude towards authority, on which see Vucinich, *Science in Russian Culture*. For examples of praise of *Nauchnye problemy* see N-skii, 'Malen'kaia knizhka bol'shogo znacheniia (Iu. Lomonosov, "Nauchnye problemy eksploatatsii zheleznykh dorog")', *Izvestiia Sobraniia inzhenerov putei soobshcheniia*, 12 (1912): 12–14; and V. Shargin, 'Iu. Lomonosov. Nauchnye problemy eksploatatsii zheleznykh dorog', *Vestnik inzhenerov*, 11 (1915): 509. See also N., 'Novaia obblast' prilozheniia truda inzhenerov putei soobshcheniia', *Izvestiia Sobraniia inzhenerov putei soobshcheniia*, 5 (1912): 3–4.

[78] See, for example, 'Vospominaniia', vol. 4, pp. 424–8, 761–2, 1225–33; 'Vospominaniia', vol. 5, pp. 163, 751–2. On cartelization in the railway engineering industry see Westwood, *History of Russian Railways*, pp. 157–9.

did, however, suggest a compromise in 1908 whereby he would do two types of experimental trip: one – designated 'cycle I' – would use his own method and the other – 'cycle II' – would be a traditional test.[79]

Lomonosov's efforts to address his critics were only partially successful. The 'cycle II' trips had the desired political effect, to judge by his repeated use of them over the years.[80] But his procedural reforms were compromised by failures of his personnel to follow rules to the letter when, as was often the case after 1908, he was absent from the experiments.[81] Moreover, the likes of Gololobov were never convinced by his responses to the major theoretical objections. In truth, Lomonosov struggled to find long constant gradients on relatively quiet lines. For example, of the six places selected for the 1908 experiments, he reported only two with a constant gradient: one had 36.75 km of level track (allowance made for curves) and the other had a very slight gradient of 1 in 2000 for 36.6 km. The other four sections all had some form of gradient change: two were level sections of 25.4 km and 25.6 km, each interrupted briefly by a gradient of 1 in 167; and the other two had 1 in 167 gradients for 14.75 km and 12.52 km respectively but were interrupted either by short level sections or by a short section of 1 in 1,000.[82] For Lomonosov such variations were admissably slight, but critics could and did reject them as incompatible with 'constant conditions'. The same problem of differing

---

[79] For greater detail about Lomonosov's responses see Heywood, 'Lomonosov and the Science of Locomotive Testing: Consolidation', pp. 272–7. For a typical example of Shchukin's disdain as perceived by Lomonosov, see 'Vospominaniia', vol. 4, p. 1233 (in July 1911).

[80] The technical value of the cycle II trips was emphasized in the following terms by J.W. Knowles in a letter to the author, 24 October 2000: 'The locomotive operates at a constant rate. Once it has been "tuned up" by "constant conditions" testing, complementary road tests on the lines on which the locomotive is to be used, aimed at most satisfactory running, are essential, even more valuable. The "satisfactory" applies to maximum load, times with various loads, duration of non-stop runs allowing for fire conditions, etc. They will include ability to draw on the capacity of the boiler on short upgradients, to be recovered on ensuing easier stretches, ability to accelerate from restricted speeds, filling of the firebox, effect on the fire of periods of easy working, etc. Many such tests were done on a comparative basis without indicating or the dynamometer car. The other, equally important aspect of road testing is examination of defects and maintenance over time.'

[81] For example, 'Vospominaniia', vol. 4, pp. 844–5, 1226–7; 'Vospominaniia', vol. 5, pp. 464–9.

[82] Iu. Lomonosov, *Sravnitel'noe issledovanie tovarnykh parovozov bol'shoi moshchnosti, vypusk II: Opyty nad parovozami 0-4-0 normal'nogo tipa 1901g. i 1-4-0 izmenennogo kitaiskogo tipa, proizvedennye v 1908 g. na Ekaterininskoi zheleznoi doroge* (Odessa, 1915), p. 51. One should note two problems with this example: first, there are discrepancies in his description of one track section: the claimed level section and the section with a 1-in-2000 gradient both involve the Apostolovo–Dolgintsevo section, albeit in opposite directions of operation; second, the gradient profile for this section shows slight undulation. See ibid., pp. 47–51.

assessments blighted his efforts to address weather variations, train resistance and other theoretical matters.[83]

Mixed results were likewise the product of Lomonosov's attempts to overcome organizational obstacles. His main bureaucratic worry was to get his proposed annual research schedule and budget approved. In 1908, for example, he sought to study six different locomotives, but financial constraints largely explain why he received only one 0-8-0 and a 'modified Chinese' 2-8-0.[84] No experiments were conducted in 1909 because the MPS did not approve a budget and the private Moscow–Kazan' Railway retracted a promise to send a class Y 0-8-0.[85] In 1910 he used some of his savings from his departmental budget to fund a small series of experiments with an oil-burning 0-8-0 locomotive, the MPS having decided to fund only some of his proposed experiments. Not until 1913–1914 did funding for his experiments rise appreciably, and here the explanation lay with the creation of the MPS Experiments Office. Yet this development did not signal his acceptance by the Russian railway-engineering establishment: it simply emerged fortuitously from a bureaucratic anomaly. During 1908–1912 his research budget was part of the budget of the traction department of his railway. However, when he moved to the Directorate of Railways in 1912, his research staff and budget had to remain with the Nicholas Railway's Traction Department because the directorate was prohibited from having any executive powers that bypassed line superintendents; from May 1913 this research group was called the Experiments Office of the Nicholas Railway. This peculiar situation was regularized in March 1914 by the office's redesignation as an MPS office within the Directorate of Railways. In his capacity as director Lomonosov thus reported to the head of the directorate, and enjoyed the same status as a line superintendent; the responsibility for approving the budget and accounting was assigned to the directorate's Committee.[86]

---

[83] For example: M.V. Gololobov, 'Sravnenie dvukh sposobov issledovaniia rabotosposobnosti parovozov – na spetsial'nykh stantsiiakh i pri probnykh poezdkakh, i organizatsiia pervoi takoi stantsii na russkikh zheleznykh dorogakh', in *XXX Soveshchatel'nyi S"ezd inzhenerov podvizhnogo sostava i tiagi v Khar'kove, 25 avgusta 1913 g.* (St Petersburg, 1914), pp. 64–73; L.M. Levi, 'Osnovaniia dlia ustanovleniia odnoobraznoi programmy ispytaniia novykh parovozov na russkikh zheleznykh dorogakh', *Inzhener*, 2 (1913): 33–7; and 3 (1913): 79–85.

[84] Lomonosov, *Sravnitel'noe issledovanie tovarnykh parovozov bol'shoi moshchnosti, vypusk I: tsel' issledovaniia i ego metod* (Odessa, 1910), preface; Lomonosov, *Sravnitel'noe issledovanie tovarnykh parovozov bol'shoi moshchnosti, vypusk II*, pp. 41–3.

[85] Lomonosov had agreed privately with Nol'tein to test one of these locomotives: Lomonosov, *Sravnitel'noe issledovanie tovarnykh parovozov bol'shoi moshchnosti, vypusk III: Opyty nad parovami 0-4-0 normal'nogo tipa na nefti, proizvedeny v 1910 g. na Tashkentskoi zh.d.* (Odessa, 1916), pp. 341–3; 'Vospominaniia', vol. 4, pp. 427–8, 761–2, 835.

[86] 'Vospominaniia', vol. 5, pp. 94–5, 99–103, 378, 565, 567. This last change was possible thanks to Lomonosov's transfer from the Directorate of Railways to the Engineering Council in early 1913. See also Iu.V. Lomonosov, *Opyty nad tipami parovozov: Opyty*

These changes also reflect Lomonosov's aptitude for promotional activity. He was able to inspire a certain *esprit de corps* among his research team as an elite; that his associates nicknamed him 'the General' by 1914 was for some of them a sign not just of his high rank and swagger, but also of pride and personal loyalty. Publicity, of course, was one aim of his publications, whether in the form of a textbook for students, a pamphlet-type 'passport' for traffic schedulers or a treatise for fellow locomotive specialists. A parallel tactic was to lobby senior officials and cultivate goodwill among colleagues.[87] In 1913, for example, he entertained the minister, the Shchukin Commission and journalists on the occasion of high-speed trips between St Petersburg and Moscow that were intended to demonstrate the practicability of a much faster passenger service between the two cities. Similarly, he offered opportunities for engineering students to do work placements, and when colleagues accepted his invitations to spend a few days observing experiments, he hosted them lavishly.[88] But while he gained a reputation – certainly among his enemies – as an energetic and talented self-publicist, his efforts did sometimes backfire. Among his worst experiences was a locomotive failure in the minister's presence during the high-speed Moscow–St Petersburg trials of 1913. This fiasco produced ridicule in the press, and Lomonosov suspected – without any real evidence – deliberate sabotage.[89]

Given his need for good contacts and publicity it is worth considering his ambivalence towards the professional associations that had become popular with Russia's railway engineers. Such associations were important in the rise of the engineering profession in Russia as elsewhere, contributing among other things to the development of professional identities and the dissemination of fresh thinking.[90] By 1910 it was common for Russian railway engineers to be involved in them, much as associations elsewhere like the Union of German Engineers, the Institution of Mechanical Engineers in Britain or the American Society of Mechanical Engineers had many railway members. Colleagues like Baidak, Gololobov and Romanov were stalwarts of the railway section of the Imperial Russian Technical Society. Lomonosov did value such associations in principle, but his record of active involvement was very patchy. Apart from his participation in the Union of Engineers in 1905, which in any case was essentially a political entity, his only major connections appear to have been his paper for the Technical

proizvodivshiesia v 1912–1914 gg. na b. Nikolaevskoi (nyne Oktiabr'skoi) zheleznoi doroge, tom 1 (Berlin, 1925), pp. 64–5.

[87]   See, for example, 'Vospominaniia', vol. 4, pp. 6, 122–3, 279–81, 758, 1293–5; 'Vospominaniia', vol. 5, pp. 577, 820–21.

[88]   For example, 'Vospominaniia', vol. 5, pp. 161–2, 425–34, 540.

[89]   For example, *Russkoe slovo*, 7 December 1913, p. 6; *Peterburgskaia gazeta*, 10 December 1913, p. 5; *Peterburgskii listok*, 11 December 1913, p. 4; *Puti soobshcheniia*, 1 (1913): 8; 2 (1913): 5; 'Vospominaniia', vol. 5, pp. 526–31, 535–48.

[90]   Balzer, 'Problem of Professions', especially pp. 188–9, and 'Engineering Profession', especially pp. 61–71.

Society's Nikolaev branch in 1899, his very active involvement with the Technical Society's Ekaterinoslav branch during 1908–1909 (he was elected as the branch's deputy chairman in May 1908), and service as railway book reviews editor of the journal *Vestnik inzhenerov* during 1915–1917.[91] Oddly, he appears not to have kept any connection with the Assembly of Engineers of Ways of Communication, which was essentially an IPS alumni association and published a respected journal.

The explanation for his activism in the Technical Society's Ekaterinoslav branch in 1908 was probably a desire to exploit a specific opportunity at a delicate moment in his career. In his memoirs he claims that he had ignored the Kiev branch of the Technical Society because it was conservative in technical matters, whereas he valued the 'lively' and 'progressive' character of the activities at Ekaterinoslav.[92] But we should note that his immediate superior on the Catherine Railway, Gutovskii, was in the branch council and that colleagues were members: Lomonosov's own involvement was surely advisable for building his credibility with them as he tried to find his feet in railway management. Further, membership of the Technical Society enabled him to submit his book about his Khar'kov experiments for the society's newly created A.P. Borodin prize for research on locomotives; he would duly give a lecture about his Khar'kov research to the society's railway section in the capital in November 1908, exploiting the occasion to promote his latest research as well. Importantly, his quest for this prize put him in competition with Gololobov, who had submitted a book about the Putilov locomotive testing plant – a fact that doubtless explains the abrupt end of their cagey friendship at this time. In this tussle the victor was Lomonosov: after lengthy deliberations the Borodin committee awarded him the prize in 1911. The judges even declared that his book had 'historic significance' for Russia's railways, criticized the neglect of his research by the railway authorities, and demanded full support for his work in the future.[93] Ironically, Lomonosov had meanwhile ceased to be active in the society. Activism in any relevant association was impracticable

---

[91]   Concerning Ekaterinoslav see 'Vospominaniia', vol. 4, pp. 111–18; *Zapiski Ekaterinoslavskogo otdeleniia Imperatorskogo russkogo tekhnicheskogo obshchestva*, 3–4 (1908): 113; 7–8 (1908): 345–6; 1–2 (1909): 14–17; 3–4 (1909): 156, 169, 171–2. His involvement with the *Vestnik inzhenerov* is explained and noted at 'Vospominaniia', vol. 5, pp. 327–38, 746.

[92]   'Vospominaniia', vol. 4, pp. 111–12. For the 1907 annual reports of the Kiev and Ekaterinoslav branches see *Zapiski IRTO*, 9–10 (1908): 362–7.

[93]   See *Zapiski IRTO*, 11 (1908): 449–50; 12 (1908): 519; 6–7 (1909): 208–9; and the sections headed 'Deiatel'nost' obshchetva' in 2 (1910): 76–7; 2 (1911): 24–5; 11 (1911): 62; 3 (1912): 22; 8–9 (1912): 56; also RGIA, f. 90, op. 1, d. 335; Lomonosov, 'Ob opytakh nad parovozami normal'nogo tipa'; and 'Piatoe prisuzhdenie premii imeni Aleksandra Parfen'evicha Borodina: Trud Iu.V. Lomonosova – "Opytnoe issledovaniia tovarnykh vos'mikolesnykh parovozov kompaund normal'nogo tipa"', *Zheleznodorozhnoe delo*, 13–14 (1912): 31–2Д. Coincidentally one of the assessors was Baidak, although the latter's disdain for Raisa as a Jewess prompted Lomonosov to avoid him from 1910: 'Vospominaniia', vol. 4, p. 916.

for him in Orenburg during 1909–1910 because no local branches existed there. In any case, he now had other possibilities for publicizing his research and influencing policy. As a traction superintendent he was again entitled to attend the annual congresses of traction engineers, which were intimately associated with the MPS. He could also exploit his closeness to Rukhlov and other influential MPS officials like Kozyrev. At heart, too, he unquestionably felt that the Russian state – distinct from the autocracy – was a legitimate sponsor and protector of scientific research, and that the relationship between state and engineer need not be adversarial. Indeed, he geared his career to power and influence in that framework.[94]

His involvement with the *Vestnik inzhenerov* did not mark any great change in this outlook. It arose more by chance than design from a conference of railway engineer-technologists that was held on Shchukin's initiative at the St Petersburg Technological Institute in February 1913, and for which Lomonosov 'reluctantly' agreed to give a paper at the request of the institute's director, D.S. Zernov. According to Lomonosov, this meeting helped to generate the idea of forming a general Union of Engineers that would be headed by Kirpichev, the erstwhile much respected director of the Kiev Polytechnic Institute; but this project collapsed due to Kirpichev's death and opposition from the Assembly of Engineers of Ways of Communication. Its two main advocates, the Moscow Polytechnic Society and St Petersburg Society of Technologists, subsequently joined forces in 1915 to publish a journal – the *Vestnik inzhenerov* – and again Zernov asked Lomonosov to get involved, now as a member of the editorial board.[95]

The overall impact of Lomonosov's road experiments on pre-war Russian locomotive development and railway operation was limited, primarily because he could not persuade Shchukin and his influential committee to take them sufficiently seriously. For example, the main finding of his 1908 research on the Catherine Railway was that the 'modified Chinese' 2-8-0 locomotive performed indifferently at all speeds, but that minor changes could make it 'ideal' for fast freight and heavy passenger traffic. Yet apparently Shchukin rejected all the criticism of this class, which continued to be used mainly for slow heavy freight.[96] Similarly, tests with a class U 4-6-0 locomotive in 1910 confirmed its suitability for fast passenger traffic on light-weight track, but did not alter Shchukin's opinion of these engines

---

[94]   Lomonosov's attitude towards the state here supports the general argument that many Russian professionals accepted the legitimacy of the state's role: see Balzer, 'Problem of Professions', pp. 187–8.

[95]   'Vospominaniia', vol. 5, pp. 327–38. *Vestnik inzhenerov* was the product of merging the *Vestnik Obshchestva tekhnologov* and the *Biulleten' Tekhnicheskogo obshchestva*. For detail about the proposed union, including opposition from the Ministry of the Interior, see Balzer, 'Engineering Profession', pp. 68–71.

[96]   The main proposals concerned the blast pipe and piston valve: Lomonosov, *Sravnitel'noe issledovanie tovarnykh parovozov bol'shoi moshchnosti, vypusk II*, pp. 336–7; 'Vospominaniia', vol. 4, pp. 231–2, 425–7.

as indifferent steamers.[97] Nor did Lomonosov's reports about shortcomings in the design of the standard Class N passenger locomotive halt the mass production of these engines in favour of better designs.[98] As for his impact on railway operations, a like-minded colleague complained in 1917 that most traffic officials were still unaware of Lomonosov's work.[99] A decade later he was not even regarded as one of Russia's leading experts on traffic costs.[100]

Given all the problems that Lomonosov encountered with his method of road testing and the existence of an apparent alternative from 1905 in the form of the Putilov test plant, one must ask why he persisted with his road tests. Although he did not address this question directly in his writings, it is certain that he did wish to work with test plants. After the proposed Kiev plant was lost he continued to support the project for a national test plant, which by 1914 was planned to be located at Liublino near Moscow, but which ultimately was never built.[101] Meanwhile, as the IPS looked for a site in the capital to build one for its own use in 1911, he used his position as Traction Superintendent of the Nicholas Railway to offer a redundant building at the line's Alexander Works in St Petersburg. This idea having been agreed by Rukhlov, Lomonosov himself approved the detailed plans on behalf of the Directorate of Railways in 1912. The building work was virtually finished by December 1915, but the plant was not tested and commissioned until after 1917 because of the war, and it was dismantled in the 1920s.[102] However, the Putilov laboratory was another matter. Lomonosov was unlikely to have access to that plant due to his conflict with Gololobov, and if the MPS paid Putilov to do locomotive research, Lomonosov would struggle to maintain his own involvement in testing. True, the MPS could not have concentrated locomotive research at Putilov because, aside from the cost, the company used the laboratory extensively for running-in new locomotives. But for good measure Lomonosov impugned the

---

[97]   'Vospominaniia', vol. 4, pp. 661–2, 776–7, 871–2.

[98]   Lomonosov blamed Shchukin for this situation. See, for instance, 'Vospominaniia', vol. 4, pp. 761–2, 1240–1241; 'Vospominaniia', vol. 5, pp. 235–41.

[99]   A.A. Persivanov, 'Institut eksploatatsii zh.d.', *Zheleznodorozhnoe delo*, 9–10 (1917): 88–92.

[100]   See V.A. Sokovich, 'K voprosu o sebestoimosti perevozok', *Zheleznodorozhnoe delo: Obshchii otdel*, 6 (1928): 8–11. It is conceivable, however, that Lomonosov was not mentioned here because of his failure to return to Russia in 1927.

[101]   'O vybore predstavitelei dorog v stroitel'nyi Komitet parovoznoi, imeni A.P. Borodina, opytnoi stanstsii russkikh zheleznykh dorog', in *XXX Soveshchatel'nyi S"ezd inzhenerov podvizhnogo sostava i tiagi*, pp. 496–8; Technical Department to Head, Directorate of Railways, 24 November 1914: RGIA, f. 273, op. 15, d. 275, l. 50.

[102]   Minutes of the Temporary Construction Committee, 1 (22 November 1911): TsGIA SPb, f. 381, op. 12, d. 5, ll. 1–5; 101 (21 December 1915): ibid., ll. 519–24ob.; 'Vospominaniia', vol. 4, pp. 1173–6; 'Vospominaniia', vol. 5, p. 24; V. Bolonov, 'Parovozoispitatel'naia laboratoriia', *Krasnyi transportnik*, 3 (1923): 13–15; Westwood, *Soviet Locomotive Technology*, p. 8. This project was kept separate from the proposed national test-plant.

integrity of the Putilov engineers by alleging that they were ignoring faults in the plant's brakes. This fact, he claimed, explained why in 1908 Gololobov refused to implement an MPS order to do experiments with an 0-8-0 locomotive at regulator and cut-off settings specified by Lomonosov.[103] There was some theoretical merit in Lomonosov's argument that his road experiments could help produce a more complete picture of the operating costs and performance of a given locomotive class: aerodynamic resistance, the frictional resistance of the train and the dynamic action of the locomotive on the track could never be studied on a test-plant.[104] But essentially he continued to promote his road-testing method mainly because of organizational politics and his determination to remain involved with locomotive testing.

*   *   *

At first glance Lomonosov's railway career during 1908–1914 seems to vindicate historians who believe that the tsarist regime did have the potential to modernize the economy for the industrial twentieth century. In just six years he appeared to achieve great success and national prominence in both railway management and research. By the time of his thirty-seventh birthday in 1913 he had already held one of the most senior administrative posts at the MPS and was a member of the ministry's Engineering Council, the appointed group of distinguished but usually venerable engineers responsible for making technical policy for the Russian railways. A year later he became the founding head of the unique MPS Experiments Office with a remit to subject all new locomotive designs to extensive road tests according to his own methodology. His energetic work on raising efficiency, it would seem, was valued and rewarded at the highest level of the MPS.

Yet on balance his experiences tend to support the negative assessments of the tsarist regime's modernizing potential, although they do not merely reflect the structural reasons that are usually emphasized. The analysis of his managerial work and locomotive research in this chapter suggests that his successes were matched and perhaps outweighed by failures. Although in 1914 he was personally

---

[103]   Lomonosov, *Opyty proizvodivshiesia v 1912–1914 gg.*, p. 7; 'Vospominaniia', vol. 3, pp. 1018–19; 'Vospominaniia', vol. 4, pp. 137, 914–15.

[104]   See, for example, *Sravnitel'noe issledovanie tovarnykh parovozov bol'shoi moshchnosti, vypusk I*; Iu. Lomonosov, *Tekhnicheskie perspektivy zheleznodorozhnogo transporta v blizhaishee vremia* (Moscow, 1924), pp. 16–17. Lomonosov's insistence on the value of his experimental method for operational purposes returns us to the relative merits of cycle I and cycle II trips (see note 75 above). The counter-argument here is stressed by Knowles: 'road tests which allowed the conditions to vary, allow the boiler to be depleted uphill (in water level and pressure), [and] to be recovered downhill, on undulating roads, would be even more important for economical working, especially of lines with undulations of say 2-3-4 miles (depending on speed; the slower the speed, the shorter the distance over which this can be done)': letter to the author, 25 April 2001.

content as a member of the Engineering Council, this position was more akin to early retirement than to a platform for advancing his managerial career; indeed there was no obvious course for him even to resume that career. Moreover, his ability to create a bureaucratic niche for his locomotive research did not mean that his views on locomotive testing, its applications and railway economics were widely endorsed; on the contrary, they were still strongly contested. His research difficulties did derive partly from funding problems related to state indifference – the traditional explanation for late Imperial Russia's poor showing in industrial research and development. However, more significant than funding problems for shaping both his research and his career as a whole were the way that he rejoined the railways at a fairly senior level and the combination of such varied phenomena as staff opposition, expert disagreements and, not least, his personality. In this light people, more than deficiencies of infrastructure, equipment and funding, emerge as the crux of the empire's railway troubles, with the implication that a leap in productivity required profound socio-cultural – and arguably hence political – change. Now, all too brutally, the war would press that point.

Chapter 6
War and Revolution, 1914–1917

The war revitalized Lomonosov's career. Not only did his research become fashionable as the MPS explored every conceivable means to boost railway capacity, but his managerial experience was mobilized too. In addition to his pre-existing duties he served as a special envoy for eliminating railway congestion and as the first MPS representative in the Special Council for State Defence – the inter-ministry committee that was created in 1915 to coordinate Russia's war effort. Also, he was appointed to the board of a major electricity-generating company as part of the extension of state control over strategic industries, in his case again to play a troubleshooting role. True, office politics continued to mar his career, contributing to his removal from congestion management in early 1916. However, he was recalled to this work in January 1917 to lead a task group to reorganize the railways of allied Romania. After the revolution of February 1917, which he actively supported, he was put in charge of the MPS mission that was urgently buying railway supplies in North America.

In the meantime tsarist Russia endured its final crisis. Just over two weeks after the declarations of war Russian armies began to advance towards Berlin and Vienna. But almost immediately they suffered appalling defeats against Germany in East Prussia, and were forced into a prolonged retreat during April–September 1915. That August, amid fears that Riga, Minsk and Kiev might fall, the Tsar defied intense pressure from liberal politicians to appoint 'a government of public confidence': he took charge of the armed forces as Supreme Commander and left his unpopular German-born wife, Tsarina Alexandra Fedorovna, to manage government affairs in the capital during his long stays at army field headquarters. Russia's western front was more stable from September 1915, but the situation remained dangerous. A major offensive against German forces in March 1916 failed so badly that 'it condemned most of the Russian army to passivity'.[1] Crisis flowed even from substantial Russian successes against Austria–Hungary in June 1916: they helped persuade Romania to join the Allies but Romania's army promptly collapsed, and this disaster increased the demand for Russia's scarce resources. As the Russian war effort faltered, public opinion turned against the tsarist regime, and the autocracy expired in the February revolution.[2]

---

[1]   N. Stone, *The Eastern Front, 1914–1917* (Abingdon, 1976), p. 231.

[2]   Good analyses of Russia's war are W.B. Lincoln, *Passage through Armageddon: The Russians in War and Revolution, 1914–1918* (Oxford, 1994) and Gatrell, *Russia's First World War.*

The place of the Russian railways in this protracted crisis has yet to be clarified.[3] The notion of a railway crisis took root in 1915 and the network attracted severe criticism throughout 1916. That criticism and especially the analysis presented by V.I. Grinevetskii in 1919 in his influential book about post-war economic prospects have shaped the view that prevails now. The essence is that the railways' infrastructure, equipment and wartime performance were inadequate, and so too were the remedial actions; railway deficiencies limited the army's ability to mobilize quickly and maintain its front lines, exacerbated shortages of vital industrial supplies, and caused the urban food crisis that eventually sparked the revolution.[4] This analysis actually suited the MPS insofar as it supported the ministry's case for more resources, but railway experts did criticize it as oversimplified, and several historians have since raised questions about it. Why, for instance, should one expect Russia's railways to have prepared for a long all-embracing war when Europe's military authorities mostly envisaged a brief conflict? Were non-railway issues more relevant for explaining the military and economic crises?[5]

Framed by this controversy and the related question of the railways' role in the February revolution, this chapter examines Lomonosov's wartime experiences up to his departure from Russia in May 1917. It looks first at the period to February 1917, relating his work to the broader debate about the railway response to the war emergency. Then it discusses his politics during the years 1907–1917, his involvement in the February revolution and his first efforts to define his place in the new Russia: can we accept his claim that he supported this revolution from conviction, what do his experiences reveal about the roles of the railways and MPS in this cataclysm, and how should we interpret his appointment as head of the MPS Mission in North America?

---

[3]   The sole, somewhat problematic, monograph about the wartime railways is N. Vasil'ev, *Transport Rossii v voine 1914–1918* (Moscow, 1939).

[4]   For instance, L.M. Levi, 'K voprosu o rasstroistva transporta: Doklad sdelannyi L.M. Levi v Sobranii I.P.S. 27 oktiabria 1915 goda', *Izvestiia Sobraniia inzhenerov putei soobshcheniia*, 25 (1915): 514–17; and 26 (1915): 535–9; and V.I. Grinevetskii, *Poslevoennye perspektivy russkoi promyshlennosti* (Moscow, 1922 edn), especially pp. 108–37. For Russian military analyses of the railway preparations for war and the railways' wartime performance see K. Ushakov, *Podgotovka voennykh soobshchenii Rossii k mirovoi voine* (Moscow and Leningrad, 1928) and Vasil'ev, *Transport Rossii v voine*. An example of an authoritative critical comment in the Western literature is A. Nove, *An Economic History of the USSR* (Harmondsworth, 1982), p. 30.

[5]   For example, A.N. Frolov, 'O zheleznodorozhnykh zatrudneniiakh', *Zheleznodorozhnoe delo*, 28 (1915): 269–75; L.A., 'O zheleznodorozhnykh zatrudneniiakh', *Izvestiia Sobraniia inzhenerov putei soobshcheniia*, 5 (1916): 73–4; L.M. Levi, 'Rasstroistvo zheleznodorozhnogo transporta, ego neposredstvennye prichiny i prosteishchie mery protivodeistviia', *Inzhener*, 2–3 (1916): 33–9; S.S. Ostapenko, *Transport i voina* (Ekaterinoslav, 1916); *O rabote zheleznodorozhnoi seti v usloviiakh nastoiashchei voiny* (Petrograd, 1916); Stone, *Eastern Front*, pp. 297–300; Westwood, 'The Railways', pp. 169–78.

Russia's railways at war

Although A.J.P. Taylor's notion of 'war by timetable' in 1914 has yet to be examined thoroughly in relation to Russia's railways, it is broadly clear that those railways were fundamental for the military decision-makers in the early twentieth century.[6] Russia's mobilization plans hinged on railway capacity for moving masses of people, horses, equipment and supplies quickly to the borders and naval bases. A great expansion of this capacity *vis-à-vis* the borders with Austria–Hungary and Germany was authorized for 1912–1917 to allow much faster mobilization, with France providing much of the capital within the framework of the Franco–Russian alliance. However, this investment alarmed Germany's military leaders, whose war strategy was to defeat France before Russia could mobilize fully. Paradoxically, then, Russia's effort to enhance her security by improving railway access to her western borderlands may have strengthened Berlin's readiness to risk war in the July crisis of 1914.[7]

Whatever the controversies about their wartime performance, there is no question that Russia's railways faced extreme challenges from the outset. In the short term they had to contend with the disruption of urgent mobilization, while in the longer term they experienced unprecedented demand for their services in exceptional circumstances. The war's impact on traffic flows illustrates the latter point well. Huge amounts of men and supplies for the army had to be funnelled through a handful of major junctions like Kiev, Minsk and Rostov. Grain that had formerly travelled to ports for export was redirected across the empire. Import/export traffic was badly disrupted by the enemy blockades of the Baltic and Black seas that closed all major rail-connected ports except Archangel and Vladivostok. Moreover, that disruption severely affected other traffic, too. For example, before the war the capital (now renamed Petrograd) had relied on British coal imported through the Baltic because that option was cheaper than getting coal from the Donbass; the Baltic blockade thus necessitated a large flow of coal from the Donbass to Petrograd. Ironically, this situation highlighted the inadequacy of the military's pre-war definition of 'strategic railway' as a line essential for

---

[6] See A.J.P. Taylor, *War by Timetable: How the First World War Began* (London, 1969); D. Stevenson, 'War by Timetable? The Railway Race before 1914', *Past and Present*, 162 (1999): 163–94; and more generally on Russia's entry into the war: D.C.B. Lieven, *Russia and the Origins of the First World War* (London, 1983).

[7] See P. Luntinen, *French Information on the Russian War Plans, 1880–1914* (Helsinki, 1984), pp. 184–96; Collins, 'Franco–Russian Alliance', p. 788; and A.J. Heywood, '"The Most Catastrophic Question": Railway Development and Military Strategy in Late Imperial Russia', in T. Otte and K. Neilson (eds), *Railways and International Politics: Paths of Empire, 1848–1945* (London, 2006), pp. 45–67.

mobilization for a short war. Equally vital, it transpired, were lines needed for the smooth operation of the economy in the rear.[8]

Despite the prevailing assumption that the war would be short, the MPS was not complacent about the possibility of a protracted conflict. True, Rukhlov told Lomonosov in August 1914 that planning should assume a maximum duration of only six months.[9] But in practice the ministry immediately began developing extensive longer-term plans to obtain more equipment, and as early as October 1914 it sought permission to place large-scale orders abroad, from which the deliveries could not be completed before mid-1915 at the earliest – a step that also marked the end of the MPS's traditional support for protecting Russia's engineering industry from foreign competition. Significantly, the government rejected the MPS argument for imports at this juncture, with the notable exception of one order for 30 Mallet-type locomotives for the vital, overloaded Archangel–Vologda Railway: the Putilov works in Petrograd wanted this contract for delivery in mid-1915, but the government allowed the MPS to give the job to the Baldwin Locomotive Works of Philadelphia because the Americans promised quicker delivery.[10]

The general realization during late 1914 that the war might continue for a long time brought an important change in procurement policy for the MPS. In January 1915 the government authorized the ministry to purchase equipment in North America that included 400 freight locomotives and 10,000 high-capacity modern freight wagons for delivery by autumn 1915. With unprecedented speed the Shchukin Commission ratified the outline specification for a 2-10-0 heavy-freight locomotive, and one of Lomonosov's IPS contemporaries, Count S.I. Shulenburg, was instructed to oversee the implementation of the contracts and delivery with a small staff based in New York.[11]

---

[8]   Among the most informative contemporary books about the wartime railways are: [Upravlenie Zheleznykh Dorog], *Kratkii ocherk deiatel'nosti Russkikh zheleznykh dorog vo vtoruiu otechestvennuiu voinu, chast' 1-aia (S nachala voiny po 1 ianvaria 1915 goda)* (Petrograd, 1916); [Upravlenie Zheleznykh Dorog], *Kratkii ocherk deiatel'nosti Russkikh zheleznykh dorog vo vtoruiu otechestvennuiu voinu, chast' 2-aia (Pervoe polugodie 1915 goda)* (Petrograd, 1916); *Otchet o deiatel'nosti Osobogo Soveshchaniia dlia obsuzhdeniia i ob"edineniia meropriiatii po perevozke topliva i prodovol'stvennnykh gruzov za period sentiabr'1915g.–sentiabr'1916g.* (Petrograd, 1916); and M.M. Shmukker, *Ocherki finansov i ekonomiki zhel.-dor. transporta za 1913–1922 gody (v sviazi s obshchimi ekonomicheskimi iavleniiami zhizni strany* (Moscow, 1923). See also Vasil'ev, *Transport Rossii v voine* and A.L. Sidorov, 'Zheleznodorozhnyi transport Rossii v pervoi mirovoi voine i obostrenie ekonomicheskogo krizise v strane', *Istoricheskie zapiski*, 26 (1948): 3–64.

[9]   'Vospominaniia', vol. 5, pp. 668–9.

[10]   A.J. Heywood, 'Russia's Foreign Supply Policy in World War I: Imports of Railway Equipment', *The Journal of European Economic History*, 32/1 (spring 2003): 77–108 (90–91). The issues concerning protection for the railway engineering industry are summarized in Heywood, *Modernising Lenin's Russia*, pp. 18–20.

[11]   Heywood, *Modernising Lenin's Russia*, pp. 28–30; Heywood, 'Russia's Foreign Supply Policy', pp. 93–102.

While that plan was being put into effect, the Russian military catastrophe of 1915 plunged the railways into a crisis from which they never fully recovered. At worst the Russian retreat measured about 200 miles. By the autumn Russian Poland had been lost, the battlefront stretched from Riga on the Baltic coast towards Chernovitsy in the south, and the cities of Minsk and Kiev were under threat.[12] Throughout this emergency the railways made a tremendous effort to evacuate people and industrial equipment. But counter-balancing their successes were severe congestion in the rear, disorganization and a worsening shortage of freight wagons. Many wagons with evacuated goods waited for months in the rear to be unloaded, and hence were effectively out of service.[13] To their credit the railways moved record volumes of traffic in both 1915 and 1916, but equally they became ever less able to cope with all the demands being made of them.[14]

Although Lomonosov later voiced the usual opinion that the railways were ill prepared, the text of his memoirs suggests that he had few if any such worries when the war started. He describes how he was supervising locomotive experiments in the south in July 1914 and how his reaction to Vienna's declaration of war against Serbia was 'very militant'. Neither he nor his partner Raisa wanted war, he claims, but they agreed about the need to teach the 'scoundrels' a lesson. He also recalls his impression – gained while hurrying back to the capital – that the railways worked like a 'well-oiled machine' during the military mobilization. His optimism was boosted by Britain's declaration of war.[15] In short, he confidently supported his government, like so many other people in the major powers, including most of those legions of socialists who had backed the anti-war stance of the Second International.

His recollections of the MPS in August 1914 suggest that the ministry did take the emergency very seriously. In particular, I.N. Borisov, head of the Directorate of Railways since April, had already designated the Experiments Office as a department working for war needs, thereby exempting its personnel from military conscription. True, Borisov put this decision to Lomonosov as a political move: to reduce the directorate's reliance on its traction department, which was just a 'Shchukin secretariat'. But the underlying aim was clearly to try to enhance efficiency by using Lomonosov's mathematical techniques for calculating locomotive capabilities, schedules and so forth despite Shchukin's objections. The practical implication for Lomonosov was that he would continue with his work at the Engineering Council, Experiments Office, Commission for

---

[12]   On the military situation see Stone, *Eastern Front*, pp. 165–91.

[13]   *Otchet o deiatel'nosti Osobogo Soveshchaniia dlia obsuzhdeniia i ob"edineniia meropriiatii po perevozke topliva i prodovol'stvennnykh gruzov*, pp. 42–9, 53–5, 61–2.

[14]   *Otchet o deiatel'nosti Osobogo Soveshchaniia dlia obsuzhdeniia i ob"edineniia meropriiatii po perevozke topliva i prodovol'stvennnykh gruzov*, pp. 165–70 and prilozheniia 6, 11.

[15]   'Vospominaniia', vol. 5, pp. 628–61.

New Railways and IPS, and would have greater influence concerning the theories that informed traffic management.[16]

That the MPS and railways became better disposed towards his ideas is evident from the projects that were directed to the Experiments Office. For example, technical information from the office was used in discussions during the autumn of 1914 about how to expand network capacity by raising speed limits without compromising safety. Similarly, individual railways sought assistance with calculating the maximum permissible length of train formations, braking distances, and so forth. Meanwhile, the Directorate of Railway Construction began requesting technical data from locomotive passports for use in designing railways, and thereby helped to justify an intensive effort during 1915–1916 to publish a series of brochures with data and results from Lomonosov's pre-war experiments.[17]

This interest is probably explained more by desperation than by any collective change of mind about the value of Lomonosov's research. Generally the tsarist regime was slow to shift the economy to a full war footing, and remained cautious about how far scientific research could help the war effort. A good illustration is its refusal to provide adequate funding for the Commission for the Study of the Natural Productive Forces of Russia, which various scientists formed in 1915 to collate and extend knowledge about the empire's strategic raw materials and resources.[18] Yet the situation of the railways was unusual in that this industry faced a full-scale emergency from the outset, which forced the MPS immediately to use every conceivable means to increase capacity at virtually any cost. Moreover, in August 1914 the MPS understood that even six months would be insufficient to obtain all the extra locomotives and wagons that were deemed essential for coping with the crisis. Thus, although it did seek permission to place substantial orders for new locomotives and wagons with foreign as well as domestic suppliers, the MPS knew that it would have to find most of the extra capacity by operating the network more intensively, upgrading core routes and speeding up the construction of potentially vital new routes.[19] Accordingly, any idea for raising efficiency was welcome, especially if it needed little or no money, and various engineers aired proposals in the railway press, including a new method of doing 'traction

---

[16]   'Vospominaniia', vol. 5, pp. 662–73.

[17]   'Vospominaniia', vol. 5, pp. 694–5. Examples of the brochures are: *Glavneishie rezul'taty opytov, proizvodivshikhsia v 1908 g. na Ekaterininskoi zh.d. nad parovozami tipa 1-4-0 Shch*, 2nd edn (Petrograd, 1915); and *Neposredstvennye dannye opytov I tsikla s parovozom M.–Kur. zh.d. 2-3-0 B17, proizvodivshikhsia v 1913 g. na Nikolaevskoi zh.d.* (Odessa, 1915). Many such brochures are listed in the Bibliography in the section devoted to works by Lomonosov, but this listing is incomplete, for it contains only those items of which I have found a copy: at least nine other brochures are likely to have been published.

[18]   Bailes, *Science and Russian Culture*, pp. 138–41.

[19]   *Kratkii ocherk deiatel'nosti Russkikh zheleznykh dorog, chast' 2-aia*, pp. 46–80.

calculations'.[20] The ministry gave particular support to 'work packing' (*uplotnenie rabot*), a technique devised by B.D. Voskresenskii (head of the Southern Railways) to raise productivity in traffic organization, wagon and locomotive maintenance, and so forth.[21] By the same token the MPS would want to employ Lomonosov's mathematical methods to clarify how locomotives and routes could cope with higher speeds and heavier and more frequent trains.

Nonetheless, the mobilization of Lomonosov's scientific ideas was not all plain sailing. One obstacle stemmed from the regime's failure to take a strategic view of skill needs throughout the economy: Lomonosov could not recruit staff to handle his office's increased workload because transport students – his main source of suitable personnel – generally decided, and were allowed, to volunteer for military service.[22] Also problematic was Lomonosov's unwillingness to make compromises that would secure a consensus for immediate but incomplete improvements. This difficulty is apparent in discussions during autumn 1914 about raising speed limits. He had become interested in this question while at Kiev. By 1911 his research had contributed to the creation of a commission of the MPS Engineering Council to consider whether to raise the speed limits for the latest passenger locomotives; his ill-fated high-speed St Petersburg–Moscow trials in 1913 were part of this investigation.[23] Concurrently another commission began studying the related matter of physical stresses in rails, including the effects of locomotive speed. Here Lomonosov argued that the interaction between the steel tyre and rail was primarily a function of the locomotive's design, not of the physical condition of the tyre and rail; hence in 1913 he began road experiments to identify the dynamic characteristics of selected locomotive classes.[24] Meanwhile, however, the idea of

---

[20]   For example: N.P. Kemmer, 'O predel'nykh skorostiakh podkhoda tovarnykh poezdov k signalam (Raschety tormzoheniia)', *Izvestiia Obshchego biuro soveshchatel'nykh s"ezdov*, 4 (1915): 292–323; Z., 'Neskol'ko slov o podtalkivaniia', *Izvestiia Sobraniia inzhenerov putei soobshcheniia*, 13 (1915): 282–3; I. Zausailov, 'K voprosu ob uvelichenii sostava tovarnykh poezdov', *Zheleznodorozhnoe delo*, 40 (1915): 420–422; and V. Shcheglovitov, 'K voprosu ob usilenii propusknoi sposobnosti odnoputnykh linii', *Zheleznodorozhnoe delo*, 41–42 (1916): 384–8; 43–44 (1916): 403–6; and 45–46 (1916): 427–9.

[21]   For example: B. Voskresenskii, 'Naibol'shee uplotnenie rabot putem primeneniia parallel'nykh i posledovatel'nykh rabot, kak osnovnoi metod eksploatatsii zhel. dorog (metod inzh. Khlebnikova)', *Zheleznodorozhnoe delo*, 16 (1915): 137–41; 'Pereotsentka odnoi tsennosti', *Zheleznodorozhnoe delo*, 16 (1915): 137; Taranov-Belozerov, 'K voprosu ob intensifikatsii raboty tovarnykh poezdov', *Zheleznodorozhnoe delo*, 24 (1915): 224–5. For Lomonosov's views see 'Vospominaniia', vol. 5, pp. 1053–64.

[22]   'Vospominaniia', vol. 5, pp. 694–5.

[23]   See Chapter 5, this volume.

[24]   For example: RGIA, f. 240, op. 1, dd. 1016, 1174, 1175; *Trudy Komissii pod predsedatel'stvom chlena Inzhenernogo Soveta d.s.s. Kunitskogo za vremia s 31-go maia po 23-'e oktiabria 1912 goda*; Kunitskii, *Kratkii istoricheskii ocherk deiatel'nosti Inzhenernogo Soveta*, pp. 99–108; 'Vospominaniia', vol. 5, pp. 699–707.

reducing speed limits became the focus of a third commission in early 1914 due to public anxiety about several big train crashes.[25] Normally several years would have been needed to pursue and ideally coordinate the diverging threads of these three enquiries. But in September 1914 Rukhlov demanded an interim directive within three months. Lomonosov was fully involved in the consequent discussions, but was incandescent about their outcome: for him the published directive was 'completely incomprehensible' owing to last-minute amendments stipulated by the Directorate of Railways.[26]

His involvement in traffic troubleshooting during the crisis of 1915 tends to suggest that, indeed, the MPS was beginning to be overwhelmed in its management role. In April he was one of the large team of MPS experts that accompanied deputy minister Dumitrashko on a tour of the railways of European Russia; they concentrated on planning improvement works and reorganizing traffic. In May he led a group that tackled these issues in the Briansk–Smolensk area (a vital district for westbound army supplies and eastbound evacuation traffic). In August, armed with the same decision-making powers as a deputy minister, he toured the South Eastern and Riazan'–Urals railways to assess progress with implementing the decisions of Dumitrashko's April tour. Next he directed the railway planning for the potential evacuation of Kiev. During October–December he investigated the congested Donbass–Petrograd route and the related traffic troubles of Moscow, Petrograd and the Nicholas Railway.[27] He claims also to have co-authored (with Rukhlov) the reform whereby the MPS liaison group at the Supreme Commander's field headquarters was replaced by a Directorate of Ways of Communication – a change that was intended to increase MPS influence over the front-area railways and improve MPS–army relations.[28] Generally his memoirs offer an upbeat picture of his work, but pending detailed investigation one suspects that in reality his impact was marginal. Fundamental deficiencies in the infrastructure were not susceptible to quick solutions, such as the difficulty that the heavy Donbass–Petrograd traffic had to traverse Moscow for want of an alternative route. Also, a plenipotentiary could find his options restricted by decisions outside his control: for example, when planning the evacuation of Kiev

[25]   This commission's papers are in RGIA, f. 240, op. 1, d. 1074.

[26]   See RGIA, f. 273, op. 6, d. 3115, especially ll. 13–14 and 26–31ob.: Minutes of meeting No. 3 of Sub-commission for working out temporary general guidelines for considerations when determining maximum axle-loadings of rolling stock in relation to the rail type and maximum traffic speed, 31 October and 11 November 1914; Draft guidelines, [December 1912]; Kunitskii, *Kratkii istoricheskii ocherk deiatel'nosti Inzhenernogo Soveta*, pp. 105–8; 'Vospominaniia', vol. 5, pp. 705, 707–20.

[27]   'Vospominaniia', vol. 5, pp. 763–807, 856–73, 935–53, 967–1015, 1064–83, 1091–139. Most of the associated official documents have not been found, but the trips are confirmed in [Iu.V. Lomonosov], *Tekhnicheskii i denezhnii otchet za 1915 g.* (Petrograd, 1916), pp. 15–17.

[28]   'Vospominaniia', vol. 5, pp. 1039–40; Kriger-Voinovskii, *Zapiski inzhenera*, p. 65.

Lomonosov was ordered to prioritize 11,000 religious relics.[29] More generally, traffic reorganization seems to have been hampered by a failure to coordinate the activity of plenipotentiary troubleshooters.

Lomonosov's service in the committee that became the Special Council for State Defence was part of a broader political dimension of the railways' war that is sometimes overlooked. His experiences suggest that however well the railways tackled their traffic challenges, the ministry lost the parallel battle for national political and public support, and became a scapegoat for a wide variety of failings. Chaired by the war minister, the council was created in May 1915 to organize artillery supplies for the army, and soon became responsible for mobilizing the whole country for the war effort.[30] According to Lomonosov, ministers were expected to represent their ministry, but the head of the MPS, Rukhlov, saw this committee's establishment as a personal attack by Duma members, and sent the professor instead. The latter thus encountered some political irritation as well as scepticism about the MPS and railway performance. In practice, therefore, his main role was to defend the MPS at council meetings, mainly with regard to MPS–army tensions, congestion on the Petrograd network and delays to the Donbass and Archangel traffic. He also had some specific associated tasks, such as chairing a conference about congestion on the Nicholas Railway.[31] After just two months, however, he was replaced by the more senior Kozyrev, a change that derived from the national political crisis of summer 1915. Under intense liberal pressure to appoint 'a government of public confidence', the Tsar reasserted his authority as autocratic tsar, not least with the reasoning that major political change might cripple the war effort. His controversial decision of August 1915 to take command at army field headquarters was one result of this crisis, as was the reform of the Special Council that led to Kozyrev's appointment.[32]

---

[29] 'Vospominaniia', vol. 5, pp. 991–2.

[30] Gatrell, *Russia's First World War*, pp. 90–98.

[31] For example: L.G. Beskrovnyi et al. (eds), *Zhurnaly Osobogo Soveshchaniia dlia obsuzhdeniia i ob"edineniia meropriiatii po oborone gosudarstva (Osoboe soveshchanie po oborone gosudarstva): Publikatsiia: 1915 god* (Moscow, 1975), pp. 57, 103, 131–2; Minutes of conference on the question of changing the existing order of train traffic on the Nicholas Railway, 11 July 1915: RGIA, f. 273, op. 10, d. 3069, ll. 30–36; 'Vospominaniia', vol. 5, pp. 874–93, 901–23, 956–60, 962–1015. According to a news report of mid-1915, he was also appointed to a shipments group that was formed under the Central War Industries Committee: 'Razlichnye izvestiia', *Izvestiia Sobraniia inzhenerov putei soobshcheniia*, 14 (1915): 309. Almost certainly this position was connected with his work in the Special Council and was short-lived.

[32] Lomonosov notes his removal at 'Vospominaniia', vol. 5, p. 1015. On the political crisis see, for instance, R. Pearson, *The Russian Moderates and the Crisis of Tsarism, 1914–1917* (London, 1977), pp. 39–64; L.H. Siegelbaum, *The Politics of Industrial Mobilization in Russia, 1914–17: A Study of the War Industries Committees* (London, 1983), pp. 69–84; D.R. Jones, 'Nicholas II and the Supreme Command: An Investigation of Motives', *Sbornik*, 11 (1985): 47–83.

Lomonosov's removal from the Special Council foreshadowed a rapid decline in his administrative influence at the MPS that was connected with the dismissal of minister Rukhlov on 27 October 1915. To a degree that sacking was occasioned by the widespread criticism of the MPS. But according to Lomonosov, high politics and a personal tragedy were also relevant. Rukhlov's relationship with the Tsarina was poor, not least because he disputed her right to 'govern' on the Tsar's behalf and disliked her notorious favourite, the monk G.E. Rasputin. Rukhlov's loss of royal favour was evident in the palace's rejection of all his nominations for appointment as deputy minister to succeed Dumitrashko, who was retiring. Also, Rukhlov lost his 'taste for political battle' after the death of his eldest daughter in September 1915. Reportedly he confided to Lomonosov in mid-October that he was a 'living corpse' as minister and that the professor was the only man to whom he could transfer his duties with a calm heart, although that change was 'impossible'. These words may overstate the closeness of their working relationship, but Lomonosov did owe his position to Rukhlov, and the latter's replacement by A.F. Trepov, a non-railwayman, left the professor somewhat vulnerable.[33]

Lomonosov's dealings with Trepov over the next few months became reminiscent of his quarrels with Shtukenberg and Ivanovskii in 1909–1911.[34] In early November the new minister accepted his argument that the difficulties of the Donbass–Moscow–Petrograd route were mainly in Moscow, Petrograd and the MPS, and appointed him to chair a new commission to clarify those problems.[35] This second investigation uncovered numerous organizational difficulties, starting with poor management and the lack of an overall work plan. Related problems included an excessive accumulation of wagons on the Nicholas Railway, poor-quality fuel, sabotage, shortages of labour and materials, and the poor physical condition of the locomotives and wagons. The solution, Lomonosov affirmed, lay not in 'miraculous super-measures', but in detailed, sustained organizational work. Apparently Trepov approved the report in December, but when the professor presented it to the inter-ministry Special Conference for Shipments, Trepov condemned him for 'shaming

---

[33]   See 'Vospominaniia', vol. 5, pp. 1035–6, 1038. The Tsarina pressed for Rukhlov's removal in letters to her husband of 8 and 10 October 1915: J.T. Fuhrmann (ed.), *The Complete Wartime Correspondence of Tsar Nicholas II and the Empress Alexandra: April 1914–March 1917* (Westport, 1999), pp. 268, 273; the significance of that pressure is unclear, for the Tsar expressed a favourable opinion of Rukhlov a year later: ibid., p. 677. Confirmation of Rukhlov's daughter's death has not been located; Shilov does not record it: *Gosudarstvennye deiateli*, p. 575. For a thoughtful, critical contemporary assessment of Rukhlov's tenure of the MPS see 'Ministr Putei Soobshcheniia S.V. Rukhlov (1909–1915 gg)', *Izvestiia sobraniia inzhenerov putei soobshcheniia*, 1 (1916): 2–7; and 2 (1916): 18–22. For an attempt to understand why the Tsarina involved herself in such matters see P. van Reenen, 'Alexandra Feodorovna's Intervention in Russian Domestic Politics During the First World War', *Slovo*, 10/1–2 (1998): 71–82.

[34]   For Trepov's career and appointment see ibid., pp. 676–8 and A.O. Bogdanovich et al. (eds), *Ministry i Narkomy Putei Soobshcheniia* (Moscow, 1995), p. 111.

[35]   'Vospominaniia', vol. 5, pp. 1085–8.

the ministry'. A more circumspect report was substituted on 4 January 1916. Now most of the findings were accepted, but Lomonosov's team was criticized for not indicating how, for instance, to strengthen the Nicholas Railway.[36] The upshot was that Lomonosov became a 'disgraced dignitary'; whether in fact he was sidelined for his association with Rukhlov is unclear.[37]

Lomonosov's activity during 1916 was thus much the same as in early 1914. He worked part-time at the IPS, gaining confirmation as a full professor.[38] He coordinated the MPS involvement in the Commission for New Railways until he was removed from that commission in the autumn.[39] Most of his time, however, went on the Experiments Office and Engineering Council. He had kept an eye on the office's locomotive research in 1915, when the three main tasks had been to solve technical problems in the Archangel line's new Baldwin locomotives, collect data for drafting the passport of the class E 0-10-0 locomotive (which was replacing the class Shch as the standard freight engine) and conduct trials in operating 2,000-tonne freight trains.[40] Now, in 1916, the projects included experiments with the new American-built 2-10-0 locomotive (called the class Ye), tests with peat as fuel in differing designs of locomotive firebox and attempts to run 4,000-tonne freight trains, which was a huge load by Russian and European standards.[41] Much of this work reflects the railways' continuing desire to use his research to increase line capacity. However, his interest in the 4,000-tonne trains seems self-indulgent: their regular operation in Russia was known to be impracticable without the fitting of automatic couplings and the extension of many sidings and refuges – both

---

[36]   *Otchet o deiatel'nosti Osobogo Soveshchaniia dlia obsuzhdeniia i ob"edineniia meropriiatii po perevozke topliva i prodovol'stvennnykh gruzov*, prilozhenie 1, pp. 64–7; 'Vospominaniia', vol. 5, pp. 1091–139. An example of the newspaper coverage is *Rech'*, 5 January 1916, p. 4 and 6 January 1916, p. 4.

[37]   'Vospominaniia', vol. 5, p. 1197. Lomonosov's focus on the whole route from the Donbass to Petrograd seems to have been dropped at this point, leaving local managers to take the strain, especially Voskresenskii on the Southern Railways and in Moscow. See, for example, N., 'Doklad o rabotakh po Moskovsko–Kurskoi zh.d. i Moskovskomu uzlu', *Zheleznodorozhnoe delo*, 19–20 (1916): 173–6.

[38]   'Vospominaniia', vol. 5, pp. 1358–62.

[39]   'Vospominaniia', vol. 5, p. 1354.

[40]   [Lomonosov], *Otchet za 1915 god*, pp. 9–17. On the weight experiments see 'Poezd vesom 100,000 pudov: Beglye vpechatleniia po opytam prof. Iu.V. Lomonosova', *Izvestiia Sobraniia inzhenerov putei soobshcheniia*, 23 (1915): 478–9; and N.A., 'Eshche o "Poezde vesom 100,000 pudov"', *Izvestiia Sobraniia inzhenerov putei soobshcheniia*, 26 (1915): 534–5 (including editorial note, p. 535).

[41]   For example: *Zhurnaly Komissii podvizhnogo sostava i tiagi*, 19 (1916): 2–11; 23 (1916): 7–8; 26 (1916): 14–16; E. Koriukin, 'Ob opytakh s parovozami Malleta i Dekapod', *Zheleznodorozhnoe delo*, 37–38 (1916): 353–4; Iu.V. Lomonosov, 'Tovarnye poezda vesom v 200,000 pudov i bolee', *Vestnik inzhenerov*, 2/19 (1916): 618–20; 'Vospominaniia', vol. 5, especially pp. 1205–12, 1254–80, 1299–1314, 1341–53. The transliteration Ye is used instead of E to avoid confusion with the class E 0-10-0.

complicated matters, and the ministry did not view even the 2,000-tonne trains of his 1915 experiments as practicable for daily service.[42]

The dismal fate of one of his more important reports to the Engineering Council exemplifies how the railways were running out of time by 1916. The redevelopment of the Petrograd network had been under discussion for years, but was rendered urgent by the traumas of 1915. The MPS tried to speed up its decision-making process, but only by giving this issue priority, not by simplifying the process itself or by focusing only on the traffic needs of the moment. Accordingly, there was protracted analysis of possible future peacetime lines into the area, with deep disagreements about likely post-war traffic flows, not to mention the construction and shipping problems posed by the need to bridge the River Neva. By January 1916 two plans were under consideration. Much work was done to compare their construction and operating costs for various traffic scenarios to 1930. The cost difference was deemed negligible, so the decision hinged on operating criteria, including post-war assumptions, and was eventually made on 25 February 1916.[43] By then, however, there was insufficient time for the detailed planning to be completed for the 1916 construction season, after which the circumstances changed beyond recognition. The extensive work of Lomonosov and many other colleagues on this project was thus wasted.

A negative verdict also seems appropriate for his role as a board member of the 1886 Company, a leading generator of electricity. The context was the fact that one of the government's responses to wartime economic problems was to extend state control over key private companies. Prime examples were the Putilov company (sequestered in 1916) and the electricity-generating industry, which hit trouble as early as the winter of 1914–1915.[44] Lomonosov became one of several state appointees on the 1886 company's board. Nominated in April 1915, he retained this post until after the February revolution. His duties again included troubleshooting: he investigated possible sabotage in Petrograd and issues concerning a power station in Moscow. But realistically he could not achieve much in the minimal time that he could devote to this work. For him the experience was interesting, but its salary was more important – hardly a resounding endorsement of the policy.[45]

Somewhat unexpectedly Lomonosov was given an opportunity to resurrect his MPS administrative career in late 1916. The chance arose from the dismissal of

---

[42]   See the editorial note at the end of N.A., 'Eshche o "Poezde vesom 100,000 pudov"', p. 535.

[43]   Minutes of Engineering Council, No. 4, 27 January 1916: RGIA, f. 240, op. 1, d. 1143b, ll. 1–14ob.; Minutes of Engineering Council, No. 14, 25 February 1916: RGIA, f. 240, op. 1, d. 1143b, ll. 15–45; E. Nekrasov, 'Pereustroistvo petrogradskogo zheleznodorozhnogo uzla', *Izvestiia Sobraniia inzhenerov putei soobshcheniia*, 21 (1916): 426–9; 'Vospominaniia', vol. 5, pp. 1171–3, 1195–6.

[44]   Grant, *Big Business in Russia*, pp. 113–35; Coopersmith, *Electrification*, pp. 99–120.

[45]   'Vospominaniia', vol. 5, pp. 822–34, 929–30, 1173–6, 1179. The papers of this company are in TsGIA SPb, f. 1243, but were inaccessible at the time of my visit.

Trepov from the MPS on 27 December, and took the form of a six-week assignment to lead the reorganization of the Romanian railways. Romania had joined the Allies in August 1916, but her military situation was always precarious and overseas Allied aid could be sent only via Russia, which was itself short of such supplies and capacity to transport them. Soon most troops in Romania's front line were Russian and the Romanian railways were struggling.[46] In December Russia's Field Headquarters decided to send an expert mission to tackle the railway problems, and the new minister, E.B. Kriger-Voinovskii, asked Lomonosov to lead it.[47] The professor left Petrograd on 6 January 1917, reached Romania four days later, and stayed there for five weeks.[48] The work was identical to his managerial tours of 1915. His team inspected the main routes and identified 11 fundamental problems. These ranged from shortages of fuel, materials and spare parts to inadequate water supplies, a lack of traffic planning and large accumulations of wagons in yards. The mission listed 165 remedies. It checked the running times for trains on each route for a range of train weights, devised a plan for reallocating the operable locomotives among eight depots by reference to their haulage capacity, and produced new calculations of line capacity, the number of engines needed and so forth. Also, it convened a Russo–Romanian conference about the supply of fuel.[49]

---

[46]   G.E. Torrey, 'Rumania and the Belligerents 1914–1916', *Journal of Contemporary History*, 1/3 (1966): 171–91; G.E. Torrey, 'The Rumanian Campaign of 1916: Its Impact on the Belligerents', *Slavic Review*, 39/1 (1980): 27–43; G.E. Torrey, 'Indifference and Mistrust: Russian–Romanian Collaboration in the Campaign of 1916', *The Journal of Military History*, 57/2 (1993): 279–300.

[47]   Given Berberova's allegation (see Chapter 4, note 22) of a friendship and Masonic connection between Lomonosov and the army Chief of Staff, General Alekseev, it is worth adding that a biographical note written in the USA in 1917 or 1918, possibly by Lomonosov, asserts that Alekseev insisted on this recall. No other evidence of any connection between the two men has been found. See Bakhmetev Archive, Columbia University, B.A. Bakhmeteff Collection (henceforth Bakhmeteff Collection), box 24, file 'April–July 1918'.

[48]   The assignment is described in Lomonosov–Chief of Staff of Supreme Commander, 19 February 1917: RGIA, f. 229, op. 2, d. 1638, ll. 2–7; 'Vospominaniia', vol. 5, pp. 1381–532. On the minister see 'Upravliaiushchii Ministerstvom putei sobshcheniia E.B. Kriger-Voinovskii', *Izvestiia Sobraniia inzhenerov putei soobshcheniia*, 1 (1917): 4; and Shilov, *Gosudarstvennye deiateli*, pp. 335–6. His surname is found as both Kriger-Voinovskii and Voinovskii-Kriger. For his gloomy view of Russia's railway situation in mid-February 1917 see his *Doklad Sovetu ministrov: o putiakh soobshcheniia i usloviiakh perevozok v tretii god voiny* ([Petrograd, 1917]).

[49]   Lomonosov–Chief of Staff of Supreme Commander, 19 February 1917: RGIA, f. 229, op. 2, d. 1638, ll. 2–7; Minutes, 9 February 1917: RGIA, f. 229, op. 2, d. 1638, l. 8–ob.; Measures which, in the opinion of the commission for investigating the railways of Romania, should be taken immediately for improving their work, [February 1917]: RGIA, f. 229, op. 2, d. 1638, ll. 9–19; Diagram of line capacity: RGIA, f. 229, op. 2, d. 1638, ll. 20–21; Calculation of requirement for working locomotives: RGIA, f. 229, op. 2, d. 1638, l. 22; Draft allocation of working locomotives: RGIA, f. 229, op. 2, d. 1638, l. 23; Table

This tour was Lomonosov's last special task for the tsarist railways. He arrived home on about 19 February, satisfied that he had been doing valuable practical work.[50] The new minister, Kriger-Voinovskii, now asked him to oversee all coal shipments. Lomonosov was dubious: 'I knew from bitter experience that the old authorities would continue nonetheless to give commands, and that nothing would come of this dictatorship except dual power and confusion'.[51] But in the event nothing came of this conversation because within a week the February revolution had started.

The February revolution

As in 1905, so in the revolution of February–March 1917 the railways were critically important. The MPS was among the first ministries to be seized because control of the railways was essential for any government; the railway telegraph system publicized the revolution throughout the empire; and revolutionary railwaymen thwarted both the Tsar's attempt to return to Petrograd from Field Headquarters and a bid to move front-line troops into Petrograd to quash the rebellion.[52] In these events Lomonosov played a prominent part, but there were doubts among at least some contemporaries about whether his support for this revolution and later the Bolshevik regime was merely opportunist.[53] This question returns us to the difficulty of identifying the politics of any given person in a society where to demand constitutional change publicly was to court arrest. Presumably like many of his compatriots in late Imperial Russia, Lomonosov compartmentalized his life, generally keeping his counsel about his politics, which explains why the available evidence about his political thinking is limited and even contradictory.

The tsarist authorities treated him as politically reliable during the decade prior to 1917. He had no trouble with the political vetting that occurred whenever a state official was appointed to a new job. Luck, however, was part of the explanation.

---

of train consists, journey time, line section capacity and water consumption per section by train locomotives, including bankers: RGIA, f. 229, op. 2, d. 1638, ll. 24–25ob.

[50]   The memoirs give the return date as 20 February: Iu.V. Lomonosov, *Vospominaniia o Martovskoi revoliutsii* (Moscow, 1994), p. 219 (first published in English as *Memoirs of the Russian Revolution* (New York, 1919)). But 19 February is shown in Lomonosov–Chief of Staff of Supreme Commander, 19 February 1917: RGIA, f. 229, op. 2, d. 1638, l. 1. An MPS deputy minister visited Romania in June 1917 and reported the transport situation as satisfactory: RGIA, f. 229, op. 2, d. 1639, l. 60–ob.

[51]   Lomonosov, *Vospominaniia o Martovskoi revoliutsii*, p. 221.

[52]   For a detailed account of the February revolution see T. Hasegawa, *The February Revolution: Petrograd, 1917* (Seattle, 1981).

[53]   In 1922, for instance, N. Meshcheriakov remarked that Lomonosov was not a Bolshevik in February 1917: 'Iu.V. Lomonosov: "Vospominaniia o martovskoi revoliutsii, 1917 g.", Stokgolm–Berlin 1921 god, 86 str.', *Proletarskaia revoliutsiia*, 5 (1922): 275–9.

The Kiev police thought that he had concealed student criminal activity and headed a Jewish self-defence group in the 1905 revolution, and so they included him in a list of KPI professors whom they recommended for dismissal in spring 1906; however, they did not consider him sufficiently dangerous to warrant a report to St Petersburg.[54] Furthermore, they evidently failed to examine his connection with Raisa, who had been arrested in Odessa in spring 1905, and to connect him with Professor Tikhvinskii, whom they investigated at length. Hence the police in the capital had no reason to object when the MPS notified them of Lomonosov's appointment to the Catherine Railway in the winter of 1907–1908. His clean police record in St Petersburg also allowed his promotion to traction superintendent in 1909 – an unambiguous sign of official trust because this post involved responsibility for royal travel.[55] Similarly, in 1912 Rukhlov could get permission for Lomonosov and his family to live at Tsarskoe Selo, the village outside the capital where the Tsar normally resided.[56]

These official assumptions about his political loyalty were probably strengthened by Raisa's contribution to the war effort. In August 1914 she volunteered to train as a nurse, as did many other women of Tsarskoe Selo, including the Tsarina and two of her daughters, the princesses Ol'ga and Tat'iana. During September–November 1914 Raisa trained at the Palace Hospital in Tsarskoe Selo, where the Tsarina and her daughters also worked. From May 1915 Raisa served at a nearby sanatorium for recuperating officers called the Community of Her Highness (*Obshchina Ee Velichestva*); this hospital was headed by the Tsarina herself. By 1916 Raisa was a senior sister and, apparently, one of the Tsarina's deputies. Whatever her politics, she found the Tsar 'charming' and reportedly worked well with the Tsarina. The Lomonosovs enjoyed dairy products from an imperial farm, privileged access to the royal parks and the Tsarina's assistance when their son Iurii contracted appendicitis. Indeed, Lomonosov may have been hoping to strengthen Raisa's position when

---

[54] Kiev Governor–OO Head, 10 April 1907; OO Head–[Kiev Governor], 16 May 1907: DAKO, f. 2, op. 44, spr. 171, ark. 2–zv, 4; Los' et al., *Iz istorii Kievskogo Politekhnicheskogo Instituta*, pp. 238–9.

[55] One surviving central police record of these inquiries is: Registration Bureau, Guard Agency Subordinate to the Palace Commandant, Tsarskoe Selo–OO St Petersburg, 20 February 1914; and reply of the same date: GARF, f. 102, 6 deloproizvodstvo, opis po 1914 g., d. 101Bt.1, ll. 129–30, 153. The Kiev gendarmes and secret police both kept files on Tikhvinskii: TsDIAU, f. 274, op. 1, spr. 1113 and f. 275, op. 1, spr. 152. For examples of police correspondence about Raisa see Head of Gendarmerie, Podol'sk–Dept of Police, 27 September 1903: GARF, f. 102, DPOO, 1903g., d. 2397, l.4-ob.; Head of Gendarmerie, Odessa–Dept of Police, 31 May 1905: GARF, f. 102, DPOO, 1905g., d. 2974, ll. 14–15ob.; Protocol [of search], 16 May 1906: TsDIAU, f. 275, op. 1, spr. 1141, ark. 14–15; Information about person searched at 1.30 am [R.I. Rozen], 16 May 1906: TsDIAU, f. 275, op. 1, spr. 1141, ark. 12–zv.; Head of Okhrannoe Otdelenie, Kiev–DPOO, 21 June 1906: GARF, f. 102, DPOO, 1905g., d. 1877, ch. 72, ll. 210–11ob.

[56] 'Vospominaniia', vol. 5, p. 83.

he accepted Sonia's terms for divorce in 1915 and married Raisa within days of the divorce being finalized in February 1916.[57]

As for non-official assessments of Lomonosov's politics, the surviving evidence indicates varying views. At least one acquaintance assumed that he firmly supported the tsarist regime. B.A. Bakhmetev (1880–1951) was a professor of hydraulic engineering at the St Petersburg Polytechnic Institute who also taught at the IPS, and was involved with the Special Council for Defence. Bakhmetev had been a Menshevik in the 1905 revolution and a member of the Social Democratic Party's Central Committee, but was evidently a liberal by 1917. The Provisional government appointed him as a Deputy Minister of Trade and Industry in March 1917 and then as extraordinary plenipotentiary and ambassador to the United States in April. A few weeks later Bakhmetev accused Lomonosov of having displayed firm support for the tsarist regime at the Special Council for Defence, and was surprised to learn about his participation in the 1905 revolution.[58]

In his capacity as a prominent state official Lomonosov was a potential target for attention from both extremes of the political spectrum. The Socialist Revolutionary Party and other radical groups killed many state officials as agents of class oppression during 1905–1917. Lomonosov must have epitomized tsarist oppression for at least some workers, whether for sacking subordinates or simply for his high rank. Indeed, Krzhizhanovskii told Lenin in 1920 that Lomonosov had been closely associated with the reactionary Rukhlov and that he was said to have supported reactionary policies.[59] No evidence has been found of any assassination plot against Lomonosov in the last pre-war years, but the possibility of murder by fellow socialists was surely a concern for him given the murders of a friend who was head of the Poltava workshops in 1905 and his own subordinate Borshchkov, head of the Lugansk workshops, in 1908.[60] Meanwhile, as we have seen, the extreme right-wing Union of the Russian People cursed him as a Kadet

---

[57]   Newspaper cutting, November 1914, at LRA, MS 717.2.244.2; 'Vospominaniia', vol. 5, pp. 673–5, 682, 690, 835–6, 1040, 1145–7, 1150–1155, 1165–6; Lomonosova–Miss Ripley, 17 September 1919: LRA, MS 717.2.257.1. Lomonosov had first asked Sonia about a divorce in March 1914: 'Vospominaniia', vol. 5, p. 573. The hospitals are described in the memoirs of an engineer whose mother, like Raisa, was a volunteer nurse there in 1914–1917: G.P. Tschebotarioff, *Russia, My Native Land* (New York, 1964), pp. 52–62. An overview of the Tsarina's nursing activities is in C. Erickson, *Alexandra, the Last Tsarina* (London, 2003), pp. 229–35.

[58]   'Vospominaniia', vol. 5, p. 915–16; vol. 6, pp. 7–10, 47–51. The council minutes neither confirm nor refute Bakhmetev's impression. For an appreciation of Bakmetev, who would play a major role in Lomonosov's life during 1917–1918, see 'Boris A. Bakhmeteff, 1880–1951', *Russian Review*, 10/4 (1951): 311–12.

[59]   Typed copy of notes by Lenin, May 1920: RGASPI, f. 2, op. 1, d. 14079, l. 7. The same source notes a remark by V.P. Miliutin that Lomonosov was rumoured to have been a member of the extreme right-wing Black Hundreds in 1908–1909: l. 5

[60]   Geifman estimates that there were about 17,000 attacks at that time: *Thou Shalt Kill: Revolutionary Terrorism in Russia, 1894–1917* (Princeton, 1993), p. 21; however,

liberal intent on destabilizing tsarism. It accused him of supporting a vendetta against its members in Aktiubinsk, and of favouring Jews and Poles over Russians. That the police suspected Russian nationalists for the attempted murder of the traction section chief in Aktiubinsk, Baron fon Tizengauzen, doubtless reminded Lomonosov of how exposed his own position had become.[61]

As a Soviet official in the early 1920s Lomonosov was very fond of saying that he had been a 'locomotive general and a footsoldier of the revolution'.[62] But the picture provided by his memoirs of his political beliefs during the last decade of tsarist Russia years is actually unclear. Overall it seems probable that he still hoped for systemic change, but only in a vague, passive way without expecting progress in his lifetime. He states that he and Raisa remained Marxist in outlook, but that they moved to the point, between the Mensheviks and Kadet liberals, that would be occupied in 1917 by G.V. Plekhanov's newspaper *Edinstvo* (*Unity*); in particular, they supported the war from 1914.[63] However, his memoirs also indicate his readiness to support a constitutional monarchy at the time of Nicholas II's abdication on 2/15 March 1917.[64] One might conclude, then, that for all his many caustic remarks about the Kadets, his position actually became quite close to theirs.

Lomonosov did have contacts with Leftist revolutionaries during 1908–1916, but none with any immediate political significance. For instance, occasional journeys to or through Kiev during 1908–1909 were used for social visits to former comrades; however, the Lomonosovs' closest radical friend there, Vera Solomon, died, Tikhvinskii moved to St Petersburg, and contact with the Ianovskiis ceased after an argument in 1909.[65] Later Lomonosov frequently sat with Tikhvinskii and Krasin while commuting by train from Tsarskoe Selo to the capital; apparently both criticized his high rank and Raisa's connection with the Tsarina.[66] Similarly he met the Bolshevik engineer R.E. Klasson through his secondment to the 1886 Company, yet their discussions about politics seem to have remained superficial.[67] Crucially, it seems unlikely that Lomonosov was the high-placed railway official whom the secret police suspected of informing SR terrorists about the movements of the royal family. Given the exhaustive, generally frank character of his memoirs and the Soviet-era questions about the sincerity of his radicalism, one would expect

---

other sources estimate that there were only eight terrorist acts by a key proponent of terror, the SR Party, during 1908–1911: see Perrie, 'Political and Economic Terror', p. 67.

[61]   See Chapter 5, this volume.

[62]   See, for example, the foreword (probably written by his Bolshevik friend A.G. Pravdin) to the first Russian edition of his memoirs about the February Revolution.

[63]   'Vospominaniia', vol. 5, pp. 648, 1660.

[64]   'Vospominaniia', vol. 5, p. 1578.

[65]   For example: 'Vospominaniia', vol. 4, pp. 11, 28, 432–7. Solomon may have committed suicide: Genis, *Nevernye slugi rezhima*, p. 75.

[66]   'Vospominaniia', vol. 5, pp. 834–5.

[67]   'Vospominaniia, vol. 5, pp. 827–8.

him to describe any such actions, yet none are mentioned.[68] He reports simply that at the Nicholas Railway he delegated his technical responsibilities for VIP travel to his first deputy. In fact, his interest was more voyeuristic than political: he enjoyed seeing the unofficial handbook maintained by saloon car attendants about the habits and foibles of their regular VIP travellers, including himself.[69]

That said, he would benefit after the Bolshevik revolution from some positive memories of his contacts with Bolshevik revolutionaries during 1905–1917. In about 1921, for example, a secret police report noted that he had been known to party leaders before 1917 and had helped to hide Bolsheviks from the authorities, find them employment and so forth.[70] In this regard Lomonosov's association with the Bolshevik A.G. Pravdin (1879–1938) is exceptionally important. A carpenter from Vladimir province and husband of one of Raisa's radical friends from Odessa, Pravdin would become a deputy head of the People's Commissariat of Ways of Communication in the 1920s. Until about 1915 he was not known to the tsarist police as a revolutionary activist and hence could work on the railways. Over the years Lomonosov found jobs for him in Lugansk, Ust-Katav, Tashkent and ultimately the capital – a move that presumably facilitated Pravdin's activities with the Bolshevik Party's Central Committee. These contacts ended abruptly in 1915 when Lomonosov rejected a plea for help. The next year Pravdin was arrested. Apparently remorseful, Lomonosov helped to get Pravdin released by asking a police acquaintance to intervene, and then helped Pravdin to get a job in Ust-Katav. Pravdin would remember this help, not least by vouching for Lomonosov when Lenin sought his opinion about the professor in May 1920, and by maintaining his friendship when the political pressure on Lomonosov became considerable in the mid-1920s.[71]

Radical or not, Lomonosov's political principles did not stop him from trying to inherit the Nikolo-niz estate upon his Uncle Vania's death in 1916 – a development that also implies that the engineer did not expect a socialist revolution in the near future. The questions in his mind were not political or social, but financial. Much of the estate had been sold, and the remnant had been mortgaged several times. However, he calculated that if he could revive a surviving peat enterprise and brick factory, the value of Nikolo-niz and Tatarka (which had been left to his son Vova) would jump from about 230,000 rubles to at least 1.5 million rubles.

---

[68]   A.V. Gerasimov, *Na lezvii s terroristami* (Paris, 1985), pp. 126–30.

[69]   'Vospominaniia', vol. 4, pp. 944, 951–2, 978–81.

[70]   OKTChK (Moscow–Kazan' Railway)–Blagonravov, [circa 1921]: RGASPI, f. 76, op. 2, d. 76, l. 27.

[71]   'Vospominaniia', vol. 4, pp. 109–10, 337, 376, 444, 1095; vol. 5, pp. 808, 1375–7; Typed copy of notes by Lenin, May 1920: RGASPI, f. 2, op. 1, d. 14079, ll. 8–9. The political pressures of the 1920s are discussed in Chapters 9 and 10, this volume. Pravdin was shot 'for anti-Soviet activity' on 3 September 1938: Memorial, 'Zhertvy politicheskogo terrora v SSSR': http//lists.memo.ru/index16.htm, citing *Moskva, rasstrel'nye spiski – Kommunarka* (last accessed 30 June 2008).

The problem was that the debt and inheritance costs might exceed 300,000 rubles and that Tatarka could not be sold before Vova came of age. Strong pressure to seize this opportunity came from N.A. Dobrovol'skii, who was now his closest collaborator in the locomotive experiments and had just married his daughter Mar'ia. But eventually Lomonosov renounced the inheritance because of the costs and practical difficulties, including Raisa's objection that he lacked the necessary business experience.[72]

In February 1917 he supported the revolution overtly long before the outcome was clear. During Monday 27 February/12 March, which he spent at the Experiments Office in central Petrograd, he remained unaware that the city's garrison was starting to mutiny, that Duma deputies at the Tauride Palace were forming a temporary committee to govern, or that the Petrograd Soviet of Workers' and Sailors' deputies was forming a rival committee. He heard about barricades and sporadic street fighting, but dismissed them as provocations organized by the regime to justify a crackdown.[73] The next day, which he spent at home preparing lectures, he realized that the situation was serious from signs of unrest in the Tsarskoe Selo garrison. At about 9 p.m. he received a telegram from the Duma deputy A.A. Bublikov on behalf of the so-called Committee of the State Duma, which requested him to come urgently to the MPS. Now, belatedly, he understood that a revolution was happening. Faced with making an immediate decision, he suspected that troops would crush the rising within a few days, yet he reacted 'like a cavalry horse to a bugle'.[74] Careerism was surely not a factor. He had no information about the situation in Petrograd, let alone elsewhere, and no certainty about the outcome. Nor did he have any prospect of personal advantage from supporting the revolution: he was already a senior ministry official, and though dubious about being a 'coal dictator', he had a reasonable relationship with the new head of the MPS, Kriger-Voinovskii. Lomonosov most likely expected only a change of ruling elite, and possibly a constitutional monarchy, without fundamental ramifications for the state bureaucracy and his own position.

Over the next few days he made an unusual contribution to the fall of tsarism. He acted as principal assistant to Bublikov, who had seized the MPS on 28 February/13 March for the Duma Committee that would become the revolutionary Provisional government. Among Lomonosov's first tasks was to establish and maintain telegraph contact with railways throughout the empire and especially the Petrograd area; he thereby helped both to keep traffic moving and send news to the provinces. A related job was to monitor the royal train's progress towards

---

[72] 'Vospominaniia, vol. 5, pp. 1201–2, 1238–52, 1259–60, 1279–80, 1282–3, 1286, 1288, 1330–36, 1340, 1371.

[73] Widespread fear of police provocations is stressed in P. Miliukov, *The Russian Revolution, vol.1: The Revolution Divided: Spring 1917* (Gulf Breeze, 1978), p. 24.

[74] Lomonosov, *Vospominaniia o Martovskoi revoliutsii*, pp. 222–4. On Bublikov see 'Vospominaniia', vol. 5, pp. 443–4, and P.V. Volobuev et al. (eds), *Politicheskie deiateli Rossii 1917: biograficheskii slovar'* (Moscow, 1993), pp. 46–7.

Petrograd and then Pskov during 1/14 March, and to help the Duma Committee to communicate with the Tsar by railway telegraph; conceivably he was instrumental in ensuring that the Tsar could not reach the capital. In due course Lomonosov organized the special train that took a Duma delegation to meet with the Tsar at Pskov. Meanwhile he helped thwart the attempt to move front-line troops into Petrograd to crush the rising. On 3/16 March he took the Tsar's abdication decree to the MPS from the Warsaw terminus, whither it had been brought by the Duma delegation, and that night he organized the printing of the abdication manifestos for the government at the MPS printing press.[75]

His recollections highlight the chaos of those days. For instance, relief, joy and confusion were his first impressions of the MPS late on 28 February/13 March. Nobody seemed to know what to do, and his roles developed more by chance than design.[76] Very troubling, in his view, was the apparent inability of the Duma Committee to make urgent decisions.[77] Thus, charged with organizing a special train to take the Duma chairman, M.V. Rodzianko, to meet with the Tsar on 1 March, he was frustrated by protracted uncertainty about whether the trip would be made.[78] Even more shocking for Lomonosov was the confusion that surrounded the printing of the abdication decrees of Nicholas and his brother Grand Duke Michael. The Duma Committee authorized the MPS to print these two statements together during the evening of 3/16 March. Lomonosov therefore took Nicholas's text to the MPS press, but could not start the printing for want of Michael's

---

[75] Lomonosov's published memoir is the main source for his role in the February Revolution; his unpublished memoirs have some additional information. Bublikov's memoirs barely mention Lomonosov, probably because the two men quarrelled in 1918: A.A. Bublikov, *Russkaia revoliutsiia (ee nachalo, arest tsaria, perspektivy): Vpechatleniia i mysli ochevidtsa i uchastnika* (New York, 1918), especially p. 29. Indeed Bublikov claims credit for keeping the MPS printers at readiness to print the abdication documents: loc. cit. On the diversion of the Tsar's train see Hasegawa, *February Revolution*, pp. 446–52. The handover of the abdication document by the Duma delegate V.V. Shulgin to one of Lomonosov's colleagues (engineer Lebedev) is described by Shulgin in his *Days of the Russian Revolution: Memoirs from the Right* (Gulf Breeze, 1990), pp. 196–7. A.I. Guchkov's less precise recollections are at P.E. Shcheglovitov (ed.), *Padenie tsarskogo rezhima: stenograficheskie otchety*, vol. 6 (Moscow and Leningrad, 1926), p. 266.

[76] Lomonosov, *Vospominaniia o Martovskoi revoliutsii*, pp. 224–34.

[77] The chaotic nature of the initial admistration of the Provisional government's affairs is recalled by its chief secretary, V.D. Nabokov, in his memoir 'Vremennoe Pravitel'stvo', in I.V. Gessen (ed.), *Arkhiv russkoi revoliutsii*, vol. 1 (reprinted edition Moscow, 1991), pp. 24–6.

[78] Lomonosov, *Vospominaniia o Martovskoi revoliutsii*, pp. 233–9. The socialist émigré historian S.P. Mel'gunov cites Lomonosov's account at length to help demonstrate that, contrary to Rodzianko's account, the committee chairman had no desire to meet with the Tsar: see his *Martovskie dni 1917 goda* (reprinted edition Moscow, 2006), pp. 66–79, and also Shulgin, *Days*, pp. 148, 164–5; Miliukov, *Russian Revolution*, p. 35. Mel'gunov also alleges that Lomonosov's account has some serious factual inaccuracies (p. 73), but does not reference his claims.

document. But then the committee demanded to see Nicholas's original document. So, accompanied by a colleague, Lomonosov took it to the Tauride Palace and gave it to Prince G.N. L'vov, leader of the new government. He then watched as ministers had a long argument about how to phrase the headings for the decrees. Not until about 3 a.m. did he reach the MPS press with the two texts.[79]

A further theme is the apparent severity of the counter-revolutionary threat posed by General N.I. Ivanov. Acting on orders issued by the Tsar on 27 February/12 March, the general planned to crush the uprising by taking several trainloads of front-line troops to Petrograd through Vyritsa and Tsarskoe Selo. Travelling ahead of his main force Ivanov reached Tsarskoe Selo on 1 March, but after a brief meeting with the Tsarina he returned that night to Vyritsa. Lomonosov spent much time on 2 March liaising with the Moscow–Vindava–Rybinsk Railway under M.E. Pravosudovich to block any attempt by Ivanov's force to resume its northward journey. Local railwaymen removed track components between Semrino and Tsarskoe Selo, and also blocked the Semrino–Gatchina section. Lomonosov thought that he himself prompted Ivanov's eventual retreat by informing him about the abdication. In fact, by that time the field headquarters had recalled Ivanov, and the general learnt about the abdication from the station commandant at Dno during the night of 3–4 March.[80]

---

[79] Lomonosov, *Vospominaniia o Martovskoi revoliutsii*, pp. 262–5; Hasegawa, *February Revolution*, pp. 557–63. Lomonosov spoke publicly about these last events on 24 November 1917 (n.s.): What has happened and is happening in Russia, US National Archives Record Group (NARG) 261, entry 24, box K18, file 'Originaly rechei'. Nicholas II, acting also on his son's behalf, abdicated in favour of his brother Michael, who then too renounced the throne. Miliukov asserts that Rodzianko arranged for the two texts to be published simultaneously: *Political Memoirs, 1905–1917* (Ann Arbor, 1967), p. 413. Lomonosov gives the two texts at *Vospominaniia o Martovskoi revoliutsii*, pp. 254–5, 264, while the original documents are now at GARF, f. 601, op. 1, d. 2100a, ll. 5, 7. I am very grateful to S.V. Mironenko and E. Chirkova for tracing a copy of the MPS-printed edition in GARF, f. 1834, op. 2, d. 11. For English translations see, for example, R.P. Browder and A.F. Kerensky (comps and eds), *The Russian Provisional Government 1917: Documents*, vol. 1 (Stanford, 1961), pp. 104, 116. Petrograd-printed MPS publications usually cited a press owned by I.N. Kushnerev, alongside the ministry at Fontanka 117.

[80] Lomonosov, *Vospominaniia o Martovskoi revoliutsii*, pp. 240–252; I. Ge, 'Ekspeditsiia gen. Ivanova na Petrograd', *Krasnyi arkhiv*, 4 (1926): 225–32 (incorporates a long, detailed statement by Ivanov); 'Dokumenty k "Vospominaniiam" gen. A. Lukomskogo', in Gessen, *Arkhiv russkoi revoliutsii*, vol. 3, p. 254; 'Dopros Gen. N.I. Ivanova, 28 iiunia 1917 goda', in P.E. Shchegolev (ed.), *Padenie tsarskogo rezhima: stenograficheskie otchety doprosov i pokazanii, dannykh v 1917 g. v Chrezvychainoi Sledstvennoi Komissii Vremennogo Pravitel'stva*, vol. 5 (Moscow and Leningrad, 1926), pp. 313–26; G. Katkov, *Russia 1917: The February Revolution* (London, 1969), pp. 400–401, 405–8, 411–15, 433–4; Hasegawa, *February Revolution*, pp. 459–86. Again, Mel'gunov claims that Lomonosov's account has some factual inaccuracies: *Martovskie dni*, pp. 128–34.

A third theme is the importance of the MPS and the railways in the revolution. One historian has concluded with some justification that Bublikov's short rule was 'all-important for the success of the revolution' and that 'if ever there was a spontaneous revolutionary departmental authority, this was the railway administration established by Bublikov'.[81] Lomonosov elucidates some of the key moments. For instance, Bublikov circulated two telegrams to all railway telegraph points on 28 February/13 March. The first, which Bublikov drafted for Rodzianko, effectively informed the country that a revolution had occurred, for it announced that a Duma Committee had taken power – an exaggeration of the actual situation – and exhorted all railway workers to save Russia by keeping the trains moving. The second telegram prohibited troop-train movements within 265 km of Petrograd, the aim being to stop any influx of counter-revolutionary forces.[82] On 4/17 March Lomonosov drafted a telegram on behalf of Bublikov and the acting commissar of Ways of Communication, A.A. Dobrovol'skii, to announce the abdications, the establishment of the Provisional government and the intention to convene a Constituent Assembly; in this way the MPS answered the many hundreds of telegraph requests for information that it had received.[83] Clearly, Bublikov appreciated that control of the railways was essential for holding national political power. Undoubtedly he, Lomonosov and their colleagues deserve some of the credit for the uninterrupted operation of most train services throughout the week. But arguably more important for the revolution as such was their ability to spread information quickly. The railway telegraph network became the means by which much of the population learnt of key events like the abdications and the creation of the Provisional government.[84]

From Lomonosov we learn too that this unique chapter in the history of the MPS was closed symbolically on 6/19 March. The noted photographer V.K. Bulla took two formal photographs of Bublikov, Dobrovol'skii, Lomonosov and other colleagues, one in Bublikov's improvised office and the other in Lomonosov's temporary office. Then an exchange of telegrams was conducted. First, Lomonosov addressed a note of thanks to commissars Bublikov and Dobrovol'skii, copied to the new minister – the Kadet N.V. Nekrasov – and deputy ministers, the head of the Directorate of Railways and all line superintendents for all staff. It declared that 'thanks to your courage, energy, knowledge and devotion to the cause of

---

[81]   Katkov, *Russia 1917*, pp. 502–3. Note, however, that Katkov controversially denies the spontaneity of the February revolution, arguing that the German authorities and Russian Freemasonry consciously did much to topple the tsarist regime.

[82]   Lomonosov, *Vospominaniia o Martovskoi revoliutsii*, p. 230. On Bublikov's action see, for example, Katkov, *Russia 1917*, pp. 396–7, 502–3; W.B. Walsh, 'The Bublikov–Rodzyanko Telegram', *Russian Review*, 30/1 (1971): 60–70; and Hasegawa, *February Revolution*, pp. 367–9. Kriger-Voinovskii's perspective is in his *Zapiski inzhenera*, pp. 91–7.

[83]   Lomonosov, *Vospominaniia o Martovskoi revoliutsii*, pp. 266–7.

[84]   See also R.W. Pethybridge, 'The Significance of Communications in 1917', *Soviet Studies*, 19/1 (1967–1968): 109–14.

freedom you succeeded not only in keeping traffic moving normally during the days of revolution but also prevented a bloody conflict in the suburbs of Petrograd'. Replying, Bublikov praised the railway family for its self-discipline, and thanked those with whom he had worked directly, starting with Lomonosov as his chief assistant.[85]

Bublikov's message conveyed a sense that the revolution was over; in reality, it was just beginning. Over the next eight months the Provisional government struggled to consolidate its power, before succumbing to the Bolsheviks in October. Among the more important obstacles to the Provisional government's survival was the disintegration of labour discipline and productivity on the railways, which caused a huge backlog of traffic and maintenance, and especially food and fuel shortages. These events were overshadowed, of course, by the drama of the government's political struggles. But the fact that railway traffic collapsed after February 1917 certainly helps to explain the Provisional government's demise in the autumn.[86]

Two points are striking in Lomonosov's description of the first revolutionary days and weeks. One is his alarm at how the radicalization of the society was disrupting the railways, which may suggest that he did not want a Marxist social revolution at this juncture. Very dangerous, in his opinion, was interference by some of the new political commissars. For example, the Vindava line's 'sensible and liberal' superintendent, Pravosudovich, found it impossible to work with his commissar, a 'staunch Bolshevik from the intelligentsia'. Lomonosov wanted the MPS to resolve these problems urgently, yet the revolutionary government seemed likely to compound them. In particular, on 5 March Nekrasov and the Minister of Justice, A.F. Kerenskii, notified line superintendents about a plan to 'democratize' the ministry's Statute and create an elected representative body there. For Lomonosov such parliamentary democracy was inappropriate for an environment that, in his opinion, required a hierarchical culture of discipline and responsibility for its smooth and safe operation.[87]

---

[85]   Lomonosov, *Vospominaniia o Martovskoi revoliutsii*, pp. 276–8; 'Vospominaniia', vol. 5, pp. 1609–12. I have not yet found copies of these messages in the Russian archives. Concerning Nekrasov see Nabokov, 'Vremennoe Pravitel'stvo', pp. 49–50; Volobuev, *Politicheskie deiateli*, pp. 231–2; A.S. Senin, *Ministerstvo putei soobshcheniia v 1917 godu*, 2nd edn (Moscow, 2008), pp. 29–31.

[86]   There is no book-length study of the railways during February to October 1917, but their collapse can be traced in, for example, A.L. Sidorov et al. (eds), *Ekonomicheskoe polozhenie Rossii nakanune Velikoi Oktiabr'skoi Sotsialisticheskoi Revoliutsii: Dokumenty i materialy, mart–oktiabr' 1917 g., chast' vtoraia* (Moscow and Leningrad, 1957), pp. 211–73. See also R.T. Argenbright, 'The Russian Railroad System and the Founding of the Communist State, 1917–1922' (unpublished PhD dissertation, University of California at Berkeley, 1990), pp. 49–65; A.S. Senin, 'Zheleznye dorogi v marte–oktiabre 1917 g.: ot krizisa k khaosu', *Voprosy istorii*, 3 (2004): 32–56; and Senin, *Ministerstvo*.

[87]   Lomonosov, *Vospominaniia o Martovskoi revoliutsii*, pp. 270–71; 'Vospominaniia', vol. 5, pp. 1603, 1641–2.

The second point is that he did not identify with any of the main political groups. He supported the Provisional government – dominated by Kadet liberals at this time – only as an interim administration pending the election of a Constituent Assembly. Indeed, he viewed this regime as merely the least objectionable option, except for its intention to continue the war, which he endorsed wholeheartedly. He doubted that this government could achieve social progress, for he saw the Kadets as socially conservative. Moreover, as a person with a strong sense of Russian identity and power he was alarmed by the new government's apparent desire to appease all dissenting groups in the society, including separatist movements and striking workers: he thought that this tactic would undermine both state and economy.[88] As for the Petrograd Soviet, he agreed only with its call for the arrest of the Tsar and the Grand Duke Nikolai Nikolaevich. He rejected an invitation from Pravdin to meet Lenin, dismissing the latter's call for civil war as insanity. He endorsed Plekhanov's *Edinstvo* newspaper, but regarded this group as politically insignificant.[89]

Some of his contempt for the Provisional government reflected his difficulties in defining his place on the post-tsarist railways; in a sense he became a political casualty of the revolution. Had Bublikov been the new minister, Lomonosov might have become a deputy minister or head of the Directorate of Railways. But Nekrasov was another matter. Lomonosov felt excluded as a 'politically alien element' and suspected that his recent association with Bublikov rendered him undesirable for Nekrasov, who, moreover, was rumoured to have grudges against both the IPS and the Engineering Council.[90] Certainly the professor's position deteriorated markedly during March 1917. He was absent from Nekrasov's flurry of new railway appointments, and his income of some 22,000 rubles fell by 7,200 rubles through his removal from the 1886 Company in a board reshuffle. If, as he began to fear, Nekrasov removed him from the Engineering Council (salary: 6,000 rubles), his combined income from the IPS (3,000 rubles) and Experiments Office (6,000 rubles) would be less than his equivalent salary on the Nicholas Railway in 1911.[91]

That said, Lomonosov was again the architect of some of his misfortune. He claims that he resigned as Main Director of Experiments after a decision by Shchukin – sensible in the chaotic circumstances – to cancel the 1917 plan for experiments on anthracite coal, but that his staff forced the minister to reinstate him as Main Director for life. No such change was announced formally, but such a clash would be in character, and if it did occur, it surely annoyed Nekrasov.[92] Furthermore, Lomonosov declined some managerial assignments. For example,

[88]   Lomonosov, *Vospominaniia o Martovskoi revoliutsii*, p. 275; 'Vospominaniia', vol. 5, pp. 1626–34, 1657–9.

[89]   'Vospominaniia', vol. 5, pp. 1620–29, 1654–6, 1660.

[90]   'Vospominaniia', vol. 5, pp. 1637, 1651.

[91]   'Vospominaniia', vol. 5, pp. 1596, 1639.

[92]   'Vospominaniia', vol. 5, pp. 1643–7.

he refused to lead an investigation of the Moscow network, reportedly because he thought that Nekrasov wanted to censure its acting director, Voskresenskii. Another offer was to head the Alexander Railway – a second-tier line based on the Moscow–Smolensk–Brest route, but one that offered a good opportunity to learn the ropes of line management; Lomonosov wanted a first-rank line, and unsurprisingly found no support in the MPS leadership. Meanwhile he dismissed the logical idea of gaining management experience as head of a regional traffic committee, simply because he saw that position as even less prestigious than head of a second-rank railway.[93]

His way forward proved to be as appealing as it was unexpected. On about 8/21 April it was suggested that he might go to North America to manage existing and new contracts there for railway equipment. A week earlier the Provisional government had approved an MPS request for 2,000 new heavy freight locomotives and 40,000 freight wagons from the United States in addition to the large wartime orders already placed there; and with an expected cost of some $220 million these contracts would become revolutionary Russia's largest planned commitment to foreign expenditure in 1917. If appointed, then, Lomonosov would be the senior MPS official in a state delegation to North America that was to be led by his IPS colleague Professor Bakhmetev as Russia's first ambassador to the United States in the post-tsarist era.[94] Succeeding Count Shulenburg as the chief MPS representative in North America, Lomonosov would have ultimate power to decide all technical questions, while Bakhmetev would oversee financial matters. Needless to say, Lomonosov was delighted: this job would solve his career dilemma, extract him from Nekrasov's MPS and transport him to the bright lights of the United States with the prestige of diplomatic status.[95]

The confirmation of Lomonosov's appointment was due mainly to Bublikov. When agreeing to help, the latter also proposed the job of office manager for the private Council of Congresses of Industry and Trade, with a salary of 15,000 rubles, and noted that Lomonosov could double that salary by keeping his IPS professorship and taking directorships in several small railway companies. The professor agreed to consider this offer, but claims that as a 'state-sector person' (*chelovek kazennyi*) he felt uncomfortable about working in the private sector. This comment, if true, contrasts with his earlier willingness to work for private railways – a difference that is perhaps explained by his new option of working abroad. After much thought he informed Bublikov about his preference for the

---

[93] 'Vospominaniia', vol. 5, pp. 1649–53, 1661–2.

[94] On Bakhmetev see the biogaphical notes in Bakhmetev Collection, box 46, and his unpublished oral history memoir in boxes 37–38. Oleg Budnitskii is preparing a biography.

[95] MPS Report to Provisional Government, 31 March 1917: NARG 261, entry 24, box K116, file 7; Zhurnal zasedaniia Vremennogo pravitel'stva, 1 April 1917, in B.F. Dodonov (ed.), *Zhurnaly zasedanii Vremennogo pravitel'stva, tom 1: Mart–aprel' 1917 goda* (Moscow, 2001), pp. 214–15; 'Vospominaniia', vol. 5, pp. 1662–3; Heywood, *Modernising Lenin's Russia*, pp. 36–7.

American assignment on about 21 April/4 May, and the deal was done within a day. Nekrasov formalized the appointment with an Order on 4/17 May.[96]

*   *   *

Lomonosov's career was rejuvenated by the outbreak of the war in 1914. Immediately the MPS displayed fresh and apparently genuine interest in his research as a means to improve efficiency, and within a few months it also brought him back into top-level operating management. For two months in mid-1915 he was the face of the ministry in the new government committee charged with coordinating the national war effort. He was even selected to serve the state in the strategic electricity-generating industry. These major changes in his work lend credence to the view that the MPS did treat the war as a serious emergency from the outset, arguably in contrast to many other parts of the Russian state bureaucracy. But the practical impact of his work was undoubtedly far less than he would have hoped. Familiar problems of bureaucratic strife helped to undermine his activities and, more generally, the railway war effort. He was basically returned to the professional wilderness in early 1916 after falling out with Trepov, the new minister, and he was not recalled to a senior railway managerial role until January 1917. Detailed analysis of the traffic on individual lines would be needed to clarify the benefits, if any, of his prescriptions for improving the utilization of locomotives, rolling stock and line capacity. The same need applies for determining the impact of his considerable work on redesigning and rebuilding major junctions. But given the length of time usually required for such improvements, even in these emergency conditions, one suspects that any such work after about mid-1915 was effectively fruitless.

Political principle was almost certainly Lomonosov's motive for supporting the cause of revolution in February 1917. In personal and professional terms he had little to gain and much to lose from rebelling against the tsarist regime. His involvement in that cataclysm reaffirms the centrality of the MPS and the railways for the overall outcome, including the Tsar's abdication. But paradoxically Lomonosov's career did not appear to benefit from the revolution, partly because of his own rejection of appropriate job offers and partly because of personal distrust on the part of the new authorities. Certainly his posting to the United States was a prestigious assignment, but it was essentially a product of his alienation, disillusionment and despair.

The Provisional government's decision to send him abroad would prove to be a key moment of his life. With the exception of just nine months between mid-September 1919 and mid-June 1920 he would spend his remaining 35 years based outside Russia. More immediately, his duties would mean that he was abroad when Lenin's Bolshevik Party seized power in Petrograd in October 1917 and

---

[96] *Vestnik putei soobshcheniia*, 13 May 1917, p. 204; 'Vospominaniia', vol. 5, pp. 1662–8, 1671–3.

founded the Soviet proletarian state. How Lomonosov fared in the United States during 1917–1919 and how he reacted to the Bolshevik revolution are the main subjects of the next chapter.

1. Vladimir Grigor'evich Lomonosov, circa 1903

2. Mariia Fedorovna Lomonosova, née Pegelau

3. The Lomonosov home at Sychevka, circa 1903

4. Iurii Lomonosov and his sister Ol'ga, circa 1892

5. Sonia (S.A. Lomonosova, née Antonovich), circa 1903

6. The Institute of Ways of Communication, St Petersburg, early 1900s (TsGAKFFD SPb, E-14371)

7. Locomotive experiment, circa 1899; Lomonosov is standing above the cylinder, third from the left

8. Professor N.L. Shchukin (J.N. Westwood Collection)

9. Engineer N.P. Petrov (Central Museum of Railway Transport, St Petersburg)

10. The Warsaw Polytechnic Institute, circa 1902

11. The Kiev Polytechnic Institute, circa 1902

12. Oksana (K.A. Zabugina), circa 1902

13. Lomonosov, Sonia and their children Marusia and Vova, Kiev, circa 1904

14. Lomonosov, Marusia and Vova with Raisa Rozen (far right) and a German friend, Kiev, May 1907

15. Lomonosov supervising the River Ros' land survey, June 1907

16. The Locomotive Room at Kiev, autumn 1907

17. A.S. Gutovskii, Traction Superintendent of the Catherine Railway, 1908

18. 'Modified Chinese' 2-8-0 No. 5529 (future Class Shch) on test, 1908

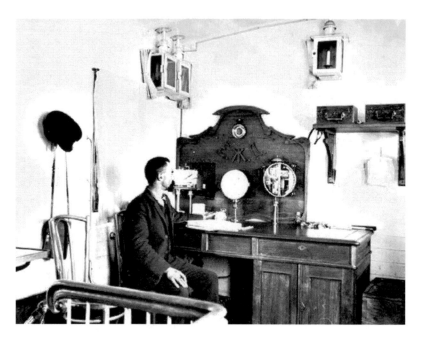

19. Interior of dynamometer car

20. Exterior of dynamometer car, with ladder for access to the locomotive's footplate via the tender

21. A.I. Lipets, 1909

22. S.V. Rukhlov, Minister of Ways of Communication 1909–1915 (TsGAKFFD SPb, E-1623)

23. A diesel engine at a water pumping station, Tashkent Railway, circa 1909

24. The Ministry of Ways of Communication, St Petersburg, early 1900s (TsGAKFFD SPb, E-14775)

25. Lomonosov with his research team, Nicholas Railway, circa 1912

26. A lecture to mark the fifteenth anniversary of Lomonosov's road research, 1913

27. Delegates to the XXIX Congress of Traction Engineers, 1912, including (seated) M.E. Pravosudovich (fourth from left), L.M. Levi (centre with cane), Lomonosov (fourth from right) and (first row standing) A.O. Chechott (behind Levi) and M.V. Gololobov (fourth from right, with trilby)

28. Iurii Vladimirovich Lomonosov, wartime troubleshooter, August 1915

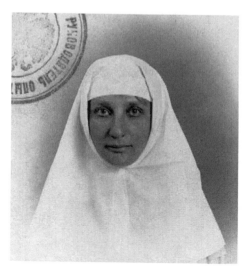

29. Raisa Lomonosova, wartime nurse, circa 1916

30. In Bublikov's temporary office at the MPS, 6 March 1917 (left to right): (seated at front right) Commissar Dobrovol'skii, Perlov; (seated behind desk) Uspenskii, Lomonosov, Bublikov; (standing) Grinenko, Tukhin, Korelin, Sidel'nikov, executor, inspector Nekrasov, Lebedev, Tiumenev, a courier, Arapov

31. Ambassador B.A.
Bakhmetev, 1917

32. Head of the MPS Mission in North America, 1917

33. Lomonosov and Lipets (fourth left) at Alco (Schenectady, NY) with a Russian decapod locomotive, 1917

34. A meeting of the Society for the Defense of Russian Democracy, New York City, 29 December 1917; Lomonosov and Bublikov sit in the front row on the platform

35. Crossing into Soviet territory, September 1919

36. L.B. Kamenev

37. A.I. Rykov

38. The Novorossiisk crash, May 1920

39. The People's Commissariat of Ways of Communication, Moscow, early 1920s

40. L.D. Trotskii

41. I.N. Borisov, head of the MPS Directorate of Railways 1914–1916 and NKPS Collegium member 1920–1928 (*Zheleznodorozhnoe delo*, No. 6 (June 1928): 1)

42. L.B. Krasin (seated right) with A.N. Shelest (seated centre), Mme
N.P. Shelest et al., summer 1920

43. A Class E^G locomotive bound
for Russia, Hamburg, 1922

44. F.E. Dzerzhinskii

45. The new elite: A.G. Pravdin with family and friends, 1922

46. The Avanesov commission, 30 November 1922 (seated left to right): Iaroslavskii, Avanesov, Medvedev

47. Lomonosov's portable locomotive test plant, Esslingen, 1924

48. Class E^G steam locomotive No. 5570 on test, Esslingen, 1924

49. Diesel-electric locomotive Iu^E No. 001, Esslingen, 5 June 1924

50. The ceremonial diesel/steam comparative tests at Esslingen, 6 November 1924; Meineke is standing third from the left, Dobrovol'skii is at the left front window, and the white-coated figure at the front door may be S.S. Terpugov.

51. The cooling tender for Iu^E No. 001

52. Iu^E No. 001 attracting crowds in Soviet Ukraine, May 1925

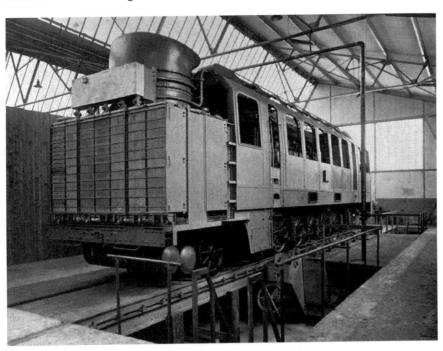

53. Iu^M No. 005 on the test plant at Hohenzollern AG, Dusseldorf, 1925

54. The diesel researchers at Esslingen, June 1924

55. S.V. Makhov and N.A. Dobrovol'skii, April 1924

56. Masha (M.I. Shelkoplasova) with her sons Zhorzh and Anatolii, Dergachi, September 1925

57. Raisa with P.L. Kapitsa (second left) and friends, Cambridge, 1928

58. Diesel traction seminar at Caltech, 1930

59. Sir Alfred Ewing (standing) and Sir Charles Inglis, Cambridge, 1931

60. Spa life in Germany: Bad Mergentheim, 1935

61. Testing the ill-fated grass-drying machine, March 1937

62. Lomonosov, Raisa, George and the family cat, circa 1945

63. Lomonosov with N.R. Crump, Montreal, May 1952 (Canadian Pacific Archives B2530-1)

Chapter 7
America and the Bolshevik Revolution

Lomonosov's assignment to North America became a nightmare.[1] Initially he enjoyed the diplomatic high life in both the United States and Canada as head of the Mission of the Ministry of Ways of Communication. But he was ostracized in mid-1918 when he became possibly the highest-ranking Russian official abroad to demand foreign diplomatic recognition of the Soviet government that Lenin created after the Bolshevik revolution of 25–26 October/7–8 November 1917. When Lomonosov left New York for Soviet Russia in 1919 he was in key respects a shadow of his former self and his country was embroiled in civil war. Meanwhile US–Russian governmental relations collapsed.[2] In March 1917 both sides were optimistic: the US government quickly recognized the Provisional government, while Russia's new Minister of Foreign Affairs, the Kadet leader P.N. Miliukov, envisaged a strong partnership with the United States founded on shared political and economic values and interests. Within a month these bonds were reinforced through the entry of the United States into the war on the Allied side. However, the mutual warmth soon faded, and it vanished after Lenin's October revolution. By spring 1919 the United States was engaged in the anti-Bolshevik foreign military intervention in Russia; for its part, the Bolshevik regime called for diplomatic and economic relations with the capitalist United States, yet still championed worldwide proletarian revolution. The way that this great crisis became a turning point in Lomonosov's life, and also his own impact on it, form the subject of this chapter.

---

[1]   Vol. 6 of his 'Vospominaniia' covers the two-year period from his departure from Petrograd in 1917 to his departure from New York for Europe in 1919. For a critical yet fairly balanced commentary about his work by one of his subordinates, who would be Latvian Ambassador to the USSR during 1923–1929, see K.V. Ozols, *Memuary poslannika* (Paris, 1938), pp. 27–32.

[2]   For example, G.F. Kennan, *Soviet–American Relations, 1917–1920*, 2 vols (Princeton, 1956–1958); D.W. McFadden, *Alternative Paths: Soviets and Americans, 1917–1920* (Oxford, 1993); D.S. Foglesong, *America's Secret War Against Bolshevism: US Intervention in the Russian Civil War, 1917–1920* (Chapel Hill, 1995); N. Saul, *War and Revolution: The United States and Russia, 1914–1921* (Lawrence, 2001). Soviet books include L. Gvishiani, *Sovetskaia Rossiia i SShA (1917–1920)* (Moscow, 1970).

America

As revolutionary Russia's first ambassador to the United States Professor Bakhmetev had two priorities.[3] One was his government's aim to build closer relations with that country. An immediate motive of the Russians was to reduce their political, military and financial dependence on Britain and France. Also relevant was a sense of common values, which included the admiration of Kadet liberals for the US achievement of forging a powerful liberal democracy within a revolutionary tradition and the belief that the United States' economic strength would offer essential post-war opportunities.[4] The ambassador's other priority was the Russian state's vast operation to purchase war supplies. Day-to-day responsibility for this endeavour in North America (including delivery to Russia) lay with a large army-dominated organization called the Russian Supply Committee in America; Bakhmetev's role primarily concerned diplomatic and financial issues, especially US loans for new contracts.[5]

Although the Provisional government continued its predecessor's wartime policy of buying supplies in North America, it changed the focus to address

---

[3]   The Bakhmetev mission's formation and tasks were defined in a report by foreign minister P.N. Miliukov to the Provisional government, 25 April/8 May 1917: A.N. Iakovlev et al. (eds), *Rossiia i SShA: diplomaticheskie otnosheniia, 1900–1917* (Moscow, 1999), pp. 666–8. See also: Zhurnal zasedaniia Vremennogo pravitel'stva, 25 April 1917, in Dodonov, *Zhurnaly zasedanii*, pp. 348–9; Interview of B.A. Bakhmetev with American newspaper correspondents, 8/21 June 1917: ibid., pp. 678–80; and O.V. Budnitskii (ed.), '*Sovershenno lichno i doveritel'no!': B.A. Bakhmetev–V.A. Maklakov, perepiska, 1919–1951, tom 1: avgust 1919–sentiabr' 1921* (Moscow and Stanford, 2001), pp. 34–6. On Miliukov see, in particular, M. Stockdale, *Paul Miliukov and the Quest for a Liberal Russia, 1880–1918* (Ithaca, 1996).

[4]   For example: Miliukov–President Wilson, 26 March/8 April and 27 March/9 April 1917, in Iakovlev et al., *Rossiia i SShA*, pp. 656–7. Studies of the Provisional government's foreign policy include A.V. Ignat'ev: *Vneshniaia politika vremennogo pravitel'stva* (Moscow, 1974) and R.A. Wade, *The Russian Search for Peace, February–October 1917* (Stanford, 1969).

[5]   On the Russian Supply Committee see, for instance, Saul, *War and Revolution*, pp. 20–31. A report of the committee's work was prepared as Otchet o deiatel'nosti Russkogo zagotovitel'nogo komiteta i ego Likvidatsionnoi komissii, [1918]: NARG 261, entry 26, box L37. Other works on Russia's wartime foreign procurement include A.L. Sidorov, 'Otnosheniia Rossii s soiuznikami i inostrannye postavki vo vremia pervoi mirovoi voiny, 1914–1917 gg.', *Istoricheskie zapiski*, 15 (1945): 128–79; K. Neilson, *Strategy and Supply: The Anglo–Russian Alliance, 1914–1917* (London, 1984); D.C. Rielage, *Russian Supply Efforts in America During the First World War* (Jefferson, 2002); F.R. Zuckerman, 'The Russian Army and American Industry, 1915–17: Globalisation and the Transfer of Technology', in S.G. Wheatcroft (ed.), *Challenging Traditional Views of Russian History* (Basingstoke, 2002), pp. 55–65; and I.R. Saveliev and Yu.S. Pestushko, 'Dangerous Rapprochement: Russia and Japan in the First World War, 1914–1916', *Acta Slavica Iaponica*, 18 (2001): 19–41.

the needs of the home front. Hitherto military supplies had accounted for most of this expenditure; the main non-military item was railway equipment, taking about 10 per cent of the money committed to date. However, most of the military goods ordered in 1915–1916 had yet to be deployed due to delays in planning, production and delivery. For the moment, therefore, the Provisional government felt able to prioritize non-military products. Its shopping list included agricultural equipment, food and, not least, civilian footwear, but the foremost items in terms of their monetary value were the 2,000 2-10-0 locomotives – commonly described as decapods on account of their 10 powered wheels – and 40,000 wagons that Lomonosov was instructed to acquire urgently.[6]

This procurement plan was given a decisive boost by the entry of the United States into the war against the Central Powers on 24 March/6 April 1917. To date the Russian state's orders in North America had been funded mostly with British credit, not least because neutral countries could not give loans to belligerents. However, the British had become sceptical about Russia's appetite for imports, which seemed to ignore the acute Allied shortages of money, industrial capacity and merchant ships.[7] They became especially worried about railway contracts, both because the MPS flouted procedures for funding and placing them and because some contracts appeared to address post-war aspirations rather than urgent war needs. Eventually, in autumn 1916, the British refused to fund any more major railway orders.[8] The resultant hiatus ended thanks to the US declaration of war: the Provisional government requested $500 million in a US scheme to lend $3 billion to Allied nations, and also approved a new MPS imports programme on 31 March/13 April 1917. Readily the US government accepted that much more railway equipment was vital for Russia's ability to stay at war and hence for the overall Allied position. Already by 7/20 April $220 million were earmarked in the proposed loan for 2,000 decapod locomotives and 40,000 wagons.[9]

The non-military emphasis of the imports plan explains the predominance of civilian officials in Bakhmetv's delegation. Apart from the ambassador and Lomonosov, the leading officials included E.I. Omel'chenko (Ministry of Trade and Industry), N.A. Borodin (Ministry of Agriculture), V.I. Novitskii (Ministry of Finances) and I.I. Sukin (Ministry of Foreign Affairs); even the main army representative, Lieutenant-General E.E. fon Ropp, specialized in transport rather

---

[6]   See, for example, lists of materials for purchase in 1917 in NARG 261, entry 24, box K12, file 12; Heywood, *Modernising Lenin's Russia*, pp. 36–47. Discussions about this change of emphasis may have started before the February revolution. For overviews of Russo–American relations during the war to February 1917, including the procurement effort, see Saul, *War and Revolution*, pp. 8–85; J.K. Libbey, *Russian–American Economic Relations, 1763–1999* (Gulf Breeze, 1999), pp. 58–71.

[7]   Neilson, *Strategy and Supply*, especially pp. 171–248.

[8]   On the MPS orders to February 1917 see Heywood, *Modernising Lenin's Russia*, pp. 26–35 and 'Russia's Foreign Supply Policy', pp. 77–108.

[9]   Heywood, *Modernising Lenin's Russia*, pp. 36–7.

than weaponry.[10] The remainder of the party consisted of support staff and relatives. Lomonosov was therefore accompanied by various railway engineers, including protégés from his MPS Experiments Office such as A.A. Postnikov and A.I. Dolinzhev, and by Raisa with their nine-year-old son. He did not, however, take his two children from his first marriage: Mar'ia was in Kaluga with her new husband Dobrovol'skii (formerly of the Experiments Office), while Vsevolod was preparing to study mathematics at Moscow University. In a typically expansive gesture Lomonosov gave Vsevolod the balance of his bank account, reportedly 1,300 rubles; but apparently Vsevolod expected more and took offence, such that 'our relations were ruined for life'.[11]

Only after his arrival in America in June did Lomonosov fully appreciate that the MPS procurement plan had a fundamental flaw: it did not allow for the acute shortages of North American industrial capacity relative to Allied supply demands and of merchant ships to carry supplies to Western Europe and Russia. Count Shulenburg, hitherto the senior MPS official in North America, had faithfully carried out his instructions to obtain and evaluate bids for the contracts for the first 500 locomotives and 10,000 wagons. He proposed to divide this initial locomotive order between the two biggest US suppliers – the Baldwin Locomotive Works (250) and American Locomotive Company (Alco; 250) – and give the wagons to the American Car and Foundry Company (6,500) and Standard Steel Car Company (3,500). Lomonosov endorsed this proposal, resolved the remaining technical matters, and negotiated with the relevant US authorities for the money, factory capacity, materials and trans-Pacific shipping to Vladivostok. He had all four contracts signed by 12/25 July. However, the deliveries could not start for a whole year because the Americans were prioritizing the material, production and shipping needs of the putative American Expeditionary Force to France. Furthermore, for these reasons among others, the US government refused to let him order the next 1,500 decapod locomotives and 30,000 wagons.[12]

Lomonosov's response – characteristically combative and single-minded, but as yet completely unsuccessful – was to lobby hard for US permission to begin the contractual negotiations for the remaining locomotives and wagons. So doing, he was of course implementing his instructions to buy this equipment urgently. However, it seems probable that he also had technology transfer in mind in the general sense that this modern equipment could serve as a physical and intellectual basis for a thorough renewal of the railway network. He certainly supported the

---

[10]   Sidorov, *Ekonomicheskoe polozhenie Rossii, chast' II*, pp. 453–4, 578; Budnitskii, 'Sovershenno lichno i doveritel'no!', p. 36.

[11]   Lomonosov–Nekrasov, 28 April 1917, and memorandum The Mission of Ways of Communication, [June 1917]: NARG 261, entry 24, box K116, file 7; 'Vospominaniia', vol. 5, pp. 895–6; vol. 6, pp. 11, 14, 24.

[12]   Heywood, *Modernising Lenin's Russia*, p. 39.

extraordinary MPS desire to continue importing these 2-10-0 locomotives as far as 1922 at the rate of 1,000 per year, to judge by one of his reports to Bakhmetev.[13]

The failure of his lobbying at this juncture led to two disputes that affected the whole Russian procurement operation in North America and highlighted serious conflicts of interest between Russia and her allies. The first dispute concerned Lomonosov's legal status *vis-à-vis* the Russian Supply Committee and the ambassador. The MPS mission had originally operated independently of Russia's other procurement missions, with Shulenburg answering directly to the MPS, but it had been absorbed into the Supply Committee in 1916.[14] Now, citing talks in Petrograd prior to his departure, Lomonosov insisted that he answered only to the Minister of Ways of Communication and that Bakhmetev had no authority over MPS personnel in North America. However, that interpretation was not reflected in the published MPS Order about his appointment, a discrepancy that Lomonosov claims to have discovered only in November 1917.[15] To be fair, there were coherent professional reasons for his stance: with some justification he believed that the Supply Committee's inefficiency was disrupting the MPS mission, especially in relation to shipping matters. He failed to grasp, however, that his departmental logic was trumped by bigger logistical, financial and diplomatic considerations, including the critical lack of merchant ships that had arisen through Germany's resumption of unrestricted submarine warfare in January 1917. Hence Lomonosov continued to baulk at restrictions imposed by Bakhmetev and the Supply Committee – a situation that would erupt spectacularly in 1918.[16]

The other dispute had the potential to compromise the Russian desire for a strong bilateral partnership with the United States. It occurred during July–August 1917 but also had aftershocks.[17] From Lomonosov's perspective there were two main difficulties. One was the reluctance of the US government to let him order the final 1,500 locomotives; the other was Bakhmetev's refusal to press for American permission. Lomonosov insisted that better railway performance depended on this equipment, whereas Bakhmetev wanted patience for diplomatic

---

[13]   See Lomonosov–Bakhmetev, No. 65, 31 May/12 June 1917: NARG 261, entry 24, box K12, file 12.

[14]   Heywood, 'Russia's Foreign Supply Policy', pp. 93–7.

[15]   See *Vestnik putei soobshcheniia*, 19 (13 May 1917): 204; 'Vospominaniia', vol. 6, pp. 3–10, 15–16, 21–2, 420–27, 475–81; Budnitskii, *'Sovershenno lichno i doveritel'no!'*, pp. 35–6. The confusion concerning Lomonosov's position is reflected in the wording of the memorandum about the Bakhmetev mission that was sent by foreign minister P.N.Miliukov to the Russian embassies in Paris, London, Rome and Tokyo: Iakovlev et al., *Rossiia i SShA*, pp. 664–5.

[16]   For example: Lomonosov–Bakhmetev, 5 October 1917, and Lomonosov–Minister, 15 October 1917: both in NARG 261, entry 24, box K197, file 22 Confidential correspondence about Professor Lomonossoff; 'Vospominaniia', vol. 6, pp. 316–35.

[17]   See Heywood, *Modernising Lenin's Russia*, pp. 39–44.

reasons.[18] Crucially, Lomonosov failed to appreciate that the US administration was now acutely worried about not just Russia's deep political crisis, but also the huge scale of Allied supply demands – so much so that it had begun reassessing the whole matter of inter-Allied supply relations. In fact, as one of the largest foreign demands, Lomonosov's procurement plan forced this review: the Treasury Secretary, W.G. McAdoo, feared a national cash-flow crisis in mid-August 1917 if he authorized the Russian locomotive and wagon contracts, which would cost roughly $143 million.

During April–July 1917 an important difference of opinion developed between the US and Russian governments in relation to the Russian railways. The MPS defined a shortage of modern equipment as the decisive issue. But as early as mid-April the Americans began to lean towards the British analysis. They assumed that some new equipment was needed, but suspected that poor organization and management were serious problems. Hence they proposed to send a fact-finding delegation to Russia. The Russians were affronted, but could not refuse. A small group of US railway specialists, led by the very experienced engineer John F. Stevens, duly arrived in Vladivostok on 19 May/1 June. During the following weeks the Russians bridled as the US representatives became convinced that labour unrest, management failings and food shortages largely explained why the railways were operating at only about 60 per cent of capacity. For Stevens, the Russian government's attitude was clear: 'They want us to put a big bag of money on their doorstep and then to run away.'[19]

Nonetheless Stevens made probably the decisive contribution to resolve this dispute in the Russians' favour. He telegraphed his recommendations to his government on 19 June/2 July:

> This whole nation [is] imbued with the one idea that additional engines and cars are necessary to maintain [the] army and navy, also to provide foods and other needs, and is the salvation of the people of Russia. [It is] expedient for moral [*sic*] effect [that] rapid action be taken to insure [*sic*] continuous operations now and through next winter, otherwise complete collapse may be expected. This commission strongly recommends furnishing from the United States immediately thirty thousand additional American type freight cars and twenty-five hundred additional [decapod] engines ... We have committed our Government to a fair price and the earliest possible delivery. A prompt confirmation of this programme by our Government is vital to the Allied cause and the aims of this commission.[20]

---

[18]   For example, 'Vospominaniia', vol. 6, pp. 261–5, 291–5, 405–18.

[19]   Quoted in L.F. Reitzer, 'United States–Russian Economic Relations, 1917–1920' (unpublished PhD dissertation, University of Chicago, 1950), p. 30; C.M. Foust is preparing a history of the Stevens mission.

[20]   Stevens–Secretary of State/Willard, 2 July 1917: NARG 39 Bureau of Accounts (Treasury), Country Series, box 177, file RS-212.20(b) 'US Railroad Commission to Russia'.

More importantly, Stevens publicized this commitment on his own initiative. He aimed to boost Russian morale, but thereby gave his superiors the unpleasant choice of authorizing the Russian orders at the risk of disrupting overall Allied borrowing and procurement, or appearing to desert Russia and risk her collapse. Prompted partly by these 'extraordinary outgivings', the Wilson administration demanded the establishment of an international organization – the Allied Purchasing Commission – to coordinate all Allied contracts in the United States on the basis of bilateral framework agreements. Moratoria were declared on all large proposed contracts until this system was created. The resultant diplomacy became Bakhmetev's priority and led to a US–Russian framework agreement of 11/24 August. Meanwhile, the Assistant Secretary of the US Treasury, O.T. Crosby, suggested telling the Russians that their railway orders could be progressed in full only at the expense of other proposed Russian contracts. Yet the Americans did not pursue this compromise. Apart from feeling bound by their promises, they calculated that these orders would contribute more to victory than if the money was used for some different purpose.[21]

It has been argued that the US government felt obliged to conceal its growing misgivings about Russia's political situation.[22] The developments noted above appear to confirm that view. However, some US officials did continue to act on their worries about the locomotive and wagon orders: the Treasury cited the impending imposition of federal price controls as a reason not to release the funds until the autumn. Indeed in September Lomonosov felt obliged to resort again to intensive lobbying. He even told Daniel Willard, chair of the US Council of National Defense, that Russia would leave the war if she did not get the decapods.[23] Nonetheless, it was late September before the professor was able formally to request US permission to proceed. At this point, ironically, objections were voiced by the Russian Supply Committee and the Allied Purchasing Commission. Lomonosov, however, used the resultant month-long delay to obtain and evaluate the bids, assigning the locomotives to Baldwin and Alco and the wagons to four companies. Thus, when the objections were resolved, he needed merely to fine-tune and sign the draft contracts and submit them for US government approval.[24]

---

[21]   Heywood, *Modernising Lenin's Russia*, pp. 43–4.

[22]   Kennan, *Soviet–American Relations*, vol. 1, p. 26.

[23]   For example: *New York Times* and *New York Tribune*, 8 September 1917 (cuttings at NARG 261, entry 24, box K20, envelope 'Gazetnye vyrezki'); 'Vospominaniia', vol. 6, pp. 405–18.

[24]   For example: Russian Supply Committee in America: Report of the MPS member of the Committee, No.10477, 27 September: NARG 261, entry 26, box L8, file 116; Minutes of Russian Supply Committee, 3 October 1917: NARG 261, entry 24, box K138, file 721; memorandum for Willard, 5 October 1917 (by Vauclain): NARG 62 Council of National Defense, entry 1-A1, box 9, file F-1-2-2 'Russia: Equipment and Material'; Crosby–Willard, 9 October 1917, and Willard–Crosby, 11 October 1917: NARG 39, Country Series, box 177, file RS-212.21(c) 'Car contracts'; Extracts from resolutions adopted at the 118th

There can be no doubt, then, that Lomonosov had a decisive impact on the US decision to allow the placement of these enormous orders. The US authorities had strong political, financial and logistical reasons for continuing to block them, and Bakhmetev was always unwilling to pressure them for permission. Lomonosov made the crucial difference with his single-minded persistence, playing ultimately on the sense of moral obligation that the Americans felt as a result, above all, of the incautious public remarks by engineer Stevens. A less forceful and determined character than him would surely not have gained the same result.

Yet it was acutely ironic, given how Lomonosov had linked Russia's future to the fate of these proposed orders, that his two-day conference with the firms to finalize the contracts coincided with the Bolshevik revolution on 25–26 October/7–8 November. He was pleased to get the overall cost down by some $10 million to $133,218,700, including associated contracts for spare parts and brake equipment. Dramatically an adjournment was declared on the second morning because of the confusion reported from Petrograd. But because Bakhmetev advised them to continue working as normal, the contracts were initialled, and were thereby made ready for the US government's stamp of final approval. The question now was whether the latest twist in Russia's political crisis would destroy all this work.[25]

Reactions to the October revolution

Although the Bolshevik slogan 'Bread, Peace, Land' caught Russia's popular mood in 1917, the October revolution provoked a civil war. Having won only about a quarter of the seats in the November 1917 election for the Constituent Assembly, the Bolsheviks shocked the populace by dispersing the deputies at gunpoint on 5/18 January 1918. By summer 1918 several anti-Bolshevik Russian armies had been assembled. The civil war was contested mainly by these 'White' armies and the Bolshevik Red Army, but it also involved many other political and national groups in countless local and separatist conflicts, plus armed intervention by British, French, German, US, Japanese and other foreign troops, a Polish invasion in 1920 and, not least, banditry. In Soviet territory, where Moscow replaced Petrograd as the capital from April 1918, life was dominated by the improvised system of rationing, food requisitions, industrial mobilization and 'Red terror' that would

---

meeting of the Russian Supply Committee, 8 October 1917: NARG 261, entry 26, box L8, file 118; Extracts from resolutions adopted at the 119th meeting of the Russian Supply Committee, 15 October 1917: NARG 261, entry 26, box L8, file 119; 'Vospominaniia', vol. 6, pp. 414–15.

25   Lomonosov, Lipets and Balkov–Bakhmetev, 10 November 1917: NARG 261, entry 24, box K57, file 'Correspondence with the Ambassador, May–December 1917'; [Memoranda of understanding], 8 November 1917: NARG 261, entry 24, box K6, file 1; Lipets–Vauclain, 16 November 1917: NARG 261, entry 24, box K4 [no file number]; 'Vospominaniia', vol. 6, pp. 439–41.

become known as War Communism. Elsewhere, similarly, life was overshadowed by the 'White terror' and requisitions. The Bolsheviks gained the upper hand in the winter of 1919–1920, although fighting continued in some areas until 1922. Directly and indirectly the conflict caused an estimated 7–10 million fatalities.[26]

International reactions to the Bolshevik revolution were an important dimension of this complex situation. In autumn 1917 Bolshevik leaders expected their initiative to inspire worker revolutions in Germany, Britain and elsewhere, and the resultant proletarian republics to send aid. Indeed, one motive for abrogating Russia's foreign debts in January 1918 was to spark those revolutions. The only other scenario that they foresaw was a victorious intervention in Russia by the major capitalist powers in league with Russian counter-revolutionary groups: Bolshevik Russia could not conceivably survive alone.[27] In reality, neither the best hopes nor worst fears of the Bolsheviks were realized. Proletarian revolution did not engulf Europe, but nor did the capitalist world unite militarily against Bolshevism. The Allied Supreme Economic Council 'quarantined' Soviet territory and various countries sent troops, especially after the Soviet–German peace treaty was signed at Brest-Litovsk on 3 March 1918 (n.s.). But this intervention was modest compared to the Bolshevik predictions – a reticence that gave Lenin's regime a chance to survive almost complete isolation.[28]

In this context a preoccupation with 'America' was a distinct continuity in policy between the Provisional government and Soviet Russia. Diplomatic recognition from the United States became a prime objective of early Bolshevik foreign policy. It was part of their controversially dualistic strategy to promote world revolution while using traditional diplomacy to seek peace with the Central Powers, end the foreign military intervention and build foreign trade relations. Bolshevik leaders like Lenin and Krasin hoped that the United States' revolutionary tradition and the perception of Russia as a potentially great market would quickly persuade the US government to confer recognition and promote trade. The Europeans would follow suit for fear of exclusion from the Russian market, the friendless White armies would collapse and the Bolshevik regime would thrive.[29]

In reality the US authorities remained hostile towards the Bolsheviks. For President Wilson in particular, Bolshevism was a challenge to his worldwide hopes for liberal democracy. However, he and his colleagues struggled to answer

---

[26] On death toll estimates see E. Mawdsley, *The Russian Civil War* (Boston, 1987), pp. 285–7.

[27] For example: N. Buharin and E. Preobrazhensky, *The ABC of Communism: A Popular Explanation of the Program of the Communist Party of Russia* ([n.p.], 1922), pp. 159, 272–3; R.B. Day, *Leon Trotsky and the Politics of Economic Isolation* (London, 1973), pp. 6–16.

[28] See, for example, R. Debo, *Revolution and Survival: The Foreign Policy of Soviet Russia, 1917–1918* (Liverpool, 1979) and his *Survival and Consolidation: The Foreign Policy of Soviet Russia, 1918–1921* (Montreal, 1992).

[29] See in particular McFadden, *Alternative Paths*, pp. 15–32.

the question of how best to deal with it. At first, like Britain and France, the US federal government expected Lenin's regime to perish quickly, and kept its ambassador *in situ* for informal contacts in the interim. In time, however, the American attitude hardened, especially after the Soviet–German peace treaty. Eventually, after resisting British and French pressure for military intervention for several months, Wilson withdrew the US ambassador in July 1918 and decided to send troops to Russia; it remains unclear whether his main concern was to fight Bolshevism, counter Japanese expansionism or extend US influence. Meanwhile the US government continued to recognize the anti-Bolshevik Bakhmetev as Russia's envoy to the United States. It halted the shipment of goods to Russia, ostensibly to avoid helping the German war effort inadvertently, and pressed Bakhmetev to pay Russian obligations by selling assets. In early 1919 Wilson authorized an informal attempt to establish contact with the Soviet regime, which was undertaken by William C. Bullitt of the American Commission to Negotiate Peace. But the US administration ignored the consequent peace proposals from Moscow, and allowed Bakhmetev to send supplies to White forces in Siberia. Neither the Bolshevik triumph in the civil war nor the growth of Soviet–Western trade from 1920 persuaded Washington to confer diplomatic recognition.[30] In fact, the United States would not take that step until 1933, the last of the major powers to do so.[31]

There was little sympathy for the Bolsheviks among the US public. Support was voiced by relatively small leftist groups like the American Communist Party, Socialist Party, Socialist Society and, in Chicago, the Hull House community founded and led by the Nobel peace laureate Jane Addams. More important, however, was fear of communist subversion, which gave rise to the panicky 'Red Scare'. One significant consequence was the creation by the New York State Legislature in March 1919 of a Joint Legislative Committee to Investigate Seditious Activities. This committee became known as the Lusk Committee after its chairman, Senator Clayton R. Lusk, and conducted a year-long investigation. It seized numerous documents in raids on various US-based radical organizations, including the American Socialist Society's Rand School for Social Science (which gave classes for immigrants on economics and history along with English, public-speaking and other practical subjects), the Industrial Workers of the World (the leading advocate of revolutionary industrial unionism), the Communist Party, the pro-Bolshevik Finnish Information Bureau and the Russian Soviet Bureau, which

---

[30]   For example: Kennan, *Soviet–American Relations*; McFadden, *Alternative Paths*, especially pp. 33–54; Saul, *War and Revolution*, pp. 177–442; Libbey, *Russian–American Economic Relations*, pp. 71–108.

[31]   US policy towards Russia through the 1920s was shaped by a memorandum of August 1920 known as the Colby note after its author, Bainbridge Colby, the then US Secretary of State. See, for instance, McFadden, *Alternative Paths*, pp. 325–35, and K.A.S. Siegel, *Loans and Legitimacy: The Evolution of Soviet–American Relations, 1919–1933* (Lexington, 1996).

was created in New York in March 1919 by L.K. Martens as the putative Soviet embassy.[32]

The attitude of the US business community towards relations with Bolshevik Russia was initially mixed, but became quite positive during the 1920s. The American–Russian Chamber of Commerce quickly became a vocal opponent of Bolshevism, yet many companies attempted to continue 'business as usual' until about mid-1918; there were also some attempts to lobby the US government to give economic assistance to Soviet Russia and keep aloof from the Allied intervention.[33] Many companies worked with non-Bolshevik Russia, which was not covered by the Allied and US embargoes.[34] From 1919 various businesses talked with the Martens bureau in New York, but only draft contracts could be signed due to the US ban on trade with Soviet organizations, and Martens had to leave the country in 1921. Meanwhile the governments of several European countries, including Britain, began meeting with Soviet trade officials in 1920 and the US federal government, though still averse to talks, reduced its trade restrictions.[35] During the early 1920s US–Soviet trade was mostly conducted via third parties and, as mutual confidence grew, direct trade increased nearly threefold in 1924–1930 despite the lack of diplomatic relations; by 1930 the USSR took 3 per cent of total US exports in value and was the United States' eighth largest export customer.[36]

Across the United States the large, diverse community of military personnel, diplomats, immigrants, exiles and others from the former Russian empire was split by the October revolution. Most of these people opposed the Bolsheviks to a greater or lesser extent. Their chief spokesman was ambassador Bakhmetev. Condemning the Bolsheviks as German agents without a popular mandate, he opposed recognition of their regime, supported foreign intervention and the suspension of trade, and cooperated with the US, British and other authorities in

---

[32]   For example, Saul, *War and Revolution*, pp. 226–31, 276–9. On the Martens Bureau and Lusk Committee see McFadden, *Alternative Paths*, pp. 270–308, and Siegel, *Loans and Legitimacy*, pp. 6–38. As noted above, the Lusk Committee's records are at the New York State Archive, Albany, NY. For an introduction to the life of Addams see V.B. Brown, 'Addams, Jane (6 Sept 1860–21 May 1935)', in *American National Biography*, vol. 1 (New York, 1999), pp. 139–41.

[33]   Libbey, *Russian–American Economic Relations*, pp. 70–71; C.A. White, *British and American Commercial Relations with Soviet Russia, 1918–1924* (Chapel Hill, 1992), pp. 39–42. Especially prominent in lobbying for economic relations with Bolshevik Russia in June–July 1918 was Raymond Robins, who had played a major role in US–Bolshevik relations as a Red Cross official in Russia until May 1918. See McFadden, *Alternative Paths*, pp. 79–124, 159–66.

[34]   White, *British and American Commercial Relations*, pp. 32–108.

[35]   McFadden, *Alternative Paths*, pp. 7, 267–335; Siegel, *Loans and Legitimacy*, pp. 12–16, 24–38.

[36]   White, *British and American Commercial Relations*, p. 228; Libbey, *Russian–American Economic Relations*, pp. 74–99.

curtailing or cancelling Russia's wartime contracts and selling assets.[37] Among the many stranded diplomats and procurement personnel who held similar views were Shulenburg and Lipets at the MPS mission.[38] Others opposed the intervention, but favoured trade or aid without diplomatic recognition. An example from this camp, at least initially, was A.A. Bublikov, the man who had seized the MPS for the revolution in February 1917: upon arriving in the United States in December 1917 he called for the shipment of goods to continue so as to block German control over Russia, though by spring 1918 he opposed US–Soviet trade.[39] By contrast, a minority of the community wanted recognition for the Bolshevik regime; leading figures here included Martens and A.F. Nuorteva (head of the Finnish Information Bureau and later a member of the Martens bureau).[40]

Lomonosov was unusual in that he seemed to switch from opposing to supporting the October revolution – a development that was especially significant due to his seniority in the Bakhmetev delegation.[41] Initially he was horrified by Lenin's gambit. For example, on 11/24 November 1917 he lamented that only Russia's enemies would benefit from this 'bloody madness' and civil war. He continued:

> I am far from thinking that the Bolsheviks or at least their leaders are German emissaries. No, they are one bone and one flesh with the Russian people, a product of our cultural backwardness and secret activity, a mixture of extreme idealism with the narrowness of fanatics, of lofty dreams with complete unscrupulousness concerning resources; they are people who place force above law, dogma above the will of the people, and their wishes above the realistically achievable, above duty to the motherland.[42]

In his opinion, the Bolsheviks could not deliver 'bread, land and peace': their revolt meant that Russia would not get further imports of locomotives and wagons, without which food could not be moved; the 'land' slogan was demagogy because the Constituent Assembly would redistribute it anyway; and a 'general democratic peace' was impossible without democracy in Germany, of which there was no inkling. No one had given Lenin and Trotskii the right to decide matters of war and peace on behalf of 182 million people. Their 'dictatorship of the proletariat' was

---

[37]   On Bakmetev's role see McFadden, *Alternative Paths*, pp. 48–50.

[38]   This difference would lead Lomonosov and Lipets to bitter conflict in June 1918, as will be shown below, and destroyed their friendship. Their last meeting for over a decade, on 5 September 1918, was dominated by their politics: 'Dnevnik', 5 September 1918.

[39]   See 'Vospominaniia', vol. 6, pp. 483–6, 497–507, 606.

[40]   Biographical notes about Nuorteva are at Lusk Committee papers, series L0038, box 2, folder 15.

[41]   This change is noted by McFadden: *Alternative Paths*, pp. 30, 273.

[42]   Iu.V. Lomonosov, What has happened and is happening in Russia, 24 November 1917: NARG 261, entry 24, box K18, file 'Originaly rechei'.

like autocracy as a principle. Instead, Russia needed unity and a national political authority (*vlast'*) created through the Constituent Assembly in a democratic republic.[43]

Lomonosov's anti-Bolshevik anger only increased during early 1918. He saw the dispersal of the Constituent Assembly as an act against 'the expression of the State Will of the Russian Land' that was 'worse' than the Tsar's dispersals of the State Duma. He even hated the Bolsheviks as users of force and traitors to their own ideals.[44] He was appalled by their repudiation of Russia's foreign debts: the high cost would include the withdrawal of essential foreign capital from Russia and possibly attempts to establish foreign control there.[45] And he was 'knocked from his saddle' in mid-March 1918 by the Bolshevik-inspired decision of the All-Russian Congress of Soviets to ratify the peace treaty with Germany. Only days earlier he had welcomed a message from President Wilson to the Congress of Soviets that promised US support for the Russian people's battle to liberate themselves from autocracy and become masters of their own lives. Lomonosov had interpreted this note hopefully as de facto recognition of the Soviet system, which could pave the way for US aid; now his optimism was dashed by the separate peace with Germany.[46]

How, then, did it transpire that Lomonosov was sacked by Bakhmetev for supporting the Bolshevik regime? The answer lies not in a Damascene conversion but in Lomonosov's definition of Russia's national interests and the fact that he did not equate Soviet with Bolshevik. Instead of accepting the ambassador's emphasis on removing the Bolsheviks from power at all costs, he prioritized the physical well-being of the Russian people within, crucially, an independent nation state. That interest required the railways to function as well as possible, not least for moving food and fuel, and for that purpose the American railway equipment was essential. Hence he always opposed the suspension of trade. When, for example, Washington announced its decision to withhold approval for the latest locomotive and wagon contracts in November 1917, he tried to appeal. He was supported by some important US decision-makers such as Samuel Vauclain and Daniel Willard at the Council of National Defense. Bakhmetev, however, insisted that foreigners should work only with the 'legal Russian government', and that the way to exploit the railway equipment bypassing political concerns was for a syndicate of private capitalists to buy it for resale or lease to the Russian railways.[47]

---

[43]   Lomonosov, What has happened and is happening in Russia, 24 November 1917: NARG 261, entry 24, box K18, file 'Originaly rechei'.

[44]   'Vospominaniia', vol. 6, pp. 527–9, 543–4.

[45]   'Vospominaniia', vol. 6, pp. 580–585.

[46]   Wilson's message is in *Papers Relating to the Foreign Relations of the United States* (henceforth *FRUS*), *1918: Russia, vol. 1* (Washington, DC, 1931), pp. 395–6. See also 'Vospominaniia', vol. 6, pp. 701–3, and McFadden, *Alternative Paths*, pp. 113–14.

[47]   For example: Lomonosov–Vauclain, 15 November 1917; Vauclain–Willard, 16 November 1917; Baker–McAdoo, 23 November 1917; McAdoo–Baker, 27 November 1917:

Similarly, Lomonosov became alarmed in 1918 by the prospect of foreign military intervention in Russia. For him, foreign troops were acceptable only to strengthen the battlefront against the Central Powers; any other military intervention was intolerable for Russian patriots, and to support it was treason.[48]

As for the future political structure, Lomonosov misread the relationship between the Bolshevik Party and the system of soviets in a way that echoed his perception of the tsarist system. From March 1917 he acknowledged the Provisional government as the legal successor of the tsarist regime on the basis of Grand Duke Michael's decree, but he never cared much for it despite the growth of socialist influence within it during 1917. By 1918 he wanted a system of elected national and local councils, which in his view could be based on the existing soviets; indeed he may have expected the Constituent Assembly to adopt that approach. Significantly, he differentiated between the Bolshevik Party and the soviets much as he differentiated between the tsarist regime and the state. Ignorant or incredulous of the Bolsheviks' ruthless determination to keep control of the soviets, he believed that other parties might win control in future elections, and he evidently thought that President Wilson was making the same distinction with his message to the Congress of Soviets in March 1918.[49]

Accordingly, he tried to influence US policy and public opinion in favour of trade with Russia and non-military assistance to the Russian people. For instance, in January 1918 he became a founding member and chair of a lobby group called the Conference of Plenipotentiaries of the New Russia for Purchases in America. Grandly titled but feeble and short-lived, this group pressed fruitlessly for the completion of the railway contracts as well as for deliveries of food, clothing and other essential civilian goods.[50] Lomonosov stressed the same points in many speeches to business, academic and other audiences, notably in Chicago, Seattle and San Francisco as he inspected stored MPS goods during January–March 1918.[51] For example, in his addresses to the University of Washington and Seattle

all in NARG 39, Country Series, box 177, file 212.21(b) 'Locomotive contracts'; [Russian representative]–US Treasury, [draft, 20 November 1917]: NARG 261, entry 24, box K138, file 722; 'Vospominaniia', vol. 6, pp. 452–9. On Vauclain see his autobiography *Steaming Up!* (San Marino, 1973) and J. Brown, 'Vauclain, Samuel Matthews (18 May 1856–4 Feb 1940)', in *American National Biography*, vol. 22 (New York, 1999), pp. 285–6.

[48]   For example: Statement made by Professor Iu.V. Lomonosov to representatives of the press, 1 June 1918: NARG 261, entry 24, box K19, file 'Rechi Iu.V. Lomonosova'; 'Vospominaniia', vol. 6, pp. 678–84.

[49]   For example: Lomonosov, What has happened and is happening in Russia, 24 November 1917: NARG 261, entry 24, box K18, file 'Originaly rechei'; 'Vospominaniia', vol. 6, pp. 526, 541–3, 702, 750–56.

[50]   The minutes of its eight meetings are at: NARG 261, entry 24, box K20, file 'Soveshchanie predstavitelei Novoi Rossii'. See also 'Vospominaniia', vol. 6, pp. 586–600.

[51]   The texts are in NARG 261, entry 24, box K18, file 'Originaly rechei'. Some are included in his memoirs, such as vol. 6, pp. 610–16.

Chamber of Commerce on 12/25 February he argued that by denying Russia help, the United States was assisting Germany.[52]

Meanwhile he complied with, but also tried to limit the impact of, Bakhmetev's instructions to curtail or cancel contracts, sell goods and downsize the MPS mission.[53] His main achievement was a concession regarding his contracts of June 1917 for 500 locomotives. An agreement of 11/24 January 1918 confirmed that the builders would accept the money already paid against these contracts as full payment for the first 100 engines, which were nearing completion and would be handed to him. The next 200 engines would be built, modified for use on US-gauge track, and sold to the American government, but the Russians would be entitled to repurchase them in 'good working order'. The final 200 locomotives would not be built, but for a payment of $500,000 the Russians could revive the contracts at any time before December 1918 at a guaranteed unit price of $55,730. To facilitate this plan the US War Department would buy Russian-owned rails and fittings worth $500,000.[54]

The eventual, decisive breach in the Lomonosov–Bakhmetev relationship in June 1918 was precipitated by Nuorteva, head of the pro-Soviet Finnish Information Bureau. The Lomonosovs had been on friendly terms with him for some months.[55] On 6 June he invited Lomonosov to oppose the military intervention and advocate recognition at a rally at Madison Square Gardens in New York City. This request came at a time when the Allies were struggling to contain Germany's huge 'March offensive' in France and were pressuring Wilson to join the intervention in Russia partly as a means to help alleviate this crisis. So, had the moment come for Lomonosov to add his diplomatic weight to the public calls for recognition? He understood that such a statement would cause a rupture with Bakhmetev and many colleagues, including his friend and deputy, Lipets, and that he might face an uncertain future. Hesitating, he confided to the diary that he started keeping at this time:

> The second addition [i.e. recognition] did not much please me, but the main thing was that there was too little time to think it through and prepare myself, not just for the speech itself but also for subsequent events. After all, this is the Rubicon.

---

[52]   Speech of Professor Iu.V. Lomonosov at a breakfast meeting organised by the Seattle Chamber of Commerce at the Masonic Lodge in Seattle, 25 February 1918: NARG 261, entry 24, box K18, file 'Originaly rechei'; 'Vospominaniia', vol. 6, pp. 648–53.

[53]   The disposal of the railway supplies in 1918–1920 is described in: Reports about the activity of the mission of the Ministry of Ways of Communication, 1 January 1918 to 30 June 1919, and 1 July 1919 to 1 May 1920: Hoover Institution Archives, Russia: Posol'stvo (US), box 102. See also, for example, 'Vospominaniia', vol. 6, pp. 531–9, 716–30.

[54]   [Memorandum], 23–4 January 1918: NARG 261, entry 24, box K6, file 1; Leffingwell–Polk, 31 January 1918: NARG 59 Dept of State, 1910–1929 decimal series, file 861.77/285; 'Vospominaniia', vol. 6, pp. 531–9.

[55]   'Vospominaniia', vol. 6, pp. 753–6.

Of course my programme has included a ceremonial statement about my views on the intervention and on the necessity of aiding the Soviet government not because it's better but because it's the sole extant government ... and [I have intended to make it] specifically at Madison Square Gardens, but ... this needs preparation, preparation and preparation ... I hesitated ... and I also hesitated because our Washington friends asked me not to speak at American meetings.[56]

The easy option was to avoid a conflict and continue drawing his large salary ($1,000 per month) as head of the MPS mission with the chance of settling in the United States. But instead he called for US recognition, first at a meeting in Chicago on 8 June and then at Madison Square Gardens on 11 June; he thereby became possibly the most senior official abroad of the ill-fated Provisional government to support Lenin's government publicly.[57]

He was right to perceive a political Rubicon here. For him this step was a logical product of his consistent interpretation of his patriotic duty to work for Russia and her people, and did not mean that he supported the Bolshevik Party.[58] But he also realized that few people shared his perception of a difference between the soviet system and the Bolshevik Party. Hence, although he misread the Bolshevik/soviet relationship, he did appreciate that his call for recognition of the Soviet government would be treated as a pro-Bolshevik statement, even though that was not what he meant. In that context, he would obviously be unable to recant without losing his credibility.

His timing hinged on his reading of the political conjuncture and Raisa's views. He believed that the US administration would either confer recognition or join the Allied intervention, and that the President was edging towards recognition. This assessment, which would prove incorrect, was based on discussions with business contacts and with friends in Washington, DC, notably a member of the US War Trade Board named Thomas Chadbourne and Grace Abbott, a former Hull House resident who was now a leading light in the US Children's Bureau.[59] Probably he knew that Chadbourne had recently signed a Trade Board report that urged the State Department to investigate how 'to bring about closer and more friendly commercial relations between the United States and Russia', recommended a mission to Russia, and suggested that the Soviet government might be approached

---

[56]   'Dnevnik', 6 June 1918.

[57]   'Dnevnik', 6 and 11 June 1918; 'Vospominaniia', vol. 6, pp. 813–51; copies of the New York speech are at NARG 59, decimal file 861.01/21½ and Lusk Committee papers, series L0032, folder 12. His salary from 1 May is confirmed in Mission Order No. 36, 22 April 1918: NARG 261, entry 24, box K2, file 7.

[58]   'Dnevnik', 8 June 1918.

[59]   For biographical information on Chadbourne and Abbott see respectively: A.C. Sutton, *Wall Street and the Bolshevik Revolution* (New Rochelle, 1974), p. 148; J. Longo and S. Van Burkleo, 'Abbott, Grace (17 Nov 1878–19 June 1939)', in *American National Biography*, vol. 21 (New York, 1999), pp. 24–6.

to request such a mission.[60] Also, Lomonosov was alarmed by the Allied pressure for intervention: as he wrote in September 1918, 'I became an ardent supporter of Soviet power only because traitors to the Motherland summoned the Varangians against it'.[61] As for Raisa's role, he claimed that she was so angry at his 'indecision' on 6 June that she intervened to accept Nuorteva's invitation on his behalf and that she persuaded him to change 'cooperation' to 'recognition' in his speech at Madison Square Gardens.[62]

This *démarche* initiated a bitter two-week battle with Bakhmetev.[63] Contending that he had been 'given absolute freedom to discharge any official in any [Russian] government department in the United States', the ambassador sent a dismissal letter on 12 June, ordered Lipets to take over the MPS mission, and notified the US Secretary of State, Robert Lansing, that 'Professor Lomonossoff's political statements were announced without my knowledge and bore a partisan character' – a stance that was popular at the MPS mission, to judge by press and US secret service reports.[64] Lomonosov consulted G.E. Klodnitskii, a Russian lawyer and representative of the Murmansk Railway, and Morris Hillquit, an American socialist lawyer recommended by Nuorteva, and decided to contest the legality of his dismissal.[65] He reasoned that if he could thereby discredit Bakhmetev, he would be better placed to campaign for recognition of the Soviet government and contest the legal ownership of Russian assets in America – an important consideration given the large amount and value of equipment still owned by the MPS mission, and his concern with accountability to the MPS. Yet his arguments were insubstantial. One was the claim that Russian law did not recognize the phrase 'is releasing [him] from his duties' (*osvobozhdaet ot obiazannostei*): it used only 'removal from duties' (*udalenie ot dolzhnostei*) and 'dismissal from [state] service' (*uvolnenie ot*

[60]   Cited in Sutton, *Wall Street*, p. 156. See also Kennan, *Soviet–American Relations*, vol. 2, p. 359.

[61]   'Dnevnik', 25 September 1918. The Varangians were the Norsemen who took charge of Slav lands in the ninth century; their leader, Rurik, made his capital at Kiev and established the dynasty that would rule the Kiev Rus and Muscovy until 1598. Lomonosov uses the term very loosely to mean unwelcome foreign invaders.

[62]   'Dnevnik', 6, 11 June 1918. See also 'Vospominaniia', vol. 6, pp. 813, 839–40.

[63]   'Vospominaniia', vol. 6, pp. 852–921.

[64]   Bakhmetev–Lomonosov, 12 June 1918: NARG 261, entry 24, box K114, [Documents pertaining to the dismissal of Professor Iu.V. Lomonosov]; Bakhmetev–Lansing, 12 June 1918: NARG 59, decimal series 1910–1929, file 033.6111/48; Russian Situation, report by operative Frank Burke, 21 June 1918: NARG 59, decimal series 1910–1929, file 033.6111/55a; 'Dnevnik', 12 June 1918; B.A. Bakhmeteff, oral history memoir, p. 298: Bakhmeteff Collection, box 38, file 3. The sacking was formalized in Supply Dept Order No. 16, 14 June 1918: NARG 261, entry 24, box K104, file 'RSC correspondence, May–July 1918'.

[65]   Klodnitskii arrived in America in December 1917: 'Vospominaniia', vol. 6, pp. 486–7. On Hillquit see 'Vospominaniia', vol. 6, p. 857 and 'Morris Hillquit': Lusk Committee papers, series L0038, box 2, folder 11.

sluzhby). Another was the assertion – at odds with Bakhmetev's mandate – that he could be sacked only by the authority that had defined his assignment, meaning the Minister of Ways of Communication. His final point was that public expression of one's beliefs was not justification for dismissal (of a state official) in the United States or contemporary Russia – a claim that, if nothing else, shows ignorance of the political, social and legal realities of Bolshevik Russia.[66]

If Lomonosov did have any chance of success, he soon killed it. Financial need, legal advice, inaccurate information about Lansing and perhaps a sense of adventure explain why he began 'military action' on 16 June. Needing to cover legal and other expenses, he withdrew the $20,000 balance of one mission bank account and tried unsuccessfully to get money from another account. Meanwhile, Chadbourne told him – incorrectly – that Lansing was sympathetic, and advised him to consult J.E. Davis, a lawyer who was reputedly a friend of the President. This lawyer stated that possession of the mission's New York premises would be half of the victory, citing a 'right of seizure' in US law. Lomonosov liked this suggestion, and although dismayed by news that Lansing was supporting Bakhmetev, he believed that Lansing would change his mind when fully acquainted with the issues; strangely, given his diplomatic experience, he seems to have been unaware that Lansing was virulently anti-Bolshevik.[67] Accordingly, aided by several supporters, Lomonosov occupied the MPS offices in the Woolworth Building on 18 June, posted guards, and required staff to pledge their loyalty to him. Over the next four days he and his comrades sorted papers, formalized dismissals and appointments, and sent a letter to Lansing with supporting documents. However, the State Department was unimpressed. Earlier an official had intimated a readiness to meet Lomonosov 'if this could be done without compromising [the State Department's] action with the Ambassador, whose authority it recognized'. But now the same official indicated that the recent steps 'forfeited ... any claims which Professor Lomonossoff might have for consideration from the Department'. Bakhmetev asked for US help with reoccupying the mission, and Lansing agreed. The Department of Justice duly evicted Lomonosov and his supporters on 21 June, and the premises were handed to Lipets.[68]

Ironically, it is conceivable that Lomonosov's occupation of the offices assisted the advocates of military intervention within the US administration. The Treasury

[66]   Lomonosov–Bakhmetev, 12 June 1918; Opinion of Mr G. Klodnitzki, 17 June 1918 and Lomonosov–Lansing, 18 June 1918: all in NARG 59, 1910–1929 decimal series, file 033.6111/53; 'Dnevnik', 13 June 1918.

[67]   Lomonosov–B, 19 June 1918: NARG 261, entry 24, box K114, file of papers pertaining to dismissal of Lomonosov; 'Dnevnik', 16–17 June 1918. In fact, Lansing had acknowledged Bakhmetev's letter on 13 June: NARG 59, 1910–1929 decimal series, file 033.6111/48. On Lansing's outlook see McFadden, *Alternative Paths*, p. 35.

[68]   B. Miles, Memorandum for the Secretary of State, 26 June 1918: NARG 59, 1910–1929 decimal series, file 033.6111/57a; 'Dnevnik', 21, 24 June 1918; 'Vospominaniia', vol. 6, pp. 877–900.

and War Industries Board were worried – understandably, given the Bolshevik repudiation of Russia's foreign debt – that Lomonosov might harm US interests by destroying papers pertaining to contracts.[69] His actions surely reminded policy-makers that the economic dimension of possible US military intervention was not just a matter of future trade, but also of the money lent in 1917, and showed the likely financial benefit of allowing the compliant Bakhmetev to control the relevant papers.

That said, it seems unlikely that Lomonosov's actions had any decisive effect on US policy. Wilson was doubtless unaware of the dispute when he told the Allies on 18 June that he would reassess his Russian policy. True, he was lobbied by the Lomonosov camp several days later. Lomonosov sent Wilson a letter that described his predicament, requested the compilation of an official inventory of the mission's property, and ended:

> I am deeply desirous of doing everything in this situation that is possible with a view to preserving the mutual confidence which ought to exist between the great democracy of America and the young democracy of Russia. I desire, however, to respectfully renew my protests against the unwarranted, and what I believe an unlawful and unwise exercise of force by the Government of the United States in this situation.[70]

Also, Davis sent a letter to the President, Chadbourne tried to see Lansing, and one of the 'White House ladies' with whom Raisa had become friendly tried to arrange an audience with Wilson for Lomonosov, Raisa and Davis.[71] Yet apart from some sympathy with the request for an inventory, Wilson was unmoved. As he replied to Davis:

> I am sorry but I am sure it would be a mistake for me to intervene on the Bakhmeteff–Lomonossoff matter. It would only lead to the Government itself becoming involved in a controversy which ought to be left entirely to the Russians themselves. Pardon the brevity of this note. I know you will understand that I am not forming this judgement in haste.[72]

---

[69] Memorandum, Department of State, 20 June 1918; Leffingwell (Treasury)–Polk, 25 June 1918; Miles, Memorandum for the Secretary of State, 26 June 1918: NARG 59, 1910–1929 decimal series, file 033.6111/54a, 57, 57a.

[70] Lomonosov–Lansing/Wilson, 27 June 1918: NARG 59, 1910–1929 decimal series, file 033.6111/58.

[71] T.T. Ansberry–Lansing, 21 June 1918, and Lomonosov–Lansing, 24 June 1918: NARG 59, 1910–1929 decimal series, file 033.6111/55, 56; 'Dnevnik', 24, 25, 26 June 1918; 'Vospominaniia', vol. 6, pp. 895–908.

[72] Wilson–Lansing, 28 June 1918: NARG 59, 1910–1929 decimal series, file 033.6111/60; 'Vospominaniia', vol. 6, pp. 908–9.

Unsurprisingly, given that the US government was already involved, this note bemused Lomonosov. Hillquit suggested that its emphasis on non-interference implied that the conflict would become an internal Russian matter.[73] In fact, it signalled the intention to continue recognizing Bakhmetev; whether the affair did influence Wilson's decision a few days later to intervene in Siberia is unclear.

Across the Rubicon to Soviet Russia

Lomonosov's worst fears had been realized: he had 'lost everything except his honour'.[74] As of mid-June he held about $24,000 of MPS money, which he used for mission-related bills and several salaries; and although he drew no salary for himself after August 1918, only about $700 was left by December 1918.[75] In principle he wanted to travel to Soviet Russia, but visa problems and a certain caution about rushing into the unknown meant that for the moment he and Raisa remained in the United States, living off their recent personal savings. They were ostracized not just by the Russian embassy and the procurement personnel, who derisively dubbed him 'the hero of our time', but also by the US administration.[76] Ironically, Lomonosov alienated even leaders of the pro-Soviet camp by withdrawing, at the insistence of Raisa and several US friends, from a lecture tour that they devised for him in July 1918.[77] His few American supporters included the Chadbournes, Alfred Scattergood (a Philadelphia banker who was one of several Quaker friends in that area), the proprietors of the small private school attended by their son at Ossining, New York, and various people associated with Hull House such as Grace Abbott, Jane Addams and Rachelle and Victor Yarros. Indeed, the Lomonosovs depended on such friends to lodge

---

[73]   'Dnevnik', 27 June, 1 July 1918.

[74]   'Dnevnik', 24 July, 23 August 1918, 6 May 1919.

[75]   'Dnevnik', 20 December 1918. Ozols implies that Lomonosov purloined this money: *Memuary*, p. 32. However, his itemized financial report is 'Otchet Glavnoupolnomochennogo Ministerstva Putei Soobshcheniia v Amerike, profa Iu.V. Lomonosova v izraskhodovanii im poluchennogo iz summ Ministerstva avansa', 15 April 1919: Lusk Committee papers, series L-0032, folder 12.

[76]   See A. Zak–M.M. Karpovich, 19 September 1918: Bakhmeteff Collection, Box 24, file April–July 1918. The name refers to the bitter and scornful anti-hero of Mikhail Lermontov's novel *A Hero of Our Time* (1841).

[77]   'Dnevnik', 26–7 July 1918, 6 April 1919. Three arguments persuaded him to withdraw: that court proceedings might be started against him as a non-official figure without diplomatic immunity; that such speeches would put their American supporters in a very difficult position; and that the speeches might prompt the US authorities to prevent the family from returning to Russia: 'Dnevnik', 27 July 1918.

at Ossining school until September and then at Hull House as they wondered what to do.[78]

Perhaps hardest to endure for Lomonosov were certain unforeseen consequences of this disaster. One was a crisis of self-confidence, which permeates his diaries. Another was the near-collapse of his marriage: recriminations abound in the surviving personal papers. He was especially pained by Raisa's recourse to such epithets as 'self-promoting turkey' and 'fat bourgeois'. By 1919 he suspected that their relationship was beyond full repair, like a 'broken vase'.[79] A particularly painful experience for him was to receive a reproachful and, to his mind, thoroughly misguided letter from Rachelle Yarros in April 1919. Apart from anything else, he was furious that Raisa had shown her his two latest, intimate letters.[80] This crisis was resolved, after a fashion, in May 1919 on Raisa's terms under pressure of their imminent departure for Russia. They agreed that only he should go there because he would probably be travelling extensively and Raisa would struggle to cope alone there with young George because, apparently, her health was poor. 'Eternal peace' was concluded on this basis on 6 May.[81] Thereafter, however, the marriage would recover its earlier warmth only occasionally, for brief intervals. Furthermore, once the family was reunited in early 1921, Raisa would remain the dominant partner – a change that would significantly affect Lomonosov's career as well as his family life.

That the Lomonosovs were helped by the US State Department on three occasions between July 1918 and May 1919 is noteworthy given the intense hostility of Western governments towards Soviet Russia. The historian A.C. Sutton has seen these contacts as evidence of official sympathy for efforts by business interests and Soviet representatives to promote US–Soviet relations.[82] But that verdict goes too far. The first intervention was merely a personal favour to help the Lomonosovs leave for Russia via Sweden. During July and August

---

[78] For example: 'Dnevnik', 25 July, 27 August, 25 September, 4–5 October, 20 December 1918, 26 April 1919. An obituary of Scattergood was published in *The Friend*, 128/5 (Ninth Month 2, 1954): 66–7 (copy at LRA, MS 717.2.271.1). The Yarroses lived and worked at Hull House for some 20 years. Victor Yarros was best known as an editorial writer for a Chicago newspaper and editor of *Literary Digest*: see 'Old Age – Without Tears', *Unity* (March–April 1954): 9–10 (copy at LRA, MS 717.2.340.30) and obituary in the *Evening Tribune* (San Diego), 9 November 1956 (copy at LRA, MS 717.2.11.1). Rachelle Yarros became the first female professor of social hygiene in the United States: obituary in *San Diego Union*, 19 March 1946 (copy at LRA, MS 717.2.340.4).

[79] For example: 'Dnevnik', 19 July, 18, 23 August, 8–9, 18 October, 8 December 1918, 3–4 February, 18 March, 2 May 1919; Lomonosov–Lomonosova, 25 August 1918, 8, 13 April, 1 May 1919: LRA, MS 716.4.1.5, 12, 15–16, 18; 'Vospominaniia', vol. 6, pp. 993–1006, 1359–71.

[80] R. Yarros–Lomonosov, 10 April 1919: LRA, MS 716.4.36.1; 'Dnevnik', 13 April 1919.

[81] 'Dnevnik', 2, 6 May 1919; 'Vospominaniia', vol. 6, pp. 1372–92.

[82] Sutton, *Wall Street*, pp. 145–61.

1918 the Assistant Secretary of State, Frank L. Polk, whom the couple knew well, persuaded Bakhmetev to issue passports to them; however, the consequent attempt by the Lomonosovs to use these passport to obtain Swedish visas was unsuccessful.[83] The two other instances involved assistance that did have a notable political dimension. Again, one concerned the logistics of departure, this time in April 1919, when the State Department apparently asked the Swedes to provide visas and interceded with the US Customs about baggage; but Polk was probably hoping that this intervention would help secure the release of US citizens from Bolshevik custody, including Xenophon Kalamatiano, who was awaiting execution for spying.[84] The other occasion is piquant given the Allied blockades of Soviet territory and gold. In 1918 Lomonosov paid for two socialist acquaintances, I. Petersen and M.N. Gruzenberg, to visit Moscow for up-to-date information and instructions.[85] Upon returning in December Petersen reported that Gruzenberg was in Scandinavia with $25,000 for Lomonosov, who was to stay in the United States as a Soviet representative.[86] With the State Department's help, Lomonosov received $10,000 of this money in April 1919, which he treated as back-paid salary. The link was Chadbourne, who had resumed his peacetime job as a Wall Street lawyer: at the bidding of Rachelle Yarros he approached Polk, who instructed the US ambassador in Stockholm to get the cash and send a cheque to Chadbourne. But since Chadbourne and Polk were led to believe that the cash was the private property of the Lomonosovs, one should not ascribe broad political significance to their roles.[87] For his part, Lomonosov did not become heavily involved in promoting US–Soviet official relations until he joined the Martens

<hr>

[83]   G. Creel (Committee on Public Information)–Lomonosova, 24 July 1918: LRA, MS 717.2.73.1; 'Dnevnik', 26–7, 30 July, 7–8, 15–16, 22 August, 16 September 1918; Passport (issued by Russian General Consulate, New York), 25 July/7 August 1918: LRA, MS 716.1.1.73; 'Vospominaniia', vol. 6, especially pp. 961–8. On Polk see McFadden, *Alternative Paths*, pp. 35–6.

[84]   Note of content of Lomonosov–State Department, 16 May 1919: NARG 59, Central Decimal File, entry 199, Name Index 1910–1944, Lomonosoff, Iu.V., record 033.6111/85; Puller–Lomonosov, 19 May 1919: NARG 59, 1910–1929 decimal series, file 033.611/85; 'Dnevnik', 1–2, 24, 26, 28 April 1919; 'Vospominaniia', vol. 6, pp. 1346–55. The idea of exchanging Lomonosov for Kalamatiano was evidently suggested by M.N. Gruzenberg without authority: see Litvinov–[Chicherin?], August 1919, in A.N. Iakovlev et al. (eds), *Sovetsko–Amerikanskie otnosheniia: Gody nepriznaniia, 1918–1926* (Moscow, 2002), p. 114. Soviet agreement to release Kalamatiano if Lomonosov had safe conduct to Russia was noted on 29 March 1919 in a telegram from the US legation in Christiana to the US embassy in London: National Archives, London, FO 371/3940/51362; I am grateful to Jon Smele for this reference.

[85]   On Gruzenberg see Iakovlev et al., *Sovetsko–Amerikanskie otnosheniia*, p. 102; Petersen is described at 'Vospominaniia', vol. 6, pp. 366–8.

[86]   'Dnevnik', 3 November, 11 December 1918; 'Vospominaniia', vol. 6, pp. 1120–1128.

[87]   For example: Chadbourne–Polk, 7 January 1919; Polk–Chadbourne, 5 February 1919; Chadbourne–Phillips, 5 April 1919: NARG 59, 1910–1929 decimal series, file

bureau in March 1919. Moreover, he was dismayed by the prospect of remaining in the United States as a Soviet representative: he wanted to resume productive work on the Russian railways. None of his papers indicate that he seriously wanted a role in bringing the two countries closer together at this time.[88]

In the meantime he busied himself with a variety of professional and personal projects. He concluded the affairs of the MPS mission as far as possible, completing a formal report in September 1918.[89] He tidied his notes of the February revolution into a short memoir for publication in English.[90] He did some reading on favourite scientific topics like maximum train speeds, although he abandoned his analysis of the 1912–1914 passenger locomotive experiments because Lipets refused to provide the raw data from the Experiments Office.[91] Much of the autumn and winter went on preparing lectures about Russia for Hull House; related activities were writing some magazine and newspaper articles and giving some speeches and lectures to radical audiences in cities like Detroit and Cleveland.[92] Also, he tried to improve his English. Partly because of his marital crisis, he wanted to reduce his linguistic dependence on Raisa, who had acquired fluent English and now often acted as his interpreter. He made little progress, however – a fact that helps explain why Raisa became so dominant in their relationship and involved in his work.[93]

The manner of his attachment to the Soviet bureau in 1919 was symptomatic of his self-absorbed, troubled state. The idea evidently came from Martens, and Lomonosov was unsure how to reply. They met in New York on 22 March, after which Lomonosov consulted Hillquit, above all concerning his legal status vis-à-vis the Moscow authorities. The upshot was his decision to resume his effort to travel to Russia, not least to clarify that question of status, and meanwhile to help Martens on an unpaid basis. Thus, having moved to New York, he divided his time between the bureau (where he was made acting head of the Railway Department) and preparations for his journey to Soviet Russia. Raisa and George stayed in Chicago until he visited them in early May, after which they all went to New York.

033.611/79, 77, 84 respectively; R. Yarros–Lomonosov, 10 April 1919: LRA, MS 716.4.36.1; 'Dnevnik', 13 April 1919.

[88] For example: 'Dnevnik', 22 January, 19, 25 March 1919.

[89] For example: 'Dnevnik', 26–7, 29 August, 3, 9 September, 1–2 October 1918; 'Vospominaniia', vol. 6, p. 981.

[90] 'Dnevnik', 11, 13, 23 August 1918; the manuscript became his *Memoirs of the Russian Revolution*.

[91] For example: 'Dnevnik', 1–2, 4, 7, 10, 13, 15, 18–23 September, 10–11 October 1918.

[92] For example: Advertisement (for Hull House lectures, November 1918–February 1919): LRA, MS 716.1.76; 'Dnevnik', 28 August, 8 September, 10, 12, 14, 18, 20, 26 October, 18–19, 25, 30 November, 20 December 1918, 5, 8 January 1919. The publications included 'A Voice Out of Russia', *The Dial*, LXVI/782 (25 January 1919): 61–6; 'Russia at the Cross-Roads', *The Nation*, CVIII/2800 (1 March 1919): 321–2.

[93] 'Dnevnik', 28 August, 20 September, 12, 14 October 1918.

He finally left the United States on 24 May 1919, hoping to reach Moscow via Scandinavia, whereas his family returned to Chicago.[94]

Offering a rare glimpse inside the Martens bureau, his papers support the view that, whatever the bureau's involvement in revolutionary propaganda and agitation, its desire for trade was genuine, and that many US engineering businesses had no objection in principle to trading with Soviet Russia. He does not discuss revolutionary activity, simply noting the creation of a 'political and intelligence' department. But he does confirm that five trade-related departments – legal, commercial, financial, railway and economic-statistical – were formed at the outset.[95] He tried to attract business in his capacity as acting head of the railway department by sending letters to engineering companies and holding meetings. His description of the responses may imply that the bureau had some potential credibility: the companies tended to cite government policy as their reason for reticence and did not – with the exception of the Baldwin Locomotive Works – display principled hostility.[96]

Particularly interesting are Lomonosov's indications of problems in the bureau's leadership, functioning and culture. So far as he could ascertain, the bureau was overseen by a group of Jewish Bolsheviks from New York's East Side, who were nicknamed the Sanhedrin. They spent much time with Martens, and incensed Lomonosov by questioning his appointment to the railway department.[97] Martens seemed concerned only with the broad political picture and prospects. He was uninterested in detail, yet unwilling to delegate power, and seemed not to have a strategy. Also, he did not display authority or try to establish and enforce work priorities, such that colleagues did not know what they were supposed to be doing. Thus, for instance, staff members were distracted by frequent instructions to attend meetings and conferences, and minor matters eclipsed major ones. Consequently, Lomonosov feared a recurrence of the same misunderstandings that had fouled his relations with Bakhmetev. Yet he found it harder to get appointments to see Martens than he had known even with tsarist ministers. As for the bureaucratic culture, he was dismayed that the colleagues were behaving like tsarist-era bureaucrats. He perceived the same formalistic attitude to work,

---

[94]   'Dnevnik', 18 March–24 May 1919; 'Vospominaniia', vol. 6, pp. 1271–392. It was thus not the case that Lomonosov left for Europe on instructions from M.M. Litvinov, the deputy People's Commissar of Foreign Affairs: see Siegel, *Loans and Legitimacy*, p. 10. Litvinov's role in Soviet–Western relations at this time is analysed in H.D. Phillips, *Between the Revolution and the West: A Political Biography of Maxim M. Litvinov* (Boulder, 1992), pp. 31–45.

[95]   'Dnevnik', 29 March 1919.

[96]   'Vospominaniia', vol. 6, pp. 1338, 1343–6. Some of the bureau's commercial papers are in Lusk Committee papers, series L-0032, folders 3–7. By spring 1920 Baldwin would be more amenable: Siegel, *Loans and Legitimacy*, pp. 29–30.

[97]   'Dnevnik', 29 March, 4 April 1919; 'Vospominaniia', vol. 6, especially pp. 1305–19.

favouritism and battles of personal pride. Wages, for example, seemed to depend on personal acquaintance more than skills and experience.[98]

Lomonosov's reaction to this situation reveals an intellectual discord that, in one form or another, would always taint his relationship with the Soviet regime. He felt that his duty was to work for the office, yet he felt out of place there; indeed, probably correctly, he sensed a view that he was a 'hired-for-money' bourgeois.[99] He strove to maintain a degree of professional independence – a response indicative of confusion and naivety about the politics of Soviet Russia. For instance:

> I am serving the Russian people, just as I served under the tsar. I confess that I have both personal honour and love for power. But I have principles too: I am serving the cause, not persons, and I do not fawn. And I feel that in current circumstances [ie. in the Martens bureau] this is just as difficult as under the tsarist satraps.[100]

And further, in his anger at the 'Sanhedrin': 'I refuse to kiss the proletarian or Jewish–Bolshevik arse, just as I refused to kiss that of the tsar and [the Kadet leader] Miliukov.'[101] That he was indeed confused about the politics is all the more clear from remarks about his own views. He now supported certain core Bolshevik ideas: the necessity of using force for social revolution, because persuasion was insufficient; the impossibility of a middle way; the need for the 'dictatorship of the proletariat' at the moment of revolution; and the importance of equality, including the right to have employment. But he did not see himself as a Bolshevik. If anything, he sympathized with Menshevik thinking about not forcing the pace of change. He 'wholeheartedly' supported the Soviets rather than anti-Bolshevik leaders like Admiral Kolchak, but he wanted a 'moderate and careful policy' within Soviet Russia – which was not a course that the Bolsheviks were likely to choose. Indeed, he was wrong to think that the Bolshevik Party was moving towards his belief that the whole Russian people should have power through the soviets. Also, he rejected the 'dictatorship of the proletariat' as the long-term basis for the Soviet state because it represented social vengeance, negated social justice, and would merely generate a new privileged class and sinecures. Above all, he had little sense of the Bolsheviks' siege mentality, their ruthlessness and their tendency to perceive counter-revolutionary opposition everywhere. For example, in criticizing

---

[98]   'Dnevnik', 26–7, 31 March, 10, 15, 17, 20 April 1919.

[99]   Lomonosov–Lomonosova, 21 March, 1, 6 April 1919: LRA, MS 716.4.7, 9, 11.

[100]   'Dnevnik', 20 April 1919.

[101]   'Dnevnik', 20 April 1919. This invective is probably not indicative of anti-Semitism on his part, given that he had stayed with Raisa, a Jewess, in defiance of pressure from Rukhlov, and it may just be an echo of contemporary popular Western attitudes to the Bolsheviks.

the murder of the imperial family in July 1918, he failed to link this act with the logic of forced social revolution and class-based civil war.[102]

Furthermore, he justified his plan to go to Russia partly in legalistic terms that assumed continuity across 1917 in the functioning of the state. In effect he reiterated and extended the bureaucratic argument that he had put to Bakhmetev: he was responsible to his minister, and wanted to check his status *vis-à-vis* the Soviet government and the MPS (now the People's Commissariat of Ways of Communication – NKPS). In particular, he wanted to know whether formally he was 'in Soviet service', whether an NKPS Order had been published to dismiss him or put him at Martens' disposal, and whether Martens had been appointed as the 'manager' of all Russian state property in North America. By contrast, Martens dismissed these worries in iconoclastic revolutionary fashion: 'we're starting everything from scratch'.[103] Ironically, Lomonosov seems to have been unaware that Lenin's government answered many of his questions in April 1919 by appointing him as Chief Plenipotentiary of the Mission of Ways of Communication in the United States.[104] The rationale for that appointment was perhaps partly to give the bureau some political and technical credibility at a time when the Soviet regime was desperate to expand its representation abroad. However, other evidence indicates that in late May 1919 Moscow was sceptical about the likelihood of any US–Soviet trade and that it wanted Lomonosov back on the Russian railways as soon as possible.[105]

That said, Lomonosov's legal concerns did have some relevance for his situation. He considered himself accountable for assets that were worth over 50 million dollars, yet were controlled by the Bakhmetev embassy.[106] Moreover, Hillquit was advising that the Soviet side would probably need to launch a court case to claim the ownership of those assets. Lomonosov, meanwhile, was in the peculiar position of being a product of the nobility who was supporting a workers' government embroiled in a class-based civil war where 'terror' and sabotage were reportedly widespread. Also, he had scant information about life in Soviet Russia that he considered trustworthy.[107] Thus it was logical that he would want to try to demonstrate his professional integrity to the Moscow regime and to Martens as its local representative, not least by trying to protect the state property in his care.

---

[102]   For instance: 'Dnevnik', 21 July 1918, 14 March, 20 April, 2 May 1919.

[103]   'Dnevnik', 18–19, 22 March, 4, 6 April 1919.

[104]   Minutes of meetings of Council of People's Commissars, 1 and 17 April 1919: GARF, f. r-130, op. 3, d. 30, ll. 3 and d. 36, ll. 1–2. Lomonosov's ignorance was probably due to the difficulties of communication between Moscow and the Martens bureau: this news may well have taken more than a month to reach Martens.

[105]   Litvinov–Martens, 27 May 1919: in Iakovlev et al., *Sovetsko–Amerikanskie otnosheniia*, p. 100.

[106]   Lomonosov–Phillips (State Dept), 2 October 1918: Bakhmeteff Collection, Box 19, file L.

[107]   'Dnevnik', 4–6 April 1919; 'Vospominaniia', vol. 6, pp. 1309–11.

Indeed, he ensured that his MPS duties were transferred formally to Martens by means of a witnessed Assignment.[108]

Overall, then, Lomonosov was unable to achieve much at the Martens bureau. Yet his involvement probably did have some broader significance. It seems very likely that Lomonosov's concern about the railway materials now controlled by Bakhmetev prompted Martens to instruct the Coudert Brothers law firm to lobby the US authorities for recognition of the bureau as the legal owner of that property. For these few weeks Lomonosov was, for the bureau, a direct connection with the affairs of the erstwhile Russian Supply Committee, and his presence gave the Soviet side at least a little hope for their argument. But Martens, of course, did not get any positive reply from the US government.[109]

Lomonosov's journey to Soviet Russia lasted four months.[110] He was delayed in Sweden for two months as he became embroiled in the efforts to secure Kalamatiano's freedom. Perturbed by the Soviet regime's apparent indifference towards himself in this affair, Lomonosov had to wait, though he did talk with several Swedish companies at their request.[111] Eventually he risked an unauthorized departure when German companies invited him to Berlin and promised to facilitate his onward journey. In Berlin he met not just businessmen but also senior officials of the Foreign Ministry, and detected strong interest in trade as soon as the Allies would permit it. He therefore took the initiative to sign some provisional contracts for railway equipment.[112] Then he travelled via Poland to the new Baltic state of Latvia. With some difficulty he crossed into Soviet territory near Daugavpils (formerly Dvinsk) on 12 September, and a week later he reached Moscow.[113]

*   *   *

Lomonosov's two years in North America became pivotal in his life because of the October revolution. Although the visit began well, it was quickly complicated by the deepening political crisis. Then Lenin's revolution effectively nullified his work

---

[108]   Hillquit–Martens, 10 May 1919, and [Assignment], 21 May 1919: Lusk Committee papers, series L-0032, folder 14.

[109]   Polk (State Dept)–Coudert Brothers, 19 April 1919: Bakhmeteff Collection, Box 9, file Coudert.

[110]   'Dnevnik', 24 May–2 September 1919; 'Vospominaniia', vol. 7, pp. 1–270.

[111]   On the idea of exchanging him for Kolomatiano see note 83 above. Apart from Lomonosov's diary and memoirs, see also *FRUS, 1919: Russia* (Washington, DC, 1937), pp. 167–90; D.S. Foglesong, 'Xenophon Kalamatiano: An American Spy in Revolutionary Russia', *Intelligence and National Security*, 6 (1991): 154–95; J.W. Long, 'Searching for Sidney Reilly: The Lockhart Plot in Revolutionary Russia, 1918', *Europe–Asia Studies*, 47/7 (1995): 1225–41. Kalamatiano was released in 1921.

[112]   Report to People's Commissar of Ways of Communications, 18 September 1919: LRA, MS 716.1.83; Heywood, *Modernising Lenin's Russia*, pp. 70–72.

[113]   'Dnevnik', 2–20 September 1919; 'Vospominaniia', vol. 7, pp. 271–408.

and forced him to choose between the pro- and anti-Bolshevik camps. His support for the Soviet system – but not yet the Bolshevik Party – was sincere, although it was founded on a misreading of the relationship between the Bolsheviks and the Soviet political system. His call for US recognition of Lenin's government was likewise based on his misreading of President Wilson's attitude towards Russia. In the event, Lomonosov alienated the anti-Bolshevik majority of the Russian community in North America without gaining the full trust and acceptance of the pro-Soviet camp, and his marriage was strained to the utmost. Writing in October 1918 he regretted three major errors: that he had not obtained signed instructions prior to his departure from Petrograd; that he had spoken at Madison Square Gardens without adequate preparation (for which he self-indulgently blamed Raisa); and that he had not removed important documents from the Mission's offices.[114] In reality, the key factor was his public support for Lenin's government, which he did not regret. He had the choice of foreign exile, but his principles took him in a different, difficult direction and ultimately to Soviet Russia.

His assignment – part of the crucial phenomenon of inter-Allied collaboration during the First World War – reflected a major shift of emphasis towards non-military products in Russia's wartime foreign procurement policy after the February revolution. His difficulties with the US government over locomotive and wagon orders highlight a rapid change from enthusiasm to caution in Washington's attitude towards Russia after March 1917. This development adds weight to the idea that Russia's allies welcomed the Bolshevik revolution as an excuse to terminate an enormous supply operation that they no longer considered worthwhile for the Allied war effort. Lomonosov's call in 1918 for foreign recognition of the Moscow regime embarrassed the Russian embassy, but if it had any impact on US policy, that effect was to harden the line against recognition. Similarly, the assistance rendered to him by the Department of State between August 1918 and May 1919 should not be construed as evidence of a pro-Soviet conspiracy in US government and business circles. Indeed, although Lomonosov's work at the Martens bureau implies that US business interest in trade with Soviet Russia was greater in spring 1919 than is normally assumed, it also suggests that the companies were generally content to follow their government's anti-Bolshevik lead.

---

[114] 'Dnevnik', 16 October 1918.

Chapter 8
Building the New Russia

The seven years that Lomonosov spent helping to build Soviet Russia were a desperate time for his country. They encompassed civil war, famine, epidemics, the consolidation of the Bolshevik one-party state, the slow revival of the economy, and Lenin's final illness and death during 1922–1924, together with much of the associated power struggle in the Bolshevik leadership. Lomonosov held various administrative and research responsibilities for the Soviet state, but basically he helped to lead the operation, repair and modernization of the railway system as a 'commanding height' of the economy. During the climax of the civil war in the winter of 1919–1920 he concentrated on railway operations and urgent repairs, and frequently consulted with Lenin. From June 1920 until April 1923 he was based abroad to lead the acquisition of a vast quantity of railway supplies – a task that echoed the Russian overseas procurement efforts of 1914–1917 and located him at the heart of both the regime's economic strategy and its foreign policy.[1] He then concentrated on the new scientific and technical challenges of building main-line diesel locomotives.

At first glance this list of high-level duties and concerns may seem to indicate that Lomonosov had little difficulty in resuming his railway career in the Soviet milieu. But this chapter argues to the contrary. Focusing on the period 1919–1923, it reveals that although he developed good relations and considerable professional influence with Lenin and certain other top party leaders, he never commanded much respect and support at the NKPS. His assignment abroad from 1920 proved significant for the repair and revival of the Soviet railways, but in career terms it marked a certain loss of favour that only increased over time. Those developments were partly a function of his own attitudes and actions, including his often high-handed treatment of subordinates, his decision not to join the Bolshevik Party, errors of judgement and his limited abilities to understand political currents and trends. But also important was the inveterate Bolshevik distrust of the community of pre-revolutionary scientific and technical specialists, whose expertise was still indispensable for the economy, yet whose politics were assumed to be bourgeois and therefore anti-Soviet. Indeed, doubts about whether the Soviet government wanted his expertise troubled Lomonosov even as he journeyed to Russia in 1919.

---

[1] The history of this procurement policy, including its significance for the domestic Soviet economy and Soviet foreign trade, is the subject of Heywood, *Modernising Lenin's Russia*.

A return to power?

Soviet Russia was tiny, weak and beleaguered in September 1919. It comprised only a modest proportion of the lands known as European Russia. The civil war was at its peak: the Bolshevik Red Army was under strong pressure from counter-revolutionary White armies, the military intervention by Britain, France, the United States and other foreign powers was continuing (albeit with some hesitation), and the Allies were still blockading Soviet territory to assist the anti-Bolshevik struggle, prevent the 'disease' of Bolshevism from spreading, and advance their commercial and other interests. Within Soviet Russia the Bolshevik regime was mobilizing the economy for its war effort through a system of state control that would later be known as War Communism. Private enterprise had been outlawed in 1918, military demands had priority for resources and industrial output, and food was being requisitioned ruthlessly from the peasantry. Yet industry and agriculture were collapsing, the transport system was barely operating and the shortages of food and fuel were acute.[2]

These events were levying an enormous human death toll. For instance, the Red and White armies suffered over two million fatalities from combat and disease between 1918 and the end of the civil war in about 1922.[3] Throughout the former empire all sides dealt harshly with political opposition, including alleged sabotage. In Bolshevik territory this task was entrusted to F.E. Dzerzhinskii's infamous All-Russian Extraordinary Commission for Combating Counter-revolution, Speculation and Misconduct in Office, usually called the Cheka or VChK. The Cheka's many victims included acquaintances of Lomonosov such as G.O. Pauker, erstwhile head of the MPS Directorate of Railway Construction, and professor M.M. Tikhvinskii, his one-time revolutionary comrade, who were executed in 1918 and 1921 respectively.[4] Millions of civilians died from disease and starvation. For instance, the typhus epidemic claimed engineers like V.I. Grinevetskii, Russia's foremost authority on diesel engines, while seven of the 45 members of the Academy of Sciences starved to death.[5]

As Lenin and his colleagues fully understood, the railways were central to overcoming Soviet Russia's disastrous predicament.[6] The civil war was essentially

---

[2] A good recent survey, with particular reference to the experiences of ordinary people, is C. Read, *From Tsar to Soviets: The Russian People and their Revolution, 1917–21* (London, 1996).

[3] G.F. Krivosheev (ed.), *Rossiia i SSSR v voinakh XX veka: Poteri vooruzhennykh sil: Statisticheskoe issledovanie* (Moscow, 2001), pp. 149–51.

[4] On Pauker see Kriger-Voinovskii, *Zapiski inzhenera*, p. 65.

[5] Bailes, *Technology and Society under Lenin and Stalin*, p. 53; Coopersmith, *Electrification of Russia*, p. 159. On the typhus epidemic see R.T. Argenbright, 'Lethal Mobilities: Bodies and Lice on Soviet Railroads, 1918–1922', *The Journal of Transport History*, 29/2 (2008): 259–76.

[6] See especially Argenbright, 'Russian Railroad System'.

a railway war, contested along and for the trunk routes because control of them was essential for asserting and consolidating political power as well as for running the economy and supporting military action. Having reformed the MPS as the NKPS in late 1917, Lenin's regime nationalized the whole transport system in early 1918. But the precipitate decline in railway traffic in 1917 continued through 1918 and was accentuated by the civil war. In due course the physical damage and destruction seemed considerable in areas such as South Russia, not to mention the desperate plight of staff.[7] As of September 1919 the Bolsheviks controlled only about one third of the pre-1917 network in terms of route mileage, and their stock of operable locomotives was shrinking rapidly. The resultant shortages of fuel, food and raw materials inevitably deepened the economic collapse. But the railways did give the Soviet regime a strategic advantage. Soviet troops could be moved relatively quickly between battlefronts because the Bolsheviks held the hub of the railway network at Moscow. By contrast, the White armies struggled with the comparative paucity of connecting lines between the arterial routes in the distant provinces.

Thanks to this advantage, among many other reasons, the war turned decisively in the Bolsheviks' favour in the winter of 1919–1920. During October the Red Army reeled from two White offensives. General N.N. Iudenich advanced from the Russo–Estonian border and reached Krasnoe Selo, only a few dozen kilometres from Petrograd. Concurrently General A.I. Denikin moved north from Khar'kov and Belgorod, and seized Orel, just 380 km from Moscow. Yet the Red Army responded successfully. Iudenich retreated to Estonia, where his army was interned and disarmed in December. Denikin's forces were repelled to Ekaterinoslav and Rostov; having gathered the survivors in the Crimea, Denikin relinquished his command to General P.N. Vrangel'. Meanwhile the other large White army fled eastwards across Siberia; its leader, Admiral A.V. Kolchak, was captured and shot near Irkutsk. Thereafter, with the foreign troops being withdrawn from European Russia, the military threat to Lenin's regime proved less acute. In April 1920 a Polish invasion made rapid progress, but after three weeks the Red Army began a counter-offensive that almost captured Warsaw; the subsequent Polish–Soviet peace treaty of March 1921 effectively confirmed Soviet Russia's western borders. Vrangel', meanwhile, had scant success: by December 1920 the remnants of his army had been evacuated abroad. The later fighting was basically localized and sporadic, and by 1923 most of the former Russian empire was in Bolshevik hands.[8]

---

[7]   The physical damage was probably less than is commonly thought: A.J. Heywood, 'War Destruction and Remedial Work in the Early Soviet Economy: Myth and Reality on the Railroads', *The Russian Review*, 64/3 (2005): 456–79.

[8]   Mawdsley, *Russian Civil War*, pp. 194–271.

With the economy, however, the Bolsheviks found resounding triumphs more elusive.[9] The situation was so threatening by December 1919 that, sensing their military victory over the Whites, Lenin and his colleagues made the economy their first priority. So doing, they tried to focus on one severe danger – an echo of their successful military tactic of concentrating their scarce resources against the army that seemed most threatening at the given time. And because they defined the transport crisis as the worst economic threat, their battle-cry 'Everything for the Front!' became 'Everything for Transport!' Yet they also expected to make rapid progress in building a modern socialist economy. Lenin, for instance, spoke of solving the economic crisis in weeks and starting to transform the economy.[10] This defiant confidence, coupled with his party's ideological commitment to technology-led industrialization, helps explain his enthusiasm at this time for the so-called GOELRO project for nationwide electrification.[11] It also helps explain why, following the end of the Allied blockade of Soviet territory in January 1920, the regime decided to try to buy a vast quantity of railway supplies abroad, including 5,000 locomotives, and why, crucially, 300 million gold rubles – about 40 per cent of the Soviet gold reserve – were allocated as initial funding for this plan in March 1920.[12] In reality, of course, the economy continued to collapse and the dream of rapid industrialization receded. In early 1921 Lenin endorsed a political compromise that included the partial revival of market forces to stimulate food production and a slower schedule for industrial recovery. Such was the core of the New Economic Policy (NEP) that, to the consternation of numerous party members, was introduced by the Tenth Party Congress to replace War Communism in March 1921. This apparent retreat could not avert the terrible famine of 1921– 1923, but it did facilitate a gradual recovery of industrial and agricultural output, with the levels of 1913 being regained in most sectors in the mid-1920s. Rapid economic modernization would not resume in earnest until the NEP's demise, the rise of Stalin and the launch of the First Five-Year Plan during 1927–1928.[13]

A fundamental paradox of the Soviet economy in the decade from 1917 was its reliance on people whom the Bolsheviks heartily distrusted. The NEP's revival of food production depended on the peasants most able to produce sufficient food to market their surplus, whom the regime saw as petty-bourgeois. An earlier, analogous case was the mobilization of former tsarist army officers in 1918 to help build and lead the Red Army. In the industrial arena, including the railways, an equivalent compromise was made with the engineers and managers whom the Bolsheviks derided as 'bourgeois specialists', or *spetsy*. These people had

---

[9]	See especially Nove, *Economic History*, pp. 46–82; S. Malle, *The Economic Organisation of War Communism, 1918–1921* (Cambridge, 1985).

[10]	Heywood, *Modernising Lenin's Russia*, p. 76.

[11]	See Coopersmith, *Electrification of Russia*, pp. 121–257.

[12]	Heywood, *Modernising Lenin's Russia*, pp. 48–82.

[13]	On the NEP-era recovery see in particular Nove, *Economic History*, pp. 83–118; and Davies, *From Tsarism to the New Economic Policy*, *passim*.

the professional expertise that was essential for their industry, but usually had a relatively privileged social background, and were therefore normally assumed to be anti-Soviet in their politics. Indeed, few engineers were Bolsheviks, two of the leading ones being L.B. Krasin and G.M. Krzhizhanovskii, both of whom Lomonosov knew from 1905. This dependence implied the possibility of a concerted technocratic challenge to the Bolsheviks' grip on political power, especially in view of the growing political influence of the engineering profession in the United States.[14] But the Bolsheviks were more worried that many of these experts would commit individual acts of sabotage. Accordingly, party members were appointed as 'commissars' to shadow specialists who held important responsibilities. A notable effect of the October revolution in industry was therefore to politicize routine administrative and technical decision-making, with supervisory power entrusted to Bolsheviks who probably lacked the requisite expertise. Clearly, as shown by the executions of Pauker and Tikhvinskii, the cost of arousing suspicion could be dire for the specialist. But generally it has been thought that the relationship between these people and the Bolshevik Party was relatively stable, if tense, until at least the mid-1920s, and that a terror campaign against specialists did not begin until 1928.[15]

Yet Lomonosov's place in Soviet society was never stable. Even as he travelled to Russia in 1919 he experienced doubts that challenged his sense of purpose. At that time his personal drive came from the way that his support for the Soviet system – if not yet fully for Bolshevism – reinforced his notion of patriotic duty:

> Not having been in Russia for two years it is hard for me to judge why economic life in Russia is being put right so slowly; I think the main reason is this absence among the defenders of Soviet power of people with a business or technical background. That is why I consider it my duty to strive to get to Russia and offer my services. That is why I considered myself justified in spending government money to achieve this. I think, I feel that I can be useful.[16]

He even showed signs of career ambition: he was intrigued by a (false) rumour of his appointment as People's Commissar of Ways of Communication.[17] But he also feared rejection, and not just by the Bolsheviks. His response to news in June 1919 that Moscow was refusing to exchange Kalamatiano for him was to conjecture that the party leadership did not want him for political reasons or that NKPS 'enemies' were opposing his return.[18] In reality, he would soon discover in Moscow that these worries were mistaken. There, however, the crux of his situation became his

---

[14]　See E.T. Layton, *The Revolt of the Engineers: Social Responsibility and the American Engineering Profession* (Cleveland, 1971).

[15]　The classic study of this relationship is Bailes, *Technology and Society*.

[16]　'Dnevnik', 12 July 1919. See also entries for 17 and 24 July 1919.

[17]　'Dnevnik', 29 June 1919.

[18]　'Dnevnik', 6, 19, 20 June, 9, 15, 24, 31 July 1919.

decision not to join the Bolshevik Party and instead to continue defining himself politically as a 'non-party Marxist' who was loyal to the Soviet political system and government. So far as he was concerned, he had nothing in common politically with the majority of the scientific and technical specialists, whom he regarded as anti-Soviet. Unfortunately for him, however, this stance ignored the reality that the Bolshevik Party was treating even Mensheviks – fellow Marxists – as politically suspect, and by 1922 it would help fuel suspicion that he was engaging in anti-Soviet activity.

Lomonosov's adoption of this semi-independent political position was surely encouraged by the warmth with which the Bolshevik Party's leaders greeted his return. Especially important for this welcome were his revolutionary activism in 1905–1907, his prominence in a vital industry, and Lenin's endorsement. Lomonosov was fortunate that the NKPS was headed by Krasin, one of the few people familiar with his revolutionary activities in Kiev. Additional support probably came from other Bolshevik acquaintances now in government like Krzhizhanovskii, Pravdin and Shlikhter. Among other things Krasin urged him to help lead the technical transformation of the railways by chairing the new NKPS Technical Committee, while Krzhizhanovskii and others pressed him to chair the Committee for State Construction Projects (Komgosor), which formed part of the Supreme Council of the National Economy (VSNKh) and oversaw the building of roads, railways and other non-military infrastructure. Given Lenin's backing for both ideas, Lomonosov was appointed to the NKPS Technical Committee and the Komgosor by 10 October 1919, and to the VSNKh Presidium by 28 October.[19] By November Lenin was consulting Lomonosov frequently about transport problems.[20] Meanwhile, friendships developed with other party luminaries such as the Politburo member L.B. Kamenev and the chairman of the VSNKh, A.I. Rykov. Kamenev was delighted that Lomonosov had known his father (the traction engineer B.I. Rozenfel'd) and became a sort of political patron for him; the friendship included their families in due course.[21]

---

[19] 'Dnevnik', 8, 10, 18, 29 October 1919; Golikov et al., *Lenin: Biograficheskaia khronika*, vol. 7, pp. 608, 610. Lomonosov does not mention whether a commissar was assigned to monitor his work; if he did not have a commissar, which is conceivable, that would be a clear sign of the party leadership's trust.

[20] For instance, 'Dnevnik', 28 November 1919.

[21] For example, 'Dnevnik', 23 May 1920; 21 May, 13, 22 June, 23 October 1922; 'Vospominaniia', vol. 7, pp. 610, 724, 731–2, 782, 798, 945, 1195–6; vol. 8, pp. 604, 1108–12, 1176–6, 1204–5, 1494–8. On Kamenev see D. Shelestov, 'L.B. Kamenev (1883–1936): Odin iz vidneishikh bol'shevikov i kommunistov...', in A. Proskurin (comp.), *Vozvrashchennye imena: Sbornik publitsisticheskikh statei v 2-kh knigakh* (2 vols, Moscow, 1989), vol. 1, pp. 211–33. A full biography of him is being prepared by Alexis Pogorelskin. On Rykov see V. Kolodezhnyi and N. Teptsov, 'A.I. Rykov (1881–1938): Predsedatel' Sovnarkoma', in Proskurin, *Vozvrashchennye imena*, vol. 2, pp. 149–75; A.S. Senin, 'Aleksei Ivanovich Rykov', *Voprosy istorii*, 9 (1988): 85–115; S.A. Oppenheim, 'Aleksei

The way that Lomonosov saw himself in an intermediate position between the party and the specialists is evident from his first letters to Raisa from Russia. These are permeated by his political alignment with the regime, yet reveal a certain professional affinity with the specialists:

> The more daring among [the intelligentsia] run to Denikin or try to help him here. The more passive element among them and perhaps the more educated of them, have buried themselves in scientific research; they write books and work in laboratories (at this time!). The most passive and the most numerous sit in their offices and drown the revolution in a sea of ink. It is a formal sabotage ... It is enough to say that in the offices in Moscow there are about five thousand civil engineers and it is impossible to make them go out and work in the factories or on the railroads.[22]

In this way, he concluded, the intelligentsia was presenting such a difficult problem for the Moscow regime that '*even I* begin to feel that there can be no compromise with them' (emphasis added); in such circumstances it was 'obvious that all specialists' would be 'looked upon with suspicion'. On the other hand, he attacked arrogance among party officials concerning technical expertise. 'Everyone' understood the necessity of using the specialists, but 'everyone' distrusted them too, and was thus trying 'semi-consciously' to assume control in the mistaken belief that they could soon gain the expertise: 'There are many narrow-minded, self-contented people who think that a communist can overreach any specialist in two weeks.' The self-defeating result, in his opinion, was to frighten the experts into working negligently and evading their duties.[23]

In this context the party leadership's trust had the ironic effect of setting Lomonosov on a collision course with party members holding important positions in the economy. At the NKPS, for example, he clashed with three Collegium members who were basically party appointees. These were P.N. Mostovenko, who headed the Technical Directorate; B.M. Sverdlov, head of the Organizational Department and inspectorate, and brother of Lenin's deceased comrade Ia.M. Sverdlov; and S.D. Markov, Deputy People's Commissar since February 1919

Ivanovich Rykov (1881–1938): A Political Biography' (unpublished PhD dissertation, Indiana University, 1972).

[22] Lomonosov–Lomonosova, 23 September 1919: LRA, MS 716.4.1.38, and translation at NARG 261, entry 24, box K170, file 52 Transfer of Mission to control of A.I. Lipets: correspondence of Lomonosov.

[23] Lomonosov–Lomonosova, 23 and 29 September 1919: LRA, MS 716.4.1.38–39 and translations at NARG 261, entry 24, box K170, file 52 Transfer of Mission to control of A.I. Lipets: correspondence of Lomonosov. See also, for example, 'Dnevnik', 22–3 September 1919.

and head of the Operations Directorate.[24] Exceptionally for a non-party engineer, Lomonosov could and did challenge them for influence. This situation became especially troubling for Mostovenko and Markov because they oversaw the matters of greatest interest to him.[25] Had Lomonosov been a party member these conflicts might have been defused easily as an issue of party discipline; in practice the tensions would become acute.

So why did Lomonosov continue to define himself as a non-party Marxist? His reasoning reflected both integrity and naivety. He quickly understood that leadership positions were reserved for party members, yet he did hope to lead the NKPS. As early as 23 September 1919 he wrote to Raisa: 'I asked the other day for the appointment of Commissar of the Tashkent Railway. Couldn't get it – I am not a communist. Just now, after the failure of the Bullitt negotiations and discovery of a series of low plots to help Denikin, all offices of responsibility are filled only with communists.'[26] But equally these comments suggest that for him the restrictions were just a temporary emergency measure; he seems not yet to have sensed the fundamental importance of party membership. Indeed he believed, wrongly, that Krasin, Sverdlov and Krzhizhanovskii were not members.[27] In any case, he still disagreed with aspects of Bolshevik policy, and he also thought – with perhaps some justification – that he could do more as a non-member to develop support for the Soviet system among specialists and foreigners.[28]

Later, as we shall see, Lomonosov would downplay such general political issues when analysing his loss of favour in Soviet Russia, and instead gave greater weight to certain professional issues. But whatever the validity of his own views it is clear

---

[24]   Mostovenko's role in 1905 is noted in H. Phillips, 'Riots, Strikes and Soviet: The City of Tver in 1905', *Revolutionary Russia*, 17/2 (December 2004): 60. Markov's appointment is noted in Argenbright, 'Russian Railroad System', p. 250; biographical notes are in V.I. Lenin, *Polnoe sobranie sochinenii*, vol. 51 (Moscow, 1965), p. 509.

[25]   'Dnevnik', 18, 28 October, 1, 27 November 1919.

[26]   Lomonosov–Lomonosova, 23 September 1919: LRA, MS 716.4.1.38, and translation at NARG 261, entry 24, box K170, file 52 Transfer of Mission to control of A.I. Lipets: correspondence of Lomonosov.

[27]   'Dnevnik', 23 September 1919; 'Vospominaniia', vol. 7, pp. 423–4.

[28]   'Dnevnik', 15 June 1920; 'Vospominaniia', vol. 7, pp. 174–83, 405–8. Igolkin cites a secret memorandum to the effect that Lomonosov was a party member by July 1921: 'Leninskii narkom'. Igolkin does not footnote his claim, but his source may be a secret police report of 21 July in which Lomonosov is called a 'communist', or a similar document from the same period which states that he joined the communist party in America: RGASPI, f. 76, op. 2, d. 76, ll. 23, 27. However, we should note that the report of 21 July also uses 'citizen', which normally implies that the person was not a party member; that Lomonosov's diary makes no reference to the American communist party, and indicates that he did not join the Bolshevik Party; and that his wife Raisa did join the American communist party (this last point is noted in a non-paginated memorandum of 15 September 1922 in Rossiiskii gosudarstvennyi arkhiv ekonomiki (RGAE), f. 4038, op. 1, d. 19). In this light Igolkin's source seems likely to be wrong; possibly it confused Lomonosov with Raisa.

now that he could not win concerning party membership. As long as he did not join the party, most people in the society would categorize him as a *spets* because of his social background and profession. Yet it is certain that even if he had joined, sooner or later he would have been accused of careerism and opportunism, as happened to countless others. Indeed, a crisis began to develop in his career during the winter of 1919–1920. He did remarkably well in consolidating his connections at the highest levels of government. But at the same time hostility towards him became widespread among colleagues and subordinates, both for political reasons and because of his own choices. These two phenomena can already be identified in his brief reign at the Komgosor in late 1919, and they came to the fore on the railways in early 1920.

The immediate sources of his Komgosor troubles were his faith in reorganization, his commitment to strict labour discipline, his readiness to act ruthlessly and his apparent lack of relevant expertise – none of which would have been such serious problems had he been a party member. Komgosor staff already had the measure of crisis management by reorganization: ordered in June 1919 to make staff cuts, they responded by creating a Main Directorate, merging two other directorates and requesting expansion.[29] But their new chief was an energetic non-party outsider who perceived widespread duplication, inadequate coordination and a tendency verging on sabotage to ignore instructions.[30] Lomonosov's strategy included a peculiar and perhaps self-serving proposal to merge Komgosor with the NKPS Technical Directorate to create a People's Commissariat of Public Works.[31] More immediately, he used potentially lethal 'revolutionary measures' against alleged sabotage and indiscipline. For instance, he consigned two specialists to a 'concentration camp' for three months for unauthorized travel.[32] Whether he deployed this tactic partly to prove his revolutionary machismo is unknown, but the idea was certainly his: neither his predecessor nor his successor used it.[33] Lenin duly commended his vigour in evicting saboteurs.[34] But Komgosor staff were sceptical. To quote one engineer:

---

[29] Report to VSNKh Presidium by Komgosor Chairman, 25 August 1919: RGAE, f. 2259, op. 2, d. 22,l. 9–10ob.; Order No. 4, 15 January 1920: RGAE, f. 2259, op. 2, d. 46, l. 4.

[30] 'Dnevnik', 12, 15–16, 18, 20, 24, 28 October, 1 November 1919.

[31] On the question of reforming the KGS: A note by Iu.V. Lomonosov, [November 1919]: RGAE, f. 2259, op. 2, d. 16, ll. 38–9ob.

[32] Komgosor Order No. 5, 17 January 1920: RGAE, f. 2259, op. 2, d. 46, l. 6ob.

[33] See the departmental orders in RGAE, f. 2259, op. 2, d. 46 and, for example, 'Dnevnik', 5 November 1919. Lomonosov was replaced at the Komgosor from 2 March 1920 by K.A. Alferov: Komgosor Order No. 17, 19 March 1920: RGAE, f. 2259, op. 2, d. 46, l. 19. Since January 1920 Alferov had been supervising the central directorates and planning their reorganization: Komgosor Order No. 1, 10 January 1920: RGAE, f. 2259, op. 2, d. 46, l. 1.

[34] 'Dnevnik', 2 November 1919.

Engineers of Ways of Communication have headed and continue to head the managements of the railways in use. This can be harmonized with everything else when the economy's situation is no higher than a certain level. But if we're aiming to achieve the maximum, a long-term differentiation of labour is essential here too. It's not appropriate for the head of a railway to lose in his past the enormous amount of work needed to acquire the knowledge and skills of construction work. In operational matters he will simply be a dilettante. He needs a completely different educational and career experience.[35]

It is thus hardly surprising that Lomonosov's initiatives were resisted, and indeed were abandoned after he left the Komgosor to concentrate on NKPS work.[36]

His prospects at the NKPS were better, despite the hostility in the Collegium.[37] His assignment to chair the Technical Committee – ideal for his seniority and expertise – gave him a notable role in technical policy-making that was similar to the chairmanship of the old MPS Engineering Council. His professional ethos and interests matched the committee's remit to employ scientific and technical research to support the survival, restoration and modernization of the railways. He could use it to develop a replacement for his Experiments Office, which evidently ceased to exist as such at some point during 1917–1919.[38] Also, he could hope to build a sense of common purpose between engineers and government. Not only had many of the staff worked on his locomotive experiments, but also he shared the iconoclastic enthusiasm of those engineers who saw the economic crisis as an unprecedented chance to start afresh – in this instance, to develop an ultra-efficient modern railway network. Revolutionary in its own way, such thinking was a significant element of shared logic between specialist engineers and the Bolshevik Party, which Lomonosov could embody.[39]

---

[35] Conclusion about the note of Iu.V. Lomonosov regarding the reform of the KGS, 6 November 1919: RGAE, f. 2259, op. 2, d. 16, ll. 62–4.

[36] Note about the question of reforming the Komgosor, 5 November 1919: RGAE, f. 2259, op. 2, d. 16, ll. 54–5ob; resolution of Komgosor Collegium about Lomonosov's report, 11 November 1919: RGAE, f. 2259, op. 2, d. 16, ll. 39ob.–40; resolution of Komgosor Collegium, 8 December 1919: RGAE, f. 2259, op. 2, d. 16, l. 15; Order no. 2/31, 1 April 1920: RGAE, f. 2259, op. 2, d. 46, l. 34; 'Dnevnik', 14 November 1919.

[37] Extracts from Lomonosov's memoirs about his NKPS activity in 1919–1920 have been published as Iu.V. Lomonosov, 'V Narodnom Komissariate Putei Soobshcheniia, noiabr' 1919–ianvar' 1920 (Otryvok iz vospominanii i neizdannaia korrespondentsiia V.I. Lenina)', edited by H.A. Aplin, *Minuvshchee: Istoricheskii almanakh*, 10 (1990): 7–63.

[38] The precise fate of the Experiments Office is unknown. As for its papers, in 1917 Lomonosov took a large selection of documents to the United States, where they remained with Lipets after the turmoil of June 1918, and these have since disappeared. The remaining papers were presumably with the NKPS Technical Committee by the early 1920s, but they do not appear to be in the NKPS collection at RGAE and may well have been destroyed.

[39] On this general point about how railway engineers saw the destruction as an opportunity for technical modernization see Heywood, *Modernising Lenin's Russia*,

That he could unlock this potential is suggested by the way that he immediately galvanized the staff to work on a remarkable variety of projects, many of which had been noted by Krasin at the committee's inaugural meeting in September 1919. Among them were the calculation of how many new locomotives were needed to replace obsolete and war-damaged stock, a plan for the temporary closure of low-priority routes, designs for high-capacity freight wagons, specifications for replacing steam locomotives with electric and diesel traction, plans for fitting automatic brakes and couplings, and outline planning for a network of 'supermainlines' for high-speed long-distance freight traffic using new types of high-power locomotive. Most of these ideas predated the 1917 revolutions; the difference now was optimism that the new regime might fund them.[40]

However, Lomonosov's primary concern at the NKPS was power. Although intrigued by the Technical Committee's projects, he pined for 'active work' – a formula that encompassed his desire to head the commissariat. He clearly expected Krasin to quit the NKPS to lead the growth of Soviet foreign trade following the end of the Allied blockade, which began to appear imminent in late 1919. In the meantime his aggressive lobbying was bound to antagonize colleagues like Markov, for he denigrated the NKPS's anti-crisis policies. For example, as the sole railwayman in a government commission that investigated the NKPS in November 1919 he proposed a Collegium of just five members – a reform that would evict Mostovenko, among others.[41] To Lenin he pressed his old complaint that managers were isolated from the lines. He praised Markov individually for making contact with lines, but damned him and more generally the Collegium for lacking a coherent plan.[42]

Lomonosov's analysis of the railway crisis was unremarkable, at least for the average engineer, but as at the Komgosor, his proposed strategy showed his assimilation of the regime's life-or-death mentality. He attributed the huge traffic delays primarily to chaos in train management; recognized that fuel shortages were disrupting even urgent military traffic; objected that the rampant use of VIP saloon carriages wasted locomotives and line capacity; and argued that the Cheka's enforcement of martial law was stirring counter-revolutionary sentiments among ordinary workers.[43] His prescription was largely an old hobby-horse: from the People's Commissar down, managers should lead by example, getting out of their offices to roam the rails and solve problems. However, he was also keen for them to shoot saboteurs.[44]

---

pp. 48–63.

[40] 'Vospominaniia', vol. 7, pp. 502–22, 643–62; Heywood, *Modernising Lenin's Russia*, pp. 61–2.

[41] 'Dnevnik', 21, 27 November 1919.

[42] 'Dnevnik', 7, 13, 28 November 1919.

[43] 'Dnevnik', 10, 13, 27–8 November, 24 December 1919, 31 March, 13 July 1920.

[44] 'Dnevnik', 13, 28 November 1919.

His appointment to the NKPS Collegium in December 1919 greatly strengthened his position.[45] He did not simply join the commissariat's leadership, but took control of the two key directorates that handled technical and operating affairs, ousting Mostovenko and defeating counter-arguments from Markov.[46] He thereby effected, at least informally, the merger of technical and operations policy-making for which he had been preaching since his time at the Kiev Polytechnic Institute: the new era was offering opportunities for radical new thinking. True, Markov was to retain overall responsibility for all NKPS directorates, for liaison with the Central Directorate of Military Communications, and for presenting NKPS reports to the key government bodies overseeing the economy, notably the Council of People's Commissars (Sovnarkom) and Defence Council (soon to be the Council of Labour and Defence – STO).[47] But Markov's authority was shaken, which fuelled Lomonosov's aspiration to become the People's Commissar.

Two events soon emphasized his prestige. First, he was instructed to prepare and deliver a major public speech: a report about the transport crisis for the Third Congress of Councils of the National Economy, which was due to convene in Moscow in late January 1920.[48] Second, at Lenin's behest he addressed the Defence Council on 24 December 1919. To both audiences he argued that without decisive initiatives the railway system might grind to a halt within weeks. Whether or not the council meeting was as dramatic as Trotskii's memoirs suggest, it seems certain that Lomonosov did much to convince the political leadership about the seriousness of the railway crisis.[49]

As a Collegium member he prioritized troubleshooting tours of the type that he had been advocating. He began with three one-day trips within Moscow. On 3 January 1920 he visited the Northern Railway's headquarters with an entourage of about 15 traffic, track and traction experts to assess the line's organization and activity. Similar inspections of the Moscow–Kazan' and Moscow–Kursk railways followed on 10 and 17 January.[50] His focus thereafter was the southern railways that had just been captured from Denikin, plus their connections to Moscow. Concentrating on the area's five large ex-private railways – the Riazan'–Urals, Moscow–Kiev–Voronezh, South Eastern, North Donetsk and Vladikavkaz railways – he argued that reorganization was urgent. The inter-railway boundaries reflected the chance pattern of private ownership, not the interests of efficient management

---

[45]　'Dnevnik', 3, 7 December 1919.

[46]　Minutes of NKPS Collegium, 16 December 1919: RGAE, f. 1884, op. 28, d. 1, l. 121.

[47]　Minutes of NKPS Collegium, 16 December 1919: RGAE, f. 1884, op. 28, d. 1, l. 121. On the government structures see T.H. Rigby, *Lenin's Government: Sovnarkom, 1917–1922* (Cambridge, 1979).

[48]　'Dnevnik', 31 December 1919; *Ekonomicheskaia zhizn'*, 24 January 1920, p. 2; 25 January 1920, p. 2.

[49]　'Dnevnik', 28 December 1919; 'Vospominaniia', vol. 7, pp. 828–30; L.D. Trotsky, *My Life* (New York, 1960), pp. 462–3.

[50]　'Dnevnik', 10–11 January 1920; 'Vospominaniia', vol. 7, pp. 887–902.

of regional and national traffic flows; and reorganization would hinder any bid to restore private ownership by anti-Soviet *spetsy* and/or other supporters of the Paris-based émigré Union of Private Railways. Thus, for example, he proposed that the main coal-loading points be grouped within one railway, to be called the Donetsk Railway; this entity could be created by adding parts of the South Eastern and Catherine railways to the North Donetsk Railway. Having piloted his proposals through the NKPS Collegium, he travelled south after the debate on his speech at the Congress of Councils of the National Economy. Unhappily his tour was terminated after three weeks when he caught typhus and was brought back to Moscow.[51] Having survived, he cut short his convalescence in early April to return to the south, where he worked until mid-May.[52]

Overall it is doubtful that his work during January–May 1920 contributed much to the key achievement of the Soviet railways at this time, which was to avert total sclerosis by stabilizing the proportion of 'sick' locomotives at about 60 per cent. True, Lomonosov acquired the responsibility for locomotive and wagon repairs, and stressed this issue to the Defence Council in December 1919. But he focused on regional and local situations rather than broad strategy and coordination. Absent from Moscow or ill, he played no part in developing the commissariat's unprecedented five-year plans for repairing the locomotive and wagon stocks, which became famous as NKPS orders 1042 and 1157 and marked the regime's first attempt at long-term economic planning.[53] His reorganizations were criticized for causing more disruption than they cured; by late April the NKPS was often countermanding his decisions in the south.[54] In any case, the decisive development was surely the capture of several thousand 'healthy' engines from Denikin and Kolchak, which could be sent to the busiest lines.[55]

In career terms, however, Lomonosov's position appeared to improve remarkably during these months. The party leadership may have treated his friction with the railway *spetsy* as a good sign: that attitude underlies some of the comments

---

[51] 'Dnevnik', 25 January to 12 February 1920; 'Vospominaniia', vol. 7, pp. 873–4, 879–881, 905–12, 919–22, 951–1049. Some of his telegraph correspondence with Moscow during this trip is collected in RGAE, f. 1884, op. 92, d. 245; some of the later messages are misdated as March 1920.

[52] Medical certificate, 1 April 1920: LRA, MS 716.1.97; Lomonosov–Lomonosova, 28 May 1920: LRA, MS 716.4.1.42; 'Dnevnik', 3 April to 19 May 1920; 'Vospominaniia', vol. 7, pp. 1084–180.

[53] For these plans see Ia. Shatunovskii, *Vosstanovlenie transporta: Puti soobshcheniia i puti revoliutsii* (Moscow and Petrograd, 1920), pp. 23, 25–31; Heywood, *Modernising Lenin's Russia*, p. 138.

[54] For example: 'Dnevnik', 1, 3 May 1920; 'Vospominaniia', vol. 7, pp. 1146–51; Typescript of notes by Lenin, circa 24 May 1920: RGASPI, f. 2, op. 1, d. 14079, l. 5; *Leninskii sbornik*, vol. 38, p. 312.

[55] Heywood, *Modernising Lenin's Russia*, p. 136.

made to Lenin in May 1920 and a secret police report of 1921.[56] Furthermore, Markov was demoted in March 1920 after his feud with Lomonosov was reported to the party leadership by V.V. Fomin, the Commissar of the NKPS Operations Directorate.[57] At the same time Lenin identified Lomonosov as potentially the best person to head the NKPS for the few months that Krasin was expected to spend abroad as head of the first main Soviet trade mission to western Europe. In practice, partly because Lomonosov was still recovering from typhus, Trotskii became acting People's Commissar from 25 March as a temporary measure. Apparently Trotskii was to be a political 'shield' for Lomonosov, who would be the de facto People's Commissar. But even this compromise did not materialize because Lomonosov returned to the south in early April and the vastly experienced engineer I.N. Borisov was installed as head of a new Main Directorate of Ways of Communication, which promptly became the railway heart of the NKPS.[58] However, Trotskii raised the question of NKPS leadership again in mid-May: the railway situation was serious, his own appointment was just for the short term, he was preoccupied with the war against Poland in his capacity as commissar for war, and Krasin seemed likely to spend at least a few more months abroad. Thus, following a Politburo discussion on 22 May, Lenin consulted over a dozen leading party and railway officials about whether Lomonosov might be the man to provide the longer-term strong leadership that the NKPS needed desperately.[59]

That this question was broached, let alone investigated, was an extraordinary political development. Lomonosov was not a member of the Bolshevik Party, and although he would have a supervisory commissar – he requested his old acquaintance Pravdin – the principle had long been established that only party

[56] Typescript of notes by Lenin, circa 24 May 1920: RGASPI, f. 2, op. 1, d. 14079, ll. 4, 6; OKTChK (Moscow–Kazan' Railway)–Blagonravov, [circa 1921]: RGASPI, f. 76, op. 2, d. 76, l. 27.

[57] Report to CC by Fomin, [March 1920]: RGASPI, f. 17, op. 112, d. 14, ll. 37–40. See also Golikov et al. (eds), *Lenin: Biograficheskaia khronika*, vol. 8 (Moscow, 1977), pp. 359, 367–9.

[58] 'Dnevnik', 19, 22 March, 3 April 1920; Lomonosov–Lomonosova, 20 March 1920: Swarthmore College Peace Collection, Jane Addams Collection, series 1: Correspondence, 1870–1937, microfilm reel 12, frame 1538; Typescript of note by Lenin to Trotskii, mid-March 1920: RGASPI, f. 2, op. 1, d. 24406. See also J.M. Meijer (ed.), *The Trotsky Papers* (The Hague, 1964–1971), pp. 114–17; Lenin, *Polnoe sobranie sochinenii*, vol. 51, p. 155; and Bogdanovich et al., *Ministry i Narkomy*, p. 178. Recent biographies of Trotskii include I.D. Thatcher, *Trotsky* (London, 2003) and G. Swain, *Trotsky* (Harlow, 2006). I am grateful to Katy Turton for checking the RGASPI reference.

[59] Politburo resolution, 22 May 1920: RGASPI, f. 17, op. 163, d. 67, l. 13; 'Dnevnik', 22 May 1920; Lenin–senior party workers, 22 May 1920: RGASPI, f. 2, op. 1, d. 14068; Typescript of notes by Lenin, circa 24 May 1920, in RGASPI, f. 2, op. 1, d. 14079; *Leninskii sbornik*, vol. 37 (Moscow, 1970), p. 209; Golikov et al., *Lenin: Biograficheskaia khronika*, vol. 8, pp. 519–20, 581–4.

members could hold the top leadership posts in the economy.[60] The fact that the Politburo even considered appointing him to the highest position in an industry vital for both the economy and the Red Army clearly shows that Lenin had the greatest respect for Lomonosov's expertise and, perhaps, that the party leadership was getting desperate.

Sadly for Lomonosov, this tantalizing development produced a serious setback from which his Soviet career would never fully recover. Most of the opinions expressed to Lenin were critical, and there was little support for the idea of making him the People's Commissar.[61] This overwhelming negativity persuaded Lenin to retreat. As a compromise the Politburo appointed Lomonosov as the commissar for a large railway district, which was specified as the Turkestan region of Soviet Central Asia. First, however, he was to visit Sweden for a few weeks at Krasin's request to create the quality control process for one of the first significant Soviet foreign trade operations, whereby the Swedish engineering company Nydqvist & Holm AB (Nohab) would supply 1,000 freight locomotives by 1925.[62] Lenin portrayed the Turkestan job to Lomonosov as a chance for him to 'rebuild' his position, and Trotskii called it an opportunity to build a political reputation, given that such senior positions were normally reserved for party members. But Lomonosov was dismayed. He had sought this sort of job back in September 1919, and had recently suggested it again, albeit flippantly; but the real possibility of becoming the People's Commissar had captured his imagination.[63] In the event, he would always enjoy Lenin's support, but he would never manage to rebuild his prestige with the wider party leadership.[64]

Lomonosov did not blame Bolshevik policies and prejudices for this reverse. He suspected that his NKPS 'enemies' had intervened, but mainly he reproached himself. He regretted that he had not cultivated supporters among subordinates, colleagues and key officials, that he had neglected to persuade workers that he was on their side, and that he had accepted Trotskii's assurances uncritically. He understood, too, that his long tours of the south and his illness had excluded him from NKPS office politics. In particular, they ensured that he was not involved in the creation of the NKPS Main Directorate of Ways of Communication in April 1920, which was important because this entity immediately became the operational

---

[60] Lomonosov himself mentioned the party-membership requirement in a letter to Raisa dated 28 May 1920: LRA, MS 716.4.1.42. His request for Pravdin is mentioned in a note by Sklianskii, circa late April 1920, at RGASPI, f. 2, op. 1, d. 14186.

[61] See Typescript of notes by Lenin, circa 24 May 1920: RGASPI, f. 2, op. 1, d. 14079, ll. 1–10.

[62] Politburo resolution, 25 May 1920: RGASPI, f. 17, op. 3, d. 82, l. 3; Politburo resolution, 8 June 1920: RGASPI, f. 17, op. 163, d. 72, l. 10.

[63] 'Dnevnik', 31 May, 1–4, 9, 12 June 1920; 'Vospominaniia', vol. 7, pp. 1217, 1220–31, 1244, 1251, 1257–8.

[64] For a later example of Lenin's support see *Polnoe sobranie sochinenii*, vol. 52 (Moscow, 1978), p. 226.

heart of the railways and its head, Borisov, became Trotskii's deputy in practice if not name.[65] Lomonosov also suspected that his decision not to join the Bolshevik Party was problematic. Above all, however, he blamed a serious train crash that occurred at Novorossiisk on 12 May 1920. That disaster was entirely his fault: he arrogantly rejected local advice to add an extra locomotive to his train to provide extra brake power for the steep descent into the city, then his train's brakes failed, and in the ensuing smash one person was killed, three locomotives were damaged (of which at least one was beyond repair) and rolling stock was destroyed. The Cheka launched a sabotage investigation, and for years afterwards Lomonosov regarded this disaster as the turning point of his Soviet career.[66]

Lomonosov's regrets were to some extent his translation, as one experienced in the ways of the Russian state bureaucracy, of Lenin's explanation that the Politburo had three main worries when considering his candidacy to head the NKPS: that he had been a very senior tsarist official; that the masses did not know him; and that he had caused the Novorossiisk disaster.[67] But although these issues did surface in the comments made to Lenin during the consultation process, they were by no means the whole story. Clearly the crash did not deter Lenin and the Politburo from considering his candidacy, and in fact only one person mentioned it to Lenin.[68] Far more important in their assessments were questions of political attitude, expertise, managerial aptitude and character. Thus, for example, Pravdin spoke of his work ethic, honesty and pre-war political assistance, and several people interpreted his friction with 'the *spetsy*' as a good sign; but Fomin damned him as a 'Pompadour' and 'petty tyrant' who was 'of little use' for transport, while Krzhizhanovskii intimated that the radical professor of 1905 had become a reactionary bureaucrat through associating with minister Rukhlov and that he behaved like a tsarist general at the Komgosor in 1919. Sverdlov, a fellow member of the NKPS Collegium, even called him a 'bad administrator' and 'impossible character' who was 'unbalanced' and ill-disciplined, and asserted that his two southern tours had produced nothing but trouble.[69]

More generally, Lomonosov underestimated the strength of ill feeling that he had caused over the previous six months at all levels of the railway hierarchy. For example, a sense of how minor party officials could perceive him as an out-of-control *spets* can be gleaned from a complaint sent to Krasin after the inspection visit to the Moscow–Kazan' Railway in January 1920. That line's former commissar, a man

[65] 'Dnevnik', 20 May 1920; 'Vospominaniia', vol. 7, pp. 1137–8, 1168–70.

[66] 'Dnevnik', 13, 16–17 May 1920; 'Vospominaniia', vol. 7, pp. 1164–72. See also the diary entry for 18 June 1920.

[67] 'Dnevnik', 18, 27 June, 13 July 1920; 'Vospominaniia', vol. 7, pp. 1160–61, 1167–8, 1225–6, 1320–23.

[68] See Typescript of notes by Lenin, circa 24 May 1920: RGASPI, f. 2, op. 1, d. 14079, l. 4.

[69] Typescript of notes by Lenin, circa 24 May 1920: RGASPI, f. 2, op. 1, d. 14079, ll. 3, 7–9.

called Piatnitskii, was furious because Lomonosov had accused him of concealing information and dismissed a reliable engineer. Having sarcastically 'admired' Lomonosov's ability to judge situations quickly, Piatnitskii concluded:

> This whole incident is merely the best proof of how much Lomonosov himself is 'completely uninformed' about the actual conditions of current work and of how difficult it is for him to renounce old capitalist ideals and customs.[70]

A few weeks later a certain I. Beliakov attacked Lomonosov publicly in a newspaper article, arguing that the railway crisis had far deeper roots than Lomonosov believed and that the recovery strategy should include measures to prevent locomotives becoming 'sick', rather than just concentrating on repair work.[71]

One issue not mentioned by Lomonosov as a possible part of the explanation was corruption. One Russian author, A.A. Igolkin, has alleged that during the winter of 1919–1920 he acted corruptly with a project to build a railway and pipeline for moving oil from Emba (on the Tashkent Railway) to the Saratov area.[72] But this allegation seems unlikely to be true: he had little to do with this project; even his sternest top-level critics like Krzhizhanovskii and Sverdlov did not mention corruption to Lenin in May 1920 – on the contrary, several people emphasized his honesty; and this case was not raised during 1922–1923 when a determined effort was made to prove that he was corrupt.

The setback generated one of Lomonosov's most controversial publications. He hoped to evade the move to Turkestan and salvage his position in Moscow by writing newspaper and journal articles that would raise his public profile, answer his critics and establish him as a strategist at one with the revolutionary times. In his view, Borisov and most of the senior NKPS experts wanted merely to restore the railways to their pre-war condition, and failed to grasp that Russia had an unprecedented opportunity for radical technical, technological and administrative change. He reckoned that he could enthuse party members, ordinary workers and

---

[70]   Piatnitskii–Krasin, circa January 1920: RGAE, f. 1884, op. 92, d. 195, l. 305–ob. Lomonosov's inspection report, dated 12 January, is at l. 306.

[71]   I. Beliakov, 'O transporte', *Ekonomicheskaia zhizn'*, 22 February 1920, p. 1. This author may be N.K. Beliakov, a member of the NKPS Collegium; if that is the case, this article would be an unusual case of a public dispute between two high-ranking Soviet officials.

[72]   A.A. Igolkin, 'Algemba: Nefteprovod v nebytie', *Ekonomicheskii zhurnal*, 1 (2001): 5–34. Lomonosov's recollections of this project are at 'Vospominaniia', vol. 7, pp. 882–6, 1080. His role in this affair was mentioned in a Russian television documentary called 'Bezumnaia afera Lenina', screened on 29 October 2002, in which he is alleged to have devised the means of taking gold abroad that Russian entrepreneurs are using in the early twenty-first century: Richard Davies–author, 31 October 2002, citing E. Chukovskaia. The idea for such a railway was studied carefully before the revolution, during 1912–1916: Spring, 'Railways', p. 55.

specialists alike by publicizing that vision. The first – and in the event only – such article appeared in mid-June in the influential newspaper *Ekonomicheskaia zhizn'*.[73] A bold manifesto for railway modernization, it presented some of his most profound convictions, including ideas from his experience of 'progressive' techniques in North America. He argued that Russia's railways had four main costly failings: their centralized administrative structure, which was bureaucratic and useless for rapid decision-making; the weakness of most of the locomotives, bridges and rails, which prevented the operation of high-speed high-capacity trains; their policy of sending freights by the shortest route – a long-established principle that was supposed to minimize state expenditure in the context of a distance-based tariff system, but was expensive to implement and often not the quickest option; and their reliance on steam traction, an inefficient technology where only about 7 per cent of the fuel's heat energy was converted into useful work. He objected that restoring the network to its pre-war condition would simply perpetuate these flaws, and that a new philosophy was needed. Thus, he advocated decentralized management whereby the NKPS would provide the strategic leadership and coordination while regions and railways managed day-to-day operations. He wanted the construction of a network of new 'supermainlines' by 1950 for high-speed heavy-weight long-distance freight trains with modern locomotives and high-capacity wagons. In the meantime, long-distance freight traffic should be concentrated on the trunk routes to minimize costs and improve delivery times. Finally, the development and use of diesel locomotives should be an urgent priority because this new technology was much more efficient than steam and more flexible than electric traction. He wanted 20 experimental main-line diesel locomotives to be ordered abroad immediately for comparative testing.

Ultimately, however, this article was significant mainly for its dismal fate that confirmed Lomonosov's isolation in the upper echelons of the NKPS. Its revolutionary agenda echoed the many pleas for modernization that railway engineers had published since 1917; indeed, supermainlines, freight routing and diesel traction had been under discussion before 1914. It was not even the first article to argue that revolutionary Russia had an unprecedented 'now or never' opportunity to create the 'ideal' railway system. However, it was the only such statement by a serving Collegium member, and its attempt to weld these ideas into a coherent vision was rare. In that sense it could be compared with the industrial strategy that Grinevetskii published in 1919 and the work being done under the leadership of G.M. Krzhizhanovskii to draft the GOELRO plan for nationwide electrification. Lomonosov's argument might have had greater resonance had he spoken out six months earlier, when his personal stock was high, the NKPS had a taste for immediate modernization and Lenin's regime launched the GOELRO

[73]   Iu. Lomonosov, 'V kakom vide dolzhny byt' vosstanovleny russkie zheleznye dorogi?', *Ekonomicheskaia zhizn'*, 3 July 1920, pp. 1–2. The idea for this article apparently came from A.I. Rykov: 'Vospominaniia', vol. 7, pp. 1253–4. Lomonosov sent a copy of the typescript to Trotskii: RGAE, f. 1884, op. 92, d. 229, ll. 192–204.

project.[74] But now his position was weak, the economic crisis was deepening, and the NKPS of Trotskii and Borisov was prioritizing repair work to guarantee survival. Almost certainly perceiving the article correctly as a bid for power and influence, the NKPS organized a conference on 14 July to discuss it. Chaired by Trotskii as the People's Commissar, this gathering rejected all Lomonosov's proposals as dangerous distractions. Trotskii attacked him for not realizing that extreme centralization was essential for transport management in a socialist society. Other speakers ridiculed the notion of large-scale reform and research at a time of economic crisis: gradualism was the only viable option. Scarcely any supportive comments were made. Lomonosov claimed a moral victory, but defeat was the reality. Afterwards each NKPS directorate drafted a list of priorities for the restoration process which did include some modernizing measures, but which assumed a gradual pace of change. Effectively the revolutionary dream of building an 'ideal' Soviet railway system was dashed, as were Lomonosov's hopes for power at this time.[75]

Nonetheless he never went to Turkestan. In mid-June, with the party's approval, the Council of People's Commissars appointed him as its plenipotentiary representative for all railway contracts abroad. This brief extended well beyond Krasin's request for him to organize the quality control for the Nohab locomotive contract, and was expected to need months of work rather than weeks. It originated with Lenin and G.V. Chicherin, the People's Commissar for Foreign Affairs, and was devised in part to help consolidate the regime's tenuous diplomatic foothold in Europe.[76] In the event, Lomonosov would retain this post until 1923. Despite his non-party status, he would be the dominant character in the implementation of the Soviet project to import a vast quantity of railway equipment.

Soviet trade plenipotentiary

The railway imports policy became a uniquely important aspect of early Soviet economic strategy and diplomacy.[77] It was devised in January–March 1920 as

---

[74] Why he did not make such a statement earlier is unknown.

[75] K., 'Vosstanovlenie nashikh zheleznykh dorog', *Ekonomicheskaia zhizn'*, 11 July 1920, p. 1; 'K voprosu o vosstanovlenii nashikh zhel. dor.', *Ekonomicheskaia zhizn'*, 25 July 1920, p. 2; 'Dnevnik', 16 July 1920; 'Vospominaniia', vol. 7, pp. 1328–51; key points of reports in response to the article: RGAE, f. 1884, op. 92, d. 229, ll. 205–16; and lists of measures for improving transport, August 1920: RGAE, f. 1884, op. 3, d. 32, ll. 9–11, 17–20. See also Heywood, *Modernising Lenin's Russia*, pp. 136–45.

[76] Heywood, *Modernising Lenin's Russia*, p. 104.

[77] Heywood, *Modernising Lenin's Russia*, *passim*, and associated articles: 'The Armstrong Affair and the Making of the Anglo–Soviet Trade Agreement, 1920–1921', *Revolutionary Russia*, 5/1 (June 1992): 53–91; 'The Baltic Economic "Bridge": Some Early Soviet Perspectives', in A. Johansson et al. (eds), *Emancipation and Interdependence: The*

the first component of an ambitious strategy for urgent economic modernization. Unprecedented anywhere in its scale, and in effect continuing the railway foreign procurement plans of 1914–1917, it envisaged the purchase of 5,000 locomotives, 100,000 freight wagons and hundreds of thousands of tonnes of other supplies.[78] Crucially, the allocation of 300 million gold rubles as initial funding in March 1920 meant that large contracts could be signed as soon as Western companies and governments were willing to trade. The results did not reach the target, but were still enormous by any standard. From over 450 contracts Soviet Russia received several hundred thousand tonnes of railway equipment by 1925, including 1,200 new steam locomotives, 1,700 oil-tanker wagons and over 80,000 tonnes of rail; the actual expenditure was about 220 million gold rubles, which equated to about 30 per cent of the Soviet gold reserve as of March 1920. No other sector of the Soviet economy enjoyed comparable investment at the time. Indeed two of the locomotive contracts may have been the largest individual Soviet foreign orders by value in the first half of the 1920s, if not the entire inter-war period. The episode was also the last time that Russia's railways enjoyed priority as the driving sector for national economic development. Not least, it showed that the introduction of the NEP a year later, in March 1921, marked a retreat from grandiose ambitions of rapid modernization.

Essentially there were three phases in the implementation of this imports policy. The first, lasting from March 1920 to March 1921, was pre-eminent in terms of the total monetary value of the contracts signed. The breakthrough was Krasin's contract of 15 May 1920 for the small Swedish firm Nohab to supply 1,000 new steam locomotives by 1925. Then came Lomonosov's appointment as Sovnarkom plenipotentiary for railway imports. The next landmarks were Lomonosov's contracts of mid-October for German companies to supply 100 locomotives and many spare parts; a contract devised by the People's Commissariat of Foreign Trade (NKVT) for Estonian factories to overhaul 230 Soviet locomotives; and

---

Baltic States as New Entities in the International Economy, 1918–1940 (Stockholm, 1994), pp. 63–85; and 'Breaking the "Window into Europe": A Case-Study of Soviet–Estonian Economic Relations, 1920–24', *Revolutionary Russia*, 12/2 (December 1999): 41–62. This policy was virtually ignored by Soviet historians, and recent Russian attention has mostly been polemical and sensationalist, such as O.G. Kuprienko, 'O "svoekorystnoi pomoshchi zagranichnym kapitalistam …"', *Lokomotiv*, 11 (2000): 46–8 and 12 (2000): 38–41; Igolkin, 'Leninskii narkom'. The latter explains the contracts with an elaborate conspiracy theory where the pivotal figure is a corrupt Lomonosov. A rare balanced view is Genis, 'General ot parovozov', *Politicheskii zhurnal*, 36 (4 October 2004): 78–80.

[78] The figures were calculated by Krasovskii in consultation with Voskresenskii. They indicated what was needed to regain the pre-war quantities of equipment, allowing for war losses and the mass withdrawal of old stock. If the regime reduced these figures, the resultant plan was expected to cover the war losses and immediate repair needs, and the old stock would simply be withdrawn more slowly. See Krasovskii, Historical note about the order abroad of locomotives and tankers, 2 March 1923: RGASPI, f. 76, op. 2, d. 76, ll. 70–ob.

Table 8.1   Principal Soviet railway contracts abroad, 1920–1922 (value above 1 million Swedish crowns at exchange rates of 15 April 1923)

Date	Item	Company and country	Cost (SEK)
27 April 1920	113 class Ye 2-10-0 'decapod' locomotives	Revalis, Estonia	see note 1
15 May 1920	1,000 class E 0-10-0 locomotives	G. Anderson, for Nohab, Sweden	see note 2
5 October 1920	overhaul of 230 class O 0-8-0 locomotives	United Mechanical Factories of Estonia	see note 3
18 October 1920	16,000 locomotive tyres	Krupp AG, Germany	4,064,060
20 October 1920	22,000 locomotive tyres	Henschel AG, Germany	3,575,071
20 October 1920	153,900 boiler tubes	Mannesmann et al, Germany	2,509,381
21 October 1920	100 class E 0-10-0 locomotives	German locomotive-building syndicate, via Nohab, Germany/Sweden	25,741,950
[11 January 1921]	overhaul of 1,500 locos	Armstrong Whitworth Co., UK	see note 4
24 January 1921	5,720 tyres	Krupp AG, Germany	1,056,000
25 January 1921	5,800 tyres	Krupp AG, Germany	1,375,000
21 February 1921	1,350 weights	Holmquist, Sweden	1,082,527
28 February 1921	600 class E 0-10-0 locomotives	German locomotive-building syndicate, via V. Berg, Germany/Sweden	178,613,400
5 March 1921	100,000 boiler tubes	Mannesmann et al, Germany	1,204,500
9 April 1921	58,510 tonnes of rails	Dortmund Union, Germany	13,442,520
13 April 1921	20,000 tonnes of rails	Wolf, Germany	4,114,000
13 April 1921	249,470 boiler tubes	Rheinmetall et al, Germany	1,554,773
25 June 1921	500 oil tanker wagons	Canadian Car & Foundry Co., Canada	8,939,242
25 August 1921	1,000 oil tanker wagons	Leeds Forge Co. and Linke-Hofmann Werke, UK/ Germany	13,027,950
15 October 1921	200 locomotive boilers	Armstrong Whitworth Co., UK	14,667,620
10 March 1922	10 steam turbines	Nohab, Sweden	4,200,570; see note 5

Notes:

1. Cancelled in 1921 due to US government opposition.

2. Originally for a maximum of 275 million Swedish crowns; order reduced to 500 locomotives in July 1922, and valued at 109,112,500 crowns in April 1923.

3. Reduced to 200 locomotives in December 1921, and valued at 4,616,220 crowns in April 1923; further reduced to about 70 locomotives in 1924.

4. Provisional agreement, worth about 60 million gold rubles; final contract not signed, and not valued in crowns in April 1923.

5. Placed by the Railway Mission for the Volkhovstroi dam construction project; used partly as means to reduce factory overheads on the class E locomotives.

Source: Iu.V. Lomonosov, *Predvaritel'nyi otchet Iu.V. Lomonosova o deiatel'nosti Rossiiskoi zheleznodorozhnoi missii zagranitseiu za vse vremia ee sushchestvovaniia (1920–1923)* (Berlin, 1923), Appendix V, pp. 1–11.

the establishment by Lomonosov of an organization called the Russian Railway Mission Abroad to negotiate contracts, monitor product quality and ship the purchases to Russia. In January 1921 Krasin signed a provisional contract for 1,500 locomotives to be repaired in Britain by the Tyneside engineering company Armstrong Whitworth, and on 28 February 1921 a further 600 new locomotives were ordered from the German locomotive-building syndicate. The value of the main contracts signed by 1 March 1921 exceeded 490 million Swedish crowns, or roughly 230 million gold rubles. Technology transfer, it should be added, had no bearing on the locomotive orders: for various practical reasons the NKPS insisted that only the Russian class E 0-10-0 freight design should be purchased in Europe.

The project's second phase, between March 1921 and April 1923, had different priorities amid changed circumstances. The Anglo–Soviet Trade Agreement of 16 March 1921 was expected to enable Moscow to develop commercial and eventually diplomatic relations with many capitalist countries. But for the moment the regime had to curtail its imports programme: most of its gold was bespoken, export income was still negligible and foreign credit remained elusive. Thus, few major railway orders were placed in this period. The largest ones, all agreed in 1921, were for the rails, the 1,700 tanker wagons and 200 locomotive boilers (the latter from Armstrong Whitworth in lieu of the locomotive repair contract). The Railway Mission dealt additionally with several hundred minor orders together with product quality control and shipment to Russia. Meanwhile, within Russia the NEP's emphasis on balanced budgets and minimal state subsidies encouraged Soviet factories to lobby for government contracts. They demanded cancellation of the foreign contracts, which, they claimed, were overpriced and delivering poor-quality products, while Soviet industry was idle for want of working capital. After much debate the Nohab locomotive order was cut to 500 locomotives in July 1922. Lomonosov, much criticized, had his assignment as Sovnarkom plenipotentiary terminated on 15 April 1923. At this point his Railway Mission was abolished and the final phase of this procurement activity began. The remaining work was overseen by an NKPS plenipotentiary attached to the NKVT mission in Berlin (A.I. Emshanov until autumn 1923, then Iu.V. Rudyi), and effectively ended with the delivery of the last Nohab locomotives in December 1924.[79]

---

[79]   Emshanov was shot in 1937 on charges of wrecking, spying and organizing a rightist-Trotskyist group on the railways: Memorial, 'Zhertvy politicheskogo terrora v SSSR': http//lists.memo.ru/index6.htm, citing *Moskva, rasstrel'nye spiski: Donskoi krematorii* (last checked 30 June 2008). Rudyi remained in the NKPS Collegium for many years, but was dismissed as chair of the NKPS Scientific-Technical Council in May 1937, arrested three months later and shot in February 1938: E.A. Rees, *Stalinism and Soviet Rail Transport, 1928–1941* (Basingstoke, 1995), p. 174; V.I. Klimenko et al. (eds), *Magistral': Nachal'niki zhelezhnykh dorog Zapadnoi Sibiri, 1896–2006: Istoricheskie ocherki* (Novosibirsk, 2006), p. 130.

Contemporaries who condemned this imports policy as an expensive blunder had a strong case.[80] It was utopian to believe in early 1920 that the economic crisis could be solved in weeks and that the rapid modernization of the economy could be started almost immediately. To keep the gold reserve as collateral for foreign loans might have been more productive, and other options such as buying supplies for Soviet industry might have given better value for money. Key aspects of the policy's implementation were dubious. For example, Krasin's decision to order so many locomotives from Nohab was motivated by the significant political and financial advantages of agreeing a large foreign contract quickly, with little thought for practicalities of delivery: not least, this small firm was bound to struggle with even the five-year schedule that was agreed, yet urgency had been fundamental to the policy's rationale. Likewise, Krasin's plan to have 1,500 engines repaired in Britain made sense only as bait to land the Anglo–Soviet trade agreement.[81] Arguably, too, the class E was an unwise choice: the design was already a decade old, and like the older Russian standard 0-8-0 it lacked the technical benefits of a front pony truck.[82]

Yet on balance the results were a valuable achievement in difficult circumstances. If the gold reserve was to be spent on imports, the priority for transport equipment could and did strike a chord with foreign governments, demonstrating that Moscow's interest in foreign trade (as opposed to spreading revolution) was serious at a time when the regime had scant credibility abroad. The vast scale of the proposed railway expenditure helped the Soviet effort to overcome foreign hostility and break the Allied blockade of Soviet gold; indeed, in October 1920 it produced the scheme whereby, with the help of Swedish intermediaries, the Russians could spend their gold on German products without any risk of this gold being confiscated by the Allies. The consequent purchases of spare parts and raw materials eased some of the most dangerous shortages during 1920–1922, and thanks to the new locomotives the railway network's material health was much better by the mid-1920s than would otherwise have been the case. It is difficult to believe that, as things turned out, these imports badly harmed Soviet industry: the engineering factories were probably hampered more by a lack of skilled, disciplined labour than by shortages of raw materials and equipment. There is no evidence for the accusation that the quality of the imports was inferior to Soviet quality; indeed, some NKPS sources reported their superiority. Nor do the foreign prices seem

---

[80] These two paragraphs have been derived from the conclusion of my *Modernising Lenin's Russia*, pp. 233–4.

[81] Whether the British government took the Armstrong contract as bait is another matter: see Heywood, 'The Armstrong Affair'.

[82] On the technical pros and cons of a front pony truck see Chapter 5, this volume. By the 1920s the tendency outside Russia was to use a pony truck on main-line freight locomotives except for locomotives intended specifically for shunting (switching, in US parlance). In other words, internationally the balance of opinion reflected Shishkin's position rather than Lomonosov's.

unduly high by international standards, especially given that intermediaries were needed for the large German contracts. Also, there were compelling reasons to select the class E: it would not need time-consuming development or special spare parts, it was familiar to workers and it weighed less than the larger American-built class Ye – important advantages in the difficult circumstances and given the prevalence of light-weight rail and weak bridges on the Soviet network.

Although Lomonosov touched almost every aspect of this affair, his impact was especially significant in three general respects. First, unlike most early Soviet diplomats and trade officials, he could inspire much-needed confidence among foreign politicians, government officials and businessmen. Very influential were his personality, experience and, one suspects, non-membership of the Bolshevik Party. To quote Adolf Maltzan, head of the Russian desk at the German foreign ministry:

> The heavy, thickset figure of Lomonosov and his calm easy-going manner are in no way reminiscent of the hitherto more or less chauvinistic and hysterical appearances of the Soviet representatives known here or passing through. He represented far more the nuances of the good, bourgeois, experienced and realistic merchant of the old Russia. His general impression is in short bourgeois, positive and cunning.[83]

Lomonosov developed close working relationships with men like Gunnar Anderson (the owner of Nohab), Otto Hagemann (a senior director of the powerful Krupp engineering conglomerate), and Ludwig Bamberger and Harold Berg (Krupp sales agents). A useful result was considerable mutual trust. For instance, in October 1920 a consortium led by the Mannesmann engineering company waived its demand for an irrevocable letter of credit, which permitted the first three large orders for locomotive parts – an extraordinary development in the context of deep Soviet–Western distrust.[84]

Second, Lomonosov's determination to obtain locomotives shaped early Soviet gold expenditure abroad. In October 1920 the regime cut the locomotive procurement target from 5,000 to 2,000 as part of a review of its foreign trade strategy prompted by the deepening domestic economic crisis. By early 1921, with 1,100 locomotives ordered, a decision not to purchase any more was increasingly likely. Yet Lomonosov wanted at least one further substantial order, and agreed terms with the German builders for 600 locomotives; Krasin signed the final document on 28 February for an expected cost of about 80 million gold rubles. However, four days earlier the NKVT had told Lomonosov by telegram not to buy more engines for the moment. Whether this instruction arrived in time is uncertain, but if it did, Krasin was not informed. In short, Lomonosov's pressure, and perhaps

---

[83] Memorandum, 16 August 1920: National Archives, London, GFM 34/3974/frame K096055.

[84] Heywood, *Modernising Lenin's Russia*, p. 133.

his deviousness, committed Moscow to this huge expense, which the regime felt obliged to approve so as to avoid harming its relations with Germany.[85]

Third, Lomonosov disrupted the entire Soviet foreign trade bureaucracy. In theory the NKVT operated the state's monopoly of foreign trade, which was instigated in 1918 through the nationalization of all Russian private trading companies. Accordingly, the NKVT opened offices in Stockholm, London, Berlin and so forth as trade relations were established with foreign countries from spring 1920. However, in November 1920 the Soviet regime authorized Lomonosov to create the separate structure that would be called the Russian Railway Mission Abroad and that would open offices in every country where railway orders were placed. This step made much sense for the effective management of the railway contracts, which were probably the largest individual component of Soviet expenditure abroad at the time. But Lomonosov used his status as a plenipotentiary of the Sovnarkom to insist that he and his mission were separate from the NKVT bureaucracy. Krasin, the head of the NKVT, was furious, but although he eventually got the Railway Mission subordinated jointly to the NKVT and NKPS, he failed to persuade Moscow to annul Lomonosov's status as a Sovnarkom plenipotentiary. Consequently Lomonosov retained considerable freedom of action. To all intents and purposes, therefore, a large proportion of the Soviet imports from Europe in 1920–1922 was not handled by the NKVT. NKPS leaders like Fomin were quietly content with Lomonosov's relative autonomy from NKVT supervision, especially his freedom of manoeuvre with cash reserves, which enabled them to resolve some urgent supply needs during 1921–1922.[86] Furthermore, this situation provided ammunition for the NKVT's critics. For instance, Stalin advised Lenin in March 1922 that Krasin was waging 'a mad battle' against Lomonosov 'not because of the interests of the job so much as for glory in Europe'. Until recently, Stalin continued, Krasin had been seen in Europe as 'the only intelligent and business-like representative of Soviet Russia'; but 'the appearance of Lomonosov with his practical approach to the job and massive contracts has undoubtedly shaken the "position" of comrade Krasin'.[87] It is likely, therefore, that the Railway Mission's existence helped to provoke the debate about the future of the NKVT that occurred within the party leadership during 1922.[88]

---

[85] Heywood, *Modernising Lenin's Russia*, pp. 135–59.

[86] Heywood, *Modernising Lenin's Russia*, pp. 150–53, 166–71. S.D. Dmitrievskii, a party member, was evidently asked by the NKPS leadership to assess such questions in late 1921, and sent a report to Dzerzhinskii on 28 December 1921: RGAE, f. 1884, op. 31, d. 1663, ll. 133–6. On Dmitrievskii, of whom Lomonosov was immediately suspicious, see 'Vospominaniia', vol. 8, pp. 797, 805–10, 886, 894–5, 915; and V.L. Genis, 'Sergei Dmitrievskii: "Nam nuzhen liberal'nyi tsezarizm"', in O.A. Korostelev (ed.), *Diaspora: Novye materialy*, vol. 8 (Paris and St Petersburg, 2007), pp. 75–172.

[87] See Heywood, *Modernising Lenin's Russia*, pp. 168–9.

[88] For an overview of this debate see O'Connor, *Engineer of Revolution*, pp. 173–87. See also Heywood, *Modernising Lenin's Russia*, pp. 166–7.

Irrespective of whether Lomonosov deserved censure in these matters of general policy, it was unreasonable to blame him for the entire strategy, which is effectively what happened from spring 1922. He had known of the planning in early 1920, but was never directly involved in it: having just joined the NKPS Collegium, he was preoccupied with his inspection trips and then incapacitated by typhus until April. In fact, in January 1920 he explicitly warned the Congress of Councils of the National Economy not to regard the planned locomotive imports as a solution for the immediate railway crisis because even in ideal circumstances the deliveries could not start before the autumn.[89] Then, touring the south in May 1920, he was oblivious of the negotiations that produced the Nohab locomotive contract: Krasin handled that process with advice from Voskresenskii as the senior NKPS official in his trade mission.[90]

So why was Lomonosov blamed for the strategy? In particular, was this because he was seen as an untrustworthy *spets*? To a degree he did become a victim of circumstances beyond his control. The crux was the party leadership's pressing need for a scapegoat when, in early 1922, the VSNKh and Soviet engineering industry began objecting that cheaper, better versions of the railway supplies being imported could be made at home if only the relevant factories had contracts and working capital. This pressure became acutely embarrassing for the Bolshevik Party in March 1922 when an outspoken critic of the leadership, A.G. Shliapnikov, gave a lengthy exposé of this situation to the Eleventh Party Congress to prove that the leaders had become divorced from the proletariat. In reality many of Soviet industry's statements about the railway contracts were fanciful, but instead of checking them and risking an investigation of the policy's origins that would implicate themselves, the party leaders chose to accommodate this protectionist pressure. As the public face of the imports policy and a non-party engineer Lomonosov was the most obvious choice for blame.[91]

Yet there were other reasons for his misfortune, too. Slow to perceive the new political conjuncture, he made his position worse by vigorously defending his contracts, contesting the claims of the industrial lobby. Furthermore, he had antagonized not just Krasin, but also other important colleagues, notably the chief political representative (ambassador) in Germany, N.N. Krestinskii, the head of the NKVT office in Germany, B.S. Stomoniakov, the chief Soviet representative in

---

[89] *Ekonomicheskaia zhizn'*, 24 January 1920, p. 2.

[90] Heywood, *Modernising Lenin's Russia*, pp. 92–103.

[91] Heywood, *Modernising Lenin's Russia*, pp. 184–99. Shliapnikov's speech was part of the culmination of a protest campaign by a group of party members about the Bolshevik Party leadership's failure to defend the interests of workers. The leadership responded by attempting at this congress to expel Shliapnikov and some of his colleagues from the party for contravening party policy. The attempt was unsuccessful in Shliapnikov's case, although the protest was largely silenced. See B. Allen, 'Worker, Trade Unionist, Revolutionary: A Political Biography of Alexander Shliapnikov, 1905–1922' (unpublished PhD dissertation, Indiana University, 2001), pp. 281–325, especially 292–3, 312.

Sweden, P.N. Kerzhentsev, and V.A. Avanesov, the chairman of Sovnarkom's gold commission and deputy head of the People's Commissariat of the Workers' and Peasants' Inspectorate (NKRKI).[92] These people now seized the opportunity to attack him. For example, Krasin condemned him as a 'political adventurist with an overwhelming urge for power and the most ridiculous conceit', and insisted that his commercial activity 'needs investigation, and not by any virtuous communists whom naturally he would dupe but by someone like former senator Garin supported by intelligent people from the VChK'.[93] Coincidentally, Lomonosov had the misfortune suddenly to lose Lenin's constant backing. In a note to Kamenev on 5 April 1922 Lenin endorsed Stalin's support for Lomonosov against Krasin, and damned the criticism of Lomonosov by White Guards and 'idiots' in Sovnarkom.[94] But within weeks Lenin suffered the stroke that largely removed him from public life until his death in 1924. True, Lenin intervened forcefully during his brief return to work in autumn 1922, not least by pressing for Lomonosov to be put in charge of the railways with Dzerzhinskii as, in effect, his political commissar.[95] But Lenin was incapable of doing anything whatsoever for Lomonosov after December 1922.

Also important was the fact that Lomonosov's interest in this foreign assignment was always as much personal as professional. He warmed to the job in June 1920 as an opportunity to avoid his unwelcome transfer to Turkestan, allow tempers to cool, enjoy some luxury, organize some research into diesel traction and *in extremis* choose exile if the Cheka seemed likely to arrest him for causing the Novorossiisk disaster.[96] In time he found further similar reasons to continue. His trade and diplomatic activities provided a pleasing sense of power, prestige and international recognition, and he found more ways to indulge his scientific interests, such as a scheme that he devised for publishing Russian-language transport-related scientific works in Germany.[97] He could largely avoid the pressures of the office politics at the NKPS where, he believed quite realistically, most people disliked him.[98] He could also avoid the practical difficulties of daily life in Russia; well paid, he could indulge his taste for exclusive hotels and

---

[92]   The People's Commissariat of Workers' and Peasants' Inspection was the early Soviet incarnation of the tsarist regime's audit ministry, State Control. See E.A. Rees, *State Control in Soviet Russia: The Rise and Fall of the Workers' and Peasants' Inspectorate, 1920–1934* (Basingstoke, 1987).

[93]   Cited in Genis, '"General ot parovozov"', p. 79. The VChK was the Cheka.

[94]   Cited in Kuprineko, 'O "Svoekorystnoi pomoshchi"', p. 38.

[95]   See Genis, *Nevernye slugi rezhima*, pp. 470–73.

[96]   'Dnevnik', 18, 22, 24, 27 June, 13, 17, 19 July 1920; 'Vospominaniia', vol. 7, pp. 1245, 1272, 1306.

[97]   'Vospominaniia', vol. 7, pp. 1696–7. An example of the output is S.P. Syromiatnikov, *Issledovanie rabochego protsessa parovoznogo kotla i paroperegrevatelia (po opytam prof. Iu.V. Lomonosova i s ego predisloviem)* (Berlin, 1923).

[98]   For example: 'Dnevnik', 23 September, 29 October 1921.

restaurants, such as the renowned Horcher's restaurant in Berlin. Not least, Raisa found Western Europe very congenial after arriving from the United States in November 1920 with 11-year-old George, and she would accept only Moscow as a place to live in Russia.[99]

Several of these interests and choices had political consequences that undermined his position. His luxurious lifestyle infuriated colleagues, as he well knew.[100] It also exposed him (and hence the leaders of his impoverished, famine-stricken country) to accusations of hypocrisy from the anti-Soviet emigration and fuelled suspicions that he was taking bribes. Neither colleagues nor émigrés respected his argument that he needed to maintain his credibility with foreign businessmen, officials and politicians.[101] Thus, for example, in July 1921 a note about him by the Transport Cheka mentioned not just the Novorossiisk 'incident', but also that there was much talk about his lifestyle abroad. A follow-up report commented that he was a 'careerist' who always loved to live grandly and drink, and that the NKVT had reportedly summoned him to Moscow for words about his lifestyle.[102] Indeed, Lomonosov believed as of December 1922 that his ostentatiously expensive lifestyle was the main thing that was incriminating him, being incompatible with his status as a Soviet government representative.[103] To say, as he did, that one lived just once and should make the most of it was no defence for a senior Soviet official at a time of famine.

Particularly troublesome was his decision to appoint Raisa as his second secretary at her insistence in February 1921. No evidence has been found to suggest that this step aroused complaints of nepotism, perhaps because this practice was common or because she had joined the American Communist Party. However, the appointment did generate concern about her influence. M.Ia. Lazerson, who was a deputy-head of the Railway Mission during 1920–1921 with particular responsibility for financial matters, later claimed that Raisa dominated her husband

[99]   For example: 'Dnevnik', 29 April, 10 December 1922; the return of Raisa and George to Europe is recorded at 'Vospominaniia', vol. 8, pp. 19–20.

[100]   'Dnevnik', 10 December 1922.

[101]   See M.Ia. Larsons (M.Ia. Lazerson), *Na sovetskoi sluzhbe: Zapiski spetsa* (Paris, 1930), pp. 58–67; M.Ia. Larsons, *V sovetskom labirinte: Epizody i siluety* (Paris, 1932), pp. 115–32. Other examples include: Liberman, *Building Lenin's Russia*, p. 115; and G. Solomon, *Among the Red Autocrats: My Experience in the Service of the Soviets* (New York, 1935), pp. 156–60. Lomonosov's recollections of Larsons (Lazerson) and Solomon became equally biting, seemingly with good reason: see in particular 'Vospominaniia', vol. 7, pp. 1485, 1489–90; vol. 8, pp. 55–65, 71–4, 124–5, 141, 145, 152–70, 173–6, 188–9, 201–4, 206, 209–10 (Larsons) and vol. 7, pp. 1653–4, 1725, 1731–3, 1739–40; vol. 8, pp. 36–40, 65–7, 223, 323 (Solomon). On Solomon see V.L. Genis, 'Nevozvrashchentsy 1920-x–nachala 1930-x godov', *Voprosy istorii*, 1 (2000): 48.

[102]   Short preliminary description of citizen engineers Neopikhanov and Lomonosov, 21 July 1921, and additional note, 2 August 1921: RGASPI, f. 76, op. 2, d. 76, ll. 23, 28.

[103]   'Dnevnik', 10 December 1922.

and his assistants.[104] Similarly, a member of Krasin's trade mission, S.I. Liberman, recalled that Lomonosov was 'patriarchal in his relations with the employees of his Stockholm office, but actually his secretary held sway over everything and everybody'.[105] One should add that although Lomonosov did perceive this problem, he felt unable to solve it – an important reflection of the couple's marital crisis of 1918–1919.[106]

Lomonosov's reaction to the policy uproar of 1922 echoed his response to his setback in May 1920, but it also differed in key respects. Once again he blamed himself – not any new political trend – for his predicament, and he persisted with his Soviet career instead of choosing foreign exile. But he was more cautious about his future. If before he had sometimes deployed a resignation letter in a bureaucratic battle, now he became wary of having his bluff called.[107] Further, he sought to postpone the abolition of the Railway Mission for as long as possible and find a new assignment that would entail his continued presence in Western Europe. In this endeavour he was remarkably successful: the mission survived until April 1923, and in the meantime he devised a new role for himself as head of a Berlin-based Soviet diesel-locomotive research bureau.[108] True, in December 1922 he bowed to pressure from Dzerzhinskii (head of the NKPS since April 1921) to return to Moscow as chair of the NKPS Technical Committee, but although this plan did hold some appeal for him, it was overtaken by the evolution of his diesel work in Germany.[109]

Guilt for a major crime, however, was the most significant difference from 1920 in his situation. Lomonosov confesses in his memoirs that he embezzled

---

[104] M. Laserson, *From Russian Skies To Southern Cross: Diary of an Intellectual* (Sydney, 1950), p. 16. See also Larsons, *Na sovetskoi sluzhbe*, pp. 66–7. On Lazerson's appointment to the Railway Mission in August 1920 see 'Vospominaniia', vol. 7, pp. 1485, 1489–90. Lomonosov claims that he sacked Lazerson mainly for involvement in a scam to lower the price of Soviet gold: 'Vospominaniia', vol. 8, pp. 152–70.

[105] Liberman, *Building Lenin's Russia*, p. 115. Liberman's memoir is less reliable than is often assumed: for example, he records the number of locomotives ordered in 1920 as 250, not 1,100 (p. 113). His description of a farewell dinner before Lomonosov became a non-returner seems fanciful, as does his claim that Lomonosov was proud of being related to the eighteenth-century savant Mikhail Lomonosov. But Lomonosov does confirm Liberman's descriptions of Raisa (see, for example, 'Vospominaniia', vol. 8, pp. 27–8, 109, 152), and he did like expensive dining, to judge by his countless references to chic restaurants.

[106] For example: 'Dnevnik', 8 May, 28 November, 1, 3, 23 December 1921.

[107] For example, Lomonosov–Krasin, [31 July] 1920: RGAE, f. 4038, op.1, d. 14, l. 177; 'Vospominaniia', vol. 7, p. 1425.

[108] Heywood, *Modernising Lenin's Russia*, pp. 205–6; on the diesel question see Chapter 9, this volume.

[109] 'Dnevnik', 18 October, 26 December 1922, 24 March 1923; Vospominaniia', vol. 8, pp. 1681–2, 1750–1751, 1867. On Dzerzhinskii (1877–1926) see A.S. Velidov et al. (eds), *Feliks Edmundovich Dzerzhinskii: Biografiia*, 3rd edn (Moscow, 1986). Formally Lomonosov headed the committee for just over a year, but he played almost no part in its work.

85,000 Swedish crowns from the Railway Mission during the operation to curtail the Nohab locomotive contract in July 1922. This substantial sum – nearly one-third of the price of a class E locomotive – was passed to Raisa in payments of 50,000 and 35,000 crowns, and was evidently invested by her in Britain and Germany. Lomonosov does not reveal precisely how he conducted these transactions, but the most likely explanation is that he recorded them as bribes for one or more foreign contacts. His aim was to render the option of exile more practicable by creating a financial reserve that, if necessary, could support him and his family in exile for at least an interim period. He regretted this crime as a sad indictment of the revolution in general and of its impact on his morals. But more practically, his action gave real substance to the accusations of corruption that he was denying vehemently and to the suspicions of party critics like Avanesov that he was planning for exile, and it burdened him with a corrosive sense of guilt and unease as he battled against those accusations and suspicions.[110]

That battle began in earnest in September 1922. Avanesov took the lead as chair of an audit commission that investigated the Railway Mission in a determined attempt to prove that Lomonosov was guilty of crimes such as bribe-taking and mismanagement.[111] Technically the inspectors were concerned with all Soviet trade delegations in Europe, but whereas they allocated, for example, about three days for the large NKVT office in Berlin, they spent six weeks at Lomonosov's mission. Among their many concerns was Lomonosov's decision to pay bonuses to staff in November 1921 to mark the Railway Mission's first anniversary, so it was especially provocative of him at this juncture to award similar bonuses for the second anniversary. Lomonosov defended these payments as vital to help retain essential foreign employees whose salaries were derisory by foreign standards. But Avanesov was furious and Krestinskii complained to Moscow. Shortly afterwards the Sovnarkom published a formal reprimand.[112] Significantly, even Kamenev distanced himself from Lomonosov and his family.[113] For the engineer to refuse to be a deferential 'errand boy' for Krasin was one thing; to be so tactless and provocative was quite another.[114]

---

[110]   'Dnevnik', 28–29 June 1922; and especially 'Vospominaniia', vol. 8, pp. 1230–31, 1313–15, 1889–90. To judge by his memoirs, he was fairly confident that his embezzlement would not be discovered.

[111]   See Heywood, *Modernising Lenin's Russia*, pp. 200–204.

[112]   Protocol of SNK meeting, 22 December 1922: GARF, f. r-130, op. 6, d. 2a, ll. 375–6; *Izvestiia*, 28 December 1922, p. 5; 'Dnevnik', 27 November, 29 December 1922; 'Vospominaniia', vol. 8, pp. 1594–1629, 1688–90. The Politburo considered a statement from Krestinskii about Lomonosov on 14 December 1922: G.M. Adibekov et al. (eds), *Politburo TsK RKP (b)–VKP (b): Povestki dnia zasedanii, tom 1: 1919–1929: Katalog* (Moscow, 2000), p. 199.

[113]   For example: 'Dnevnik', 1, 6, 22 January, 26–28 March 1923; 'Vospominaniia', vol. 8, pp. 1679, 1683–5, 1696, 1699–1701, 1704–6, 1742, 1866, 1872–3, 1875–6.

[114]   This phrase can be found at, for example, 'Vospominaniia', vol. 8, p. 48.

Submitted in February 1923, the Avanesov commission's report was seriously flawed. It accused Lomonosov of many grave faults and mistakes. For instance, he had 'hypnotized' the Politburo and Sovnarkom, knowingly ordered locomotives that were unnecessary, agreed high prices and impossible delivery schedules, used intermediaries without good reason, accepted poor-quality products, constantly exceeded his authority, interfered in the work of Soviet ambassadors, kept 'fictitious' accounts and not least, maintained a den of White Guards and dubious foreigners.[115] Some of the criticisms were certainly apposite. The need for at least some of the locomotives was debatable, certain delivery schedules were unrealistic, the mission's accounts were not maintained with the NKVT accounting system, the financial documentation was incomplete, Lomonosov's powers and autonomy did have disruptive effects, the mission had few communists, and so forth. It is conceivable too that some staff did accept bribes. But important points were omitted, and some key extrapolations and conclusions were exaggerated or spurious. The report ignored the policy's origins, which were pivotal for assessing whether the engines were needed. It also ignored the inconvenient fact that the NKPS was content with the quality of the imports; indeed, in one engineer concluded in 1925 that the imported class E locomotives were more economical than their Russian-built fellows because they were better built.[116] Ludicrously, the Avanesov commission criticized Lomonosov for the first Nohab delivery timetable and for ordering the Armstrong boilers without competitive tendering. It did not substantiate its allegation of anti-Sovietism among the mission's staff, apart from recording their non-party status. And, ironically, it failed to reveal the embezzlement that Lomonosov had perpetrated: he had covered his tracks well.

For Lomonosov, then, the crucial question was how this report and his responses would be treated. Conscious of his financial crime, he was nervous at the meeting of the three-man commission appointed by the Politburo to evaluate the report in March 1923. Yet the commission was impressed by his replies. Indeed, one member, Dzerzhinskii, lambasted the inspectors for exaggeration, carelessness, incompetence, delays, exhuming matters that had been closed and forming judgements from incomplete data. He asked:

---

[115]   For the allegations and Lomonosov's responses see 'Dnevnik', 21, 23 March 1923; 'Vospominaniia', vol. 8, pp. 1836–7, 1839–40, 1845–56.

[116]   P. Stepanov, 'Sravnitel'nye dannye o raskhode para i topliva parovozami serii E postroiki russikh zavodov i E^G, E^{SH} postroiki zagranichnykh zavodov', *Zheleznodorozhnoe delo*, 8 (1925): 37–40. Boiler explosions on two class Eg locomotives in 1925 and 1926 were not attributed to the quality of design and construction: V.I. Taranov-Belozerov, 'Vzryv parovoznogo kotla serii E^G na Syzrano–Viazemskoi zh.d. 10 iiunia 1925g.', *Zheleznodorozhnoe delo: Podvizhnoi sostav, tiaga i masterskie*, 4–5 (1926): 13–16; V.G. Idel'son and Sh.V. Korbelov, 'Vzryv kotla parovoza serii E^G No. 5213 na Zakavkazskikh zhel. dorogakh', *Zheleznodorozhnoe delo: Podvizhnoi sostav, tiaga i masterskie*, 3–4 (1927): 3–4.

> For what should [Lomonosov] be put on trial? For the fact that Sovnarkom gave him extensive powers? For that fact that, in someone's opinion, he could have fulfilled his mission better? For the fact that you don't like him? For the fact that you suspect him of something, without having proof for your suspicions? For the fact that our transport [system] is returning to life without the sacrifices which everyone assumed were essential (the Gosplan calculations)?[117]

However, these reservations did not carry the day. The paperwork was missing for about 30,000 Swedish crowns that, so far as Lomonosov could recall, were used for bribes and so forth. Thus, although the commission did not see enough justification for a trial, it had to recommend an investigation by the NKRKI, which was to begin after Lomonosov submitted his book-length preliminary report later in the year.[118]

Paradoxically, Lomonosov's response to this setback over the next two years highlights his genuine commitment to the Soviet system. He reasoned that his future in Soviet service depended on clearing his name, for otherwise the likes of Avanesov would never leave him in peace. He continued to hope and believe that that vindication would come (and hence too that his embezzlement would remain hidden). His preference was for the Sovnarkom and Politburo to reject the accusations against him by approving his written report, which he submitted in October 1923.[119] Increasingly he expected to be sent for trial, but he remained confident that his documentary evidence would rout his accusers. Hence, surprising as it may seem, he pressed explicitly for either ratification of his report or a trial. That was his plea in, for example, a meeting in November 1924 with his pre-revolutionary acquaintance Pravdin, who was now in the NKPS Collegium, and a letter to Dzerzhinskii's successor at the NKPS, Ia.E. Rudzutak, in February 1925.[120]

Seen in this context, the result of the NKRKI investigation in 1925 was upsetting for Lomonosov. The report covered the same ground as the Avanesov commission, drew similar conclusions and recommended the rejection of Lomonosov's report, implying that criminal charges were appropriate. As before, the missing paperwork

---

[117]   Cited in Genis, "'General ot parovozov'", p. 80.

[118]   Minutes of Politburo troika meeting, 24 March 1923; Politburo resolution, 26 March 1923: RGASPI, f. 17, op. 163, d. 326, ll. 13–14, 12; 'Dnevnik', 23, 25–28 March 1923; Vospominaniia', vol. 8, pp. 1854–64, 1879, 1886–7, 1901–2.

[119]   Iu.V. Lomonosov, *Predvaritel'nyi otchet Iu.V. Lomonosova o deiatel'nosti Rossiiskoi zheleznodorozhnoi missii zagranitseiu za vse vremia ee sushchestvovaniia (1920–1923)* (Berlin, 1923).

[120]   'Dnevnik', 18 November 1924; 'Vospominaniia', vol. 9, pp. 941; Lomonosov–Rudzutak, 8 February 1925 (draft): LRA, MS 716.1.136. Rudzutak is yet another major Bolshevik figure for whom there is no biography in English. He was executed on charges of being a Trotskyist. See I. Donkov and A. Nikonov, 'Ia.E. Rudzutak (1887–1938): Emu doverial Il'ich', in Proskurin, *Vozvrashchennye imena*, vol. 2, pp. 131–47.

for some 30,000 Swedish crowns was a key problem.[121] The Politburo, however, accepted a compromise proposed by Rykov and Rudzutak: Lomonosov's report was neither ratified nor rejected, and the matter was closed with a reprimand.[122] Thus Lomonosov was spared the danger of a trial, but had no scope to clear his name. Briefly he was tempted to quit his Soviet career; but again he desisted, now because of his commitment to his diesel research, his desire to resist his enemies and, incredibly, his stubborn faith in the possibility of victory. That optimism, however, finally deserted him during the following autumn.[123]

That Lomonosov was not prosecuted remains surprising. There were certainly relevant precedents. The two major investigations of the Railway Mission in 1922–1924 coincided with a mini-wave of symbolic trials – 'show trials' – of economic managers, some of whom were even party members. Particularly notorious was the case of A.M. Krasnoshchekov, a member of the VSNKh Presidium and head of the Bank of Industry and Trade. Arrested in 1923, he was the first high-ranking party official to be convicted publicly for corruption and mismanagement, and was sentenced to six years in prison.[124] At least one of his investigators was involved in Lomonosov's case (E.M. Iaroslavskii was in Avanesov's audit commission), the charges against Krasnoshchekov were similar to, and arguably less serious than, some of the main accusations against Lomonosov, and the latter regarded the 'method of denigration' as identical to his own case.[125] Moreover, Rudzutak and Rykov accepted that the allegations against Lomonosov merited a trial and that he was a 'bad dealer and businessman'. Their argument against a trial was that the errors were not self-interested or mercenary, and that Lomonosov could contribute much to improving the transport sector.[126] This last point was presumably a reference to the much-publicized success of Lomonosov's first diesel locomotive at precisely this time (early 1925), and may imply that that success saved his liberty and conceivably his life. Yet such reasons seem flimsy against the principal allegation that many millions of gold rubles had been squandered. To speculate, perhaps a trial was considered inexpedient because Lomonosov was

---

[121] Report to Sovnarkom, [circa January 1925]: RGASPI, f. 5, op. 1, d. 149, ll. 201–529.

[122] Report by Rykov to the Politburo, January 1925, and Politburo resolution, 27 January 1925: RGASPI, f. 17, op. 163, d. 473, ll. 90, 89; 'Dnevnik', 16 March 1925; Vospominaniia', vol. 9, pp. 1105–6; Genis, '"General ot parovozov"', p. 80. Lomonosov's secretary, S.V. Makhov, told Raisa in a letter dated 20 March 1925 that the report had been ratified: LRA, MS 717.2.196.1; however, Lomonosov raised the question of ratification again with Rudzutak in the autumn: 'Dnevnik', 13 October 1925.

[123] See Chapter 9.

[124] On Krasnoshchekov see R.T. Argenbright, 'Marking NEP's Slippery Path: The Krasnoshchekov Show Trial', *The Russian Review*, 61/2 (2002): 249–75.

[125] 'Dnevnik', 22 January 1925. Iaroslavskii became notorious as one of Stalin's henchmen by 1927: see A. Vatlin and L. Malashenko (eds), *Piggy Fox and the Sword of the Revolution: Bolshevik Self-Portraits* (New Haven, 2006), pp. 78–9, 136.

[126] See Genis, '"General ot parovozov"', p. 80.

in the international spotlight, or because Lenin was central to the initial decision-making. Perhaps Dzerzhinskii again challenged the allegations. If nothing else, Lomonosov was lucky.

*   *   *

Given that Lomonosov did so much to antagonize superiors, subordinates, colleagues and even friends, it is difficult to judge overall whether his Soviet career between 1919 and 1923 should be characterized as an opportunity missed or as an inevitable disaster. He was hungry for power and influence in Lenin's Russia in 1919, but he did not put his aim of heading the NKPS above his political principles: he preferred to remain a 'non-party Marxist' despite realizing that party membership was required for such top-level posts. Nonetheless his ambition cannot be dismissed as ridiculous or naive because Lenin did propose him for the post of People's Commissar at the NKPS. The rejection of that proposal was indeed partly Lomonosov's responsibility, although he overestimated the significance of the Novorossiisk train crash and underestimated the extent to which he had alienated party and NKPS colleagues. With hindsight, we can see that spring 1920 marked the peak of his power and influence in Soviet service. For all its high diplomatic profile and glamour his posting to Western Europe tended to isolate him from the NKPS and the railways. Abroad his non-party status, social and professional background and opulent lifestyle made him susceptible to the derogatory label of 'bourgeois specialist', especially when Soviet industry challenged the imports policy. Having started to fear for his life in 1920, he became more cynical about his position as the political pressure on him mounted. Yet contrary to the suspicion of party officials like Avanesov, he never wanted foreign exile at this time; he stole state funds to make exile more practicable as a last resort, not preference. He remained proud of his citizenship, committed to his Soviet career, contemptuous of his critics and very determined to rebuild his professional and political reputations.

As discussed elsewhere, several basic issues of early Soviet history may need reassessment in the light of his work during 1919–1923. It would appear that industrial modernization, not mere recovery, headed the Bolshevik economic agenda in early 1920 within the framework of War Communism, albeit only briefly, and that the railways were the main target for the initial investment, not electrification. The railway imports policy confirms that for all the rhetoric of international revolution the Bolshevik leadership did want foreign trade to play a large role in the Soviet economy, while the abandonment of this railway-led modernization strategy and the contraction of the railway imports policy during 1920–1921 reinforce the view of the NEP as a retreat from War Communism. As for the place of the pre-revolutionary scientific and technical intelligentsia in Lenin's Russia, Lomonosov's experiences suggest several conclusions. It was possible for the regime to consider appointing the most senior non-party specialists to the key position of People's Commissar in 1920, at least in the economic arena; of course, Lomonosov was known to have helped the Bolsheviks in the 1905 revolution and he claimed to support the Soviet system, but it would still be interesting and useful to establish whether his candidacy was unique. On the other hand, the party's ability to monitor the work of a senior engineer like him was quite weak, and in the intense atmosphere of class war this weakness greatly encouraged suspicions about possible deception and sabotage – a situation in which the dividing line between substantiated allegations and demagogy was easily crossed. The frustrated anger of party officials like Avanesov may well indicate that the inclination for a thorough, ruthless purge of technical specialists already existed in the early 1920s. It is also worth noting, finally, that *spets* was, like the term *kulak* for a rich peasant, a label over which the individual had little control. Lomonosov never considered himself a *spets* because of his support for the Soviet system. But what mattered was how other people saw him, and they were quick to define him as a *spets* because he did not join the party. Try as he did, he could not discard that identity.

Chapter 9

The Diesel Revolution

From April 1923 Lomonosov's main professional concern was to design, build and test main-line diesel locomotives for the NKPS in collaboration with German and Swiss engineering firms.[1] His endeavour was one of several contemporaneous Soviet projects to develop this technology to replace steam locomotives, and it was paralleled by various foreign projects, notably in Germany, Italy, Great Britain and the United States. Lomonosov's team, however, attracted international attention during 1924–1925 for producing a locomotive that seemed to prove the technical viability of this concept. The premier British engineering periodical, *The Engineer*, called his machine 'a milestone in the history of the internal combustion locomotive', while the US journal *Oil Engine Power* lauded Lomonosov in this regard as 'one of the most important authorities' in railway engineering.[2] Some 30 years later locomotive engineers in Britain still remembered Lomonosov's project. For instance, J.M. Doherty felt that 'Lomonossoff and those associated with him deserve the greatest credit for what must rank as one of the most outstanding experiments in the history of diesel traction'; and the head of the British Railways locomotive testing plant, D.R. Carling, declared:

> It was undoubtedly the energy and enthusiasm of Dr Lomonossoff, armed not only with his extensive experience as a locomotive and railway engineer but also with authority from Lenin himself, and to the ability and persistence of his colleagues, notably the late Dr Meineke, that this epoch-making work was brought to a successful conclusion. It was this work which paved the way for almost all later developments in main line diesel traction.[3]

Lomonosov's international prominence in diesel-traction research in the 1920s, and more generally the extensive Soviet interest in developing this technology, would appear to confirm the common view that the decade was a golden age of enquiry and enlightenment in Soviet science and technology thanks to the Bolshevik

---

[1]   For a thorough discussion of the diesel research for the Soviet railways until the 1950s see Westwood, *Soviet Locomotive Technology*, especially pp. 14–22, 34–91, 105–24, 142–62, 215.

[2]   'A Diesel-Electric Locomotive', *The Engineer*, 138 (14 November 1924): 554; 'Diesel-Electric Locomotive Surpasses Steam on Test', *Oil Engine Power*, 2/12 (1924): 633.

[3]   J.M. Doherty, 'Evolution of Internal Combustion Locomotives', *Journal of the Institution of Locomotive Engineers*, 46/251 (1956): 251, 283.

enthusiasm for socio-economic modernization.[4] Equally, the fact that overall the inter-war Soviet diesel-locomotive research programme yielded limited results seems to support the contention that the early Soviet momentum of industrial and technological development was lost by the late 1930s; indeed it may endorse Loren Graham's argument that much of the USSR's chronic trouble with technological development was due to the misuse of technology.[5] In the terms of a popular three-stage model for analysing technological change – invention; development for mass production; and innovation (introduction to common utilization) – the inter-war Soviet diesel-locomotive research effort collapsed during the second phase.[6] Only about three dozen locomotives had been built, and mass production remained a distant prospect when the Soviet government halted this work in 1937 – ironically, just as the technology was making its commercial breakthrough in the United States.[7] Diesel-traction experts were among the thousands of people whom the Bolshevik authorities imprisoned (and in some cases executed) during purges of the pre-revolutionary scientific and technical intelligentsia from 1928.[8] Ultimately their inter-war research played only a marginal role when railway dieselization – the replacement of steam by diesel power – began in the USSR in earnest in the late 1940s, that process being launched with US equipment imported through the wartime Lend-Lease arrangement.[9]

How, then, did Lomonosov regard his diesel research, and how do these broader issues appear through the prism of his experiences? This chapter outlines the technical challenge of railway dieselization, traces the roots of the Soviet interest in diesel traction, and then examines Lomonosov's project. Why did he pursue this new line of work, what was he trying to achieve, how does it fit in his overall career, how far does the regime's support explain his achievement, and above all, why did his involvement in this project end just as his second locomotive was ready for testing?

---

[4]   For an overview of the relationship between the regime and scientists in 1917–1932 see Graham, *Science in Russia*, pp. 79–98. An example of the official interest is the support for popularizing science, on which see Andrews, *Science for the Masses*, pp. 17–118.

[5]   Graham, *Executed Engineer*, pp. 2–3.

[6]   On this model see J.M. Staudenmaier, *Technology's Storytellers: Reweaving the Human Fabric* (Cambridge, MA, 1985), pp. 40–61.

[7]   On American dieselization in the 1930s see, in particular, R.C. Bingham, 'The Diesel Locomotive: A Study in Innovation' (unpublished PhD dissertation, Northwestern University, 1962), pp. 1–366; A.J. Churella, *From Steam to Diesel: Managerial Customs and Organizational Capabilities in the Twentieth Century American Locomotive Industry* (Princeton, 1998), pp. 37–74.

[8]   Westwood, *Soviet Locomotive Technology*, pp. 66, 81–5, 88–9.

[9]   See Rakov, *Lokomotivy*, 2nd edn, pp. 364–7, 371–84; Westwood, *Soviet Locomotive Technology*, pp. 160–62.

Russia and the technical challenges of dieselization

The regular use of diesel engines to power main-line railway locomotives required the development and integration of several technologies. The main technical challenges were to create a high-power low-weight engine for the specific demands of the railway environment, an effective engine-cooling system and – as with automobile transport – a reliable means of enabling the engine's power output to drive wheels. Engineers from several countries, including tsarist Russia, began tackling these issues in earnest just before the First World War, and Lomonosov's first locomotive of 1924 did much to demonstrate the concept's feasibility. The transition to common utilization came in North America in the late 1930s and elsewhere from the late 1940s.

Fundamental for this transition was the development of commercially viable high-efficiency internal combustion engines in the nineteenth century. In a steam engine the combustion occurs in the firebox so as to heat the water in the boiler, thereby producing the steam that is fed into the cylinder to move the piston and, via a connecting rod, a shaft or wheel. But with internal combustion engines of the types used in railway locomotives the combustion occurs within the cylinders. The norm is for each piston to compress air within a cylinder, which generates a temperature sufficiently high to ignite fuel injected into that cylinder, which in turn causes the piston to move. Thus, for instance, Alphonse Beau de Rochas theorized the four-stroke piston cycle for such an engine in 1862. The first complete piston stroke would draw air and fuel into the cylinder; the second (return) stroke would compress the air; ignition would occur at the dead point, with expansion during the third stroke; and the burned gases would be expelled from the cylinder during the third and fourth strokes. The technical and economic attraction was that the thermal efficiency of the cylinder (the 'indicated thermal efficiency') was higher than for a steam engine; in other words, more of the heat energy was converted into useful work. For example, early gas engines enjoyed a thermal efficiency of about 14 per cent, whereas the value for an efficient railway steam locomotive was typically under 10 per cent.[10]

The contribution of the German scientist Rudolf Diesel was a type of internal combustion engine with potentially far higher levels of thermal efficiency and power output. He aimed to eliminate heat loss by achieving combustion and expansion at a constant temperature. The essence of his invention, patented in 1892 and developed with the MAN engineering company of Bavaria, was to use much more air than was necessary just for combustion, and to inject the fuel after compression had heated this air far above the fuel's ignition temperature. In practice he had to accept a certain rise in temperature during combustion, and hence also to use a cooling system to prevent the engine from overheating. Nonetheless, he and

---

[10]   For the history of the internal combustion engine up to the first engine of Rudolf Diesel see C.L. Cummins, Jr, *Internal Fire* (revised edn, Warrendale, 1989).

MAN claimed a thermal efficiency of over 30 per cent for the first small engine that they unveiled for commercial application in 1897.[11]

This achievement had potentially far-reaching implications for steam-powered railways. If diesel engines of sufficient power and reliability could be developed and if their power could be harnessed within a locomotive to move trains, they might yield huge economies for railways. Fuel costs could be significantly lower than for steam despite the higher unit price of fuel oil, and the removal of the steam boiler might reduce maintenance costs substantially. The infrastructure requirements might be much simpler and cheaper than for either steam, which needed good-quality water, or electric traction, which needed a high-voltage electricity supply from an overhead wire or ground-laid contact rail. To cite Diesel:

> We consider the new motor especially applicable to railways, to replace ordinary steam locomotives, not only on account of its great economy of fuel, but because there is no boiler. In fact the day may possibly come when it may completely change the present system of steam locomotion on existing lines of rails.[12]

And as he assured the American Society of Mechanical Engineers in 1912 shortly before his death: 'one thing is certain: the Diesel locomotive will come sooner or later, according to the perseverance with which the problem is followed.'[13]

Yet the technical and economic barriers to high-power main-line diesel traction were still significant, notwithstanding the popularity of locomotives with small petrol or diesel engines for shunting purposes and narrow-gauge lines, including military field railways. Vital was a powerful yet relatively light engine that would remain reliable in this stressful working environment, and likewise an effective, reliable lightweight cooling system. Also essential was an efficient, durable means of harnessing the engine's power to turn the locomotive's wheels. The ideal, given that the engine's pistons would rotate an axle-type shaft, was a form of direct drive, for that would minimize the losses of energy. However, to implement this idea would be technically difficult, if not impossible. The alternative – less efficient but more practicable – was to use a power transmission system. The main options were seen as electric transmission (the engine shaft drives an electricity generator, from which the current powers electric motors attached to the driving axles), mechanical transmission (gears connect the shaft to a driving axle, as in an automobile, and side rods connect this axle to the other axles) and hydraulic transmission (a fluid-based system). The commercial adoption of diesel technology also posed challenges: such locomotives were

---

[11]   M. Grosser, *Diesel: The Man and His Engine* (Newton Abbot, 1980); D.E. Thomas, *Diesel: Technology and Society in Industrial Germany* (Tuscaloosa, 1987), pp. 84–198.

[12]   Cited in D.S.L. Cardwell, *Technology, Science and History: A Short Study of the Major Developments in the History of Western Mechanical Technology and their Relationships with Science and Other Forms of Knowledge* (London, 1972), p. 169.

[13]   Cited in J.B. Garmany, *Southern Pacific Dieselization* (Edmonds, 1985), title page.

expected to cost more to build and be less robust than steamers, and be thermally less efficient than electric locomotives.[14]

Although main-line dieselization began in the United States during the 1930s, many of the early experimental main-line locomotives appeared in Europe. The first one was constructed by the Sulzer Company of Winterthur, Switzerland, for the Prussian State Railways in 1906–1913, but its direct drive system was unsuccessful.[15] The next two locomotives, which both appeared in 1924, were built for the Soviet railways. Both had electric transmission and ex-marine engines. One was Lomonosov's first locomotive and the other was completed in Petrograd to a design of Professor Ia.M. Gakkel'.

Soviet authors tended to imply that this Soviet contribution was mainly a product of the Bolshevik fascination for science, technology and economic modernization, and especially of the enthusiasm that Lenin displayed after seeing a newspaper article about diesel traction in December 1921.[16] To a degree, as we shall see, they had a point. Yet as several of those accounts also reveal, these developments had deep roots in the tsarist era. As a large oil-producer tsarist Russia was an obvious home for diesel engines, and soon it had many in industrial use.[17] The technology was especially attractive for replacing steam locomotives on the railways given the lack of good-quality water in many districts, and several research projects were begun.[18] One was headed by Grinevetskii, the Moscow professor who became Russia's foremost specialist on internal combustion engines. He focused on developing a high-power engine specifically for railway locomotives. In 1909 he had a prototype engine built by the Putilov works in St Petersburg, but Putilov halted the tests in 1912 for want of funds and, presumably, success. Meanwhile, the German-born chief draughtsman of the Kolomna locomotive-building works, Felix Meineke, sketched designs for three locomotives of 40, 300 and 1,600 h.p. with gas, mechanical and electric

---

[14]   For a comprehensive, clear overview of the technical issues see Ransome-Wallis, *Concise Encyclopaedia*, pp. 25–106.

[15]   'The Sulzer–Diesel Locomotive', *Engineering* (5 September 1913): 317–21; M. Rutherford, 'When Britain was a Contender: Some Pioneer Diesel-Electrics before La Grange, Part 1: The Background', *Backtrack*, 14/6 (June 2000): 355. Lomonosov notes this project in, for example, his *Diesellokomotiven* (Berlin, 1929), pp. 6–7; and 'Vospominaniia', vol. 9, pp. 512–14.

[16]   For example, Iakobson, *Istoriia teplovoza*, pp. 24–6; Lovtsov, 'Iz istorii bor'by Kommunisticheskoi partii za sozdanie otechestvennogo teplovozostroeniia', pp. 100–133. See also Westwood, *Soviet Locomotive Technology*, pp. 18–22.

[17]   Thomas, *Diesel Technology*, p. 188.

[18]   I.V. Zuev and V.O. Reingardt, 'O neftianykh dvigateliakh sistemy Dizelia', in *Protokoly zasedanii XXVII soveshchatel'nogo s"ezda inzhenerov sluzhby podvizhnogo sostava i tiagi russkikh zheleznykh dorog, v Varshave, v avguste 1909 g.* (St Petersburg, 1910), pp. 477–515.

transmission respectively, but the company terminated this work in 1913.[19] On the Tashkent Railway Lomonosov himself conceived a design for a 1,000 h.p. locomotive in 1909 (see Figure 9.1). Here the technical priority was to develop an effective transmission system. Mechanical transmission was preferred to electric so as to save weight, and work began on making a suitable clutch. Other early Russian ideas included a gas-transmission system using a steam-locomotive chassis, which was patented in 1913 by Grinevetskii's student A.N. Shelest, and a 1,000 h.p. diesel-mechanical design that was published by an engineer called E.E. Lontkevich in 1915.[20]

Indeed, Lomonosov's pre-revolutionary association with diesel railway traction was quite substantial. Apparently his interest was first stirred in 1906 by research seminars about internal combustion engines at the Kiev Polytechnic Institute.[21] More importantly, in 1909 he quickly saw the advantages of diesel traction for the Tashkent Railway's desert conditions, where indeed diesel engines were powering the water-pumping stations. The line's superintendent, Shtukenberg, was already considering the matter, but Lomonosov took the initiative to create a research project. Astutely, he began by seeking and winning the minister's support. Then, aware that his former student and colleague A.I. Lipets was interested in the topic and unhappy at the Kiev Polytechnic Institute, Lomonosov arranged for him to work on the Tashkent Railway. He helped Lipets to begin research, and ensured that this project was in the railway's budget for 1910.[22] In 1912–1914 he used his senior positions at the MPS to steer this work towards the policy limelight by getting Lipets and his equipment moved to the Nicholas Railway, with funding to build and test the prototype clutch, and to visit Germany to examine the Sulzer

---

[19]   On Meineke (1878–1955) see the obituary 'Prof. Dr.-Ing. Meineke †', *Glasers Annalen*, 79 (1955): 254–6, and 'Vospominaniia', vol. 5, pp. 362–3.

[20]   For details see V.T. Mikhailov, 'Primenenie dvigatelia vnutrennego sgoraniia na zheleznodorozhnykh lokomotivakh', *Vestnik Obshchestva tekhnologov*, 9 (1913): 284–90; 10 (1913): 318–24; 11 (1913): 349–55; and 12 (1913): 384–9; Iu.V. Lomonosov and E.E. Shveter, *Proekty teplovozov zakazannykh v 1921–1925 gg. v Germanii dlia SSSR* (Berlin, 1927), pp. 7–15; Iakobson, *Istoriia teplovoza*, pp. 5–23; Rakov, *Lokomotivy*, pp. 352–5. See also V.P. Goriacheva and P.A. Shelest, *Aleksei Nesterovich Shelest: Pioner teplovozostroeniia, 1878–1954* (Moscow, 1989); Westwood, *Soviet Locomotive Technology*, pp. 14–17; and Grinevetskii's major treatise, written during the First World War but published posthumously: *Problema teplovoza i ego znachenie dlia Rossii* (Moscow, 1923).

[21]   'Vospominaniia', vol. 3, pp. 1482–3. There is no substance to M. Evseev's claim that Lomonosov was the first to think of using an internal combustion engine in a railway locomotive in the late nineteenth century: 'Sozdanie teplovoza', *Oktiabr'skaia magistral'*, 1 October 1986.

[22]   See Tashkent Railway–Head, MPS Directorate of Railways, 15/16 June 1912: RGIA, f. 273, op. 15, d. 271, ll. 3–29; 'Vospominaniia', vol. 4, pp. 383–5, 480–481, 664–5, 681, 748, 763–4.

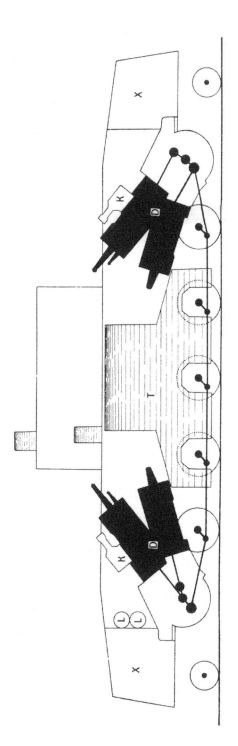

Figure 9.1   The Lipets–Lomonosov diesel locomotive (1909). Source: Iu.V. Lomonosov and E.E. Shveter, *Proekty teplovozov razrabotannye v 1921–1925 gg v Germanii dlia SSSR* (Berlin, 1926), p. 10.

Notes for Figures 9.1–9.6

A   exhaust pipe and silencer
B   auxiliary dynamo starter system
D   diesel engine
E   electric motor
F   fuel tanks
G   dynamo, generator

K   compressor
L   air reservoir
N   coupling; clutch
T   water tank (fig. 1); fuel tank
U   accumulator battery (for starting the dynamo)
X   engine cooling unit
Z   gears

locomotive and other work.[23] However, Lomonosov was less involved with the technical minutiae than he later implied: the associated patents were awarded to Lipets alone.[24]

Ultimately this project, like the tsarist regime, fell victim to the First World War and Russian revolutions. The drawings for the test bed for the clutch were completed in 1915, but no further work was attempted.[25] After 1917 the project was doomed for want of a champion, whatever its technical merits: Lipets remained in the United States with other concerns, while the engineers still in Russia had their own pet schemes. Lomonosov was ideally situated to promote it from 1919 as chair of the NKPS Technical Committee and as a member of the VSNKh Presidium and NKPS Collegium. But as the next section will show, by then he had changed his mind about how diesel traction should be developed.

Lomonosov and Soviet diesel traction

Research into diesel traction, like the parallel work on electric traction, became a core part of the Soviet railway interest in science and technology during the 1920s. A small band of diesel proponents soon began trying to exploit Bolshevik enthusiasm for scientific and technological progress. For example, Shelest published an article about his gas-transmission invention in mid-1918.[26] The idea of the diesel-electric locomotive – one with a diesel engine and electric transmission – was often presented as a type of electric traction. This stance had a certain technical logic, and could also tap Lenin's well-known passion for nationwide electrification, while stressing the advantage that diesel-electrics did not need costly, dangerous and vulnerable electrical contact rails or overhead wires.[27] Unsurprisingly given

---

[23]   Tashkent Railway–Head, MPS Directorate of Railways, 15/16 June 1912; Lomonosov–Plakid, 26 June 1912; Lomonosov–Technical Dept (Directorate of Railways), 17 July 1912; Lomonosov–Head, Nicholas Railway, 7–9 November 1912; Nicholas Railway–Lomonosov, 29 November 1912; Lomonosov–Head, Nicholas Railway, 22 January 1913; Technical Dept (Directorate of Railways)–Head, Nicholas Railway, 29 January 1913; Minutes of the Committee of the Directorate of Railways, 30 March 1914: RGIA, f. 273, op. 15, d. 271, ll. 3–29, 43, 45–6ob., 47-ob., 49; 'Vospominaniia', vol. 5, pp. 556–8, 603, 622. The report by Lipets about the trip to Germany, dated 23 January 1914, is at RGIA, f. 273, op. 15, d. 271, ll. 119–22. See also Lomonosov and Shveter, *Proekty teplovozov*, pp. 7–12.

[24]   Privilege, 31 July 1911, No. 19578 (declared February 1910) and Privilege, 31 October 1912, No. 22,822 (declared May 1910): RGIA, f. 273, op. 15, d. 271, ll. 1–2, 39–40.

[25]   See correspondence at RGIA, f. 273, op. 15, d. 271, ll. 174–6.

[26]   *Vestnik putei soobshcheniia*, 11–12 (1918): 31–2.

[27]   For instance, the term *dizel'-elektrovoz*, meaning electric locomotive with a diesel engine, was preferred by the congress of Russia's railway traction engineers in 1925: *Trudy XXXIV soveshchatel'nogo s"ezda inzhenerov podvizhnogo sostava i tiagi v Moskve, s 1-go*

the general national crisis of civil war, this initial lobbying had little effect. But by 1923 there were several Soviet projects in addition to Lomonosov's efforts. Most notably, Professor Gakkel' was working on a diesel-electric with Petrograd factories, Shelest was developing his concept with the British company Armstrong Whitworth, and Professor E.K. Mazing had a similar project with the VSNKh's GOMZA engineering group.

Lomonosov's inaugural public salvo in the Soviet diesel-traction debate was revolutionary in both spirit and scope, reflecting a certain ideological and technological confidence.[28] The medium was his ill-fated article on railway reconstruction in June 1920 (see Chapter 8). Hitherto Russian engineers had concentrated on promoting their own designs of diesel locomotive, whereas now Lomonosov advocated a comprehensive development programme for all the likely types of transmission, beginning with electric as the least difficult option and including, of course, extensive road tests. He advocated an order for as many as 20 diesel-electric locomotives and suggested that, given Russian industry's temporary inability to handle such a large project, they should be purchased abroad with some of the money recently allocated for importing steam locomotives. These experimental locomotives would allow scientific comparison of differing types and combinations of engine, electricity-generator, cooling system and so forth. The data, he implied, would indicate how best to use electric transmission to begin the dieselization process, pending development of other transmissions that might prove superior eventually. In other words, he wanted his government to fund a research and development programme that far exceeded the capability of any one engineering firm or private railway in the capitalist world and that could study all the technical issues quickly and thoroughly. With the Bolshevik commitment to technological change, Russia's expertise and bold leadership, the nationalized Soviet railways could revolutionize the world of railway traction – for Lomonosov now a matter of political as well as professional and national pride.

When the NKPS dismissed his article as utopian, he quietly organized some initial design work on a modest scale that nonetheless preserved his core idea of investigating a range of transmission systems. He had spotted an opportunity to do this with his appointment as Sovnarkom plenipotentiary for railway imports in June 1920. Needing two technical assistants for that work, he selected Shelest as one of them partly because of his diesel interests, with the idea that Shelest could make time to prepare an outline design for a locomotive with his system of gas transmission. Further, he asked Shelest to ensure that his locomotive had the same power and tractive effort as the latest standard Russian-built steam freight

---

po 9-oe aprelia 1925 goda (Moscow, 1926), pp. 1–2. Lomonosov sometimes took this view: compare 'Vospominaniia', vol. 8, pp. 1656–8 and vol. 9, p. 1721–2. See also Westwood, *Soviet Locomotive Technology*, p. 57.

[28] As noted in Chapter 8, diesel traction was one of the topics studied by the NKPS Technical Committee under Lomonosov's leadership during the winter of 1919–1920, but it is unknown whether he was personally involved in that discussion.

locomotive, the class E 0-10-0.[29] Next, Lomonosov exploited his presence in Western Europe during the second half of 1920 to contact key firms like Brown Boveri, the Swiss manufacturer of electrical equipment. He also approached the former Kolomna designer Felix Meineke, who was now a professor at the Charlottenberg Technical Higher School in Germany. Meineke agreed to take leave of absence from his academic post to head the Railway Mission's Locomotive Department, with a brief to use some of his time to design a diesel-electric locomotive; this design, too, was to match the class E steamer (see Figure 9.2). Needless to say, for now Lomonosov did not report these developments to Moscow.[30]

An apparent opportunity for official support was the reason why he did reveal his hand during a visit to Moscow in May 1921. He formally requested permission to order two diesel locomotives abroad – one with the Shelest-type transmission and the other to Meineke's diesel-electric design.[31] He was almost certainly reacting to news of parallel developments within Russia, where other engineers had continued to press their interest. One such approach had recently prompted the VSNKh Presidium to establish a diesel-locomotive commission in its Scientific–Technical Department. The NKPS also showed interest in spring 1921, particularly through Dubelir as head of its Technical Committee. It too created a diesel-locomotive commission, which studied documents from engineers Zalkind, Gakkel' and Popov, and possibly others. Furthermore, the Gakkel' project was examined by the Power Section of the new State Planning Commission, Gosplan.[32] In short, the authorities seemed interested, and Lomonosov needed to catch the moment.

---

[29]   Lomonosov and Shveter, *Proekty teplovozov*, pp. 16–17; 'Vospominaniia', vol. 7, pp. 1194, 1249, 1259–61, 1266–7, 1659. These two engineers had clashed in print in 1917: A.N. Shelest, 'Issledovanie raboty teplovoza Br. Zul'tser v Shveitsarii', *Vestnik inzhenerov*, 3/3 (1917): 71–80; Iu.V. Lomonosov, 'Po povodu stat'i g. Shelesta "Issledovanie raboty teplovoza br. Zultser'", *Vestnik inzhenerov*, 3/7 (1917): 206–7; and A.N. Shelest, 'Po povodu zametki prof. Iu.V. Lomonosova o stat'e "Issledovanie raboty teplovoza Br. Zul'tser v Shveitsarii"', *Vestnik inzhenerov*, 3/15 (1917): 336–8. The phrasing of Lomonosov's memoirs may suggest that a fancy for Mme Shelest also influenced his decision to take Shelest.

[30]   Lomonosov and Shveter, *Proekty teplovozov*, p. 17; Vospominaniia, vol. 8, pp. 1791–2.

[31]   Lomonosov–Dzerzhinskii, 17 May 1921: RGAE, f. 4038, op. 1, d. 46, l. 50; Lomonosov and Shveter, *Proekty teplovozov*, p. 17.

[32]   Z.N. Zalkind, 'K voprosu o vosstanovlenii transporta', March 1920: RGAE, f. 3429, op. 7, d. 726, ll. 25–30ob. (sent to VSNKh by Sovnarkom for comment in February 1921); VSNKh Scientific–Technical Department–Sovnarkom (Gorbunov), 25 February 1921: RGAE, f. 3429, op. 7, d. 726, ll. 19–ob.; Popov–Chair, NKPS Diesel Locomotive Commission, 16 April 1921: RGAE, f. 1884, op. 43, d. 11, l. 1–ob.; Dubelir–Borisov, 31 May 1921: RGAE, f. 1884, op. 31, d. 81, ll. 50–52; Extract from minutes of meeting of Gosplan Power Section, 4 June 1921: RGAE, f. 1884, op. 31, d. 81, l. 48; 'Vospominaniia', vol. 8, pp. 378–9, 393.

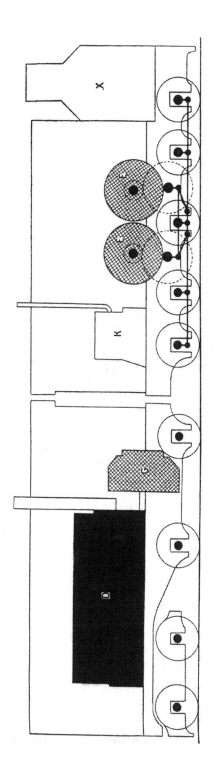

Figure 9.2   Diesel-electric locomotive Iuᴱ No. 001 as initially conceived in 1920–1921. Source: Iu.V. Lomonosov and E.E. Shveter, *Proekty teplovozov razrabotannye v 1921–1925 gg v Germanii dlia SSSR* (Berlin, 1926), p. 19.

His rapid success in this sponsorship race was due to his persistence and unique position abroad. Like all the other supplicants, he failed to obtain funds immediately, even though the NKPS Technical Committee approved his request on 27 May 1921.[33] This setback is unsurprising given the drastic contraction of the state budget that now occurred as part of the shift to a mixed state–private economy under the New Economic Policy of March 1921. But Lomonosov alone overcame this hurdle quickly. When he repeated his request during his next Moscow visit in August 1921, he obtained an NKPS instruction to 'try to order [the two diesel locomotives] in lieu of [some of] the steam locomotives approved for purchase' abroad.[34] The point was that this course did not require the NKPS or VSNKh to obtain money from the Sovnarkom – a tactic that was effectively impracticable for Lomonosov's diesel rivals within Russia. Even Gakkel', who had good connections in Gosplan, failed to get money during 1921.

This permission allowed Lomonosov to begin the negotiations that would yield, eventually, his pair of diesel locomotives. He would have preferred to build these locomotives in Germany, but that was impossible for financial reasons: the German locomotive-builders had already ordered – and hence required payment for – all the materials needed to build all the 700 class E steamers that the Soviet government had ordered from them in October 1920 and February 1921. By contrast, the Swedish firm Nohab was still in the early stages of fulfilling its enormous Soviet contract of May 1920 for 1,000 steam locomotives, and so Lomonosov could sign a memorandum of understanding with Nohab's chairman, Gunnar Anderson, on 20 October 1921 about substituting some of those steamers for diesels.[35] Hence the main tasks for the next few months were to finalize the technical specifications for these diesel locomotives and negotiate the formal contract – a very fluid situation that became, for Lomonosov, the context of Lenin's unexpected intervention in railway diesel research at the turn of the year.

It seems certain, as Westwood has argued, that Lenin's attention boosted the collective prospects of the dieselizers. It strengthened their political position *vis-à-vis* the advocates of railway electrification, who were the main rivals for funding to develop new forms of railway traction. This gain was especially important because of Lenin's much-publicized support for the GOELRO plan for nationwide electrification that the regime adopted in December 1920.[36] Additionally, the dieselizers gained a specific opportunity and incentive to redouble their effort: at Lenin's behest the STO passed a resolution about diesel traction on 4 January 1922

---

[33]   Note about an order abroad for 2 diesel locomotives, [circa 1921]: RGAE, f. 1884, op. 31, d. 81, l. 8; Lomonosov and Shveter, *Proekty teplovozov*, p. 17.

[34]   Resolution by Emshanov, 19 August 1921: RGAE, f. 4038, op. 1, d. 46, l. 50.

[35]   Iu.V. Lomonosov, *Opyty nad teplovozami: Teplovoz Iu^{e} No. 001 i ego ispytanie v Germanii* (Berlin, 1925), p. 25.

[36]   On this plan see Coopersmith, *Electrification*, pp. 151–265.

that, among other things, authorized an international competition for the best idea, complete with a reward of one million gold rubles to fund its development.[37]

One might expect that Lomonosov welcomed Lenin's intervention, yet actually it disconcerted him. Evidently because he had numerous concerns that were more urgent, he had not tried to exploit his favour with Lenin to advance the diesel case during 1920–1921. Now, learning of these developments on about 27 January 1922, he was alarmed by the draft rules for the competition, which were to be applied, in the interests of scientific comparability, to all locomotives being proposed and/or built. One rule required these locomotives to be equivalent in power output to the class Shch steam locomotive – a weaker engine than the class E that Lomonosov had specified. Another rule demanded a relatively high maximum speed of 75 kph, whereas the Railway Mission's designs were oriented to slow heavy freight traffic. Lomonosov thus had to decide whether to have his projects redesigned to conform or request an exemption.[38]

Irritated and convinced that he knew best, he sought exemption. During his visit to Moscow in January 1922 he persuaded the NKPS leadership to support this response.[39] But Gosplan ignored his concerns when approving the competition rules on 2 February, and it was supported by the STO on 10 March, albeit with caveats. Meanwhile Lomonosov initialled a draft contract with Nohab on 28 February for three (not two) locomotives. As expected, one would be a diesel-electric and another would have Shelest's transmission. The third one was earmarked for mechanical transmission by the NKPS leadership at the insistence of Lomonosov's old acquaintance P.I. Krasovskii, who was now one of the senior traction specialists at the NKPS. But no company wished to try to design and build the requisite high-power gearbox, and hence hydraulic transmission was substituted; it was to be designed by Meineke, who had just returned to his academic job and was retained as a consultant.[40] Happily for Lomonosov, in May

---

[37] See in particular A. Naporko (comp.), *Zheleznodorozhnyi transport SSSR v dokumentakh Kommunisticheskoi partii i Sovetskogo pravitel'stva* (Moscow, 1957), pp. 158–9; Westwood, *Soviet Locomotive Technology*, pp. 18–22.

[38] Lenin–Lomonosov, 27 January 1922: Lenin, *Pol'noe sobranie sochinenii*, vol. 54, pp. 144–5; 'Dnevnik', 3 February 1922; Lomonosov and Shveter, *Proekty teplovozov*, p. 19; 'Vospominaniia', vol. 8, pp. 904–8, 912.

[39] Minutes of meeting about Railway Mission affairs, 30 January 1922: RGAE, f. 1884, op. 28, d. 2, l. 248; 'Dnevnik', 3 February 1922; Lomonosov and Shveter, *Proekty teplovozov*, p. 19; 'Vospominaniia', vol. 8, p. 912.

[40] N.I. Rogovskii et al. (eds), *Protokoly Prezidiuma Gosplana za 1922 god*, vol. 1 (Moscow, 1979), p. 66; Gosplan–Lenin, 3 February 1922: RGASPI, f. 5, op. 1, d. 339, ll. 7–11; Lomonosov–Lenin, 28 January and 7 February 1922: RGASPI, f. 5, op. 1, d. 339, ll. 18, 37; STO resolution, 10 March 1922: RGAE, f. 1884, op. 31, d. 81, l. 24; Lomonosov and Shveter, *Proekty teplovozov*, pp. 19–20 and prilozhenie I, p. 3 (Report to Sovnarkom, 25 October 1922); 'Vospominaniia', vol. 8, pp. 916–18, 963, 983–4, 1312. Meineke's return to academia is noted at 'Dnevnik', 11 December 1921.

1922 Gosplan relented concerning the competition rules.[41] Indeed, Lomonosov believed that the whole competition was simply abandoned so that Gakkel' could have the money for building his locomotive design in Petrograd, which was agreed by the STO in July 1922. Certainly the competition never became the vehicle for progress and propaganda that Lenin envisaged.[42]

Over the next two years Lomonosov encountered political, financial and technical problems that each threatened to terminate his project. The first came from the decision of June 1922 to cut the orders for foreign-built steam locomotives. Lomonosov anticipated that the reduction would affect Nohab alone since the cost of compensating the Germans for the materials already bought would be prohibitive; hence his Nohab diesel-locomotive contract was also threatened. He saved it by obtaining NKPS permission to use 1.75 million Swedish crowns from the Railway Mission's 'savings' to pay for new diesel contracts, and by seeking alternative contractors; in particular, he agreed terms in August for the Hohenzollern company of Dusseldorf to build Meineke's hydraulic transmission system.[43] This reassignment of some of the Railway Mission's money was challenged in Moscow. However, Lomonosov prevailed by securing Lenin's support. Chairing the Sovnarkom on 31 October 1922 for what would be the last time due to his declining health, Lenin recommended approval for all Lomonosov's diesel decisions.[44] On this basis, and with Nohab's consent, Lomonosov moved his project to Hohenzollern. He signed a contract on 15 December 1922 for two locomotives – one diesel-electric and one diesel-hydraulic – with an option to add a Shelest 'turbine-type' locomotive later. Related contracts went to MAN for two diesel engines and the Swiss company Brown Boveri for the electrical equipment for both locomotives. The cost of these items, as valued in April 1923, just exceeded half a million Swedish crowns, which was slightly more than Nohab's price for two complete class E steam locomotives.[45]

[41]   Lomonosov and Shveter, *Proekty teplovozov*, p. 19.

[42]   'Vospominaniia', vol. 9, pp. 1339–40. On Gakkel' and his locomotive see Ia.M. Gakkel', 'O teplovoze Gᴱ 001', in *Trudy XXXIV soveshchatel'nogo s"ezda inzhenerov podvizhnogo sostava*, pp. 29–52; Rakov, *Lokomotivy*, pp. 356–9; Zenzinov and Ryzhak, *Vydaiushchiesia inzhenery*, pp. 242–55; Westwood, *Soviet Locomotive Technology*, pp. 35–9, 42–4.

[43]   Lomonosov and Shveter, *Proekty teplovozov*, p. 19, prilozhenie I, p. 3 (Report to Sovnarkom, 25 October 1922) and prilozhenie III, p. 23 (Minute of meeting, 11 August 1922); 'Dnevnik', 12 August 1922; 'Vospominaniia', vol. 8, pp. 1312–13, 1348–51.

[44]   Report to Sovnarkom, 25 October 1922; Lomonosov and Shveter, *Proekty teplovozov*, prilozhenie I, p. 3; Minutes of Sovnarkom meeting, 31 October 1922: RGASPI, f. 19, op. 1, d. 529, ll. 2–3; 'Dnevnik', 2 September 1922; 'Vospominaniia', vol. 8, pp. 1474, 1527–40.

[45]   Lomonosov and Shveter, *Proekty teplovozov*, p. 20 and prilozhenie III, pp. 23–5 (Minutes of meeting, 5 December 1922); 'Dnevnik', 5, 13, 15 December 1922; Lomonosov, *Predvaritel'nyi otchet*, Prilozhenie II, p. 10; 'Vospominaniia', vol. 8, pp. 1404–5, 1520–1521, 1636–9, 1659. The contract papers are in RGAE, f. 4038, op. 1, dd. 519, 521–2.

The reason why Lomonosov ordered only two locomotives was the collapse of his relations with Shelest. The latter suspected a Lomonosov plot to ruin his reputation and steal his idea, whereas Lomonosov began to regard Shelest as paranoid and his invention as hopeless. Later events confirmed the latter technical analysis, but there is some truth in the claim that Lomonosov tried to undermine his colleague's reputation. The final break occurred in autumn 1922 when Shelest refused to provide the blueprints necessary for the contract. Amid a flurry of mutual recriminations Shelest complained to the NKPS and was recalled to Moscow.[46] Thereafter Shelest had an intriguing career. From 1923 he was allowed to work on developing his idea in Britain with Armstrong Whitworth, but even this major firm struggled with the technical challenge. In 1926 the NKPS decided to try to repatriate the project, and in 1927 it found a suitable excuse when the UK broke off diplomatic relations. Shelest spent the remainder of his life – 27 years – working on his ideas at the Moscow Higher Technical School. Alas, he never built a locomotive, but fortunately he was never arrested in the various Stalinist purges of pre-revolutionary 'bourgeois' engineers despite at least one period of political disgrace – a remarkable tale of persistence, failure and survival.[47]

The Lomonosov–Shelest conflict contributed directly to the growing distrust of Lomonosov among Moscow decision-makers, though evidently not Lenin. Particularly unfortunate for Lomonosov was the fact that his conflict with Shelest came to a head in autumn 1922, which meant that it coincided with the Avanesov commission's hostile investigation of the Railway Mission, the Sovnarkom's reprimand for the payment of bonuses to mission staff and the final departure from the political scene of the ailing Lenin, who had shielded him from so much criticism. By February 1923, then, the head of the NKPS, Dzerzhinskii, was adamant that the main way forward for diesel traction was to establish a diesel-locomotive-building industry in Russia, and he had ever less time for the likes of Lomonosov and Shelest: 'I will not speak of political motives ... Let Lomonosov build two locomotives abroad and Shelest one at home ... [The construction of a diesel-locomotive-building industry in the USSR] would be not simply a matter of Shelest or some other engineer, but of the whole of Gosplan, the NKPS, VSNKh and our party from one day to the next. Whereas abroad, I fear, it would be a matter

---

[46] 'Dnevnik', 2 August, 1 November 1922; 'Vospominaniia', vol. 8, pp. 6–7, 261–2, 514–15, 1319–32, 1494–5, 1527–38, 1572–7; Westwood, *Soviet Locomotive Technology*, pp. 46–7. Key letters and reports from 1921–1923 are reproduced in Lomonosov and Shveter, *Proekty teplovozov*, Prilozhenie II, pp. 9–22. Shelest's perspective is described in, for example, *Dva dostizheniia v oblasti sovetskoi zheleznodorozhnoi tekhniki: Doklady VI Plenumu Biuro pravlenii zheleznykh dorog* (Moscow, 1925), pp. 13–14 (report by Shelest); Goriacheva and Shelest, *Aleksei Nesterovich Shelest*.

[47] Lomonosov and Shveter, *Proekty teplovozov*, prilozhenie I, p. 4 (Sovnarkom resolution, 30 April 1923); Report about inspection of work on building the Shelest-system diesel locomotive, 2 August 1926: RGAE, f. 1884, op. 43, d. 75, ll. 312–24; Westwood, *Soviet Locomotive Technology*, pp. 47–8.

of struggle and intrigue between two [engineers], Shelest and Lomonosov.'[48] For Dzerzhinskii, therefore, as for various other senior party figures, personalities and engineering arguments had become intertwined, and party leadership and control within Russia were essential to drive the process forward as a national economic and political priority. Lomonosov was basically an untrustworthy *spets*.

Also significant for Lomonosov was the fact that the uproar about the Railway Mission during 1922 gradually helped to change his outlook so that he regarded his diesel project as a means to remain abroad in Soviet service for some months and then years. Back in 1920 he had suggested ordering diesel locomotives abroad merely to bypass Russia's economic crisis; not expecting to be abroad for long, he had undoubtedly assumed that he would lead any experimental research in Russia as an NKPS Collegium member. As of early 1923 he remained committed to Soviet Russia: though grateful to be offered a consulting position by Hohenzollern, he declined it on the grounds that he was still in government service. Yet he was uncertain and confused about what to do next. One possibility, as he saw it perhaps naively, was to retire from government service and either live quietly in Russia or work abroad in a private capacity; indeed, he had long since been aiming to retire in about 1925.[49] But for now he wished to keep a state salary, and so he weighed various proposals and aspirations, both separately and in combinations. These included membership of the NKPS Collegium with responsibility for leading the diesel-locomotive research; chair of the NKPS Technical Committee (this was Dzerzhinskii's preference); deputy People's Commissar; a senior position on the lines if not at the NKPS (Lenin's preference); a return to the Kiev Polytechnic Institute; 'hiding himself' for a while in Germany with his two diesel locomotives (which he knew would cause protests in Moscow); and bizarrely, Soviet ambassador to Sweden. He humoured Dzerzhinskii by taking the Technical Committee, but by February 1923 his preference was reassignment to diesel research based in Germany, both for the scientific interest and the distance from Moscow's 'moods'.[50]

Surprising as it may seem given the furore over the Railway Mission, Lomonosov got his way. The key was probably Dzerzhinskii's reluctant acquiescence, which seems to have been secured through Lomonosov promising to keep his position

---

[48]   Dzerzhinskii–Krzhizhanovskii, 18 February 1923, in F.E. Dzerzhinskii, *Izbrannye proizvedeniia v dvukh tomakh*, vol. 1 (Moscow, 1977), pp. 314–15. See also the NKPS Collegium minutes for 17 February 1923: RGAE, f. 1884, op. 28, d. 52, l. 59. Dzerzhinskii's vision here proved impracticable due to the technical complexity of Shelest's project and the financial weakness of the NEP economy.

[49]   His aim to retire in 1925 is noted at 'Dnevnik', 3 September 1930.

[50]   In particular: 'Dnevnik', 2 August, 6, 30 September, 18–20, 26, 30 October, 2–3, 19 November, 9, 26–8, 31 December 1922, 2–6, 8, 13, 22–7 January, 7, 16, 28 February, 4 March 1923; Lomonosov–Lenin, 8 October 1922: RGASPI, f. 5, op. 1, d. 1166, l. 7. 'Vospominaniia', vol. 8, pp. 1332, 1678, 1682, 1691–3, 1700–1701, 1742–3, 1750–1751, 1782, 1836–7.

at the Technical Committee and return to Russia by about the end of 1923. The regime duly agreed in March 1923 that Lomonosov could stay in Germany to write his Railway Mission report and work on his diesel locomotives; he could have a small staff and keep his Sovnarkom plenipotentiary powers for the diesel work. On that basis he immediately established his so-called Bureau for Diesel-Locomotive Construction in Berlin. Initially it had a general office with three staff, a diesel department manned by E.E. Shveter (helped by Meineke as a consultant) and a temporary office to help with the Railway Mission report. Needless to say, that report took far more time than expected and the diesel work only increased. In due course more people were attached to the bureau, including Lomonosov's protégé Dobrovol'skii, a communist veteran of the Railway Mission called S.S. Terpugov and, as technical secretary, a veteran of the erstwhile Experiments Office, S.V. Makhov. Also, a subsidiary office was established near the German factory for product acceptance work and the locomotive testing.[51]

Not everyone in the party leadership was content with this outcome. At about this time the Central Committee member K.E. Voroshilov expressed the opinion to Stalin that Lomonosov did not merit the party's trust and that 'we have enough *spetsy* without such "ultra-dealers"'. Stalin's response, however, revealed a different view that helps explain why Lomonosov got his way:

> A very good technical man, he drew up a plan for building diesel locomotives (in place of steam locomotives), which can effect a revolution (*perevorot*) throughout our transport. These diesel locomotives are being built in Germany for testing and we can't complete this plan without Lomonosov. I think that for the time being there's no basis for removing Lomonosov from *technical* work.[52]

Nonetheless as early as May 1923 Dzerzhinskii began pressing Lomonosov to return to Russia, and confronted him in June and December about his continuing presence abroad. Meanwhile the ambassador in Berlin, Krestinskii, became so

---

[51] 'Dnevnik', 26–8, 31 December 1922, 22–7 January, 25 February, 21, 24 March 1923; Lomonosov and Shveter, *Proekty teplovozov*, pp. 20–21 and prilozhenie I, p. 4 (Sovnarkom resolution, 29 March 1923); 'Vospominaniia', vol. 8, pp. 1691–3, 1744, 1772–3, 1780, 1868–9, 1908; vol. 9, pp. 1–2, 15–6, 159, 178–83, 193–4, 503, 536. Terpugov, who was born in Rostov in 1885 and became a deputy head of the assembly shops at the Kolomna works, perished in the purges; his death date is unknown: Memorial, 'Zhertvy politicheskogo terrora v SSSR': http//lists.memo.ru/index19.htm, citing *Kniga pamiati Moskovskoi oblasti* (last checked 30 June 2008). Makhov joined Lomonosov's research team before the world war: 'Vospominaniia', vol. 5, p. 365. Having worked with Lomonosov until January 1931, he died suddenly on 31 October 1931: 'Dnevnik', 24 January, 31 October, 3, 7 November 1931. See also Genis, *Nevernye slugi rezhima*, pp. 468–9.

[52] Voroshilov–Stalin, Stalin–Voroshilov, [circa spring 1923]: in RGASPI, f. 74, op. 2, d. 39 (dated by contents; emphasis in the source). I am grateful to James Harris for this reference.

exasperated that, asked one day for instructions, he advised Lomonosov to die so that a monument could be erected and his colleagues could at last work in peace.[53]

Lomonosov was normally at odds, too, with the head of the NKPS office at the NKVT delegation in Berlin. With the first incumbent, Emshanov, these difficulties began fairly soon and mostly concerned affairs of the former Railway Mission. Especially bad for Lomonosov was the tenure of Emshanov's successor, Iu.V. Rudyi, from November 1923 to January 1925: even Emshanov told Lomonosov to treat this person as one of his 'most evil enemies'.[54] Lomonosov had no particular problems with Rudyi's replacement: P.N. Kirsanov was an old friend. For a time, too, Lomonosov was on reasonable terms with V.G. Chirkin, who became the fourth head of the NKPS office in spring 1926; but by January 1927 Chirkin was a 'scoundrel', not least because he pressed the Lomonosovs to vacate their government apartment in Berlin.[55]

That said, events outside Soviet control posed the main danger to Lomonosov's diesel project during 1923. From 13 January France occupied the Ruhr region of Germany in retaliation for alleged non-fulfilment of German obligations in the Treaty of Versailles. The diesel work was badly affected both by the general difficulties of life in the occupied zone and by the close French scrutiny of the Ruhr engineering industry due to its military significance. Eventually Hohenzollern allowed Lomonosov to transfer the diesel-electric locomotive to the Esslingen Machine-building Works near Stuttgart. The agreement was signed on 8 June 1923 and the locomotive was built between August 1923 and June 1924. For the time being no decision was taken about the second locomotive.[56]

For all the later praise of Lomonosov's project there has been criticism that his two diesel locomotives constituted a technological dead-end, particularly in relation to their chassis, engine and cooling system.[57] There is some justification for this opinion. Both locomotives had fixed frames, like a conventional steam

---

[53] 'Dnevnik', 14–17 May, 6, 8, 15 June, 18, 24 December 1923, 15 April, 25 June 1924; 'Vospominaniia', vol. 9, pp. 23–7, 46–8, 56–9, 167–9, 449–55, 558–9, 636–9.

[54] For example: 'Dnevnik', 3 June, 25 July, 3 August 1923, 20 April, 23–4 July 1924; 'Vospominaniia', vol. 9, pp. 35, 214, 326, 380, 404, 495 (between 405–6), 504, 628, 741–2, 771–2, 780–785, 792–4, 828–30. Rudyi's scepticism about Lomonosov is evident in, for example, Rudyi–Rudzutak, 31 March 1924: RGAE, f. 1884, op. 31, d. 1674, ll. 133–6.

[55] For example: 'Dnevnik', 28 June, 8 July, 4 October, 30 December 1926, 4 January 1927; 'Vospominaniia', vol. 9, pp. 91, 506, 1753–4, 1796, 1958–61, 2119–21.

[56] 'Dnevnik', 7 February, 2, 21, 24, 26 May, 9 June 1923; Lomonosov and Shveter, *Proekty teplovozov*, pp. 20–21 and prilozhenie III, p. 30 (Minutes of meeting, 8 June 1923); 'Vospominaniia', vol. 8, p. 1724; vol. 9, pp. 30–31, 42–3, 176, 195. A full technical description of the diesel-electric is at Lomonosov, *Teplovoz Iu^g*, pp. 34–64. A small amount of related paperwork survives in the factory archive: Deutsches Museum, Archiv, FA Maschinenfarbrik Esslingen, folders in boxes 013, 016 (listing kindly provided by the Archive; papers not consulted).

[57] For example, Westwood, *Soviet Locomotive Technology*, p. 40.

locomotive, instead of a chassis with bogies of the type that was becoming popular in first-generation main-line electric locomotives, that Gakkel' had chosen in Petrograd, and that would become the worldwide norm for main-line diesel power. Also, both his locomotives used an engine that was designed for submarines and was too heavy for general railway use. Finally, both locomotives had a large yet inadequate engine-cooling system. Indeed, from 1925 the first locomotive had additional cooling equipment mounted on the chassis of a steam locomotive's tender, which meant that it had to be turned at each destination and was too long for the usual Soviet depot stall and turntable (see Figures 9.3 and 9.4).

However, it should be remembered that at this stage Lomonosov was concerned with getting scientific data and that he envisaged a lot more research before any designs were prepared for mass production. He saw these two locomotives as the first of many experimental aggregates – his favoured term was 'mobile laboratories' – that would collectively reveal the best design ideas by enabling thorough experimentation on all combinations of the major components. To that end he found two reasons to use fixed frames: the need to cut the overall weight became critical, and in any case his aim to study the interactions of the power system's major components required a simple, reliable and stable chassis.[58] Of course, much of the weight problem came from the engine. The conundrum was that a light-weight high-power unit did not exist. Pending its development, he sought the best available compromise, and chose a 26-tonne MAN submarine-type engine as the lightest (and cheapest) option; it was no coincidence that Gakkel' also used an engine designed for submarine use.[59] Lomonosov likewise made compromises with the cooling system.[60]

His concern with getting accurate performance data also informed his response to the challenge of checking that these locomotives actually worked. Because considerable modifications to the locomotives might prove necessary, it was preferable for these tests to be completed near the German factory. But how could this be done with Soviet-gauge locomotives? A short length of track could be laid, but more imaginative and useful was Lomonosov's idea to build a Soviet-gauge locomotive test plant for both running-in and formal experiments. Uniquely, this plant was made portable so that eventually it could be shipped to Russia. It made such an impression on the Esslingen management that they built a version for themselves.[61] Also, further to an agreement of January 1925 with the German State Railways about road experiments in Germany, the second

---

58 See Lomonosov, *Teplovoz Iu*^{E}, pp. 59–64.

59 Gakkel', 'O teplovoze G^{E} 001', p. 29.

60 For technical details see Lomonosov, *Teplovoz Iu*^{E}, pp. 45–51.

61 For example: Lomonosov, *Teplovoz Iu*^{E}, pp. 85–91; 'Diesel Locomotives for Main Line Traffic', *The Railway Engineer*, 47/561 (October 1926): 363, 368; W. Messerschmidt, 'The Esslingen Roller Test Plant', *Diesel Railway Traction* (December 1956): 477–9.

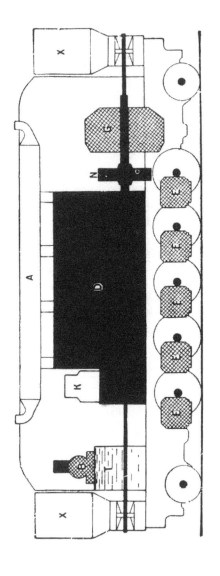

Figure 9.3   Iu[E] No. 001 as first built in 1924. Source: Iu.V. Lomonosov, *Opyty 1925 goda nad teplovozom Iu[E] No. 001 na zheleznykh dorogakh SSSR* (Berlin, 1927), p. 2.

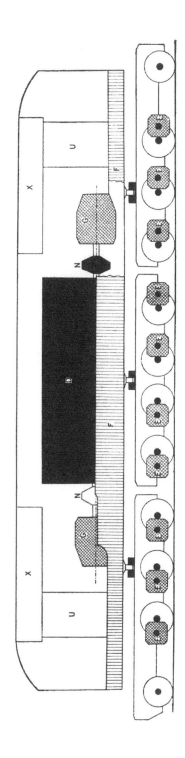

Figure 9.4   Diesel-electric locomotive Gᴱ No. 001 as designed by Ia.M. Gakkel'. Source: Iu.V. Lomonosov, *Opyty 1925 goda nad teplovozom Iuᴱ No. 001 na zheleznykh dorogakh SSSR* (Berlin, 1927), p. 193.

locomotive was redesigned to allow the use of German-gauge as well as Soviet-gauge wheelsets.[62]

The transmission types explain why these two locomotives were designated Iu[E] No. 001 and Iu[M] No. 005 when the NKPS introduced a unified system for identifying its diesel locomotives in December 1923. This system was derived from the 1912 alpha-numerical system for steam locomotives. The letter 'Iu' was, in effect, the class letter for diesel locomotives, and the suffix indicated the transmission type. The result as applied to the five locomotives that were under construction or planned at this time is shown in Table 9.1. The suffixes E, SH, K and M designated electric, Shelest, compressor and magnetic transmission respectively. The designs of Shelest and Professor E.K. Mazing were regarded as similar and hence were numbered consecutively.[63]

Table 9.1   Soviet diesel locomotives under construction or planned, 11 January 1924

	Design-team leader and transmission type	Main builder	Date of main contract	Road number
1	Lomonosov: diesel-electric	Esslingen	1922, 1923	Iu[E] No. 001
2	Gakkel': diesel-electric	Baltic	1922?	Iu[E] No. 002
3	* Shelest: gas/steam turbine	Armstr. Whitworth	1923	Iu[SH] No. 003
4	* Mazing: gas/air compressor	GOMZA	authorized 11 January 1924	Iu[K] No. 004
5	Lomonosov: diesel-magnetic (mechanical)	Hohenzollern/ GOMZA	1922/authorized 11 January 1924	Iu[M] No. 005

Note: * never completed.

Lomonosov's Iu[E] No. 001 was a qualified success in trials at Esslingen during 1924. It first moved under power on 5 June, then underwent a four-week programme of running-in and experiments. It enjoyed an indicated thermal efficiency of about 26 per cent at best, developed 1,020 h.p. at the wheel rim, and consumed about three times less fuel than the class E steamer that was kept in Germany for comparative experiments on the test plant. As for problems, the auxiliary motor for starting the excitation equipment – with which the driver controlled the magnetic field in the main generator – was noisy, unreliable, dangerous and a cause of much vibration. The cooling system was heavy yet inadequate, as expected. Also, the weight was higher than planned at 124.8 tonnes, with a maximum axle-load of almost 20 tonnes, which compared badly

---

[62]   Lomonosov and Shveter, *Proekty teplovozov*, p. 179.

[63]   P.Ia., 'Teplovozy dlia russkikh zheleznykh dorog', *Zheleznodorozhnoe delo*, 5 (1924): 24–8; Lomonosov, *Opyty 1925g. nad teplovozom Iu[E] No. 001 na zheleznykh dorogakh SSSR* (Berlin, 1927), prilozhenie I, p. 1 (Minutes of meeting of 28 December 1923). There were several renumberings during the next few years; the position at the end of the 1920s is shown in Heywood and Button, *Soviet Locomotive Types*, p. 11.

with the class E's 80-tonne weight (excluding tender) and 16-tonne axle-load.[64] Summer modifications yielded better results. The auxiliary motor was discarded (its job was reassigned to the main engine), the cooling system was rebuilt, the engine–generator connection was improved, and the excitation equipment was moved for better weight distribution. The overall weight dropped to 118.3 tonnes. The benefit of these changes was shown by experiments in September 1924, but so too was the need for more weight redistribution and changes to the cooling system. These amendments having been made, 20 satisfactory experiments were done during October (see Figure 9.5).[65]

These experiments incorporated a modification to Lomonosov's locomotive testing methodology that reflected the different nature and possibilities of diesel traction compared to steam. In 1908 he had devised cycle I and cycle II regimes for road tests, mainly to deflect criticism from the likes of Shchukin. Now he introduced cycle III trips for diesel locomotives on the test plant. The idea was to keep the engine regime constant (specifically, the number of engine-shaft revolutions per minute and the fuel pump setting) and to use the brakes to vary the speed. This tactic, he hoped, would give perhaps 10 different readings for tractive effort during one 'trip' – a much quicker and cheaper way to obtain enough data. Dobrovol'skii and some of the factory engineers expressed misgivings, but Lomonosov persuaded himself about its practicability during the June experiments and used it thereafter.[66]

The way that Lomonosov exploited this progress for publicity did much to boost his international profile and reputation. He invited many European railway experts and journalists as well as Krestinskii and Rudyi to Esslingen for a formal comparison of the diesel locomotive and the class E steamer.[67] The minutes, drafted by Lomonosov, noted the diesel-electric's fuel economy and smooth movement, and ended triumphantly:

> To judge by the results of the experiments on the diesel locomotive Iu[E] No. 001, the creation of this diesel locomotive and the experiments with it have taken the idea of the diesel locomotive out of the stage of academic research, and have embodied it in forms suitable for effecting regular freight work. The

---

[64] 'Dnevnik', 5 June–7 July 1924; Lomonosov, *Teplovoz Iu[E]*, pp. 65–173; 'Vospominaniia', vol. 9, pp. 703–58.

[65] For example, 'Dnevnik', 2, 13, 17 September, 22–30 October 1924; Lomonosov, *Teplovoz Iu[E]*, pp. 174–231; 'Vospominaniia', vol. 9, pp. 758–61, 774–5, 789–90, 832–41, 888–913.

[66] See Lomonosov, *Teplovoz Iu[E]*, pp. 65–114; 'Vospominaniia', vol. 9, pp. 699–707, 714–21.

[67] 'Dnevnik', 3, 7 November 1924; Lomonosov, *Teplovoz Iu[E]*, pp. 231–5; 'Vospominaniia', vol. 9, pp. 913–30.

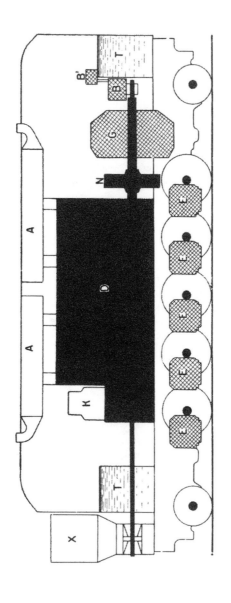

Figure 9.5   Iu^E No. 001 as rebuilt in summer 1924. Source: Iu.V. Lomonosov, *Opyty 1925 goda nad teplovozom Iu^E No. 001 na zheleznykh dorogakh SSSR* (Berlin, 1927), p. 193.

latter point deserves to be mentioned in the pages of the history of railway equipment.[68]

Enthusiastic reports duly appeared in prestigious foreign technical journals, and were noted in Russia.[69] By contrast, the first steps of the Gakkel' locomotive were reported only in Soviet publications, although probably foreign publicity was not sought.[70]

The official minutes may seem presumptuous given the technical criticisms noted above. Indeed Westwood is doubtless correct to describe them as 'a reminder to future historians that the locomotive and its designers deserved a chapter, not a footnote'.[71] But we must also consider them in the light of Lomonosov's immediate need to promote his project amid not just uncertainty about diesel traction but also increasing nationalism in Soviet politics as Stalin, Trotskii, Kamenev and others vied for power following Lenin's death in 1924.[72] The 'parallel experiments', as Lomonosov termed the Esslingen event, were not needed for gathering performance data: separate testing of the class E steamer in early 1924 provided enough information for the steam–diesel comparison. Rather, their main aims were to woo the sceptical majority of railwaymen both in Russia and abroad, and to forestall ruses by such 'friends' as Krestinskii and Rudyi; the presence of the 'impartial world authorities' was to confirm and legitimate his demonstration

---

[68] Lomonosov and Shveter, *Proekty teplovozov*, prilozhenie III, pp. 66–71 (Minutes of the comparative tests of class E^G steam locomotive No. 5570 and 1-5-1 diesel locomotive Iu^E No. 001 on the Temporary Russian Test Plant at Esslingen, 3–6 November 1924); 'Vospominaniia', vol. 9, p. 913.

[69] For example: 'A Diesel-Electric Locomotive', *The Engineer*, 138 (14 November 1924): 552–4; 'Diesel-Electric Locomotive Surpasses Steam on Test', *Oil Engine Power*, 2/12 (1924): 633–7 (also reported in *Mechanical Engineering*, 47/3 (1925): 212); F. Meineke, 'Vergleichversuche zwischen Diesel- und Dampflokomotive', *Zeitschrift des Vereines Deutscher Ingenieure*, 69/10 (1925): 321–2; and W. Willigens, 'Russische 1E1 Diesel-elektrische Lokomotive Bauart Professor Lomonossoff–Hohenzollern AG Düsseldorf', *Der Waggon- und Lokomotiv-Bau*, 6 (19 March 1925): 81–6. For examples of the Soviet coverage see V.E. Glazenap, 'Teplovoz sistemy prof. Iu.V. Lomonosova', *Elektrichestvo*, 2 (1925): 105–7; G. Trinkler, 'K voprosu o naivygodneishem tipe teplovoza', *Zheleznodorozhnoe delo*, 1 (1925): 8–14.

[70] For example, 'Pervyi teplovoz v SSSR', *Ekonomicheskaia zhizn'*, 27 November 1924, p. 3; 'Pribytie pervogo russkogo teplovoza', in *Ekonomicheskaia zhizn'*, 17 January 1925, p. 3; G.N., 'Pervyi teplovoz – v Moskve', *Tekhnika i zhizn'*, 2 (1925): 6.

[71] Westwood, *Soviet Locomotive Technology*, p. 43.

[72] On these political developments see, for example, E.H. Carr, *Socialism in One Country, 1924–1926*, 3 vols (London, 1958–1964); Day, *Leon Trotsky*; S.F. Cohen, *Bukharin and the Bolshevik Revolution: A Political Biography, 1888–1938* (Oxford, 1980), pp. 160–242; G. Gill, *The Origins of the Stalinist Political System* (Cambridge, 1990), pp. 113–98; R. Service, *Stalin: A Biography* (London, 2005), pp. 208–29, 240–250. See also note 107 below.

of diesel traction's superior economy. Moreover, he was conscious of growing Russian antipathy towards Iu[E] No. 001 as a foreign product – a development that was surely encouraged by Stalin's popular deployment in 1924 of the nationalist-revolutionary slogan 'Socialism in One Country' to strengthen his position in the power struggle. A member of the party's Central Committee, A.D. Tsiurupa, even told Lomonosov in November that his diesel project was seen as almost 'counter-revolutionary'. But here too Lomonosov reckoned that 'science' would prevail. He hoped that Iu[E] No. 001's technical success would restore his credibility in Russia as both an engineer and loyal Soviet citizen and bring permission to order more locomotives abroad. Equally, he felt that good international publicity would help him find private employment abroad should the need arise.[73]

His confidence in his locomotive's technical capabilities was vindicated by the first road tests. During late 1924 Iu[E] No. 001 was partially disassembled and sent by rail to Daugavpils workshops in Latvia for reassembly and running-in trips on the Soviet-gauge lines of that area; those trips began on 9 January 1925.[74] Meanwhile Lomonosov visited Moscow to finalize arrangements for the NKPS acceptance trials and subsequent road experiments. The examination of both Iu[E] No. 001 and the Gakkel' locomotive was entrusted to an NKPS commission under M.E. Pravosudovich, whereas the subsequent experiments on both locomotives would be managed by a Moscow office of the Berlin diesel bureau under Lomonosov's overall control, using rules drafted by Lomonosov and ratified by Pravdin as NKPS Deputy People's Commissar.[75] Some technical problems arose during the running-in trips, but they were not fundamental, and Iu[E] No. 001 left Latvia for Moscow under its own power on 19 January, while Lomonosov returned to Berlin. The acceptance trials of both locomotives began shortly afterwards and were successfully concluded for Iu[E] No. 001 as early as 3 February. In the latter's schedule was a journey with Rudzutak and other top NKPS officials on 28 January, after which Rudzutak and his colleagues sent an enthusiastic telegram to Lomonosov in Germany:

> The diesel locomotive Iu[E] No. 001 completed an experimental trip today with an 1,800-tonne train. The results were outstanding. It has been confirmed that the diesel locomotive is truly a practicable locomotive that has the broadest

---

[73] 'Dnevnik', 29 June, 11 November 1924; Lomonosov, *Teplovoz Iu[E]*, p. 231; Iu. Lomonosov, *Parovozy E, E[SH], E[G]* (Berlin, 1924), pp. 173–243; 'Vospominaniia', vol. 9, pp. 514–15, 526–7, 693–4, 746, 890, 902, 905–8, 926, 932.

[74] 'Dnevnik', 10, 18 December 1924; Lomonosov, *Opyty 1925 g.*, pp. 4–10; 'Vospominaniia', vol. 9, pp. 988–1003, 1008–9, 1018–19.

[75] For example: Minutes of NKPS Collegium, 22 November 1924: RGAE, f. 1884, op. 28, d. 74, ll. 170–176; 'Dnevnik', 30 September, 19, 28 October, 18–21 November, 6 December 1924; 'Vospominaniia', vol. 9, pp. 955–8, 963–76, 983–4. The main documents about these matters were published in Lomonosov, *Opyty 1925 g.*, prilozhenie I, pp. 1–33.

prospects. We congratulate you as the initiator of the world's most powerful and economical engine on wheels. We greatly regret that you are not with us now.[76]

Thus began a long career for this locomotive that lasted until 1954. By contrast, the Gakkel' locomotive was troubled. Its bogie-chassis was successful, but its electrical equipment and engine proved so unreliable that the acceptance was delayed for months. It was withdrawn in December 1927, and was later used as a mobile generator.[77]

Needless to say, Lomonosov's absence from the acceptance trials was a sign of trouble. Ostensibly he needed to be in Germany to discuss shortcomings in Iu^E No. 001, resolve problems with the second locomotive, and make decisions about the technical specifications for follow-on orders. In reality, he knew that these matters could wait. His diary confirms that he desperately wanted to accompany his locomotive. He was 'terribly upset' at seeing it leave Daugavpils for the Soviet border, and found that to be abroad as it reached Moscow was 'extremely painful'.[78] So why did he stay away?

Political and professional pressures did form part of the explanation. He sensed that 'for some reason' virtually everybody in Moscow was against him and that there was a 'great risk' that 'anything could happen' if he went to Russia. Part of this problem was the recent rejection of his Railway Mission report of 1923 by the audit authority, the Commissariat of Workers' and Peasants' Inspection: if the Sovnarkom and party's Central Committee concurred, he would probably be prosecuted for squandering state gold, with the risk of a death sentence.[79] But he also thought that his difficulties stemmed from general political developments.[80] As noted in Chapter 8, he believed that he was enduring the same 'method of denigration' that was being employed against party members like Trotskii and A.M. Krasnoshchekov. Equally, aware that many people considered him a bourgeois *spets*, he linked his plight with the repression that he thought the *spetsy* were beginning to suffer:

> It goes without saying that [in 1918–1920] Soviet power was very lenient towards the intelligentsia and plotters. But now, after they have lived and worked peacefully for five years, the squeezing of them for old sins has the character of injustice. On the other hand, those [specialists] who genuinely worked honestly with us are not being touched now. But 'you can't make an omelette without breaking eggs'. It's difficult to maintain justice for individuals

---

[76] Lomonosov, *Opyty 1925g.*, pp. 10–11 and prilozhenie I, pp. 33–9 (quotation: p. 37).

[77] The withdrawal dates are in Rakov, *Lokomotivy*, pp. 358–9, 361. Options for the Gakkel' locomotive's future were considered in a report of circa January 1928: RGAE, f. 1884, op. 43, d. 530, ll. 42–6.

[78] 'Dnevnik', 19 January (15.00), 22 January 1925.

[79] For example: 'Dnevnik', 15, 21, 23, 29 December 1924.

[80] 'Dnevnik', 13, 21 December 1924.

during mass actions. I am myself partly a victim of that mood, but I reconcile myself with this.[81]

He was incensed by Gosplan's pressure for a 'neutral' research bureau to direct all the experiments: this stance showed distrust of him as a non-party specialist and – worst of all in his eyes – as the biased creator of a machine, not an objective scientist. Additionally, he was struggling with 'treachery' in his camp. Possibly because of the political clouds gathering over him, even close colleagues like Terpugov and Dobrovol'skii had begun criticizing his type of locomotive research in favour of the old-style 'practical road trips' that he deemed unscientific. The construction of more experimental diesel locomotives would be absurd, he thought, if they could not be studied scientifically.[82]

For now his mood over these concerns tended to be combative. As would happen with countless party members in the 1930s purges, he blamed himself for most of the distrust. In particular, he still saw the Novorossiisk crash of 1920 as the turning point of his Soviet career. That experience had taught him that the revolution had raised the stakes of professional decisions and disagreements to life or death for engineers. Now his own confidence waxed and waned about whether a trial would be conducted fairly, but he remained hopeful, not least because – apart from a residual worry that his embezzlement might be discovered – he thought that his documentary evidence was incontrovertible. As we have seen, he pressed explicitly for the Sovnarkom to approve his Railway Mission report or commit him for trial. Doubtless correctly, he feared that otherwise the gossip and suspicions would linger and that his enemies would make his life impossible. His feelings about his research were more variable, though he remained loath to quit the experiments after so much hard work.[83] Above all, he seems still to have thought that Iu^E No. 001 could banish the hostility towards him within the NKPS, especially if Rudzutak made a good long journey with it.[84]

The crux was that Raisa did not share this confidence. In December 1924 she declared that she would leave him if he accompanied the locomotive to Russia, and she justified her ultimatum by asserting that his enemies were poised to attack. She expected them to arrange 'a second Novorossiisk' and hit him with the NKRKI's allegations. The diesel locomotive, she claimed, would lead only to shame for him. Also, she no longer wanted to endure the anxiety that was associated with his trips to Russia. Lomonosov was surprised and angry, but unwilling to call her bluff. He could not contest her logic: 'I'm out of fashion, that's a fact. One can expect any sort of nastiness; I have plenty of enemies.' But he thought that

---

[81] 'Dnevnik', 6 December 1924.

[82] For example, 'Dnevnik', 8 October, 18 November, 15, 29 December 1924, 4 January (10.00), 24 January 1925.

[83] For example: 'Dnevnik', 29 December 1924, 4 January, 16 March, 8 April, 25 May 1925.

[84] 'Dnevnik', 27 November 1924.

she exaggerated the dangers. Also, he suspected that she was protesting against an NKPS decision to ban women from the experimental train, which she viewed with considerable justification as an attempt by her husband to prevent her from accompanying him.[85] That said, he understood the need to clarify his attitude to his marriage, and he realized that he did not want a separation. So far as Iu^E No. 001 was concerned, then, he found just one consoling thought, itself a measure of the shadow over his career at a time of triumph: that his enemies could not accuse him of influencing the locomotive's trials unfairly.[86]

By January 1925, then, Lomonosov was approaching a professional, political and personal crossroads. In certain respects his situation was rosy. His professional reputation was becoming genuinely international. Unlike Gakkel' and most other Soviet scientists and engineers, he had ample scope for interacting with Western experts in his field. He had an extremely comfortable existence by Soviet standards. Based abroad with a large official apartment, he could even afford – financially if not politically – to send his son to an English private school.[87] But in Russia he had few professional and political allies and numerous enemies. Some of the latter, notably Rudyi, were still seeking evidence of crimes.[88] His scientific priority – experimental research on diesel traction – appeared to be threatened. Not least, his marriage was in crisis. As 1925 progressed, then, he would feel increasing pressure to decide which job he wanted to do, where he wanted to live, and with whom he wanted to live. His answers to these questions would shape the rest of his life.

---

[85] 'Dnevnik', 13, 21, 23, 29 December 1924 (19.30); 'Vospominaniia', vol. 9, pp. 974–6, 994–7, 1005–9. Lomonosov arranged the ban for several reasons: the experimental trains had acquired, according to Pravdin, a reputation as mobile brothels; he was irritated by Raisa's outspoken criticism of some of his recent decisions, which he viewed as interference in his work; he was 'never keen' on having her in the experimental train, evidently because she 'ruled the roost' there; and he foresaw an awkward situation when the train went through Khar'kov, where he expected to see Masha and his sons. In practice the ban was ignored.

[86] 'Dnevnik', 13, 23, 26, 29 December 1924, 8, 11, 29 January 1925; 'Vospominaniia', vol. 9, pp. 995–7, 1005–7, 1031–2.

[87] George attended a Quaker boarding school, Leighton Park, in Reading from 1921 to 1926: Aplin, *Catalogue*, p. xxvii. His parents were well disposed towards the Quakers from their time in North America. Leighton Park was founded in 1890 to prepare boys for Oxford and Cambridge universities, and had a high reputation for sciences: G. Cantor, *Quakers, Jews, and Science: Religious Responses to Modernity and the Sciences in Britain, 1650–1900* (Oxford, 2005), p. 60. I am grateful to Ben Marsden for this reference.

[88] See, for example, Sverchkov–Rudyi, 9 May 1924 and Rudyi–Sverchkov, 23 May 1924: RGAE, f. 1884, op. 31, d. 1674, ll. 83–4.

Defeat

Although Lomonosov did not leave Soviet service until January 1927, his career on the Soviet railways effectively ended in late 1925. That year there was good progress with his diesel project: he collected most of the experimental data for Iu^E No. 001 in Russia during March–July, his second locomotive was completed by Hohenzollern in Dusseldorf as Iu^M No. 005 and most of his plan to order more locomotives in Germany was accepted. However, he was sidelined from this research programme in October 1925. Obligated to write four technical and financial reports for the NKPS by October 1926, he undertook to rejoin the Kiev Polytechnic Institute thereafter.

Lomonosov's decision to visit Russia in March 1925 shows the strength of his desire to clear his name and remain in Soviet service. The journey was instigated by a summons from Rudzutak. Lomonosov hoped to exploit the success of Iu^E No. 001 to get the Railway Mission report approved, secure appointment to the NKPS Collegium with particular responsibility for diesel locomotives, conduct the planned road experiments on No. 001, and ensure that more locomotives would be ordered abroad. Crucially, Raisa did not reiterate her ultimatum: evidently accepting that the only alternative was to break with Moscow, she even agreed to accompany him.[89] In retrospect it seems extraordinary that he held such hopes for his career. But his diary entries remind us that the purges of the future were unknown for this man who, for all his indiscretions and troubles, was still well connected in the ruling elite. Was he naive, or was the situation more finely balanced?

Over the next four months two political setbacks revealed how far he had been deluding himself. The first was the Politburo's compromise concerning the Railway Mission report, which confirmed that unsubstantiated and even nonsensical allegations could carry at least as much weight as specialist expertise, and that it was inexpedient fully to support a non-party engineer against seemingly principled Bolshevik concern for state interests. Unsurprisingly, Lomonosov was distressed by this outcome, which was communicated to him in March. Months later, in autumn 1925, he tried again to get his report accepted, but again he got nowhere.[90]

Lomonosov's second setback was to lose his influence in diesel research as wider political issues made themselves felt. In December 1924 he had helped to persuade the STO to reject Gosplan's proposal for a 'neutral' diesel bureau and instead to create an inter-departmental commission under A.D. Tsiurupa, a member of the party's Central Committee and deputy chair of both the STO and Sovnarkom.[91] Lomonosov scented progress in March 1925 with his own appointment as deputy chair of this commission because Tsiurupa had no relevant

[89]   'Dnevnik', 2, 6, 16, 25, 28 February, 4 March 1925.

[90]   'Dnevnik', 4, 16 March, 13 October 1925; 'Vospominaniia', vol. 9, pp. 1536, 1551, 1574ff., 1584, 1586–7.

[91]   'Dnevnik', 6 December 1924; Lomonosov and Shveter, *Proekty teplovozov*, prilozhenie I, p. 6 (STO resolution, 5 December 1924); 'Vospominaniia', vol. 9, pp. 980, 986.

technical expertise and was well disposed towards him.[92] However, Tsiurupa soon spotted the danger of becoming entangled in the ongoing NKPS–VSNKh struggle for control of the research, and his desire to avoid controversy was a recipe for inaction. In June 1925, therefore, the STO transferred all railway diesel affairs to a new NKPS commission.[93] Lomonosov lost his special plenipotentiary status and separate budget, which also increased his vulnerability to possible investigation by the Cheka's successor, the GPU – an ominous situation as long as his Railway Mission report remained in limbo. Further, he could not build influence in the new NKPS group because the supervisory responsibility for it lay with Rudyi as the head (from January 1925) of the NKPS Central Directorate for Railways. In short, although Lomonosov kept his positions as Chief Director of Experiments and head of the Berlin diesel bureau, he was effectively subordinated to his arch-enemy. To make matters worse, he lost his main ally in the NKPS Collegium: Pravdin was transferred away from the NKPS.[94]

In this context Lomonosov's main source of professional satisfaction was the largely successful operation of No. 001, which he led in person during March–June 1925. The main aims were to obtain general operating experience, test the cooling system in warm weather and begin doing road experiments. Problems in organizing and staffing this work were mostly solved by April. The earlier technical snags were generally either overcome or bypassed. Of the new technical hitches that appeared, most concerned the engine, associated parts like the compressor, and the cooling system (the cooling tender was fault-ridden), while the traction motors became the other significant concern. Except for the motor faults, these problems were eclipsed by No. 001's achievements, which included a successful return trip from Moscow to the Caucasus during March–April. Especially pleasing were the large crowds of onlookers at stations: the locomotive played its part in popularizing Soviet science and technology.[95]

The technical difficulties with the traction motors, which proved less serious than feared, show how fine the line had become between routine technical issues and political crisis. The rear motor burnt out and the other four motors were damaged during a second Caucasus trip in late April. The repairs and installation of a replacement motor at Tiflis took several weeks, during which time an enquiry

[92]   Tsiurupa–Chair of STO, 18 March 1925, and Politburo resolution, 19 March 1925: RGASPI, f. 17, op. 163, d. 480, ll. 39, 38; 'Dnevnik', 12, 20 March 1925; Lomonosov and Shveter, *Proekty teplovozov*, prilozhenie I, p. 6 (STO resolution, 20 March 1925); 'Vospominaniia', vol. 9, pp. 1092–3, 1108.

[93]   'Dnevnik', 8 April, 2, 4–6, 24–5 June, 1925; Lomonosov and Shveter, *Proekty teplovozov*, prilozhenie I, p. 6 (STO resolution, 24 June 1925); 'Vospominaniia', vol. 9, pp. 1155–60, 1174–6, 1329, 1331, 1333–4, 1337–41, 1355–6, 1372–82.

[94]   For example: 'Dnevnik', 27 June, 13, 19 July, 9 September 1925; 'Vospominaniia', vol. 9, pp. 1380–1381, 1385, 1407, 1496.

[95]   Lomonosov, *Opyty 1925 g.*, *passim*. Photographs of crowds of onlookers appear on pp. 10, 17–18, 22–3, 33.

blamed the failures on poor design and assembly. Then the replacement motor burnt out during the return journey to Moscow and operations had to be halted for several months pending delivery of a new armature from Switzerland. When inspection of the motor in Moscow revealed a short circuit, the hypotheses included sabotage. This suspicion – symptomatic of the party-state's siege mentality – was misplaced: the faults were traced to oil leaks.[96]

For Lomonosov the most important technical outcome of the initial work with Iu^E No. 001 was a clear sense of the next engineering priority. By November 1924 he was sure that electric transmission was practicable if not theoretically ideal and that it was vital to develop an engine specifically for railway use.[97] The fact that in July 1925 he told the Brown Boveri company that No. 001's failures were discrediting electric transmission was merely a (successful) ploy to get some of the damaged motor parts replaced without charge, for actually his short-term plan concerned diesel-electrics exclusively.[98] So from about January 1925 he concentrated on the need for a much cheaper, 'flexible' engine with an effective cooling system. That month he told his German partners that No. 001 with its submarine-type engine was analogous to one of the earliest steam locomotives, whereas a locomotive with a purpose-built diesel engine would be the diesel equivalent of Stephenson's *Rocket*. The new engine would, among other things, be much simpler, develop as many as 840–1,200 revolutions per minute and not require a compressor.[99] In short, he identified the technical challenge that would be the key to the commercial success of diesel traction in the United States a decade later.

In the meantime he faced two other major technical issues. One was the delayed completion of his second locomotive, Iu^M No. 005. Here the transmission question was highly problematic. During 1923 the task of designing a reliable hydraulic transmission proved difficult, whereas mechanical transmission began to seem realistic again because a seemingly suitable gearbox and magnetic clutch had been developed. Lomonosov, Hohenzollern and the other partner companies therefore agreed to defer the diesel-hydraulic and prioritize a diesel-magnetic (later called diesel-mechanical) locomotive. However, in January 1924 the STO

---

[96]   For example: 'Dnevnik', 11, 13, 15 June 1925; Lomonosov, *Opyty 1925 g.*, pp. 25–40, 240–246; 'Vospominaniia', vol. 9, pp. 1219–37, 1248–57, 1305–7, 1323–5, 1345–8, 1354–5, 1360–63, 1366–7.

[97]   Minutes of NKPS Collegium, 22 November 1924: RGAE, f. 1884, op. 28, d. 74, ll. 170–173.

[98]   Lomonosov and Shveter, *Proekty teplovozov*, prilozhenie III, p. 95 (minutes of meeting in Esslingen, 23 July 1925); 'Dnevnik', 24 July 1925.

[99]   Lomonosov and Shveter, *Proekty teplovozov*, prilozhenie III, p. 76 (minutes of meeting in Augsburg, 28 January 1925); Lomonosov, *Teplovoz Iu^E*, p. 235. See also, for example, 'Dnevnik', 22, 27 July, 1 August 1925; Lomonosov and Shveter, *Proekty teplovozov*, prilozhenie III, pp. 93–5 (minutes of meeting in Augsburg, 21 July 1925); 'Vospominaniia', vol. 9, p. 1418.

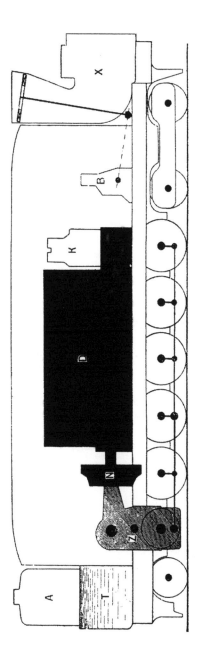

Figure 9.6   Diesel-mechanical locomotive Iu^M No. 005, Source: Iu.V. Lomonosov, *Teplovoz Iu^E No. 001 i ego ispytanie v Germanii* (Berlin, 1925), p. 32.

resolved to have this locomotive built in Russia by GOMZA.[100] By way of partial compensation Lomonosov ensured that Hohenzollern did much of the design work. The basic concept followed No. 001 in using the same engine and fixed frames, although the enlarged cooling system required an extra supporting axle. In December 1924 the STO retreated from its January decision on advice from GOMZA and the NKPS, and ultimately the contract returned to Hohenzollern.[101] Unfortunately the transmission remained troublesome. Above all, the clutch did not fit properly and seemed too complex for railway use. In July 1925 it was decided to design a new clutch, and to modify and use the existing one in the interim so as not to delay the experiments further. The locomotive was completed in this form at Dusseldorf in November 1925.[102]

As for his other major technical concern in spring 1925 – contracts for further locomotives – Lomonosov developed two plans that had long-term potential to shape both his own career and the overall diesel-locomotive research programme. Despite his fear that NKPS enemies would halt his experimental work, he campaigned for four more diesel-electrics to be built in Germany. This foreign focus was ostensibly to maximize the available expertise and facilitate comparison with his first two locomotives, but really it was mainly to give him a legitimate reason to continue working abroad in Soviet service. He instructed his Berlin bureau to prepare four diesel-electric designs as variations of Iu^E No. 001 and submitted them to the NKPS Collegium on 7 April.[103] Meanwhile he devised another plan that was, to say the least, different. He attempted to foster a large long-term NKPS demand for diesel locomotives by persuading the Krupp company to request a concession to design and build a proposed 2,000-km desert line to Chardzhou in Soviet Central Asia as a diesel-powered railway. This unusual idea, which recalled his initial diesel interest on the Tashkent Railway, would make excellent promotional sense if the locomotives could succeed in the desert conditions. The cost, for which Krupp would seek US investment, was estimated at $20 million for infrastructure and $5–8 million for the first 100 diesel locomotives. Having obtained a formal

---

[100] 'Dnevnik', 6–7, 29 November, 15, 18, 29 December 1923, 10 January 1924; Lomonosov and Shveter, *Proekty teplovozov*, pp. 21,167, prilozhenie I, p. 5 (STO resolution, 11 January 1924) and prilozhenie III, p. 35 (minutes of meeting in Esslingen, 6 November 1923); 'Vospominaniia', vol. 9, pp. 392–8, 429–30, 464–6.

[101] Minutes of NKPS Collegium, 22 November 1924: RGAE, f. 1884, op. 28, d. 74, ll. 170–176; 'Dnevnik', 27 November 1924; Lomonosov and Shveter, *Proekty teplovozov*, pp. 21, 172–9 and prilozhenie I, p. 6 (STO resolution, 5 December 1924); 'Vospominaniia', vol. 9, pp. 962–3, 979–80, 984–7.

[102] For a concise technical description of this locomotive as built see N.A. Dobrovol'skii, *Teplovoz Iu^M 005 i ego ispytanie v Germanii* (Moscow, 1927), pp. 3–21.

[103] 'Dnevnik', 12 February, 4, 8 April 1925; Dobrovol'skii, Report to Temporary Management Bureau of NKPS Diesel Commission, 16 December 1926: RGAE, f. 1884, op. 43, d. 11, ll. 104–7; 'Vospominaniia', vol. 9, pp. 1056–7, 1170–75.

proposal from Krupp in February 1925, he forwarded it to Rudzutak, the head of the NKPS since January 1924.[104]

The fates of Lomonosov's two initiatives help to show that the Soviet diesel research programme stumbled at this juncture for political and financial reasons. The Krupp proposal was welcomed by Gosplan, but still foundered: it stirred malicious gossip and suspicion about espionage and mercenary interests, which probably reflected the political support for Stalin's isolationist slogan 'Socialism in One Country'.[105] As for the Berlin bureau's latest designs, the NKPS Collegium approved Lomonosov's proposal (albeit for three instead of four locomotives) and even endorsed his request to order them abroad, notwithstanding VSNKh pressure to use only domestic resources.[106] However, the implementation of this plan was delayed for several years. The broad context was the political debate about economic strategy and industrialization that became entangled with the power struggle within the Bolshevik leadership after Lenin's death. At issue was whether the NEP's mixed state–private economy and gradual pace of development should be replaced with rapid centrally planned industrialization. This battle would be won during 1927–1928 by the advocates of forced growth led by Stalin.[107] Meanwhile the diesel research was one of many casualties as budgets were cut. The plan for ordering more locomotives abroad was postponed, and the same problem (together with a decision to concentrate gas-transmission research at GOMZA) explains the NKPS desire in 1926 to cancel Shelest's contract with Armstrong Whitworth. That said, the principle of ordering diesel locomotives abroad stayed intact. Indeed by September 1926 there was discussion about adding two diesel-mechanicals to the plan.[108] By that time, however, the most likely motive was technology transfer from the West: the main advocate of foreign orders, Lomonosov, had been removed from the programme a year earlier.

---

[104] 'Dnevnik', 7, 8, 10, 20 February, 4 March 1925; Hagemann–Lomonosov, 14 February 1925: LRA, MS 716.1.137–138; 'Vospominaniia', vol. 9, pp. 494–5, 1051–2, 1066, 1069, 1072, 1077.

[105] 'Dnevnik', 12 March, 25, 30 May, 7, 24 July 1925; 'Vospominaniia', vol. 9, pp. 1389, 1399, 1418.

[106] See 'Dnevnik', 8 April 1925; [Dobrovol'skii?], Programme for Diesel Locomotive-Building, [circa November 1925]: RGAE, f. 1884, op. 43, d. 528, ll. 15–16; Dobrovol'skii, Report to Temporary Management Bureau, 16 December 1926: RGAE, f. 1884, op. 43, d. 11, ll. 104–7; 'Vospominaniia', vol. 9, pp. 1170–1172, 1419.

[107] On this debate see, for example, A. Erlich, *The Soviet Industrialization Debate, 1924–1928* (Cambridge, MA, 1960); Carr, *Socialism in One Country*; E.H. Carr and R.W. Davies, *Foundations of a Planned Economy, 1926–1929*, 3 vols (London, 1969–1978); M. Lewin, *Political Undercurrents in Soviet Economic Debates from Bukharin to the Modern Reformers* (London, 1975).

[108] See: The situation with the question of building diesel locomotives in the USSR, [April 1926]: RGAE, f. 1884, op. 43, d. 75, ll. 48–54; Dobrovol'skii, Report to Temporary Management Bureau of NKPS Diesel Commission, 16 December 1926: RGAE, f. 1884, op. 43, d. 11, ll. 104–7.

This removal rapidly became inevitable after June–July 1925, when the changes in the top-level management of Soviet diesel-locomotive research basically subordinated Lomonosov to Rudyi, now head of the NKPS Central Directorate for Railways. Almost immediately the new NKPS diesel committee began making plans that left little room for Lomonosov. For example, it drafted proposals to close the Berlin office and restructure the organization of the experiments without consulting him as Chief Director of the diesel experiments. Also, it became clear that Terpugov, who was one of Lomonosov's closest associates by 1924, had switched allegiance to Rudyi.[109] Lomonosov seems to have interpreted these developments as a turning point and to have become demoralized. True, during August 1925 he resisted pressure from Raisa to cease all diesel work, such was his desire to finish the experiments. But one of the main reasons why he went to Russia in September, and why he rejected Raisa's last-minute request to accompany him, was his desire to say a proper farewell to the women there with whom he had been close in recent years: his former mistress Masha near Khar'kov, whom he had seen occasionally since autumn 1919 with their two sons; and two 'mistresses' in Moscow – a distant relative named Ania Pegelau, and a nurse whom he nicknamed Kuma, who had cared for him during his typhus attack in 1920. He also wanted to find Marfusha, briefly a family servant in Kiev nearly two decades earlier, possibly because he felt responsible for her daughter Sonia.[110] He was seriously, albeit reluctantly, thinking about quitting Soviet Russia for good.

Characteristically, quarrels with key colleagues precipitated his formal removal from the diesel research in October 1925. He fell out with the chair of the NKPS diesel commission, A.M. Postnikov, over basic matters like strategy and foreign involvement. But worst, in Lomonosov's eyes, was a professional and personal 'betrayal' in October: his protégé and colleague Dobrovol'skii sided with Terpugov and Rudyi over the planned road experiments. The available circumstantial evidence suggests that Dobrovol'skii now regarded his association with Lomonosov as a liability. Be that as it may, this 'treachery' stunned Lomonosov: 'an internal enemy is the most frightening'. He lost all desire to continue his struggle and relinquished his remaining positions. Rudzutak did offer to make him deputy chair of the NKPS diesel commission, but neither Lomonosov nor Postnikov saw any future in that idea, although Lomonosov did agree to serve as an ordinary member of the diesel commission and NTK council. His only real consolation was that Rudzutak allowed him to spend the next year in Western Europe. He was to prepare four

---

[109]   Draft resolution of the Temporary Management Bureau of the NKPS Diesel Commission, circa July–August 1925: RGAE, f. 1884, op. 43, d. 11, ll. 33–4; 'Vospominaniia', vol. 9, pp. 1474–7.

[110]   'Dnevnik', 19–20 July, 8, 10, 13, 24–25, 29, 31 August, 2, 7–9, 13 September 1925, 30 January 1926; 'Vospominaniia', vol. 9, pp. 1458, 1463–4, 1478, 1480, 1498. Kuma is not mentioned in his description of his planning but they did meet in Moscow during this trip.

reports: a financial account for Sovnarkom; a report about the experiments with No. 001 in the USSR; a survey of diesel research in Italy and Germany; and a review of all his work on diesel traction since 1920. He was allocated 10,000 gold rubles, engineer Makhov as his assistant and a deadline of 1 October 1926. Thereafter he would take up a professorship at the Kiev Polytechnic Institute.[111] Formally his participation in the Soviet diesel traction programme ended on 19 December 1925 with the transfer of Iu^M No. 005 and the associated staff, spare parts and documents to the NKPS Delegation in Germany. Lomonosov consoled himself with the thought that, while he had high hopes for this transmission type in the long term, he had never expected a triumph with this locomotive.[112]

The task of writing four reports indirectly determined his fate in Russia. First to be submitted in December 1925 was the financial report. Then came a supplement to the financial report in March 1926 and his analysis of foreign progress in June 1926. The review of his diesel work was co-authored with his colleague E.E. Shveter as the book *Proekty teplovozov* and submitted several weeks after the October deadline. Finally, his report about the experiments with No. 001 in the USSR was printed as the book *Opyty 1925 g.* in December 1926 and posted to Moscow on 5 January 1927. He had obtained a deadline extension until 1 December 1926, but Rudzutak treated the further delay as prevarication. Indeed, he castigated Lomonosov for not having submitted any reports or an explanation, and ordered him to return to Moscow immediately to deliver his reports and discuss his future – an instruction that Lomonosov appeared to ignore.[113]

*   *   *

Lomonosov became interested in the concept of main-line railway diesel traction upon seeing the difficulty of operating steam locomotives in desert conditions

---

[111] 'Dnevnik', 19–20, 22–30 September, 1–3, 5, 7, 9–11, 13, 16–18, 22, 26, 28–31 October 1925; Minutes of meetings of Temporary Management Bureau of the NKPS Diesel Commission, 15 October, 11 November 1925: RGAE, f. 1884, op. 43, d. 528, ll. 6–10, 11–14; Rudzutak–Lomonosov, 31 October 1925: LRA, MS 716.1.152; Lomonosov and Shveter, *Proekty teplovozov*, prilozhenie I, p. 8 (NKPS order 7909, as published in *Vestnik putei soobshcheniia*, 21 November 1925); 'Vospominaniia', vol. 9, pp. 1489–98, 1506–22, 1539–67, 1573–1603.

[112] In particular, 'Dnevnik', 11, 18–19 December 1925, 5, 11, 15 January, 22 March 1926; 'Vospominaniia', vol. 9, pp. 1567, 1664–6, 1682–6, 1697, 1700–1702, 1724–8, 1770–1771. The subsequent experiments on this locomotive in Germany to May 1926 are described in Dobrovol'skii, *Teplovoz Iu^M 005*, pp. 22–47. The locomotive proved vulnerable to transmission problems, and was withdrawn in the mid-1930s: Rakov, *Lokomotivy*, p. 364. For an example of Lomonosov's earlier confidence in this transmission see 'Dnevnik', 16 April 1924.

[113] Makhov–Lomonosov, [circa 13 October 1926]: LRS MS 716.1.179; Rudzutak–Lomonosov, 7 January 1927: LRA, MS 716.1.183; 'Dnevnik', 13 October 1926, 11, 15 January 1927; 'Vospominaniia', vol. 9, pp. 2006–7, 2135–7.

in 1909. His principal contribution before 1917 was to facilitate research by his former student and colleague Lipets. But from 1920 he played an exceptionally important if controversial role in leading the work of Soviet engineers to develop this technology. He alone argued that the Bolshevik revolution gave the Soviet railways a unique opportunity to tackle this technical challenge in a systematic, comprehensive way. He lost that argument, yet he still contrived to launch a research programme in 1920. Lenin's intervention here in the winter of 1921–1922 was initially a mixed blessing for Lomonosov. However, the latter's ability to win Lenin's support during 1922 was crucial for confirming the funds to build several diesel locomotives in Germany. By that time, however, the allegations against Lomonosov of misconduct in affairs of the Railway Mission were encouraging the engineer to treat his diesel research primarily as a means to stay abroad on official Soviet business for a few more years. His use of West European firms helped him to make fairly rapid progress in building his locomotives and attracted interest from foreign engineers, yet paradoxically it undermined his credibility in Russia. To a degree his removal from the Soviet diesel research programme in 1925 was a product of several such political and cultural circumstances that were beyond his control. Personal problems played a role, too, notably Raisa's increasingly outspoken criticism of his work. And his own character and choices were yet again critically important: his stubborn determination to chart his own course not only enabled him to start his diesel research in 1920, but also caused much of the distrust and suspicion that underlay his removal in 1925.

The inter-war Soviet research in main-line railway diesel traction was rooted in the tsarist era but was begun mainly by engineers seeking individually to capitalize on the enthusiasm of Lenin's regime for technological change. Clearly there was scope for these experts to use their initiative in the early 1920s. Yet Lomonosov's experiences cast doubt on the idea that the 1920s was a golden age in Soviet science and technology. The chronic lack of money for research in Soviet Russia was always a serious obstacle, and in this case it helps to explain why the advocates of diesel traction lost most of their hard-won momentum in the mid-1920s. A further problem, at least for Lomonosov, was Russian nationalism, encapsulated in the slogan 'Socialism in One Country': by 1924 there was a tendency to criticize his diesel-electric locomotive for the simple reason that it was built abroad. Also, his experiences provide indications that the regime's political truce with engineers may have begun to collapse sooner than we tend to believe. As early as 1923 leading party figures like Dzerzhinskii and Voroshilov were asserting party control over the diesel research project as a matter of political principle, and they were supported by lower-ranking party members in key positions of the technical bureaucracy, such as Rudyi. The fact that for several years Lomonosov managed to resist their pressure was a cause of much frustration and anger for them, which may explain in turn why allegations of sabotage appeared so quickly after the failures of No. 001's traction motors in 1925. Here, it would seem, were some of the roots of the concerted repression of scientists and engineers that began in 1928.

Chapter 10
'A Free Soviet Citizen Abroad'

The Lomonosovs never saw their homeland again after January 1927. Initially the couple stayed in Europe, mostly in Italy, Germany and the UK, while their son attended Cambridge University as George Lomonossoff. In 1929 all three sailed to the United States with a view to settling there, but they soon returned to Europe and made their home in Britain, where they acquired citizenship. For the most part, including much of the Second World War, Lomonosov and Raisa lived in rented apartments in London. They were back in North America with George from 1948, and again they considered settling there. In 1950, however, Raisa returned to Britain, where she remained until her death in 1973, whereas Lomonosov went to live with his son and daughter-in-law in Montreal, and died there in 1952, predeceasing Stalin by a few months. George died suddenly in Montreal at the age of 45 in 1954.

Russians both in and outside the USSR described the family as 'non-returners' – Moscow's epithet for Soviet citizens who, in its terms, committed treason by refusing to return home from their assignment abroad.[1] In Russia it was usually assumed that the Lomonosovs had joined the anti-Bolshevik Russian diaspora – the veritable society in exile that has been called Russia Abroad.[2] Yet by examining why they stayed abroad and how they faced the formidable financial, political, social and cultural challenges of doing so, this chapter reveals a more complex picture that also raises questions about our understanding of the inter-war Russian emigration.

A 'non-returner'?

Terms such as 'Russian émigré' are routinely employed to describe people from tsarist Russia and the USSR who lived abroad for political reasons during the nineteenth and twentieth centuries, yet these labels require caution. Historians identify two 'waves' of emigration between 1917 and Lomonosov's death in 1952: one provoked by the October revolution that encompassed perhaps two million people by 1921, and another involving displaced Soviet citizens at the end of the

---

[1] On this term and the associated Soviet law of November 1929 see Genis, 'Nevozvrashchentsy', pp. 46–63. The law specified the death penalty for this form of treason, the sentence to be carried out within 24 hours of confirmation of the accused's identity.

[2] See in particular M. Raeff, *Russia Abroad: A Cultural History of the Russian Emigration, 1919–1939* (Oxford, 1990).

Second World War.[3] However, 'Russian' was used as an all-encompassing term for many ethnicities. For example, the 1921 census in England and Wales treated 'aliens' from the former tsarist empire as Russian unless born in Poland and (sometimes) Finland.[4] Similarly, 'émigré' was only one of several words used for such people – terms which are not synonymous, but which may overlap in meaning and may reflect efforts by, for example, states and individuals to form identities for political, legal, economic and other purposes. Examples include exile, refugee, immigrant, non-returner and, mainly from the Cold War, defector. Further, these terms tended to have specific political connotations. According to the historian Marc Raeff, the 'one political notion shared by practically all émigrés who gave a moment's thought to politics was an absolute rejection of the Bolshevik regime'.[5] By contrast, the politics of an exile were not necessarily anti-Soviet. One might contend, for instance, that Trotskii became an exile upon his deportation from the USSR in 1928 but not an émigré.[6]

This political dimension profoundly affected the question of Russian citizenship, as can be seen with the first-wave émigrés. In international law a Russian living abroad during the inter-war years was either an expatriate Soviet citizen, stateless or a naturalized citizen of a host country. For the émigrés who planned to stay abroad until the Soviet regime collapsed Soviet citizenship was neither possible nor desirable. Equally they tended, at least initially, to avoid assimilating into their host society in the belief that they might soon be going home. Hence they were generally treated by host societies as stateless refugees, for whom the so-called Nansen passport was devised under the auspices of the League of Nations in lieu of a regular passport.[7] That said, the number of these refugees decreased over time, and not solely because of deaths. Their children born abroad normally became citizens of their country of birth, and eventually many émigrés did take foreign citizenship, albeit often continuing to consider themselves as émigrés.[8]

Lomonosov's case is complex and idiosyncratic, not just because he was one of the relatively small number of 'non-returners', but also because his political views remained genuinely pro-Soviet – a fact that helps to highlight an intriguing anomaly in Soviet policy. Unsurprisingly, Soviet institutions soon viewed him

---

[3]   For example, Raeff, *Russia Abroad*, pp. 3–7, 24.

[4]   *Census of England and Wales, 1921: General Tables* (London, 1925), pp. 181, 186–8; *Census of England and Wales, 1921: General Report with Appendices* (London, 1927), pp. 152–60. Armenia was treated as part of the former Ottoman empire.

[5]   Raeff, *Russia Abroad*, p. 8.

[6]   The terminology is discussed in, for example, M.A. Miller, *The Russian Revolutionary Emigrés, 1825–1870* (Baltimore, 1986), pp. 6–10.

[7]   On the Nansen passport see Raeff, *Russia Abroad*, p. 36.

[8]   For example: E.B. Kudriakova, *Rossiiskaia emigratsiia v Velikobritanii v period mezhdu dvumia voinami* (Moscow, 1995), pp. 6–9.

as a 'non-returner', and Russian émigrés followed this lead.[9] Some years later, during a two-year campaign to denigrate the pre-revolutionary Russian tradition of railway science and promote a socialist science of railway design, traction and operations, the NKPS newspaper *Gudok* and NKPS periodicals even applied the term 'White émigré' to brand Lomonosov as an anti-Soviet enemy.[10] But abroad these terms carried no legal weight for defining his citizenship. Moreover, even while executing thousands of alleged 'enemies of the people' during the 1930s, the Soviet authorities ignored their own rhetoric about Lomonosov in that they repeatedly approved his annual applications via Soviet consulates to renew his Soviet passport, and they did not strip him of his Soviet citizenship until 1945. For his part Lomonosov sought to keep his Soviet citizenship during the 1930s, even though the annual renewal of his passport involved a 'very unpleasant' interview at the consulate.[11] He never wanted nor had refugee status, and never regarded himself as an émigré; the same was true for Raisa and George. Initially from 1927, then, the Lomonosovs lived abroad as Soviet citizens with foreign-visitor visas that they renewed as necessary. However, George took British citizenship in 1932, and for reasons discussed below his father followed suit in 1938, from which time he allowed his Soviet passport to lapse. Thus, from 1938 until Moscow cancelled his Soviet citizenship in 1945, Lomonosov was a British citizen under British law and a Soviet citizen under Soviet law. To complicate matters further, until 1938 Lomonosov called himself a 'free Soviet citizen abroad' – a personal, perhaps unique response to the lack of a legal term for Soviet citizens like himself who were living abroad independently of the Soviet state. Unfortunately, the reasons why Moscow revalidated his passport during the 1930s are unknown. Conceivably it was hoped that he would return to Soviet territory, at which point he could be arrested. Presumably the Soviet authorities allowed at least a few other people to live outside the USSR in a similar way.[12]

---

[9]   For example, in August 1928 Lomonosov featured in a report by the People's Commissariat of Trade about non-returners: V.L. Genis–author, 17 May 2004. An example of an émigré source is *Poslednie novosti*, 7 October 1931, cutting at 'Dnevnik', 8 October 1931.

[10]   For example: 'Reaktsionery parovoznogo khoziaistva prigvozhdeny', *Gudok*, 9 January 1936, p. 1; N. Morozov, 'O predel'shikakh–parovoznikakh i ikh "ideologakh"', *Sotsialisticheskii transport*, 1 (1936): 25–37; and 'Godovshchina stakhanovsko-krivonosovskogo dvizheniia', *Sotsialisticheskii transport*, 6 (1936): 10–24. Also V.L. Genis–Author, 17 May 2004. The campaign against 'bourgeois' railway science was launched in August 1935, part of a concerted effort to raise railway productivity. See, for example, [L.M. Kaganovich], 'Ob uluchshenii ispol'zovaniia parovozov i organizatsii dvizheniia poezdov: prikaz Narodnogo komissara putei soobshcheniia No. 183/Ts, 7 avgusta 1935 g.', *Parovoznoe khoziaistvo*, 8 (1935): 4–7. See also Rees, *Stalinism*, pp. 106–59, and Westwood, *Soviet Locomotive Technology*, pp. 129–37.

[11]   The quotation is from 'Dnevnik', 16 September 1931.

[12]   The phrase 'free Soviet citizen abroad' (*svobodnyi sovetskii grazhdanin za-granitsei*) can be found at, for example, 'Vospominaniia', vol. 9, p. 2074. Raisa had a Soviet passport until at least 1938, but the date when she acquired British citizenship is unknown.

The roots of Lomonosov's peculiar situation developed through the 1920s. A key question was whether, as a Soviet official until 1927, he ever planned to emigrate or enter foreign exile. At issue, in short, was his political position at a time when class identity and loyalty were life-or-death matters for the Bolsheviks. Crucially, as noted in Chapter 8, most Bolsheviks (and Lomonosov too) suspected that Russia's several thousand pre-revolutionary or 'bourgeois' scientific and technical experts detested the Soviet system. For Bolsheviks like Avanesov and Rudyi, who saw Lomonosov as a bourgeois specialist, it was self-evident that he intended to use his foreign assignment as an opportunity to stay abroad. Yet in reality most of the 'bourgeois specialists' were probably not anti-Soviet, and the majority may well have shared engineer Pal'chinskii's aim to assist his country by increasing its industrial strength and the welfare of its people.[13] The same paradox applied with Lomonosov: he had no wish for emigration or exile, but the suspicions, aspersions and gossip prompted him to have doubts that otherwise might not have arisen. In fact, apart from his strong patriotism, he assumed that to find and keep a regular job abroad would be so difficult and time-consuming as to prevent him from writing books – his primary desire. In any case, he had vowed to support the Soviet system in 1918, and he would always view that decision as fundamentally correct despite his subsequent troubles. Accordingly, he was adamant that he would never become a stateless émigré with a Nansen passport. However, he was spurred to contemplate foreign exile as early as 1920 by the Cheka investigation of the Novorossiisk train crash: he fully agreed that the regime should punish deliberate sabotage, but he did not relish having his own mistake misinterpreted as sabotage. Ironically, the same worry that he might be accused of sabotage, and specifically of 'wasting' Soviet gold on 'unnecessary' foreign contracts, led him into deliberate criminality in mid-1922, when he embezzled state funds to make foreign exile a practicable option (see Chapter 8).

Consistently, however, he treated exile as a last resort. From about 1922, as we have seen, he wanted to stay abroad on Soviet business for as long as possible before spending his retirement in Russia. He exploited his diesel project successfully to this end until late 1925. Then, with Rudzutak's reluctant approval, he obligated himself to leave the NKPS and rejoin the Kiev Polytechnic Institute from October 1926 (in other words, after completing his four diesel reports for the NKPS), and he kept Kiev as his official destination despite being invited in 1926 to rejoin his old institute in Leningrad (formerly Petrograd).[14] Other things being equal, his personal preference was to retire immediately and spend

---

On George's citizenship see Aplin, *Catalogue*, p. xxviii. No information has been found about whether the Soviet authorities annulled Raisa's and George's Soviet citizenship.

[13]   See Graham, *Executed Engineer*, especially pp. 41–8.

[14]   Lomonosov–KPI, 25 February, 10 November, 3, 31 December 1926: DAK, f. 18, op. 2, spr. 159, ark. 237, 253, 256–7; KPI–Lomonosov, 3 December 1926: LRA, MS 716.1.181; N.M. Beliaev–Lomonosov, 3 May 1926 and Lomonosov–Beliaev, 30 May 1926: LRA, MS 716.1.167, 174; 'Dnevnik', 1 January, 2, 25 February, 3 June, 10 August, 11 December 1926;

his declining years in the countryside near Moscow writing books, especially a major textbook on locomotives.[15] But other things were not equal, and throughout 1926 he continued seeking permission from Moscow and Kiev to spend more time abroad. One motive was Raisa's wish to live in England near George, who expected to begin reading engineering at Cambridge in October 1926, although Lomonosov postulated in his more wildly optimistic moments that he might persuade her to live in Moscow.[16] Another reason was Raisa's poor health: he felt that for the moment he could not subject her to Russia's tough living conditions or leave her alone abroad. Above all, he assumed that for the moment NKPS meddling and his enemies would deny him a quiet life in the USSR, and that he would be unable to follow foreign developments in diesel traction if living in Russia because he would be banned from travelling abroad. He recognized, too, that he might be imprisoned.[17]

He was convinced that his plan for staying abroad was practicable. There was a precedent, he thought, with the long-term assignment to Cambridge University of the Soviet physicist P.L. Kapitsa, whom he met in 1926.[18] His own analogous idea was for the NKPS to give him a fixed-term secondment to the German State Railways as a diesel-traction consultant. Moscow, he reasoned, would welcome the technological benefits of this connection and was more likely to permit a job with a foreign state organization than with a private company. In early 1926, therefore, he offered his services to the Germans for five years at $10,000 per year, again to start in October 1926 upon the completion of his NKPS reports. His monetary demand was huge, yet he did receive a formal proposal from Berlin, which he accepted on condition – crucially – that the NKPS agreed.[19] In the meantime he tried to bolster his position by writing articles for major Western journals like *The Engineer* and

---

'Vospominaniia', vol. 9, pp. 1769–70, 1775, 1793, 1906, 1928, 1970, 1989, 2080. His idea to rejoin the KPI dated from about September 1923: 'Vospominaniia', vol. 9, p. 534.

[15] For examples of his desire to be in Russia see 'Dnevnik', 6 November 1925; 12 May 1926. The textbook project is noted at, for example, 'Dnevnik', 7 November 1926.

[16] For example: 'Dnevnik', 16 November 1925; 'Vospominaniia', vol. 9, pp. 1650–1651, 1667.

[17] For example: 'Dnevnik', 4 April 1924, 2, 10 July 1925, 15 September, 29 October, 4 November, 31 December 1926; 31 January 1927; 'Vospominaniia', vol. 9, pp. 1668, 2008–9. Raisa's health had been a matter of concern for some years. By the mid-1920s the Lomonosovs were convinced that she had a weak heart.

[18] For example: 'Dnevnik', 13 April, 2, 10 July, 26, 29 October 1925, 1, 5–8, 14 January, 23 March, 15 September, 31 December 1926, 31 January 1927; 'Vospominaniia', vol. 9, p. 2098. On Kapitsa see, for instance, Badash, *Kaptiza, Rutherford, and the Kremlin*; D. Shoenberg, 'Piotr Leonidovich Kapitza, 9 July 1894–8 April 1984', *Biographical Memoirs of Fellows of the Royal Society*, 31 (1985): 327–74; D. Shoenberg, 'Kapitza, Fact and Fiction', *Intelligence and National Security*, 3/4 (1988): 49–61.

[19] U. Kumann–Lomonosov, 23 August 1926: LRA, MS 716.1.176; 'Dnevnik', 18 November 1925; 28 June, 10, 30–31 August, 3 September 1926; 'Vospominaniia', vol. 9, pp. 1754–6, 2003–9.

the *Zeitschrift des Vereines Deutscher Ingenieure*.[20] Useful kudos came with the award of an honorary doctorate by the Charlottenburg Technical Institute, which was arranged by his friend Meineke.[21] Also, he tried to develop other employment possibilities by approaching contacts like Samuel Vauclain of the Baldwin works and Sir Henry Thornton, president of the Canadian National Railway.[22] He knew too that he and Raisa had enough money to live abroad for a while without a job if he could not get an approved assignment.

There was less confidence and consistency in his assessment of attitudes towards him within the USSR. The scandals and difficulties of 1922–1923 prompted him to export the main possessions that he would want if permanently resident abroad: as of late 1923 he felt wary about risking the vagaries of a Russian court and did not want to die 'stupidly', and he felt much the same during April–May 1924.[23] By late 1924 he was more hopeful thanks to the success of his diesel locomotive Iu^E No. 001. But during 1925, as we have seen, he became worried by the political troubles of prominent party members like Trotskii and Krasnoshchekov. He was especially pessimistic during his autumn 1925 visit to Russia. There he selected a few books and packed a trunk of papers from 1905–1917 to take to Germany, but sadly he could not find a suitable vehicle to take the trunk to the station.[24] As for farewells with relatives and friends, his mother and daughter had died in 1921, and he did not record seeing his sister, his first wife Sonia or son Vsevolod. But as noted in Chapter 9, he did try to take leave of the women in Russia with whom he had been intimate in recent years. He went to Ukraine to see Masha, give her some money and initiate arrangements to be certified legally as the father of their two sons, and he also gave money to Ania and possibly Kuma.[25]

By this juncture his state of mind was chaotic as a result of the chronic personal, professional and political uncertainty and especially from the way that

---

[20]   See for example: 'Dnevnik', 5, 8, 15 January 1926; 'Vospominaniia', vol. 9, pp. 1733–4, 1749–50, 1794.

[21]   Lomonosov–KPI, 14 May 1926: DAK, f. 18, op. 2, spr. 159, ark. 239–40; 'Vospominaniia', vol. 9, pp. 1779, 1863.

[22]   'Dnevnik', 3 December 1925; 'Vospominaniia', vol. 9, pp. 1654, 1656–7, 1850–1851.

[23]   'Dnevnik', 29 December 1923, 4 April, 18 May 1924.

[24]   'Vospominaniia', vol. 9, p. 1634. He especially had in mind papers from the Catherine and Tashkent railways, and some of Raisa's letters: 'Dnevnik', 30 January 1936.

[25]   The death of his daughter is noted at 'Dnevnik', 15 July 1921 and 'Vospominaniia', vol. 8, pp. 512–14. He last saw his mother on 6 November 1921, a few days before her death: 'Dnevnik', 6 November 1921 and 'Vospominaniia', vol. 8, pp. 733, 759, 763. Concerning his 1925 visit see in particular 'Dnevnik', 13, 20–3 September, 3 October, 4–15 November 1925; and 'Vospominaniia', vol. 9, pp. 1502–5, 1509, 1534–5, 1553, 1572, 1585, 1633–6. His sister died in mid-1926: 'Dnevnik', 28 June 1926; 'Vospominaniia', vol. 9, pp. 1951–2. Masha's sons decided not to complete the legitimation process and to retain her surname – a decision that probably saved their lives in the 1930s. That Lomonosov provided money in 1925 is remembered in the family: information from N.S. Skachkova.

he perceived his future as a three-part choice of work, female partner and location, with his personal safety as a subsidiary question. This chaos is evident with the extraordinary crisis in his private life that erupted near the end of his autumn trip to Russia in 1925. He had arrived at the view that Raisa would be the best partner for him abroad, whereas a permanent return to Russia would probably lead to divorce from her and marriage to Masha or Ania.[26] Shortly before his departure from Moscow for Berlin in mid-November he lost his 'self-control'. Suddenly obsessed with Ania, he proposed marriage and was accepted; he then wrote to Raisa in Europe to request a divorce. But by mid-January 1926 he had changed his mind again and made peace with Raisa. He accepted virtually all her demands, which included the withdrawal of his proposal to Ania, a sperm test (partly to prove that he was impotent and therefore not responsible for Ania's pregnancy) and complete sexual abstinence.[27]

Oddly, given the way that he had been analysing his personal and professional dilemmas, this 'peace' did not mark a conscious decision to quit Soviet Russia, although it presumably did reflect a subconscious step in that direction. Notwithstanding his affair with Ania and his desire to live in Russia, he had travelled from Moscow to Germany in November 1925 feeling certain that he would never see his homeland again.[28] Having recommitted himself to Raisa, he noted in March 1926 that 'in all conscience' he wanted only to work on diesel research abroad, and the following September, shortly before the deadline for his diesel reports, he wrote that for the first time in his life he had 'absolutely no desire' to go to Russia.[29] Indeed, throughout 1926 he clung to the hope that his scheme for an NKPS-approved consulting job with the German railways would enable him to stay abroad for a few more years in Soviet service.

This determined hope helps to explain how the issue of his employment and location came to a head in late 1926. On 3 September, having just received the formal German offer, Lomonosov wrote to Rudzutak asking for his deadline to be deferred until 1 December and for permission to accept the German offer. He also sought support from L.P. Serebriakov, a member of the NKPS Collegium with whom he had become friendly.[30] In late September, not having received any replies, he sent Makhov to Moscow to press the two requests and, if unsuccessful, to find a job there for himself.[31] This step did evince a reply from Rudzutak, but not one that

---

[26]   For example, 'Dnevnik', 4, 6 November, 18 December 1925.

[27]   In particular: 'Dnevnik', 4–21 November, 3–26 December 1925, 1–3, 9–11, 15–17 January 1926; 'Vospominaniia', vol. 9, pp. 1606–23, 1633–4, 1639–41, 1646–54, 1657–63, 1673–6, 1687–8, 1693–9, 1705–6, 1718–21, 1751, 1758–60. Since his memoirs indicate that impotence was not proven, he may have been responsible for Ania's pregnancy.

[28]   'Dnevnik', 19 November 1925; 'Vospominaniia', vol. 9, p. 1636.

[29]   'Dnevnik', 25 March, 8 September 1926.

[30]   On Serebriakov see N. Limonov, 'L.P. Serebriakov (1888–1937): Ostanetsia navsegda', in Proskurin, Vozvrashchennye imena, vol. 2, pp. 203–19.

[31]   'Dnevnik', 3, 8, 22 September 1926; 'Vospominaniia', vol. 9, pp. 2003–9, 2033–4.

Lomonosov wanted: the People's Commissar ordered him to return to Moscow for discussions about his future before any decision could be made about the German project.[32] Realistically Lomonosov saw this summons as 'the beginning of the end': he understood that it gave him the choice to risk being confined to Russia (and possibly worse) or break with the regime. This interpretation was supported by information, in a letter from Litvinov received on 17 October, that getting police permission to make trips abroad was now very difficult – a hint about the need for caution, for which Lomonosov was grateful.[33] Yet he still wanted to believe that he could persuade Rudzutak to approve the German plan without going to Moscow, and he perceived a basis for this hope in a report from Makhov that Rudzutak was 'conciliatory' and that Postnikov and Dobrovol'skii were 'hostile'.[34]

Lomonosov's next steps suggest that he accepted Makhov's analysis, effectively preferring yet again to treat his situation as a function mainly of professional jealousies, and not of the political distrust that was undoubtedly important for Rudzutak. He replied to the latter's summons by repeating his request for his final deadline to be deferred to 1 December, and Rudzutak agreed on condition that Lomonosov reached Moscow by 10 December.[35] At the same time Lomonosov redoubled his efforts to get NKPS approval for his German plan without visiting Moscow. He asked Rykov, Tsiurupa and Litvinov to intercede with Rudzutak on his behalf and he wrote again to Rudzutak with the request to approve the German proposal or release him from NKPS service.[36] The revised deadline came and went, no replies arrived from Moscow, and Lomonosov remained in Germany. By 31 December 1926 he was at last beginning to lose hope:

> In a few days I will have to move to the status of a fugitive to a certain extent. It's hard. But now there's no other way out. A year ago it would have been possible to remain as an NTK member, adapt myself, hide, move away from real work and make myself an ordinary *spets*. It's impossible to work, do things and be creative in the conditions that have developed. That's as clear as marmalade. One can only curry favour and go with the flow. For my private personal life Moscow and generally Russia are a more congenial environment for me … if they would leave me in peace there; but that's very doubtful.[37]

---

[32]   Rudzutak–Lomonosov, 2 October 1926: LRA, MS 716.1.178; 'Dnevnik', 8 October 1926; 'Vospominaniia', vol. 9, pp. 2042–3.

[33]   'Dnevnik', 4 October 1926; 'Vospominaniia', vol. 9, pp. 2050–2051, 2064.

[34]   'Vospominaniia', p. 2045.

[35]   'Dnevnik', 8, 13, 29 October 1926; Rudzutak–Lomonosov, 26 October 1926: LRA, MS 716.1.180; 'Vospominaniia', pp. 2050–2051, 2072.

[36]   'Dnevnik', 18 October, 10, 12, 29 November 1926; 'Vospominaniia', vol. 9, pp. 2076–8.

[37]   'Dnevnik', 31 December 1926.

To his dismay, the eventual reply – a letter dated 7 January 1927 – was uncompromising. Plainly angry, Rudzutak issued an unambiguous ultimatum: the engineer was to return to Moscow to present his reports and discuss his future, or else he would be 'considered as no longer serving in transport, with all the consequences that flow from this'.[38]

It was Lomonosov's apparent decision to ignore this ultimatum that earned him the label 'non-returner'. In reality he did respond to Rudzutak, sending a letter that was intended to be conciliatory. He reasserted his association with the NKPS, confirmed that he had submitted all his reports, reiterated his explanations for the submission delays and portrayed the German offer as a chance for him to continue his specialist work in a way beneficial for the commissariat.[39] In effect, he did not accept that he was being recalled, and he tried to negotiate, believing that meanwhile he was neither agreeing nor refusing to go to Russia. He may have misunderstood Rudzutak's demand for him to submit his reports: the point was surely for him to submit them in person. Be that as it may, his reply appeared evasive, and Rudzutak ignored it. Hence Lomonosov could no longer resist pressure from the Berlin embassy to vacate his official apartment there, which he and Raisa did on 29 January 1927.[40]

Lomonosov, then, did not choose as such to become a 'non-returner' in January 1927. He had deluded himself both about Rudzutak's position and more generally about the regime's attitude towards residence abroad, and indeed he would continue to do so for at least another year. He thought – without any justification – that if Moscow was angered by the German proposal, he could make amends simply by declining it. In that event, he would stay abroad for a while as 'a free Soviet citizen', by which he meant a person loyal to the regime who was living independently without state employment. Yet Moscow was already persuaded that Soviet citizens who disobeyed an instruction to return home were disloyal, and a suspicion that Lomonosov's evasiveness masked an intention to become a 'non-returner' surely motivated Rudzutak's ultimatum. So it would be more a matter of luck than judgement that Lomonosov was abroad when the regime began a systematic purge of Russia's pre-revolutionary engineers and scientists in 1928, including many railway engineers.

Intriguingly, Moscow did not simply severe contact with Lomonosov in early 1927. No great change of official attitude was evident to the man himself for over a year, which encouraged his delusion that his conduct was deemed acceptable. Soviet diplomats continued to accept him and Raisa as Soviet citizens. For example, during April–June 1927 the couple consulted the Rome embassy's resident GPU representative on several occasions, and in October 1927 the

---

[38] 'Dnevnik', 10–12 November, 31 December 1926; 11 January 1927; Rudzutak–Lomonosov, 7 January 1927: LRA, MS 716.1.183; 'Vospominaniia', vol. 9, pp. 2074, 2077–80, 2090, 2100, 2113–15, 2130.

[39] 'Dnevnik', 15 January 1927; 'Vospominaniia', vol. 9, pp. 2135–7.

[40] 'Dnevnik', 23 September 1926; 28, 31 January 1927; 'Vospominaniia', vol. 9, p. 2152.

Paris embassy renewed their Soviet passports.[41] Also, Lomonosov continued to have contact with NKPS officials and friends. For instance, Serebriakov, still a Collegium member, met with him in Berlin in September 1927.[42] Colleagues and friends like R.P. Grinenko and Mme Pravdina corresponded with him, as did Dobrovol'skii despite their estrangement.[43] His publications were still being cited in NKPS-published books and periodicals as late as 1931.[44] For his part, Lomonosov was mainly concerned until at least mid-1928 to avoid provoking a summons to Moscow: still he did not accept that Rudzutak had recalled him already. Thus, he eschewed jobs that might arouse objections in Russia, such as consulting work for private companies. Above all, he deferred the start of his consulting work for the German railways until eventually he withdrew altogether in January 1928 – an extraordinary and clear sign of his desire to appease Moscow, for he had no other salary in prospect, let alone such a lucrative post, and he had already drawn considerably on his capital.[45]

A few months later, in spring 1928, the commencement of the purge of the pre-revolutionary scientific and technical specialists finally made him start to sense a change. A dramatic political, social, economic and cultural revolution had begun with Stalin's emergence as the principal party leader during 1927 and the decision of the Fifteenth Party Congress (December 1927) to abandon the New Economic Policy in favour of forced industrialization with central planning and, in the countryside, the replacement of private agriculture with collective farms. The purge of scientific and technical specialists was publicized in May–June 1928 with the 'Shakhty trial', in which 55 mining engineers were convicted of sabotage and collusion with 'international capital'.[46] Less obvious, but equally ruthless, were similar trials throughout the economy, including trials of NKPS personnel. Lomonosov became aware of this convulsion during April–June 1928: one of his remaining Moscow

---

[41]   'Dnevnik', 14 May, 6 October 1927; 'Vospominaniia', vol. 10, pp. 16, 19–20, 45, 75–7, 208–9, 213, 320. The GPU was the successor of the Cheka. The Lomonosovs arranged these meetings to discuss whether to withdraw George from Cambridge following the rupture of British–Soviet diplomatic relations in 1927.

[42]   'Dnevnik', 20 September 1927; 'Vospominaniia', vol. 10, pp. 187–8.

[43]   For example: Vannevskii–Lomonosov, 14 March 1927: LRA, MS 716.1.185; 'Dnevnik', 30 May, 3 August 1927; 'Vospominaniia', vol. 10, pp. 48, 450, 609.

[44]   For example: V.S. Markovich, 'Sila tiagi parovozov serii E', *Zheleznodorozhnoe delo*, 9 (1929): 33–5; A.B. Dul'nev, 'K voprosu o naivygodneishem sostave tovarnykh poezdov', *Zheleznodorozhnoe delo*, 7–8 (1931): 38–40; M.N. Zhitkov, 'Ob osnovnykh prichinakh umen'sheniia rabotosposobnosti nashego parovoznogo parka i meropriiatiiakh dlia predel'nogo podniatiia etoi rabotosposobnosti', *Zheleznodorozhnoe delo*, 9 (1931): 24–8.

[45]   For example: Lomonosov–Hammer, 2 January 1928, and Hammer–Lomonosov, 27 January 1928: LRA, MS 716.1.186–87; Lomonosov–Lomonosova, 25 January 1928: LRA, MS 716.4.1.52; 'Dnevnik', 23–8 January 1928.

[46]   On the repression of the technical specialists during 1928–1931 see Bailes, *Technology and Society*, pp. 69–159.

contacts asked him not to correspond with NKPS people for their safety, and Grinenko sent news of many NKPS arrests and a rumour that Lomonosov was to lose his Soviet citizenship for not having submitted his reports.[47] The NKPS victims included Krasovskii, who was convicted in July 1928 and probably executed for buying unnecessary equipment abroad and agreeing disadvantageous contracts with foreign suppliers – the same allegations that Lomonosov had faced five years earlier.[48] Borisov, the head of the Main Directorate of Ways of Communication since 1920, died suddenly in his mid-sixties in June, and despite the publication of warm obituaries, there were rumours of suicide and murder.[49] During the same month a senior VSNKh acquaintance, S.A. Khrennikov, gave Lomonosov detailed information about the elimination of the pre-revolutionary engineers and scientists as a group, which made Lomonosov realize that his own troubles had been 'child's play'; Khrennikov, who was visiting Berlin on business, was fearful for his own life and was duly arrested in 1929.[50] Nonetheless, even in April 1929 Lomonosov did not believe news that Grinenko had been arrested.[51] Probably not until later that year, when senior railway engineers were among those executed for involvement in a 'counter-revolutionary organization in transport', did Lomonosov realize that he might need to stay abroad for many years for his own safety.[52]

Yet even that insight did not change his politics fundamentally: his opinion of the Soviet system remained staunchly positive, just as many Bolsheviks would still support their party while languishing in prison in the 1930s. Lomonosov's

---

[47] 'Dnevnik', 15, 27 April, 5 June 1928; 'Vospominaniia', vol. 10, pp. 409, 425–6, 429–30. Further research is needed to determine whether any of these arrests were directly connected with Lomonosov's loss of favour and apparent defection.

[48] K. Nagatsuna, 'A Utopian Ideologue in Soviet Industrialisation: S.A. Bessonov and Transport Reconstruction Debates, 1928–1930', paper presented at the Soviet Industrialisation Project Seminar, University of Birmingham, January 1989, pp. 9–10.

[49] V. Tolstopiatov, 'Ivan Nikolaevich Borisov', *Zheleznodorozhnoe delo: obshchii otdel*, 6 (1928): 1–2; Iu.V. Rudyi, 'Ivan Nikolaevich Borisov', *Zheleznodorozhnoe delo: obshchii otdel*, 6 (1928): 2; 'I.N. Borisov (1860–1928)', *Transport i khoziaistvo*, 5 (1928): 3–5; 'Vospominaniia', vol. 10, p. 481. See also *Poslednie novosti*, 7 October 1931. In 1937 he was exposed as a 'wrecker': Rees, *Stalinism*, p. 171.

[50] Lomonosov–Lomonosova, [25 June 1928]: LRA, MS 716.4.1.56; 'Dnevnik', 25 June, 3 July 1928; 'Vospominaniia', vol. 10, pp. 476–81, 488–90. Khrennikov (1872–1929) was a former director of the Sormovo machine-buliding company who knew Lomonosov through the VSNKh involvement with the work of the Railway Mission in 1922. He was arrested on 12 July 1929, accused of counter-revolutionary activity and sentenced to six years, five months of prison. He died in prison hospital on 25 December 1929. See http//lists.memo.ru/index22.htm, citing *Kniga pamiati Nizhegorodskoi oblasti* (last accessed 30 June 2008).

[51] 'Dnevnik', 14 April 1929; 'Vospominaniia', vol. 10, p. 738.

[52] On the 1929 purge see Rees, *Stalinism*, pp. 17, 24. In 1934 Lomonosov wrote of going to Russia on the off-chance of getting a professorship; but frustrated homesickness rather than serious intent was probably the explanation: 'Dnevnik', 12 October 1934.

political aversion to the anti-Soviet emigration was such that there was perhaps only a little exaggeration in his 1931 remark that he would rather be shot in the USSR than take a Nansen passport.[53] More tangibly, he made provision in a will of 1930 for some of his capital in the United States to be used to fund scientific or technical visits to the United States by staff and students of his Alma Mater in Leningrad.[54] Also, he renewed his Soviet passport annually until he took British citizenship in 1938. Even as a British citizen he could still write in 1942 that 'I support the ideas of [the] October [revolution] with all my soul'.[55] And he bridled at the idea of appeasing prospective foreign publishers by giving his memoirs an anti-Soviet tone in the late 1940s.[56]

Above all, his support for the principle of purging the pre-revolutionary scientific and technical specialists in Russia shows how his pro-Soviet politics were embedded in his scientific and technical thinking as well as his moral code. He interpreted those purges not as an indiscriminate orgy of paranoia and violence, but as a rational, albeit somewhat counter-productive response by the party to sabotage. He still believed that most of these specialists were passively anti-Soviet, and that at least some were saboteurs. Hence he was receptive to Moscow's claim that these people were threatening the revolution. In his opinion, for example, such sabotage had disrupted the Soviet diesel research programme, in such forms as Krasovskii's preference for steam technology and Dobrovol'skii's 'bureaucratic' sabotage of the diesel experiments. Thus, following their arrests in 1928 and 1930 respectively, Lomonosov could believe the charges against them, just as he accepted the charges of sabotage that were made against technical specialists in the spurious 'Industrial Party' show trial in 1930. However, he did baulk at reports of the imprisonment and execution of engineers without trial, including his former student A.F. Velichko.[57]

Further, he agreed that the interests of the revolution could justify the murder of innocent people in this process – a view that he had expressed *vis-à-vis* himself in the winter of 1924–1925 (see Chapter 9) and that anticipated the despairing logic of *Darkness at Noon*, Arthur Koestler's graphic novel about an imprisoned Bolshevik trying to comprehend the Stalinist terror.[58] The problem with these purges, Lomonosov thought, was that 'not having technical expertise, [Soviet

---

[53] Lomonosov–Lomonosova, 16 January 1931: LRA, MS 716.4.1.101.

[54] Draft deed, 1930: LRA, MS 716.7.10.

[55] 'Dnevnik', 7 November 1942.

[56] For example: 'Dnevnik', 22 June 1948.

[57] See 'Dnevnik', 25 June, 30 October 1928, 9 January, 25 May, 6 August 1929, 16 March, 30 August, 13 November, 4, 8, 11 December 1930, 6, 14 January 1931; 'Vospominaniia', vol. 10, pp. 869–70, 1405–9. His outlook reflects the argument that the 1930s purges were essentially an attempt to eliminate a potential 'fifth column': see especially O. Khlevnyuk, 'The Objectives of the Great Terror, 1937–1938', in J. Cooper et al., *Soviet History, 1917–53: Essays in Honour of R.W. Davies* (Basingstoke, 1995), pp. 158–76.

[58] A. Koestler, *Darkness at Noon* (London, 1940).

power] could not clarify who was right and who was guilty, and unleashed mass terror against all specialists in general'. As a result, innocent people were suffering, fine specialists were rejecting Soviet power, and communism's prestige was undermined. He did not write at length about his own case in this regard, but it was surely in mind. The irony was that he could not see how else the Soviet government could defend the revolution against sabotage by technical experts.[59]

In short, then, Lomonosov did not really understand what was happening to the community of pre-revolutionary engineers and scientists in the USSR. Despite his good connections and detailed information from the likes of Khrennikov, he failed to see the reality that the actions of these people had little if anything to do with their arrest. Continuing to see himself as a loyal supporter of Soviet power, and to see the *spetsy* collectively as fundamentally disloyal, he did not link his own situation with theirs. In his mind the most likely source of problems for himself within the USSR was not his social and professional background but his enemies at the NKPS.

There is little doubt that Lomonosov's and Raisa's politics were sincerely radical during these years abroad. True, they may have had a financial motive to maintain their Soviet citizenship in the 1930s: they believed that their British tax liability was less if they lived in rented furnished rooms with a 'visitor' visa than if they had unfurnished rooms and/or British citizenship.[60] Also, their socialism had never been egalitarian in practice, to put it mildly, and it was even forgotten in their snobbish disdain for George's English girlfriend Bessie Walker, a factory clerk whom he married in 1938.[61] On the other hand, the political contacts that they developed in Britain were generally on the far left of the spectrum. Specifically, they became friendly in 1930 with a leading light of the Independent Labour Party (ILP) named Fenner Brockway, who was also a Member of Parliament for the Labour Party. They attended ILP functions in the early 1930s, and they continued to associate with the ILP even after it marginalized itself by disaffiliating from the Labour Party to form a united front with the British Communist Party in 1932. Their friendship with Brockway lasted long after the ILP had collapsed; indeed Brockway may have influenced Lomonosov to join the British Labour Party in 1944 – probably the only occasion that he became a paid-up member of any political party.[62]

---

[59]   For instance: Lomonosov–Lomonosova, 10 July, 15, 29 August 1930; 14 January 1931: LRA, MS 716.4.1.66, 71, 77, 98; 'Dnevnik', 30 October 1928; 6 August 1929; 16 March, 30 August, 13 November, 4, 8, 11 December 1930; 6 January 1931; 'Vospominaniia', vol. 10, pp. 478, 810–811.

[60]   'Dnevnik', 12–13 August 1932.

[61]   For examples of their hostility towards Bessie see 'Dnevnik', 13, 19 October 1927 and Lomonosova–Lomonosov, 13 June 1932: LRA, MS 717.2.1.85.

[62]   For example: 'Dnevnik', 8, 14 August, 28 December 1931, 21 February 1932; 23 August 1940. On Brockway and the ILP see, for example, F. Brockway, *Inside the Left: Thirty Years of Platform, Press, Prison and Parliament* (London, 1942) and *Outside the*

So why did Lomonosov support the Soviet system so staunchly for so long? Above all, he was determined to remain intellectually consistent with his public political choice of June 1918 – his 'Rubicon'. Related to this point is the way that he always differentiated, before and after 1917, between the Russian state and the ruling regime: almost certainly he reasoned that if Russia had to have a repressive regime, the state structure should reflect a progressive ideology like Marxism, not a reactionary model like tsarism. Also, he strongly distrusted Western reports about the USSR. This scepticism stemmed from his experiences of the foreign press spotlight in 1917–1923, and never disappeared.[63] Instead, he preferred to trust news in letters from Sonia, his son Vsevolod and other correspondents in the USSR, until the Soviet regime made such correspondence impossible in the second half of the 1930s. Unfortunately, like much of his personal correspondence, these letters have not survived; but his diary indicates that, for example, Sonia provided news about family and friends, including Masha's sons.[64] It seems highly unlikely, given Soviet censorship, that Sonia openly criticized the purges; and it is possible that Lomonosov naively read her letters at face value in that sense.

Although he always kept his pro-Soviet outlook, his attitude towards the purges did change, albeit gradually. He was concerned by Sonia's request in July 1936 for him not to send any more money, but whether he understood that she was worried for her safety is unclear.[65] During the following months he surely became more concerned as the terror in Russia intensified and he stopped receiving letters from Russia. Presumably, too, he was upset to see most of his intellectual legacy condemned in print during the NKPS campaign of 1935–1937 to replace the pre-revolutionary tradition of railway science with a new socialist science of railway design, traction and operations: his ideas about traction calculations, rail stresses, traffic costs, railway design and railway operations were all subjected to fierce criticism, and he was branded a 'White guard' and ideologue of the railway 'limiters' – the new label for NKPS people steeped in the intellectual tradition of the pre-revolutionary specialists, as opposed to genuinely Soviet specialists who could extract far more productivity from equipment.[66] Lomonosov wished

---

Right: A Sequel to 'Inside the Left', with a Lost Play by G. Bernard Shaw (London, 1963); D. Howell, 'Brockway, (Archibald) Fenner, Baron Brockway (1888–1988)', in *Oxford Dictionary of National Biography*, vol. 7 (Oxford, 2004), pp. 765–6. Lomonosov notes that he joined the Labour Party in Lomonosov–Mikardo, 16 January 1947: LRA, MS 716.1.195. On the ILP–Labour relationship in the 1930s see, for example, B. Pimlott, *Labour and the Left in the 1930s* (Cambridge, 1977).

[63] For example: 'Dnevnik', 4 December 1930.

[64] For example: 'Dnevnik', 2 October 1931; 17 September, 22 October, 9 November 1933; 26 June 1934.

[65] 'Dnevnik', 29 July 1936; Lomonosov–Lomonosova, 3 August 1936: LRA, MS 716.4.1.264.

[66] For example: G.P. Vasil'ev, 'A.M. Babichkov. "Teoriia tiagi poezdov i tiagovye raschety"', *Parovoznoe khoziaistvo*, 2 (1935): 42; *Gudok*, 9 January 1936, p. 1; Morozov,

'for the first time' to renounce his Soviet citizenship in August 1936 following the 'monstrous' and 'shameful' executions of the main defendants in the first big show trial of Bolshevik leaders, who included his erstwhile patron and friend Kamenev. Yet with the next major show trial in February 1937 he could somehow still credit the allegation that the accused, including his friend Serebriakov, had joined a 'wrecking' conspiracy, and one can assume that he took the same view about other NKPS victims of the 1937 purges such as Emshanov, Rudyi and M.E. Pravosudovich.[67] He was probably unaware that he was mentioned as a leading NKPS 'wrecker' at the plenary meeting of the party's Central Committee in February–March 1937, because the minutes were not published.[68] But events of June 1937 did change his mind: he scorned the treason charges against Marshal Tukhachevskii and other top military leaders, and was appalled by the 'nightmare' of their execution by firing squad.[69]

Lomonosov's application for UK citizenship in April 1938 could be seen as the legal step that marked him as a 'non-returner', yet he was still not rejecting the Soviet system as such. Revulsion at the purges did partly explain his decision, but more important was a practical concern about his Soviet citizenship. From about 1929 he was constantly worried about how the Soviet authorities would handle the next annual renewal of his passport, for any refusal to renew it would render him stateless and force him to have a Nansen passport like an émigré.[70] His passport worries were heightened by the mass purges of the mid-1930s, and the execution of Rykov, N.I. Bukharin and other former Bolshevik leaders in March 1938 became the final straw. His decision to seek British citizenship was 'difficult', yet preferable to the likely alternative of statelessness. The UK authorities were evidently content with his application, despite his colourful past, for he swore the British oath of allegiance on 17 August 1938.[71]

'O predel'shikakh–parovoznikakh'; S.M. Kucherenko, 'Povysim konstruktivnye skorosti', *Zheleznodorozhnaia tekhnika*, 1 (1936): 7–11; M. Lazarev, 'Protiv "predela", za perevooruzhenie fronta proektirovaniia', *Sotsialisticheskii transport*, 2 (1936): 36–48; E. Raaben, 'Za bolshevistskuiu nauku o sebestoimosti perevozok', *Sotsialisticheskii transport*, 5 (1936): 38–44; 'Godovshchina stakhanovskogo-krivonosovskogo dvizheniia'; 'Velikaia godovshchina', *Zheleznodorozhnaia tekhnika*, 10 (1936): 3–6; 'Ot redaktsii', *Sotsialisticheskii transport*, 6 (1936): 94–5; and 'Sobranie aktiva Narodnogo komissariata putei soobshcheniia: Doklad tovarishcha L.M. Kaganovicha ob itogakh Plenuma TsK VKP (b)', *Zheleznodorozhnaia tekhnika*, 3 (1937): 4–7.

[67]   'Dnevnik', 22 January 1935; 29 August 1936; 11 February 1937; Rees, *Stalinism*, pp. 171, 174.

[68]   See Rees, *Stalinism*, pp. 167–72.

[69]   'Dnevnik', 29 April, 12 June 1937; Lomonosov–Lomonosova, 12 June 1937: LRA, MS 716.4.3.287.

[70]   See, for instance, Lomonosova–Lomonosov, 15, 19 January 1931: LRA, MS 717.2.1.65, 69.

[71]   For example: Lomonosov–Lomonosova, 12, 16 January 1931: LRA, MS 716.4.1.96–97, 101; 'Dnevnik', 29–30 April, 17 August 1938; Certificate of naturalization,

Intriguingly, there is some fragmentary evidence to show that Moscow did not necessarily sever contact with individuals in such cases. During the Czechoslovakian crisis of March 1939 Lomonosov agreed to attend a meeting at the Soviet consulate in London at the consul's request. Another request came in May 1939, which Lomonosov declined or ignored. In April 1942, with considerable trepidation, he again visited the London consulate in response to their repeated requests. The consul evidently wanted further contacts, noting that Moscow still recognized him as a Soviet citizen. But while Lomonosov confirmed his support for the Soviet system and war effort, he emphasized that he had been a British citizen since 1938 and that British law did not acknowledge dual citizenship. He was therefore required to surrender his Soviet passport.[72]

This last development, combined with Lomonosov's refusal to apply to Moscow to regularize his residence outside the USSR, prompted the Soviet authorities to revoke his Soviet citizenship – a slow bureaucratic affair that formally marked him as an enemy of the state. This process began with a report by the consular department of the People's Commissariat of Foreign Affairs. The People's Commissariat of State Security voiced agreement in December 1943, noting that Lomonosov was suspected of having contacts with foreign intelligence services during his time abroad. During 1944 the People's Commissariat of Internal Affairs visa department was consulted; and eventually the consular department sent a formal recommendation to the Citizenship Commission of the USSR Supreme Soviet in February 1945. The Presidium of the USSR Supreme Soviet confirmed the withdrawal of Lomonosov's Soviet citizenship in a decree of 14 April 1945.[73]

Employment and domicile, 1927–1937

Had Lomonosov been simply an opportunist, caring only for his own well-being, he could easily have relaunched his career abroad in 1927 by becoming a well-paid consultant with the German railways. Following their proposal of 1926, the German State Railways made a further offer in September 1927. They promised him RM6,000–12,000 per year for five years for monitoring diesel research

---

15 August 1938: LRA, MS 716.7.17; *The Times*, 30 April 1938, p. 1. A subordinate reason may have been financial: to be eligible for interest on their Dawes loan stock: Lomonosova–Lomonosov, 26 August 1937: LRA, MS 717.2.1.275.

[72] 'Dnevnik', 1–2 April 1942; V.N. Zonov–Lomonosov, 25 May 1939: LRA, MS 716.4.39.1; Lomonosova–Lomonosov, 30 May, 3 June 1939: LRA, MS 717.2.1.309, 313; Beliaev and Churilin (NKID Consular Dept)–USSR Citizenship Commission, 12 February 1945: GARF, f. r-7523, op. 59, d. 290, l. 2.

[73] People's Commissariat of Internal Affairs Visa Dept briefing note, 29 January 1945: GARF, f. r-7523, op. 59, d. 290, l. 1–ob.; Beliaev and Churilin–USSR Citizenship Commission, 12 February 1945: GARF, f. r-7523, op. 59, d. 290, l. 2; information from Catalogue department, GARF; Mikheev–author, 13 August 2002.

worldwide and testing a locomotive that the MAN company was building for them. Concurrently he could assist other private German companies keen to develop this market: Hohenzollern offered $300 per month plus travel costs and sales commission, and similar deals were discussed with Esslingen and the Henschel engineering company. Yet incredibly, as noted above, Lomonosov rejected these offers in January 1928, such was his desire for official approval from Moscow.[74]

This action overshadowed the rest of his life abroad. It demolished his job prospects in Germany, the foreign country where he was best known and where his command of the language was best (though far from perfect): his top-level contact on the state railways, Gustav Hammer, was mortally embarrassed, and the payments of $5,000 and $3,000 that were offered by Hohenzollern and MAN for previous advice had an air of finality.[75] Accordingly Lomonosov had to seek work elsewhere. Britain was one option: not least, it was home for George after six years of boarding school. Italy and North America were also considered because railway diesel research was especially vigorous there. Lomonosov had fewer contacts in these countries compared to Germany and had only a poor command of the languages. Yet he was confident that he would cope. He still expected to return to Russia in due course and he had no pressing financial worries: the family money was invested in seemingly solid stocks like US Liberty bonds, the Dawes Plan loan to Germany, and significant German and American companies (especially the German chemicals giant IG Farben); and although George's first year at Cambridge cost over half of that year's investment income, this expense was not long-term.[76] Also, Lomonosov had more ideas for supplementing the family budget. However, as will be shown below, these calculations were shattered over the next few years, and he never managed to relaunch his career.

In terms of alternatives to Germany for employment, Lomonosov concentrated on Britain during 1927–1928 for reasons that were largely non-professional. Italy had professional, climatic and other attractions, but the Lomonosovs abhorred its fascist politics. North America evoked too many unhappy memories. In favour of Britain was George's likely presence there until at least 1929. Lomonosov did have some UK contacts like J.R. Glass, a sales agent for Armstrong Whitworth with whom he had worked during 1920–1922. But his preference was for a salary in higher education, one of the few realms that in his opinion Moscow would tolerate. Here he sensed the possibility of help from P.L. Kapitsa, the young Soviet physicist at Cambridge University, which was why Lomonosov headed for Cambridge. Soon he was introduced to the head of the engineering department, Professor Charles Inglis, and gained the friendship of the Russian-born Cambridge-based mathematician A.S. Bezikovich and the retired doyen of Cambridge engineering,

---

[74]   'Dnevnik', 10 September 1927; Lomonosov–Lomonosova, 25 January 1928: LRA, MS 716.4.1.52; 'Vospominaniia', vol. 10, pp. 176–8, 181, 183.

[75]   Lomonosov–Lomonosova, 28 January 1928: LRA, MS 716.4.55; 'Vospominaniia', vol. 10, pp. 117, 127–8, 176–8, 181, 188, 192–7, 263–4, 267, 300–301, 345, 398, 437–8.

[76]   'Vospominaniia', vol. 10, p. 87.

Sir J. Alfred Ewing, whose classic book on steam machines Lomonosov had had translated into Russian in 1904. The fact that Lomonosov subsequently focused his hopes on Cambridge was simply because his contacts were clustered there, not because of any enthusiasm for the particular approach to engineering teaching that Ewing had developed.[77]

In professional terms, however, Britain was a poor location for a Soviet Russian railway engineer like Lomonosov. The rupture of British–Soviet diplomatic relations in 1927 over alleged subversive activity was unhelpful, but Lomonosov's main problem here concerned the relation of his professional skills, interests and experience to British railway engineering traditions and needs. His command of English was inadequate for teaching, and he felt unable to take a junior post, so his only possibility in the education sector was a senior research position in transport engineering. Sadly for him, such jobs were virtually non-existent. Unlike Russia and other major European countries, which had dedicated institutes and university departments for transport engineering, the UK merely had general university departments of engineering and technical colleges, and their connections with the railways were not especially close.[78] Furthermore, Lomonosov embodied the continental European tradition of stressing mathematics-based theory in engineering education for transport, whereas British railway engineers valued practical experience above all – a difference encapsulated in the retort to him by a senior British engineer that 'in this country trains are pulled by locomotives, not by differential equations'.[79] There was even some feeling in Britain that the academic world had too much influence over locomotive development in Russia. To quote E.S. Cox, who became one of the leading figures in British locomotive design in the 1950s, 'there always seemed to be a considerable gulf between the professor in his laboratory, and the man in the depot who ran the engines'.[80] It did not help that Lomonosov somehow alienated Inglis. One difficulty, perhaps,

---

[77] On engineering at Cambridge see Hilken, *Engineering at Cambridge*, especially pp. 107–28 for the years 1890–1903 when Ewing was Professor of Mechanism and Applied Mechanics and pp. 145–76 for 1919–1943 when Inglis headed the department. On Bezikovich (usually rendered as Besicovitch in English) see C. Domb, 'Besicovitch, Abram Samoilovitch (1891–1970)', in *Oxford Dictionary of National Biography*, vol. 5 (Oxford, 2004), pp. 509–11; for Inglis see J.F. Baker, rev. J. Heyman, 'Inglis, Sir Charles Edward (1875–1952)', in *Oxford Dictionary of National Biography*, vol. 29 (Oxford, 2004), pp. 258–9.

[78] The railway industry is virtually absent from the discussion of the inter-war years in M. Sanderson, *The Universities and British Industry, 1850–1970* (London, 1972), pp. 243–313, in contrast to, for example, aeronautical and wireless engineering.

[79] Cited by D.R. Carling in a letter to the Editor, *Railway Gazette*, 5 December 1952, p. 622. Unfortunately Carling withheld the speaker's identity.

[80] E.S. Cox, *World Steam in the Twentieth Century* (London, 1969), p. 162.

was that Inglis's interest in the dynamics of railway vehicles was also one of Lomonosov's particular interests.[81]

For these and other reasons the British railway environment was also uncongenial for Lomonosov's thinking on railway economics during the 1920s and 1930s. Ironically, the British tradition of private railway ownership coupled with competition from road transport and declining profits did not mean that the British railways had any more interest in systematic cost-cutting than Lomonosov had seen on the state-dominated Russian railways.[82] His work on the costs of designing and building new railways was scarcely relevant because the UK network was essentially complete.[83] But operating costs were another matter, and here there was much complacency in the UK. This problem has yet to be explored fully, but in part the explanation may stem from the fact that in the early 1920s the government grouped almost all the many private companies into four large private companies on a territorial basis: the Great Western Railway (GWR); Southern Railway; London, Midland and Scottish Railway (LMSR); and London and North Eastern Railway (LNER). For all its strengths this structure did not stimulate systematic demand for clear, reliable data about operating costs.[84] That said, such interest would probably not have helped Lomonosov. His experience and recipes, including his formula for calculating operating costs, were generally deemed irrelevant because they emanated from a very different environment of heavily subsidized state railways under centralized government control.[85]

The scope was no greater for him in locomotive testing. Hitherto, this work had been a poor relation in British railway science, which was mostly geared to the testing of materials. But interest had been growing steadily since the First World War, mainly as a means to reduce fuel costs, much as Lomonosov's work had started in the 1890s. There was even some inter-company collaboration, and the LMSR created a research department in 1932 that reflected a corporate commitment to scientific enquiry. But Lomonosov's type of road experiment was impossible because it would disrupt revenue-earning traffic and because of the lack of suitable gradients. There was just one locomotive testing plant, at the GWR's Swindon works, and despite railway lobbying for a government-funded national facility, the construction of a second plant did not begin until 1936; a private LNER–LMSR joint venture at Rugby, it was delayed by the Second World War and did

---

[81] On Inglis's interests see Hilken, *Engineering at Cambridge*, p. 170.

[82] On the British railway context see, for example, D.H. Aldcroft, *British Railways in Transition: The Economic Problems of Britain's Railways Since 1914* (London, 1968), pp. 1–88.

[83] For an overview of the development of the British railway network see J. Simmons and G. Biddle (eds), *The Oxford Companion to British Railway History* (Oxford, 1997), pp. 492–8.

[84] A good introduction to this reorganization and its aftermath is M.R. Bonavia, *The Four Great Railways* (Newton Abbot, 1980).

[85] Early examples include: 'Dnevnik', 1, 24 December 1930, 2 February 1931.

not open until 1948. In any case, until the early 1930s the principal demand of managers concerning locomotive research was merely very simple comparative data about the relative economy of the many different classes of locomotive that they had inherited through the government-imposed grouping.[86] Also, there was much scepticism – all too familiar to Lomonosov – about whether elaborate and expensive research could yield worthwhile savings. Reviewing British locomotive-testing practice in 1927, Lomonosov concluded that the methodology was 30 years behind Russia and that the few committed researchers faced the same traditionalist resistance that he had faced in Russia.[87] Indeed, even some twenty-five years later, the Engineering Department of the nationalized British Railways was viewed derisively by other railway staff 'as being full of airey-fairy, head-in-the-clouds scientists' without any practical railway knowledge.[88]

Diesel power was the one realm where the sometimes rather parochial British railway engineers acknowledged Lomonosov as an exceptional contributor at this time. Yet even he was unsurprised by the absence of British offers of work here. He still reasoned that diesel power's railway potential was greatest where good water was scarce and/or good coal was expensive. Neither problem troubled the British railways. Interestingly, he scarcely considered labour costs, perhaps reckoning that economies on footplate wages would be offset by increases in other costs. He concluded that in British terms main-line diesel power made economic sense primarily for colonial railways in arid locations. Thus, he thought that any British demand for his diesel expertise would come from independent locomotive-builders with export ambitions, and he became more receptive to the idea of working with such private firms as his worries about Russia increased during 1928. Yet few of these companies were exploring diesel technology. Among them, one of the most

---

[86] A participant in these developments, E.S. Cox, provides a good overview (with special reference to the LMSR) in three books: *Chronicles of Steam* (London, 1967), pp. 57–76; *Locomotive Panorama*, vol. 1 (London, 1965), pp. 20–131; and *Speaking of Steam* (London, 1971), pp. 86–100. See also C. Divall, 'Down the American Road? Industrial Research on the London, Midland & Scottish Railway, 1923–1947', in J. Armstrong, C. Bouneau and J. Vidal Olivares (eds), *Railway Management and its Organisational Structure: It Impact on and Diffusion into the General Economy* (Seville, 1998), pp. 131–40; A.O. Gilchrist, 'A History of Engineering Research on British Railways: IRSTH Working Paper No.10' (York, 2006), pp. 6–12; and J. Crosse, 'The Scientific Research Department of the LMS', *Backtrack*, 22/4 (April 2008): 236–7. On the Swindon and Rugby facilities see, in particular, Carling, 'Locomotive Testing Stations (Part I)', pp. 117–21; and P. Atkins, 'On Test: Rugby Locomotive Testing Station', *Backtrack*, 3/4 (September–October 1989): 148–54. More generally on the connections between economics and technical change on British railways see C. Divall, 'Technical Change and Railway Systems', www.york.ac.uk/inst/irs/irshome/papers/hulltext.htm (accessed 18 May 2004).

[87] 'Dnevnik', 25 December 1927.

[88] A. Rimmer, *Testing Times at Derby: A 'Privileged' View of Steam* (Usk, 2004), p. 45. Cox likewise recalled the scepticism of 'hard-headed operators' at that time: *Chronicles of Steam*, p. 55.

committed was Armstrong Whitworth. Glass helped to get George a job there in 1931, but for whatever reason the company did not wish to use Lomonosov himself.[89]

These unpropitious circumstances encouraged Lomonosov to try redefining himself as an inventor. He had two main ideas. One was a magnetic clutch for high-power diesel-mechanical locomotives, and the other was an electromagnetic system for operating train brakes. Initially he saw this work as insurance in case of failure in his negotiations with the German State Railways in 1927, in the belief that Moscow would consider inventing as a suitable occupation for a Soviet citizen abroad. In 1928 Kapitsa agreed to do the main development work – notably a new solenoid for the brake system – if Lomonosov would underwrite the costs, administer the patent applications and share any profits. With much effort they obtained patents in the UK, United States, Germany and even the USSR. But the products never reached the market. Kapitsa was unable – and by 1929 unwilling – to devote enough time to developing them. The inventions themselves faced stiff competition: Lomonosov's own work had helped to confirm electric transmission as the best prospect for rapid progress with main-line diesel traction, and the Westinghouse air-brake system was widely established. Without prototypes to prove their economic significance, Lomonosov could not sell his inventions to would-be manufacturers. Unable to continue bearing the development costs, he abandoned these ventures by the end of 1929.[90]

---

[89]   Lomonosov's fruitless contacts with Armstrong Whitworth are covered extensively in his diary between 14 June 1931 and 23 January 1932; George's appointment is noted in entries for 30 January and 1 February 1931. No relevant information has been found in the few surviving company papers at the Tyne and Wear Record Office, Cambridge University Library and National Railway Museum, York. On early British attitudes towards dieselization with particular reference to foreign developments see C. Divall, 'Learning from America?', *Railroad History, Millenium Special: The Diesel Revolution* (2000): 124–42; M. Rutherford, '"Export or Die!": British Diesel-Electric Manufacturers and Modernisation, Part One: Roots', *Backtrack*, 22/1 (January 2008): 52–60. Armstrong Whitworth's diesel-traction work at this time is described in its house journal in several articles by the head of its diesel department, C.J.H. Trutch: 'Diesel Rail Traction', *Armstrong Whitworth Record*, 1/2 (1930): 13–17; 'Armstrong Whitworth Diesel Traction Progress', *Armstrong Whitworth Record*, 2/2 (1932): 2–9; and 'Armstrong Whitworth Diesel Traction Development', *Armstrong Whitworth Record*, 2/4 (1933): 2–11. See also Rutherford, 'When Britain was a Contender', Parts 2 and 3, *Backtrack*, 14/7 (July 2000): 420–421 and 14/8 (August 2000): 479–83. For background concerning the company see K. Warren, *Armstrongs of Elswick: Growth in Engineering and Armaments to the Merger with Vickers* (Basingstoke, 1989), especially pp. 205–9. It would seem that Lomonosov did not contact the Beardmore company of Glasgow, which was the leading British manufacturer of diesel engines for railway use until the early 1930s; the importance of the Beardmore engines is stressed in B. Webb, *The British Internal Combustion Locomotive, 1894–1940* (Newton Abbot, 1973), p. 12 and Rutherford, 'When Britain was a Contender': 416–20.

[90]   Extensive correspondence about the inventions survives in NARG 241, US patent files 1,886,692 and 1,927,457; and in Kapitsa's archive: Lichnyi fond P.L. Kapitsy, Dom-

Lomonosov's attitude to employment in North America changed during 1928–1929 as he became more apprehensive about Russia. He and Raisa began discussing a visit in about July 1928 as a tactic to extract George from Cambridge. They thought that, having failed his examinations, George was simply indulging passions for his motorcycle and his girlfriend Bessie. However, by January 1929 job-seeking headed Lomonosov's American agenda. In his favour, still, was his professional reputation: Samuel Vauclain wanted him to investigate problems with the two diesel-electric locomotives that Baldwin had built since 1925, and Lomonosov was now willing to work with private industry in this way, not least as a means to promote his inventions.[91] There were recent precedents, too, for Russian engineers to establish themselves in US industry and academe. Their number included Lipets, who joined Alco in 1920 as their European sales and technical representative and additionally became a non-resident professor of locomotive engineering at Purdue University in 1927. Another example was the former IPS professor S.P. Timoshenko, who by 1929 was a research professor in mechanics at the University of Michigan and an adviser to the Westinghouse company.[92] Furthermore, Lomonosov was receiving encouraging letters from friends like Grace Abbott, who was now at the US Department of Labor. Eventually he decided to obtain a visitor's visa (without the right to take paid work) as the simplest option, adjust his immigration status if and when required, and meanwhile advertise himself by offering technical advice on an unpaid basis.[93]

He had two major opportunities, but his hopes were quashed by suspicion of him as an outsider together with bad luck and his own typically intemperate conduct. During March–July 1929 he spent weeks on diesel-related work at Baldwin's Eddystone factory near Philadelphia, but was appalled by the uncooperative attitude of staff at all levels. His abrasive manner was part of this problem, but so too was dislike of him as a consultant: his involvement was seen as criticism of the workforce. Also relevant was Vauclain's replacement as company president

---

muzei P.L. Kapitsy pri Institute Fizicheskikh Problem imeni P.L. Kapitsy, Moscow. See also E.S. Sharikova and L.Iu. Pokrovskaia, 'K istorii teplovozostroeniia v SSSR (Novye materialy ob izobretatel'skoi deiatel'nosti prof. Iu.V. Lomonosova)', *Voprosy istorii estestvoznaniia i tekhniki*, 4 (1986): 186–8. For early expressions of Lomonosov's fading hope see 'Dnevnik', 5, 9 October 1928.

[91]   'Dnevnik', 8 July, 5 November 1928. On the Baldwin locomotives see, in particular, G.W. and S.L. Dolzall, *Diesels from Eddystone: The Story of Baldwin Diesel Locomotives* (Milwaukee, 1984), pp. 5–10, and J.F. Kirkland, *The Diesel Builders, Volume 3: Baldwin Locomotive Works* (Pasadena, 1994), pp. 11–17.

[92]   See Westwood, *Locomotive Designers*, p. 232 (Lipets); '[A.I. Lipetz']', *Railway Mechanical Engineer* (June 1936): 236; and Timoshenko, *As I Remember*, pp. 278–9. Lomonosov wrote to Timoshenko: 'Vospominaniia', vol. 10, pp. 960, 994.

[93]   'Dnevnik', 23 May, 26 November 1928; Abbott–Lomonosova, 2 January 1928: LRA, MS 717.2.5.13.

at this time by George H. Houston, a sceptic about railway diesel power.[94] Equally, Lomonosov was disgusted by the company's penny-pinching, which he considered typical of the private sector and counter-productive for such research.[95] Having failed meanwhile to interest corporations in his inventions, he looked to academe.[96] In August 1929 some old friends from Hull House contacted the Nobel-prize-winning physicist Robert Millikan, who chaired the Executive Committee of the California Institute of Technology (Caltech) at Pasadena, and who agreed to the idea of a one-year unpaid research fellowship.[97] Lomonosov divided his time at Caltech between postgraduate teaching and drafting a development plan for a dedicated railway diesel engine. But the students disliked the highly mathematical nature of his classes and his poor English. As ever, he quarrelled with some colleagues, notably V.M. Zaikovskii, a Russian émigré and former Captain in the Russian Supply Committee. Crucially, the Executive Committee did not want a faculty of transport studies, and even Millikan, a fund-raiser par excellence, could not persuade railway companies to sponsor the proposed diesel research. Eventually in March 1930 Lomonosov confronted Millikan with an ill-judged ultimatum for a large salary. The notoriously parsimonious 'Boss' called his bluff, and Lomonosov left under a cloud. Shortly afterwards, with no other openings in America, the couple returned to Europe, where Lomonosov was scheduled to attend a conference.[98]

This outcome was a missed opportunity in the development of railway diesel technology. Lomonosov's aim was to develop a diesel engine specifically for railway service. He wanted high power combined with low weight in a format that could match steam locomotives capable of developing 4,000–6,000 h.p. at the rail, without making excessive combustion smoke and without overheating the cylinders and pistons. The project proposals that he wrote at Caltech reveal that, among other ideas, he envisaged an engine of roughly 2,400 h.p. (2,000 h.p. at the rail) with up to 12 cylinders, a two-stroke combustion cycle (lighter and more

---

[94] *Baldwin Locomotives*, 8/1 (1929): 3; 'Dnevnik', many entries from 23 February to 15 July 1929; 'Vospominaniia', vol. 10, pp. 714–832, 889–963.

[95] 'Dnevnik', 20 April 1929.

[96] For instance: 'Dnevnik', 12, 15 March, 2, 14–19, 22 May, 5 June 1929.

[97] Lomonosov's work with Caltech is the subject of A.J. Heywood, 'Barriers to East–West Technology Transfer: Iu.V. Lomonosov and Diesel Railroad Engineering in the Interwar Period', *The Russian Review* (forthcoming). Much information survives in a file of the Caltech Archives: Robert Andrews Millikan Collection, Series IV: Millikan correspondence re. Caltech, Lomonossoff George et al., 1929–1930 (folder 29.26) (henceforth Caltech folder 29.26). On Millikan and Caltech see J.R. Goodstein, *Millikan's School: A History of the California Institute of Technology* (New York, 1991) and R.H. Kargon, *The Rise of Robert Millikan: Portrait of a Life in American Science* (Ithaca, 1982).

[98] Lomonosov's stay is detailed in 'Dnevnik', 29 September 1929–10 April 1930, and 'Vospominaniia', vol. 10, pp. 1082–1451. Relief at Caltech about his departure is evident in A. Noyes–Millikan, 3 April 1930: Caltech folder 29.26, frames 788–9.

efficient than four-stroke), and a maximum weight of about 40 US tons, which could be operated in multiple by one driver.[99] This concept was broadly similar to the General Motors 1,350 h.p. model 567 engine of 1938 – the machine developed specifically for railway use that indeed would power US railway dieselization. It is thus conceivable that, had the circumstances been different, a suitable engine would have been built by the early 1930s, and that Vauclain might have avoided the blame – somewhat unfair, in the light of Lomonosov's evidence – for Baldwin's ultimate failure to master this new technology.[100]

Lomonosov's personal conflicts and ultimatum to Millikan help to explain why railway diesel engines did not bring Caltech the fame that it achieved with its inter-war excellence in aeronautics and jet propulsion.[101] The fundamental obstacles, however, lay elsewhere. Even Millikan, with his legendary fundraising prowess, was thwarted by the aversion of the railways and engineering industry to sponsor such research. This problem was partly due to terrible luck – the Wall Street Crash occurred at this time – but was also systemic. The California-based Southern Pacific Railroad was among the biggest corporations in the United States, with assets worth over two billion dollars in 1927, and became a leading operator of diesel-powered railcars by the mid-1920s thanks to its indomitable chairman Julius Kruttschnitt. But like the other US railways, it wanted the locomotive-builders to fund the research and development work. Further, Kruttschnitt's successor as chairman from the mid-1920s distrusted diesel power.[102] As for the two main locomotive-builders, Baldwin under Houston's leadership was cautious about diesel research, whereas Alco had a diesel programme in which Lipets was

---

[99]   In particular: Lomonosov–Millikan, 4 October 1929; Lomonosov, Memorandum, 23 October 1929; Memorandum No.2, 1 November 1929: Caltech folder 29.26, frames 724–6, 731–8. He also outlined his view to the American Society of Mechanical Engineers: *Transactions of the American Society of Mechanical Engineers*, OGP-51-5 (1929): 49–51.

[100]   On GM's engines and locomotives, and the parallel troubles of Baldwin, see Churella, *From Steam to Diesel*; Bingham, 'The Diesel Locomotive'; and J.P. Lamb, *Evolution of the American Diesel Locomotive* (Bloomington, 2007), pp. 1–66. Vauclain's interest is evident with his article 'Internal Combustion Locomotives and Vehicles', *Baldwin Locomotives*, 5/1 (July 1926): 43–9. Reviewing the contemporaneous work of the British firms Beardmore and Armstrong Whitworth, Rutherford argues similarly that 'it is just possible, had circumstances been a little different and investment capital more readily available, that British industry could have established a comprehensive worldwide diesel traction market before the Electromotive Division of General Motors began production in 1937': '"Export or Die!"', p. 53. His article 'When Britain was a Contender' makes the same point in greater detail.

[101]   On Caltech's role in aeronautics and jet propulsion see Goodstein, *Millikan's School*, pp. 153–77, 262.

[102]   D.L. Hofsommer, *The Southern Pacific, 1901–1985* (College Station, 1986), pp. 25, 90, 98, 115–20; Garmany, *Southern Pacific Dieselization*, especially pp. 7, 23–5, 34, 36, 40, 46–7, 53–89; P. Storey (SPRR)–Millikan, 27 February 1930: Caltech folder 29.26, frame 776.

involved – an impossible partnership for Lomonosov. After the latter's departure Millikan did contact Lipets, who showed interest, but Caltech abandoned the idea in 1931 because of the financial crisis on the US railways.[103]

The disappointments of this US visit persuaded the Lomonosovs that they had to decide once and for all whether to settle in Europe or North America, and contributed to their worst arguments since their 'peace' of December 1925. Typical of Lomonosov's perspective are the following remarks from a letter written in September 1930:

> You ask what I have decided about the USA. Risa, you must be a little fair here. I have long since lost the habit of deciding anything. Like a *spets* in the USSR, I merely coordinate, extricate myself, and devise schemes. In any case, how can I decide the question about America when the main question about our life is still in the air. I tried sincerely to discuss it fully in New Orleans, New York, Berlin, but nothing came of this. Well, I thought, conversations aren't working, so I will write. I waited until I was in a good mood in Karlsbad, and wrote: from all my heart and soul. Well, I got two replies, and I don't know which is the better one. These letters crushed me ... I have the impression that you are waiting for a miracle from somewhere. Alas, my dear Risa, miracles don't happen any more. We must somehow reach agreement ourselves.[104]

In practice, their decision was largely dictated by events outside their control. They considered Germany, Italy, Czechoslovakia and France, and deemed them all impracticable for various reasons, including fascism.[105] They were attracted to Britain because of its Labour government and especially George's intention to settle there, but they feared that the damp climate would harm Raisa. So by 10 October 1930, when Lomonosov and Raisa arrived in Britain to visit George, they had decided to return to the United States, where the employment prospects seemed best and they had many friends and contacts. However, their departure for New York was postponed because of major problems in George's recovery from a severe injury sustained in a motorcycle crash. By January 1931, still in Britain, they had to decide whether to extend their existing US visas for one final year.[106] They chose

---

[103]   'Dnevnik', 5 December 1929; Lipets–Millikan, 11, 22 April, 22 December 1930; Millikan–Lipets, 14 January 1931: Caltech folder 29.26, frames 790–791, 795–6.

[104]   Lomonosov–Lomonosova, [12 September 1930]: LRA, MS 716.4.1.89. Lomonosov had taken a cure at Karlsbad most years since 1921. The subjects of the couple's arguments were very varied, but particular favourites included Lomonosov's lack of employment, financial worries, George's future and the question of where to settle. Risa was Raisa's family nickname.

[105]   'Dnevnik', 6 January 1928, 31 August, 3, 10 September, 9 October 1930; 'Vospominaniia', vol. 10, pp. 1776–80, 1790, 1804, 1811–13, 1824–5, 1827–8, 1830.

[106]   'Dnevnik', 10–14 October 1930; Lomonosov–Lomonosova, 12, 14, 15, 16, 19, 23 January 1931: LRA, MS 716.4.1.96–99, 101, 104, 107; 'Vospominaniia', vol. 10, pp. 1831,

to stay in Britain, both because George's recovery would need many more months and because help with job-hunting was promised by Ewing, the emeritus professor of engineering at Cambridge. After spending a year in Cambridge, Lomonosov and Raisa moved to London. During 1932–1936 they had furnished rooms in Artillery Mansions, Westminster, after which they rented apartments on Lyndhurst Road in the affluent north London district of Hampstead, where they would stay until after the Second World War.[107]

Lomonosov used a variety of old and new tactics to try to relaunch his career in Britain during the 1930s. As well as relying on friends and existing contacts, he broke virtually the habit of a lifetime in his quest for new opportunities by becoming an active member of a major professional body, the Institution of Mechanical Engineers. For years he was a frequent speaker in the Institution's debates. Likewise, he attended the annual summer conferences of the British Association for the Advancement of Science.[108] He became acquainted, though not always on good terms, with many luminaries of British railway engineering, including the Chief Mechanical Engineers H.N. Gresley (LNER, 1923–1941) and W.A. Stanier (LMSR, 1932–1944).[109] Additionally, he gave priority to writing. In just four years he produced two papers for the Institution, several journal articles, an introductory textbook on railway mechanics that was published by Oxford University Press in 1933, and a large textbook called *Mechanics of a Train*. The Oxford book, which was derived from his Caltech lectures, argued the centrality of the inter-relationship between mechanics, design, operation and economics for a railway enterprise and its constituent parts. Above all, it made his formula for calculating operating costs accessible to an English-speaking audience.[110] His *Mechanics of a Train* – 'the report of my life's work' – was the first of several planned books that were to expand the themes of his Oxford book; also in mind were a book on diesel engines and another book called *Dynamics of the Rail and Wheel*.[111]

1860–1862, 1882.

[107] In particular: 'Dnevnik', 13 October 1931, 3, 13 August 1932, 22 February 1936, 23 September 1936, 15 March 1949.

[108] For example: 'Dnevnik', 2, 8 July, 22–30 September, 20, 28 November 1931, 7, 11 September 1935, 10, 12 September 1936; his many contributions to debates are recorded in the Institution's *Proceedings*, and are listed in the Bibliography. For reasons unknown he had little or no contact with the Institute of Locomotive Engineers.

[109] For example: 'Dnevnik', 2, 25 July 1931, 5, 7 September 1932, 21 January 1943. Recent biographies of Gresley and Stanier are: G. Hughes, *Sir Nigel Gresley: The Engineer and his Family* (Usk, 2001); J.E. Chacksfield, *Sir William Stanier: A New Biography* (Usk, 2001).

[110] G.V. Lomonossoff, *Introduction to Railway Mechanics* (Oxford, 1933). He was put in contact with Oxford University Press by an acquaintance, T.G. Crouther, probably because Oxford University Press had a reputation for engineering science: 'Dnevnik', 9 April, 15 May 1932.

[111] For example: 'Dnevnik', 10 January, 11 February 1932.

The professional recognition generated by these efforts was significant. His first paper for the Institution (about railway mechanics) was awarded a T. Bernard Hall Prize in 1932 as an original contribution to mechanical engineering; this award was not top-rank, in Lomonosov's view, but a 'good start'.[112] In 1933 the North-West branch of the Institution of Mechanical Engineers invited him to give a paper on diesel traction, which he treated as his first 'responsible' project in Britain; later he was asked to read it to the North-East branch too.[113] Meanwhile he was lauded in E.L. Diamond's seminal history of locomotive testing in *The Railway Gazette*; soon reprinted by the International Railway Congress Association, this publication did much to develop Lomonosov's international reputation for locomotive testing.[114] Additionally, his textbook attracted critical acclaim. Reviewers were impressed by the novelty of his approach, including his notion that each section of line was effectively a factory producing ton-miles, for which one could identify the cost of production.[115] For instance, *The Railway Engineer* welcomed the book as the first 'treatise dealing comprehensively with the railway as a single machine organized for the economical production of transport' and as a product of 'vast experience and a well-trained mind'.[116]

Yet prizes and praise did not mean acceptance: generally Britain remained an unreceptive environment for his thinking and skills. He felt hurt by the Institution's failure to invite his diesel paper to London: he felt that he had lost his 'last chance to get some sort of salary' in Britain.[117] His continuing marginality was evident in the fact that he associated primarily with younger members of the Institution such as Diamond and Dennis Carling, a future head of the Rugby test plant.[118] Ewing

---

[112]   G.V. Lomonossoff, 'Problems of Railway Mechanics', *Proceedings of the Institution of Mechanical Engineers*, 120 (1931): 643–59; 'Annual Report of the Council for the year 1931', ibid., 122 (1932): 163–4; 'Dnevnik', 30 January 1932.

[113]   G.V. Lomonossoff, 'Diesel Traction', *Proceedings of the Institution of Mechanical Engineers*, 125 (1933): 537–613; 'Dnevnik', 21 March, 8, 22 September, 2–5 October 1933.

[114]   Diamond, 'Horse-power of Locomotives', pp. 150–188.

[115]   *Engineering*, 137 (1934): 475; 'Introduction to Railway Mechanics', *Railway Gazette* (24 March 1933): 412; 'Introduction to Railway Mechanics', *The Railway Engineer*, 54 (April 1933): 123.

[116]   'Introduction to Railway Mechanics', *The Railway Engineer*, 54 (April 1933): 123.

[117]   'Dnevnik', 8 September, 16 November 1933.

[118]   For example: 'Dnevnik', 15, 20 November 1931, 16 November 1933, 8 June 1934, 3 December 1936; George Lomonossoff–D.R. Carling, 19 December 1952: LRA, MS 718.1.4.1. Having become a student member of the Mechanicals in 1920, Diamond became an associate member in 1927 and a full member in 1948, but died in 1961: K. Dike–author, 7 July 2008. The Institution's library has a small collection of papers (ref. PRP 3/12) concerning Diamond's research in the 1920s, on which see also Cox, *Speaking of Steam* (London, 1971), pp. 67–8. No obituary of Carling has been traced, but his books and an unsorted collection of his papers are kept at the library of the Institution. Fred Rich, a junior colleague at Rugby in the 1950s, has described Carling as 'a bespectacled, easy-going intellectual who personified Rugby Test Plant and all that it stood for in the world

dented his confidence and pride by refusing – for reasons not recorded – to write a preface for the textbook.[119] And the book's reviewers in Britain and North America expressed reservations that reflected rival approaches to railway engineering: the book was too mathematical for *The Railway Engineer*, and it upset *Engineering* by following the theory-based educational tradition of continental Europe, not the practical tradition of Britain and America.[120] Only 192 copies of the book were sold by August 1934: clearly it had not won popular endorsement as either textbook or handbook. Oxford therefore refused to publish *Mechanics of a Train* without a subsidy. Unable to comply, Lomonosov withdrew that manuscript in despair. It was never published, and its sequels were never completed.[121]

As in 1927–1928, Lomonosov's basic difficulty was the lack of a suitable place for an experienced outsider whose message and character were ill-aligned with prevailing local interests, attitudes and personalities, and who could not speak the language well. In relation to diesel power, for instance, Lomonosov still explained the paucity of British interest by reference to the local fuel and water situation. This point was relevant, but he harmed his cause by continuing to advocate research into other transmissions (including direct drive, which was now widely seen as impracticable) instead of supporting electric transmission as an easy, viable compromise. Evidently this stance alienated the head of diesel research at Armstrong Whitworth, C.J.H. Trutch, a strong proponent of electric transmission, possibly to the extent that he organized George's dismissal from his junior position in the company's diesel department.[122] Moreover, for reasons unknown, Lomonosov fell foul of a leading British expert on internal combustion engines, A.E.L. Chorlton, who served as President of the Mechanicals in 1933.[123]

---

of steam locomotive technology': 'You'll Go to Jail, Young Man', *Steam World*, 215 (May 2005): 8.

[119] 'Dnevnik', 3 July 1932.

[120] 'Introduction to Railway Mechanics', *The Railway Engineer*, 54 (April 1933): 123; *Engineering*, 137 (1934): 475.

[121] 'Dnevnik', 3 July 1932; 25 July, 3 August, 11, 30 September, 2, 23 November 1934; 5 February 1935. Oxford University Press retains the original contract and figures for the textbook's costs and sales. The print-run was 1,550, of which 48 were distributed as review copies, 317 were sold by 1943 and the remainder (1,185) were 'wasted' during 1940–1943. I am grateful to Martin Maw, Oxford University Press archivist, and Paul Dukes for this information.

[122] For example: 'Dnevnik', 28 June 1930, 12 April, 16 October, 17, 23 December 1931, 5 October 1933, 9 March 1934.

[123] For example: 'Dnevnik', 15 March 1932, 5, 28 October, 16 November 1933. On Chorlton (1874–1946) see, for example, F.O.L. Chorlton, 'Alan Ernest Leofric Chorlton, CBE', *Proceedings of the Institution of Mechanical Engineers*, 156 (1947): 245, and 'Obituary: Mr A.E.L. Chorlton, CBE', *Engineering*, 162 (1946/2): 354. The tension with Chorlton may have derived from the fact that Chorlton was the key man in diesel-engine development by the Beardmore company in the 1920s.

A similar situation still applied with locomotive testing. British engineers, led by Gresley and Stanier, the chief mechanical engineers of the LNER and LMSR companies, were becoming more convinced that locomotive performance should be studied systematically. But what did 'locomotive testing' mean? As noted above, Lomonosov's method of road research was considered impracticable for Britain's traffic and topographical conditions. Stanier and especially Gresley prioritized the construction of a national locomotive testing plant.[124] In any case there was controversy about how much precision was feasible and necessary. According to Lomonosov, a representative of the successful Manchester locomotive-building firm Beyer, Peacock declared at a meeting of the Mechanicals in 1933 that the only vital information about any steam locomotive was its haulage capacity and that he had never been interested in theory. That only a minority of the audience challenged this statement made Lomonosov suspect that he was wasting his energy in Britain.[125] Also, technical and personal feuds made their mark, as ever. For instance, Gresley was a forceful defender of steam power, and perhaps for that reason Lomonosov saw him by 1935 as one of his 'most evil ill-wishers'.[126]

The difficulty of finding a gainful niche in this environment is well illustrated by Lomonosov's attempt to work as a self-employed consulting engineer in association with his son and C.A.J. Elphinston, formerly one of George's colleagues at Armstrong Whitworth. Lomonosov regarded this project – a brainchild of Raisa – as profoundly uncongenial for various reasons. However, Lomonosov and his son needed an income, and Lomonosov also wanted to placate Raisa, who was complaining – unfairly, in his view – that she was having to look after him instead of doing her own work in literature and music.[127] The concept was to provide advice about the design, purchase, inspection and testing of diesel locomotives; to advise on railway design, especially for arid districts; and to analyse existing railways for improving their efficiency – issues which permeated Lomonosov's career. The perceived market was not Britain, where the railway network was complete and interest in diesels was small, but British colonies, where many lines were needed and diesels could be useful. Lomonosov thought that places like Palestine and Australia held the best prospects, but was worried because colonial railways normally hired UK consulting engineers only for inspection work. Hence he saw the first challenge as building contacts with colonial railway administrations.[128] The enterprise was launched in January 1935 in premises

---

[124]   On the lobbying by Gresley and Stanier for a test plant see, for example, Cox, *Speaking of Steam*, pp. 86–8, and Chacksfield, *Sir William Stanier*, p. 161.

[125]   'Dnevnik', 22 October 1933.

[126]   'Dnevnik', 22 February 1936. For a good insider's account of inter-war British locomotive testing see Cox, *Chronicles of Steam* (London, 1967), pp. 57–76 and his autobiography, *Locomotive Panorama* (London, 1965), vol. 1, pp. 20–131.

[127]   'Dnevnik', 14, 17, 20, 23 August, 18, 30 September, 24 October, 23, 29 November, 8, 21 December 1934; 3 January 1935.

[128]   'Dnevnik', 14, 20 August, 24 October 1934.

leased from the consulting engineers Hitchens, Jervis and Partners. However, it never covered its costs. A small, obscure office in London was unlikely to inspire interest among the colonial managers, and Lomonosov suspected that, like many managers that he had encountered, these people were inclined to treat ideas from outside their ranks as criticism to be resisted. The office's one major project was a grass-drying machine for Lord Forres, which proved unsatisfactory on test.[129] Lomonosov suggested closure as early as December 1935, and this finally occurred in 1937, though the affairs took several more years to wind up.[130]

In this light, it is difficult to judge how far Lomonosov harmed his prospects in Britain, and for that matter North America, by not mastering spoken English. This failure contrasts with, for instance, the success of his former IPS colleague Timoshenko, who made a distinguished career in American academe. Lomonosov did realize that he could not cope with the linguistic demands of a teaching position. Yet even if his English had been flawless, he would still have had to confront the lack of teaching opportunities in his fields and of empathy for his technical worldview. In the commercial arenas his poor English was more clearly a peripheral matter compared with that lack of empathy, though one might speculate that it strengthened the inclination to treat him as a foreigner with eccentric, inappropriate ideas. Interestingly, he made a belated effort to improve his English systematically in 1933 by starting a Berlitz course and watching feature films. But while cinema-going became a hobby for life, he quickly abandoned the classes. The absence of any real improvement is clear from an embarrassing incident later in 1933, when he misunderstood a request to join a prize-giving committee as an invitation to write a book.[131] One suspects that he found a pressing personal reason not to persevere. It seems that Raisa relished her scope to influence him by working as his translator and interpreter, and resented the language-learning efforts that he did make. Lomonosov was frequently annoyed by her interventions in his work, but perhaps elected not to improve his English in the interests of domestic harmony.[132]

It should be added that although Lomonosov attracted some professional attention from continental Europe in the 1930s, his prospects for employment there never became serious. In 1937, for instance, he was awarded honorary membership of the Scientific Institute of Transport in France, but did not become

---

[129]   For example: 'Dnevnik', 19 September, 12 November 1936; 29 April 1937.

[130]   For example: 'Dnevnik', 10, 22 December 1935; 3 January, 18 May 1936; 6 May, 26 July, 2, 7 November 1937; 2 February, 18 November 1938; 24 February, 29 March 1939.

[131]   Lomonosov–Lomonosova, 17, 22 November 1933: LRA, MS 716.4.1.171, 172; 'Dnevnik', 28 May, 11–12, 19 October, 17–18 November 1933. Such evidence from Lomonosov himself contradicts the assertion of an English family friend, Brian Reed, that Lomonosov had fluent spoken and written English: Reed, 'Lomonossoff', p. 47.

[132]   For example: 'Dnevnik', 13 September 1931, 28 May 1933. It should be added that he kept his distance from the Russian-speaking émigré community in London, rejecting their politics.

involved in its work.[133] Potentially more remunerative were three approaches about collaboration with locomotive-builders. One came from the Frichs company of Denmark in 1932, concerning experiments on a diesel-electric locomotive; another came from MAN in 1935; and the third came from the German firm Humboldt–Deutz in 1937. For various reasons, however, none of these projects progressed beyond preliminary talks.[134]

Lacking any regular salary, the Lomonosovs survived financially during the late 1920s and 1930s thanks to occasional income (including the $8,000 paid by MAN and Hohenzollern in 1928 and Raisa's earnings from translating some Russian literary texts into English), income from investments and a mixture of careful planning and economies. The investments, which were concentrated in Germany and the United States, were presumably made with savings and the 85,000 Swedish crowns that Lomonosov embezzled in 1922; it is also probable that the couple invested the proceeds from selling some diamonds that Lomonosov bought in Moscow in June 1920, and possible that he had illegally kept some money from a fund for publishing books about his experiments.[135] Be that as it may, the Lomonosovs held some RM100,000 (about £8,000) in Germany as of September 1934, with Farben shares accounting for at least 80 per cent of this figure. In the United States they had roughly $9,000 of 'free capital' in August 1932, which probably included some $6,000 invested in liberty bonds since about 1920, but not $5,000 loaned to an American friend in about 1929.[136] These dollars presumably constituted the balance of personal savings made in 1917–1918 and the $10,000 received from Moscow via Chadbourne in 1919, plus accrued interest. Thus, if the Lomonosovs were in a third country and needed money, they arranged an international money transfer. For instance, they received about $2,000 in Britain in 1928 by selling some US liberty bonds.[137] Conceivably, they could have lived modestly in this fashion for some years. But that scenario was precluded by side-effects of the Depression and Nazism, notably new restrictions on cross-border currency movements. The Lomonosovs did receive some of their dollars after the

---

[133]   Lomonosov–Lomonosova, 3 August 1937: LRA, MS 716.4.1.288.

[134]   'Dnevnik', 25 May, 4 June 1932; Lomonosov–Lomonosova, 16 July 1932: LRA, MS 716.4.1.130; 'Dnevnik', 26, 27, 31 August 1935; Lomonosov–Lomonosova, 7, 14 August 1935: LRA, MS 716.4.1.226, 231; 'Dnevnik', 7 August, 8 September 1937; Lomonosov–Lomonosova, 5, 13, 17 August, 6 September 1937: LRA, MS 716.4.1.289, 292, 294, 300.

[135]   The diamond purchase is mentioned in 'Vospominaniia', vol. 7, pp. 1260, 1268. As for the publication fund, he concealed its existence during an official inspection of his accounts in 1926: 'Vospominaniia', vol. 9, pp. 1875, 1977, 1980–1983.

[136]   'Dnevnik', 3 August 1932; 11 September 1934; 13 April 1948; Bank statements, 1935–1946: LRA, MS 716.7.16. The free-mark rate for sterling was about 12.55 in early September 1934: *The Economist*, CXIX/4750 (8 September 1934): 448, 465.

[137]   G. Abbott–Lomonosova, 6 January 1928: LRA, MS 717.2.5.14.

early 1930s, such as $1,400 in September 1935.[138] But they were badly were hit by German laws of 1933–1934 that prohibited the export of Reichmarks and curtailed payments of dividends to foreigners. With only about £600 banked in the UK in September 1933, they tightened their belts.[139] Nevertheless they had a large bank debt in Britain by 1935. By then they could withdraw RM2,000 each per month within Germany, but could not take this money to London.[140] Accordingly, Lomonosov spent several months in Germany each year during 1935–1939. He could live comfortably on his monthly allocation of Reichmarks and take a spa cure (usually at Bad Mergentheim) while simultaneously reducing the pressure on the family's British budget. Mostly he was alone in Germany, but Raisa and George made brief visits. For example, in September 1935 Raisa came in connection with a successful scheme to export RM5,000 to Britain in the form of household goods.[141] Old friends in Germany assisted as they could. For instance, the former Railway Mission accountant F.F. Perno gave financial advice, and Otto Hagemann of Krupp interceded with bank staff.[142]

The financial pressure receded in 1937 thanks to George's appointment as an inspection engineer with Rendel, Palmer and Tritton, a leading British engineering company.[143] He would remain with this firm until 1950 except for war service with the Royal Army Ordnance Corps during 1942–1946.[144] The fact that the company assigned him to tasks in Germany during 1937–1939 meant that he could easily withdraw money from the family's account at the Dresdner Bank. It thus seems likely that George lived off family Reichmarks in Germany, and gave his parents some of his sterling salary. The sums must have been modest, but they did allow Lomonosov to consider himself retired at the age of 61.

*   *   *

Lomonosov avoided the vicious purge of scientists and engineers that the Soviet regime unleashed in 1928, but otherwise his decade from 1927 was replete with

---

[138]   Lomonosova–Lomonosov, 19 September 1935: LRA, MS 717.2.1.229; see also George Lomonosov–parents, 15 March 1933: LRA, MS 718.1.1.408; Lomonosova–Lomonosov, 20 August 1937: LRA, MS 717.2.1.271; and A.G. Scattergood–Lomonosova, 31 January 1940: LRA, MS 717.2.269.4.

[139]   Lomonosova–Lomonosov, 28 September 1933: LRA, MS 717.2.1.140.

[140]   'Dnevnik', 25–26 August 1935; Lomonosova–Lomonosov, 22 July, 8 August 1935: LRA, MS 717.2.1.206, 214.

[141]   For example: 'Dnevnik', 28, 31 August, 18–19 October, 21 November 1935.

[142]   For example: 'Dnevnik', 25–26, 31 August 1935.

[143]   Lomonosov–Lomonosova, 19, 28 August 1937, 6, 7 September 1937: LRA, MS 716.4.1.295, 298, 300, 301. On this company's origins and its inter-war operations see M.R. Lane, *The Rendel Connection: A Dynasty of Engineers* (London: Quiller Press, 1989), especially pp. 89–119.

[144]   Aplin, *Catalogue*, pp. xxviii–xxix.

disappointment. Overall he appears more fortunate than prescient in staying abroad. Not only did he fail to anticipate the general purge, but he misread the links between politics and citizenship in the Soviet milieu. Almost incredibly, he believed – utterly wrongly – that he could resign from state service without implying any political disloyalty on his part. Indeed, he felt that he remained faithful to the political choice that he had made so publicly in 1918. Describing himself as a 'free Soviet citizen abroad', he kept his Soviet citizenship, avoided mixing with anti-Soviet Russian émigrés and, resident in Britain in the 1930s, associated with the extreme left-wing Independent Labour Party. He eventually took British citizenship only to avoid the real danger that the Soviet regime would make him stateless by refusing to revalidate his passport. His politics were the underlying reason why he failed to rebuild his career in Western Europe: had he not been so concerned to avoid upsetting Moscow in 1927, he could have accepted the lucrative German offers of consulting work. By the time that he realized the error of that judgement, he had ruined his chances of working in Germany and he faced serious obstacles elsewhere that included a lack of suitable jobs, different cultures in academic and railway research, personal conflicts and language difficulties. Never overcoming those barriers, he endured a long, demoralizing anti-climax to his high-flying career.

These experiences provide a useful reminder that politicized terminology like émigré, exile and non-returner must be used with great care. In particular, one should not simply assume that the politics of so-called non-returners were anti-Soviet. Did any more of these people think of themselves as 'free Soviet citizens abroad'? Clearly, too, official Soviet attitudes towards non-returners were more varied than has been thought. Despite vitriolic criticism of these people in the Soviet press and special laws to define them as traitors, the regime did not necessarily ostracize them. In fact, it even allowed them to maintain their Soviet citizenship by renewing their passport at Soviet consulates. How many people availed themselves of that system in addition to the staff of embassies, trade missions and so forth? Finally, the challenge of building a new life abroad was hard even for people like Lomonosov with an international professional reputation, excellent contacts and, at least initially, a solid financial base. Some coped successfully, but others did not. So in relation to the scientists and engineers of the former Russian empire who settled abroad in inter-war years we should not simply assume that the West benefited considerably from their resettlement: their expertise was not necessarily wanted.

Chapter 11

Retirement and Remembrance

In 1942 Lomonosov grumbled that 65 per cent of his time went on looking after their apartment and the family cat.[1] More seriously – for he was fond of the cat – he felt unsettled throughout his retirement. Nor was this just because of declining health and the Second World War. Homesickness for Russia persisted. Quarrels with Raisa were frequent and often bitter. He never fulfilled his dream of unpacking all his boxes of books and papers: their apartments were small and there was perpetual talk about leaving Britain. For example, the couple considered moves to the United States and France in 1939 and would have sailed to the United States in 1941 had berths been available.[2] The idea of a permanent move was inherent in the family's 1948 expedition to the United States, although at heart Lomonosov neither wanted nor expected to forsake Britain. In his last years, living in North America, he was distressed by Raisa's recriminations and pining for London, his books and papers, and the cat.[3]

Yet in many respects he was comparatively active and contented in retirement. Until his departure for the United States in 1948 he often spoke in the discussions at London meetings of the Institution of Mechanical Engineers, and he also attended functions of the Royal Institute of International Affairs after his election as a member in 1942.[4] His relationship with George improved, aided to some extent by a gradual acceptance of the latter's long-standing girlfriend Bessie; Lomonosov and Raisa recognized the couple's marriage in 1938 and were genuinely appalled by Bessie's death in a motorcycle crash in 1939.[5] Above all, liberated from job-hunting by George's steady employment from 1937, Lomonosov indulged his desire to write. Apart from continuing with his memoirs, he prepared a third paper for the Institution of Mechanical Engineers and another textbook on mechanics. True, the outbreak of war in 1939 and the resultant loss of the family's German investments caused more financial worries, but there was no crisis thanks to George: withdrawn from Germany by his British employer in August 1939, George stayed at Rendel,

---

[1] 'Dnevnik', 10 July 1942.

[2] Lomonosova–Lomonosov, 20 August 1937: LRA, MS 717.2.1.271; Lomonosov–Lomonosova, 18–26 May 1939: LRA, MS 716.4.1.319–29; Lomonosova–Lomonosov, 23, 24, 29 May 1939: LRA, MS 717.2.1.301, 303, 308; Lomonosov–Lomonosova, 1 September 1939: LRA, MS 716.4.1.353; Dnevnik, 7 August, 14 September 1940; 14 March 1941.

[3] In particular: 'Dnevnik', 21 December 1947; 7, 23 January, 6, 9 February 1948; 18 May 1950; 26 May 1951.

[4] His election to the RIIA is noted at 'Dnevnik', 5 March 1942.

[5] He noted Bessie's death in 'Dnevnik', 3 June 1939.

Palmer and Tritton with several promotions until 1950, except for war service with Britain's Royal Army Ordnance Corps during 1942–1946, when he earned the rank of Lieutenant-Colonel.[6] Additionally, Lomonosov did some occasional paid consulting work in the early post-war years, and the family gained unrestricted access to their dollar savings by travelling to the United States.[7]

Though not especially remunerative, his consulting projects during 1941–1948 merit some attention. The first provides a footnote to the history of the British–Soviet wartime alliance. On several occasions during 1938–1940 he offered his services to the British government for war work, not just in the hope of some extra earnings but also from a sense of patriotism for his adopted country.[8] Since these offers led nowhere, he was both surprised and delighted to be called to the War Office in July 1941, shortly after the Nazi invasion of the USSR. The Department of Military Communications requested him to compile a description of the Soviet railways, and then specified several other related tasks, possibly connected with Lend-Lease shipments of railway supplies to the USSR. Pleased with his new sense of purpose and with the new British–Soviet alliance, he embraced this challenge. Later, however, he was disappointed to hear that he could not undertake salaried war service because of his age, and could be paid only 'expenses' of £35 for all this work. A further small sum was probably paid in 1942 for some work on railway statistics.[9]

His other commissions encompassed an intriguing episode in the history of British railway diesel traction and an indication of how he might have developed a modest professional niche in the 1930s. In January 1943, at the recommendation of an acquaintance at the Institution of Mechanical Engineers, he was asked to prepare a survey of the literature about diesel traction for Oliver Bulleid, Chief Mechanical Engineer of the Southern Railway. He obliged, and was paid £75.[10] Bulleid's reaction to the paper is unknown, but one can speculate that it informed his work, several years later, on two 1,750 h.p. main-line diesel-electric locomotives, which were among the first such units in Britain and which influenced the design of several hundred locomotives for the nationalized British Railways in the mid-1950s.[11] Meanwhile, Lomonosov earned a certain reputation

---

[6]   Concerning George see Aplin, *Catalogue*, p. xxix.

[7]   For example: 'Dnevnik', 8 September 1937; 9 September, 11 October 1939; 27 October 1946, 2 April, 10 September 1947; 16 February 1948.

[8]   'Dnevnik', 27 September 1938; 25 August 1939; 4 June, 16, 23 August 1940.

[9]   'Dnevnik', 31 July, 16 August, 6, 25 September, 25 October, 3 December 1941; 3 January, 6 March, 29 August, 14, 21 September 1942. The Lend-Lease connection is mentioned in several obituaries but is not confirmed by other sources. See, for example, 'Built First Diesel Locomotive, Dr G.V. Lomonossoff Dies Here', *The Gazette* (Montreal), 21 November 1952, p. 27 (copy at LRA, MS 716.1.205).

[10]   'Dnevnik', 21 January, 8 March, 20 April, 8 September 1943.

[11]   According to S. Day-Lewis, Bulleid was a steam man with 'no overwhelming enthusiasm' for this diesel-locomotive project: *Bulleid: Last Giant of Steam* (London,

as an expert witness in lawsuits against the railway companies for fire-related damages. For a small fee and expenses he would prepare a report for the plaintiff and, if required, undergo formal cross-examination in court; how he coped with speaking in English here is unknown.[12]

One other project, though fruitless, deserves mention as a political postscript and distant echo of his Soviet career. In 1946 his ex-ILP friend John Aplin persuaded him that he had a 'party duty' to help Clement Atlee's Labour government with its project to nationalize transport services. Aplin organized a gathering on 11 January 1947 to introduce Lomonosov to Ian Mikardo, the left-wing Labour Member of Parliament for Reading. Lomonosov then wrote to Mikardo offering his railway expertise 'as a life-long socialist'. He was not seeking any 'position of responsibility', but believed that he could help train executive staff and advise on the technical challenges of post-war railway reconstruction. In particular, he warned with Bolshevik-style disdain for non-socialist technical specialists:

> The Government must reckon on opposition to the point of sabotage from conservative railway officials. Paradoxically, that opposition is most dangerous, not when openly advanced, but when officials appear to give full allegiance to the new regime. The answer, in my opinion, is an anglicized development of the Russian 'political commissar' system; this would work, I suggest, in the form of training selected junior grades of railways staffs and workers to become executives of a nationalized railway system. In my experience, six months' training of the right type of man is sufficient.[13]

Lomonosov had a point about potential resistance, for company loyalties were deeply ingrained among British railway managers. But he was ill-informed about the British situation. Those same loyalties were also entrenched among most other railway staff, with many older workers still cherishing the memory of the company that they had served before the Grouping of 1923. Generally the UK's railway situation bore little resemblance to Lomonosov's Soviet experience, and unsurprisingly his Soviet-style rhetoric did not persuade the British government to seek his help in preparing the 1947 Transport Act.[14]

In the mid-1940s Lomonosov became known among British engineers as an expert on condensing locomotives – a system to raise a steam locomotive's efficiency by reusing steam that was normally exhausted through the chimney. This fame arose from his one major technical publication during this decade: a paper co-authored with his son for the Institution of Mechanical Engineers, which

---

1964), p. 204.

[12]   In particular: 'Dnevnik', 3 June 1945; 27–28 February, 4 March, 12–13 November 1946; 14 January 1948; correspondence at LRA, MS 716.1.196–202.

[13]   Lomonosov–Mikardo, 16 January 1947: LRA, MS 716.1.195.

[14]   'Dnevnik', 21 December 1946; 11, 17 January 1947; Lomonosov–Mikardo, 16 January 1947: LRA, MS 716.1.195.

was also published in *The Engineer* and for which the Institution awarded its prestigious George Stephenson Prize.[15] Ironically, Lomonosov had little interest in this topic: he wrote the paper to aid George's application for associate membership of the Institution, and its critical success was mainly a function of his ability to analyse the relevant Soviet literature, which was important yet inaccessible to most British engineers as non-Russian speakers. He was annoyed, moreover, when *The Engineer* published criticisms from a Russian émigré named Pochebradskii. Having at first ignored them as unfounded, Lomonosov later concluded that they were basically correct, and vowed not to write any more technical works.[16]

Nonetheless, he did attempt one last major technical project. For much of 1947–1950 he concentrated on *Terrestrial Mechanics*, a university-level textbook. His text, of which only the first chapter survives, had an historical survey followed by chapters on basic conceptions, vectorial algebra and the mechanics of a particle, a system, a rigid body and a rigid body being deformed. However, the manuscript was rejected by John Wiley Ltd and Chicago University Press during 1950. Among the failings which Lomonosov himself identified were his book's closeness to Herbert Goldstein's textbook *Classical Mechanics* (first published in 1950 and widely used for decades), its inattention to the fundamental work in mechanics by the eighteenth-century Italian mathematician Joseph-Louis Lagrange, and linguistic faults. He spent much time revising it after he moved to Montreal with George in 1950, but abandoned it in March 1951 in favour of completing his memoirs.[17]

These memoirs were the dominant project of his retirement. He began writing them in August 1934 with the intention of describing his life until 1927; but having done that much by October 1943, he added a volume about his life abroad to 1930. He started revising his text at an early stage and was completing his revisions at the time of his death. He first put pen to paper in 1934 because Raisa had long been pressing him to do so, he had nothing else to do at that moment and, not least, he considered his tale important. Certainly he wanted to give his account of matters

---

[15]   G.V. Lomonossoff and G. Lomonossoff, 'Condensing Locomotives', *Proceedings of the Institution of Mechanical Engineers*, 152 (1945): 275–88; reprinted in 10 parts in *The Engineer*, 7 July to 22 September 1944; 'Annual Report of the Council for the Year 1944', *Proceedings of the Institution of Mechanical Engineers*, 152 (1945): 117. See also, for example, Ransome-Wallis, *Concise Encyclopaedia*, p. 500.

[16]   'Dnevnik', 21 September 1942; 21 June 1943; 7–9 January, 16 February, 19 May, 12 June, 12 September, 13 November 1944; *The Engineer* (26 May 1944): 408, 410–412; (9 June 1944): 447; (11 August 1944): 107–8; (1 September 1944): 166.

[17]   The manuscript remnant is at LRA, MS 716.5.2.1. See also, for example: 'Dnevnik', 1 December 1946; 17 April, 18 October, 21 December 1949; 6 May, 21 August, 10 September 1950; 3 January, 3 February, 7 March 1951; Lomonosov–Paul Epstein, 18 September 1949: Caltech Institute Archives, Papers of Paul Sophus Epstein, 1898–1966, Section I: Correspondence, Part A: General Correspondence, Box 5, Folder 44; Lomonosov–V. Yarros, 5 March 1951 (draft): LRA, MS 716.4.37.1; H. Goldstein, *Classical Mechanics*, 2nd edn (Reading, 1980).

that he thought were significant. His intended audience was Russian: he even sent a publication proposal to Moscow in 1935, which was duly rejected. However, he also rewrote the first parts for translation into English and evaluation by Western publishers. Yet he did not want publication at any cost. Unwilling to cut the length drastically, he declined Secker & Warburg's offer of a contract for 120,000 words in 1938. Also, he refused on principle to insert an anti-Soviet tone even if, as he believed, that refusal made his work unpublishable in the West. For him the project's value lay in its honesty as well as its story and exhaustive detail, and in that sense he treated the text as his most important legacy for George.[18]

When Rendel, Palmer and Tritton sent George to an assignment in the United States in 1948, Lomonosov accompanied his son. His main aim was to reorganize the family's US financial affairs with Alfred Scattergood, their banker in Philadelphia, but the idea of settling in the United States was also present: Raisa, who joined them after a few months, was very keen. Lomonosov hoped vaguely that he could do some consulting about diesel traction, but really he did not want to move again.[19] It was more by chance than design that he spent his last years in Canada. In spring 1950 they were expecting to return to Britain, Raisa by air and Lomonosov by sea with George and his new Canadian wife Peggy. In May, however, George accepted a job at the Montreal Locomotive Works. Because Lomonosov could no longer travel alone, it was agreed that he would spend the summer in Montreal while Raisa went to London to 'liquidate' their rented flat, pending a final decision in the autumn about where to live. One week later, Raisa announced that she planned to stay in Britain. Irritated by this unilateral announcement, and assuming from experience that soon she would change her mind, Lomonosov went to Montreal as planned. Raisa, however, did not change her decision.[20] Their correspondence from this period has not survived, but his diary indicates that it quickly became recriminatory. For Lomonosov this situation was desperate: he felt unwanted in London and a hindrance in Montreal. He hoped to spend July and August 1952 in London with Raisa while George and Peggy holidayed in Europe, but Raisa's reaction was so negative that he went instead to a favourite spot at Lake Echo, some 40 miles from Montreal.[21]

---

[18] In particular: 'Dnevnik', 2 September, 7 December 1934; 30 October, 3 November 1935; 24 October 1943; 23 January, 8 February 1944; 9 February, 22 June 1948; 16 March 1949; 3 February 1951; Lomonosova–Lomonosov, 25 June, 4, 15 July 1935: LRA, MS 717.2.1.191, 196, 206; correspondence with Secker & Warburg, April–June 1938: LRA, MS 716.4.35. English translations and synopses of parts for 1876–1901 are at LRA, MS 716.2.2.1–8.

[19] 'Dnevnik', 7, 23 January 1948; Lomonosov–Lomonosova, 20–21, 26–7 February 1948: LRA, MS 716.4.1.376, 378.

[20] 'Dnevnik', 21, 23 March, 11–14, 23, 31 May, 1–2, 4, 9–11 June 1950.

[21] For example: 'Dnevnik', 20 August, 2, 10 October, 18 December 1950; 2 January, 2 February, 26 May, 4–7 June, 15–17 August 1951; 18 February, 26 March, 26 May 1952.

His health declined gradually. As early as 1941, in his mid-sixties, he found his War Office work draining. He was quite ill for several months after suffering a stroke in February 1949, and was generally unwell through 1951.[22] But he felt much better during the summer and early autumn of 1952, such that George described him as looking quite fit at this time. He continued to revise his memoirs and make his weekly cinema visit to watch, as he liked to joke, the 'dancing girls'. He delighted in his grandson Nicholas, born that October. But he had a heart attack on 17 November, and died at Montreal General Hospital two days later. George wrote afterwards that the death was 'a great shock to us all'.[23]

The funeral, held in Montreal on 22 November 1952, became a professional as well as family event. Raisa could not travel from London at such short notice, but there were delegations from the Canadian Pacific Railway and the Montreal Locomotive Works, and flowers were sent by two of the major companies with which he had worked during 1917–1922 – the Canadian Car and Foundry Company and the Inspection Department of Alco's works at Schenectady, New York.[24]

Peripatetic to the last, Lomonosov did not reach his final resting place for another 40 years. His ashes and possessions were shipped to Raisa in London shortly after George's death from pancreatitis in January 1954.[25] Peggy then moved to Britain to raise her two sons with Raisa's help, and inherited the remaining possessions of both Lomonosov and his wife upon Raisa's death in March 1973. In 1982 she deposited most of the papers, photographs and books of Lomonosov, Raisa and George in the Brotherton Library, University of Leeds, where they became foundation collections of the Leeds Russian Archive.[26] Eventually, on 29 June 1992, Lomonosov's ashes were strewn at Cambridge City Crematorium, left of Hewitson's fir tree, where Raisa's ashes had been scattered 19 years earlier.[27]

*   *   *

---

[22]   For example: 'Dnevnik', 1 August 1941, 6 March, 2 June 1949, 1 March 1951.

[23]   For example: 'Dnevnik', 27 June, 30 August 1952; interview with G.M. Browning, 25–6 September 1995; George Lomonossoff–Lomonosova, 20 November 1952: LRA, MS 718.1.1.782; George Lomonossoff–O.R. Henschel, 31 May 1953: LRA, MS 718.1.7.4.

[24]   George Lomonossoff–Carling, 19 December 1952: LRA, MS 718.1.4.1; list of floral tributes (in family papers); George Lomonossoff–Lomonosova, 29 December 1952: LRA, MS 718.1.1.783. See also 'Built First Diesel Locomotive', *The Gazette*, 21 November 1952, p. 27.

[25]   Certificate of cremation, 25 November 1952: LRA, MS 716.7.32; Intermediate Shipping Company–Lomonosova, 7 April 1954: LRA, MS 717.2.140.

[26]   Aplin, *Catalogue*, p. xxxi; not deposited were six albums of mainly family photographs and a very few papers, notably funeral papers and obituaries.

[27]   Interview with G.M. Browning, 6 April 2002; J. Prew et al. (Cambridge City Crematorium)–author, 2, 23–4 July 2007.

Different ways of remembering Lomonosov surfaced within his family immediately after his death. In a letter to her son, Raisa called her husband a 'traditional' Russian. What she meant by this phrase is unknown, but George firmly rejected it, noting that his father had been 'a revolutionary and a colleague and admirer of Lenin'. George also condemned his mother's reaction to the death, using Lomonosov's family nickname: 'your bitterness during the last 2½ years, and your terrible remorse now that it is too late, are both pointless. What our dear Monster would have liked and what I still advise is moderation and a more rational attitude'.[28] Writing separately to his friend Dennis Carling, now at the British Railways Locomotive Testing Plant at Rugby, George agreed that his father's greatest achievement was the development of 'scientific methods' of testing locomotives, but he rather harshly dismissed the diesel work as 'merely the application of test data to certain conditions'.[29] As for Raisa's suggestion to Carling that a memorial service should be organized, George noted that his father had not wanted such an event – a preference that Carling respected.[30]

In the wider world the Soviet media ignored his death, as was their custom for 'non-returners', but obituaries did appear elsewhere. Technical periodicals provided the most substantial accounts, including one by Meineke. Largely descriptive and sympathetic, these tended to focus on his career, but could also include references to his revolutionary experiences. The longest article, in *The Engineer*, was evidently written by a staff member who knew him quite well and remembered him warmly; it concluded by recalling his extensive technical knowledge, careful attention to detail and genial personality.[31] Likewise Carling, in a letter to the *Railway Gazette*, not only stressed Lomonosov's professional eminence but also remembered him as 'a friend and a delightful host who gave us the benefit of his wide experience in many spheres of life'.[32] The majority of the newspaper and magazine reports and obituaries were in Canadian and US publications, and the tribute that Carling submitted to *The Times* of London was not published.[33] Generally short, factual and garbled, the published articles tended to stress his seniority on the Russian railways and his diesel work. This slant reflects the newspaper accounts of Lomonosov's last public appearance in May

---

[28]   George Lomonossoff–Lomonosova, 29 December 1952: LRA, MS 718.1.1.783.

[29]   George Lomonossoff–Carling, 19 December 1952: LRA, MS 718.1.4.1.

[30]   Carling–Lomonosova, 7 January 1953: LRA, MS 717.2.57.7.

[31]   'Professor G.V. Lomonossoff', *Railway Gazette*, 97/22 (28 November 1952): 594–5; 'Obituary: Dr G.V. Lomonossoff', *The Engineer* (5 December 1952): 763; 'Obituary: Dr G.V. Lomonossoff', *Engineering*, 174 (1952/2): 733; F. Meineke, 'Professor Lomonossff †', *Glasers Annalen* (January 1953): 24.

[32]   D.R. Carling, letter to the Editor, *Railway Gazette* (5 December 1952): 622.

[33]   See, in particular, 'Built First Diesel Locomotive, Dr G.V. Lomonossoff Dies Here', *The Gazette* (Montreal), 21 November 1952, p. 27; 'G.V. Lomonossoff, Engineer, was 76', *New York Times*, 21 November 1952, p. 25:4 and Carling–Lomonosova, 1 January 1953: LRS MS 717.2.57.6. No copy of Carling's text has been found.

1952, which was arranged by N.R. Crump, an admirer of his diesel work and a vice-president of the Canadian Pacific Railway.[34] To judge by their vocabulary and some confusion about his duties, those earlier reports were probably based on notes by Lomonosov himself that had some loose translations and summaries. Be that as it may, he was not responsible for the inability of reporters to distinguish between his son George, grandson Nicholas and first son Vsevolod, who was now a professor of electrical engineering in Moscow. Always cynical about the Western press after 1917, Lomonosov would probably have been amused and not surprised to read in Montreal's *The Gazette* and *The New York Times* that the infant Nicholas was a second son by a first marriage living in Moscow.[35]

In the West Lomonosov's work on locomotive testing and railway economics retained a modicum of influence for some years after his death. In particular, it helped to shape projects on two regions of the state-owned British Railways during the 1950s and 1960s. On the Western Region S.O. Ell regarded his own technique of Controlled Road Testing as a descendant of the testing done in Russia by Lomonosov and in inter-war Poland by one of Lomonosov's former colleagues, Chechott, who returned to his homeland after 1917.[36] Also, Ell was intrigued by Lomonosov's formula for costing traffic operations, and used it to help cost proposals for dieselization. He described this work to the Institute of Locomotive Engineers in 1958, and although criticisms were made of the formula, the minutes of the discussions also note warm praise from the institute's immediate past president, E.S. Cox: 'The Lomonossoff formula given so prominently in Fig. 14 of the paper had been available for 25 years, but it had to wait for the present Author as its true interpreter. Mr Cox felt that Fig. 14 should be enlarged to poster size and pinned up on the office walls of everybody who had anything to do with train

---

[34]   See 'Vice-President Meets Scientist Who Designed and Built World's First Diesel-Electric Locomotive', *The Spanner* (July 1952): 8 and, for example, 'Builder of First Diesel Meets "Student" Here', *The Gazette* (Montreal), 31 May 1952, p. 7.

[35]   Vsevolod published a textbook in electrical engineering, called *Elektrotekhnika*, in 1947, which was republished in many editions until 1990, long after his death in October 1962. An obituary appeared as 'Vsevolod Iu'evich Lomonosov', *Elektrichestvo*, 12 (1962): 88; his father was not mentioned. Almost certainly Vsevolod had no contact with his relatives abroad after the mid-1930s because of the political conjuncture. The fates of his mother Sonia and Ania Pegelau are unknown. Masha never married and died in Dergachi near Khar'kov in 1943, while her sons Iurii (Zhorzh) and Anatolii died in 1978 and 1988 respectively: information from N.S. Skachkova.

[36]   S.O. Ell, 'Developments in Locomotive Testing', *Journal of the Institute of Locomotive Engineers*, 43 (1953): 561–91 and discussions 591–633; Ell, 'Testing of Locomotives', pp. 404–5; Rutherford, 'Quantification and Railway Development', pp. 44–6; 'Measurement not Mystification', pp. 443–4. Chechott described his technique in A. Czeczott, 'A Description of the Method of Carrying Out Locomotive Tests on the Polish State Railways', *Monthly Bulletin of the International Railway Congress Association*, XII/7 (1931): 575–605.

operation or train timing.'[37] That suggestion was not followed as such, but Ell did present the formula to a very wide audience in 1959 by including it in his chapter on locomotive testing in P. Ransome-Wallis's important *Concise Encyclopaedia of World Railway Locomotives*. Moreover, building on the work of Lomonosov among others, Ell's team developed a 'Cost of Energy Diagram' to show how much it would cost for a given locomotive to move a given load at a given speed; and devised a manual called a Passenger Train Computer to enable non-engineering staff to draft timetables based on experimentally proven and economically sound data – an approach that is now common practice in many countries.[38] Finally, reference was made to Lomonosov in a project on the North Eastern Region to raise the efficiency of freight train operation in the 1960s; again, the aim was to apply a scientific approach to railway operation so as to reduce operating costs.[39]

Even in his homeland, where he was *persona non grata* for so many decades, ramifications of his work continued – and in some important respects continue – to be felt long after his death. As noted above, the technical legacy of the inter-war Soviet research into diesel traction was only marginal for Soviet railway dieselization from the 1940s. But many of the steam locomotives that Lomonosov ordered in the United States and Europe during 1917 and 1920–1921 were still working in the 1960s and the final withdrawals occurred as late as the 1980s, admittedly after some years of storage.[40] The system of locomotive classification and numbering that he introduced in 1912 is still used in much of the former USSR despite attempts by the Soviet railways to replace it in the 1920s and 1980s. Likewise railways of the former Soviet republics still use the system of technical passports that was devised with his experimental work before the First World War. Above all, the Soviet railways decided after Stalin's death to revive the tradition of scientific locomotive testing that Lomonosov had helped to found, and that tradition is being maintained in post-Soviet Russia, principally by the Moscow-based All-Russian Scientific Research Institute for Railway Transport.[41]

---

[37] S.O. Ell, 'The Mechanics of the Train in the Service of Railway Operation', *Journal of the Institute of Locomotive Engineers*, 48 (1958/1959): 528–61 and discussions pp. 561–90, especially p. 563. Clearly Cox was referring to the formula's publication in Lomonosov's *Introduction to Railway Mechanics* (1933), not to its first appearance in print in tsarist Russia.

[38] Ell, 'Testing of Locomotives', p. 410; Rutherford, 'Measurement not Mystification', p. 444.

[39] See Heywood, 'Lomonosov and the Science of Locomotive Testing', p. 283 and D.S. Hellewell, 'Lomonosov and British Railways', unpublished research note, 30 January 2001. For this information I am indebted to Mr Hellewell, a participant in the project.

[40] An example is Nohab's No. 4444 of 1924, which had its last boiler inspection in 1982: Technical Passport of Steam Locomotive E^{sh} No. 4444, Class E^{sh} (courtesy of the former Ministry of Ways of Communication of the Russian Federation).

[41] For a rare example of Soviet-era recognition of Lomonosov's contribution to this tradition see N.A. Fufrianskii and A.N. Dolganov, *Opytnoe kol'tso Vsesoiuznogo Nauchno-*

Lomonosov did not record his attitude to the prospect of a tangible memorial after his death, notwithstanding Krestinskii's ironic suggestion for a statue. At the time of writing no statue exists, nor are there commemorative plaques on his places of residence. A major part of his career was memorialized – in a manner of speaking – through the use of the Nohab locomotive contract as the backdrop for a Soviet–Swedish romantic feature film in the 1960s, although this film did not name him because he was still *non grata* in the USSR.[42] Unfortunately, his diesel locomotives were scrapped after their withdrawal from service, in contrast to the less successful Gakkel' locomotive, which is preserved in St Petersburg. On the other hand, about 10 of the foreign-built steam locomotives are preserved in the former USSR, including an Alco-built decapod from 1917 at the Central Museum of Railway Transport in St Petersburg and one of the Nohab locomotives at Ladozhskoe Ozero station, where it honours the role of the railways in relieving the German siege of Leningrad during 1941–1944.[43] Also, six of the decapods stranded in the United States after the October revolution survive in US museums.[44] Finally, his name has been revived at the erstwhile Institute of Ways of Communication, now the St Petersburg State University of Ways of Communication: his portrait hangs with the many other portraits of distinguished former professors, and the library has produced a short bibliography.[45]

*   *   *

Lomonosov's international reputation as a leading expert in locomotive engineering and railway economics was well earned, but the evidence now available also casts some long shadows over his achievements. Key aspects of his work were recognized and to some extent assimilated abroad as well as in Russia; and the eminence that he eventually enjoyed in Europe and North America was primarily a product of his successful diesel locomotive, his technical papers and his 1933 book on railway mechanics. Yet his professional life, including all his main areas of accomplishment, was also replete with disappointments – setbacks, missed opportunities, debilitating rivalries and so forth, culminating with his demoralizing failure to obtain salaried work abroad after leaving Soviet service in 1927.

This ambiguity pervaded even his much-vaunted work on high-power main-line diesel traction. The successful public debut of his diesel-electric locomotive

---

Issledovatel'skogo Instituta Zheleznodorozhnogo Transporta, 2nd edn (Moscow, 1977), pp. 8–10.

[42]   This venture was the subject of Iu. Iakhontov, 'Tysiacha parovozov dlia Lenina', *Pravda*, 27 January 1969, p. 5.

[43]   Heywood and Button, *Soviet Locomotive Types*, p. 135.

[44]   For the later history of the stranded decapods see W.D. Edson, 'The Russian Decapods', *Railway and Locomotive Historical Society Bulletin*, No. 124 (1971): 64–75; Heywood, *Modernising Lenin's Russia*, pp. 85–92.

[45]   *Iurii Vladimirovich Lomonosov (1876–1952): Biobibliografiia* (St Petersburg, 1993).

Iu^E No. 001 in November 1924 and its generally good performance in subsequent road operations were very much Lomonosov's triumphs: this locomotive and its somewhat temperamental diesel-mechanical sibling Iu^M No. 005 would never have been built without his vision, resourcefulness and tenacity. The success of the initial trials with Iu^E No. 001 did much to persuade railway engineers worldwide that this technology could be developed for regular main-line railway use, which gives this machine a strong claim for recognition as the world's first operationally successful main-line diesel locomotive. Yet ironically these successes coincided with the messy, premature end of Lomonosov's career on the Russian and Soviet railways. We see now that for him this diesel work was increasingly a means to stay outside Russia without quitting Soviet service and that ultimately, essentially for political reasons, he could not persuade Moscow to extend his assignment beyond the winter of 1926–1927. Thereafter, for a variety of reasons, he never secured a salaried job outside Russia despite his impressive record in this emerging field. At Caltech, for example, he proposed a research strategy that anticipated much of the decisive work by General Motors in the mid-1930s, but this bid for paid employment soon ended in mutual recriminations. The Soviet authorities, meanwhile, squandered their opportunity to build on the strong technical foundation that he had helped to provide.

His contribution to the development of locomotive testing was significant, but his testing method was never as important for the Russian and Soviet railways as foreign engineers later assumed. He devised his method as a compromise response to the specific technical, financial, bureaucratic and topographical circumstances in which he was working during 1898–1901. But it was never fully accepted even in Russia and was not copied as such elsewhere, not least because of its need for long, constant gradients and its capacity to disrupt commercial traffic. Indeed, his fiercest critics objected that even Russia did not have the requisite gradients. Eventually, the NKPS abandoned his type of test in the politically inspired campaign of the 1930s against 'bourgeois' railway science. In reality, then, Diamond's description of the Lomonosov method as 'the Russian method' was an exaggeration. Overall, Lomonosov's contribution was important mainly for his emphasis on system and comparability for road tests; for helping to develop a tradition of rigorous enquiry in locomotive road testing in Russia that could be revived after the excesses of the Stalin era; and for creating in that context an intellectual basis for developments elsewhere by the likes of Chechott in Poland and Ell in Great Britain. Sadly for Lomonosov, but perhaps unsurprisingly, his great ambition to develop a theory of the locomotive remained unfulfilled.

The long-term legacy of his work on railway economics was modest. Having drawn him to locomotive engineering as a student, this topic yielded his special formula for calculating railway operating costs. But again his work in this field was always controversial in Russia, and it took the war emergency of 1914 to persuade the top-level operating staff at the MPS to view it as potentially an important means to help reduce costs and expand traffic capacity. The subsequent Soviet denigration of Lomonosov's economic thinking in the late 1920s and

1930s was purely political. More surprising is the cool reception of his work as summarized for the English-speaking world in his 1933 textbook. Despite broadly favourable reviews the dismal sales figures confirm that private railway ownership in Britain did not imply a much greater emphasis on efficiency than with state ownership in Russia. True, his formula would strike a chord with certain British engineers in the late 1950s, but they appear to have been a small minority. Given that the nationalized British Railways eventually undertook radical cost-cutting in the 1960s, one should perhaps conclude that in this area Lomonosov was ahead of his time.

In bureaucratic terms Lomonosov's career up to the 1920s seemed spectacularly successful. He reached the upper ranks of the MPS while still in his thirties, he became by far the youngest member of the MPS Engineering Council in 1913, and he was seriously considered for appointment as People's Commissar at the NKPS in 1920 despite not being a member of the Bolshevik Party. However, the details behind these headlines suggest a different conclusion. Initially he wanted an academic career, and was soon well set on that course, but those hopes seemed dead as early as 1907, when he resigned from the Kiev institute to rejoin the railways; his part-time job at the IPS from 1911 was only ever a marginal part of his work, while his commitment to rejoin the Kiev institute in the 1920s was a product of despair about his situation within the NKPS. His meteoric rise through the ranks of railway management from 1908 is only partly explained by managerial achievement: also crucial were the high level at which he rejoined this hierarchy, his lobbying and especially his apparent inability to co-exist with his immediate superiors. His moves to Orenburg and St Petersburg were clearly promotions, but his 1912 transfer to the MPS was made against the minister's wishes and his move to the Engineering Council in 1913 sidelined him from the managerial career path. The wartime revival of his managerial career was somewhat marred by his 'disgrace' during 1916. After the February revolution the possibility of going to North America was, for him, mainly an attractive means to escape from the MPS and his 'enemies' there. The same was true of his assignment to Western Europe by Lenin's government in June 1920, following the resounding failure of his bid to become People's Commissar at the NKPS. From that point he became a marginal figure in the senior ranks of the Soviet railway bureaucracy, and of course he had no formal salaried position whatsoever after leaving Soviet service at the age of just 50.

Much of Lomonosov's professional activity was routine administrative work of the type that is rarely noticed, often ephemeral and yet essential for the railway industry's continued functioning. His efforts to display initiative met with mixed results, a good example being his cost-cutting measures on the Tashkent Railway: they helped him win promotion, but also disrupted MPS norms and calculations. Overall, we can suggest that the Russian and Soviet railways gained some substantial long-term benefits from three aspects of his administrative work. His locomotive classification system, introduced in 1912, is still used, with only slight modifications, in much of the ex-USSR. Many of the locomotives and wagons that were ordered and delivered on his watch in North America during 1917–1918 gave years of useful service, and the same applied for the 1,200 Class E 0-10-0 locomotives imported under his supervision from Sweden and Germany in the 1920s. Critics have complained that the latter locomotives were a waste of scarce Soviet gold, but if, as was the case, the Soviet government was determined to use gold to purchase machinery abroad, these engines surely represented a good deal. This emphasis on buying railway equipment helped to build much-needed international credibility for the Soviet government as a trading partner because it matched the foreign assessment that to repair the transport system was a top priority; and the locomotives themselves ensured that the physical condition of the Soviet railways was far better in the mid-1920s than would otherwise have been the case.

As with any Russian in the revolutionary epoch, a crucial question with Lomonosov is how far he made his own fate. Clearly his life was shaped by circumstances beyond his control on many important occasions. For example, as a youth he had to find a career because of his family's lack of money. Much of his frustration at Kiev stemmed from the bureaucratic barriers between his institute and the railways and from the severe disruption caused to academic life by the state's financial woes and the revolutionary upsurge of 1904–1907; his career and life might have taken a rather different direction had the proposed national locomotive testing plant been built at his institute, for he would have had a strong incentive to remain there. The outbreak of war in 1914 terminated a cosy routine, while the October revolution three years later nullified much of his wartime work in both Russia and North America. Crucially, the Bolshevik revolution forced him to define his politics publicly, with far-reaching consequences for not just his career but also his marriage. His Soviet career was overshadowed by the way that his social and professional background defined him in the regime's terminology as a 'bourgeois specialist', despite his rejection of that label for describing himself. Indeed, although he did not immediately see the matter in such terms, he was effectively forced to live abroad for his own safety from the later 1920s, and even there he could not escape from stereotypical labelling.

Yet by no means was he simply a victim of the chronic social, political and economic turmoil of revolutionary Russia. Even in old age Lomonosov blamed himself for causing much of his trouble with Moscow in the 1920s; he thought that his responsibility for the Novorossiisk crash in May 1920 was especially damaging. And although he almost certainly overestimated the personal impact of that crash, his life and career were unquestionably shaped decisively by his own choices and actions, influenced time and again by his complex character, ambition for power, prejudices, naivety, personal life, politics, rivalries and even misunderstandings. He chose, for example, to have affairs and illegitimate children, to engage in terrorism in 1905, to confront superiors like Shtukenberg and Bakhmetev and seek high office, to go to America in 1917 and Soviet Russia in 1919, and to enjoy an extravagant lifestyle abroad from 1920 that was bound to antagonize Bolshevik Party activists at a time of mass poverty and famine in Russia. Equally, he chose not to join the Bolshevik Party, not to go to Moscow in 1926–1927 and, crucially for the final third of his life, to reject the German offers of consulting work in 1928. Paradoxically the combination of ruthless ambition and belligerence, which colours so much of his professional life, explains not only much of his success but also to a large extent why he became an outsider in both the tsarist milieu and Soviet milieu despite having, in differing ways and for differing reasons, a big stake in each of these societies.

Accordingly, circumstances, his character and his choices are all fundamental for understanding the patterns and balance of continuity and change between the tsarist and Soviet eras in his life and career. The October revolution forced him to side with either ambassador Bakhmetev or the Soviet regime. Having opted for the Soviet side, he found a strong element of continuity in his work. During the first decade of the Soviet era the tasks of running and developing the railway system were substantially unchanged from the last pre-revolutionary years, albeit with the added complication of resolving wartime neglect, damage and destruction. The railway bureaucracy continued to function in much the same way as before 1917, many of the senior people of the tsarist MPS remained influential, and Lomonosov even spent three years on procurement work that echoed his role in implementing the tsarist wartime policy to import railway equipment. However, the Bolshevik revolution did bring rapid, significant and enduring change to his personal life and even his personality. His public call for foreign recognition of the Soviet system led to a marital crisis and a crisis of self-confidence that seem to have rendered him much less decisive in both professional and private matters. Conceivably these crises redirected his marriage to the course that produced his de facto separation from Raisa in 1950.

This intensely personal note reminds us that, for all the important continuities that historians now identify between the tsarist and Soviet eras, the October revolution rapidly brought profound change to the individual lives of most of the inhabitants of the ex-tsarist empire, not least through the ensuing nightmare of civil war. Lomonosov's life was eventful and in many respects unusual, but ultimately his fate after 1917 was typical for a professional of late tsarist Russia: there was no long-term place for him in revolutionary Russia despite his education, expertise and even, paradoxically, his genuine support for the Soviet system. The upshot for so many of those people, including many of his friends and acquaintances, was imprisonment and perhaps execution; Lomonosov, by contrast, was among those fortunate to die peacefully in foreign exile.

Select Bibliography

Unpublished sources

Archival collections

Arkhiv vneshnei politiki Rossiiskoi Federatsii, Moscow
Fond 04   Sekretariat narkoma inostrannykh del SSSR G.V. Chicherina
Fond 082   Referentura po Germanii
Fond 140   Referentura po Shvetsii

Bakhmeteff Archive, Columbia University, New York
B.A. Bakhmeteff Collection

Berne State Archives, Berne
Berne University collection

California Institute of Technology, Pasadena, CA
Papers of Paul Sophus Epstein, 1898–1966
Papers of Robert A. Millikan, 1897–1953

Cambridge University Library
Vickers Archive

Derzhavnii arkhiv Dnipropetrovs'koi oblasti, Dnipropetrovsk
Fond 566   Upravlenie Ekaterinenskoi zheleznoi dorogi, 1897–1919

Derzhavnii arkhiv m. Kyieva, Kyiv (DAK)
Fond 18   Kievskii Politekhnicheskii Institut, 1898–1918
Fond 163   Kievskaia gorodskaia uprava

Derzhavnii arkhiv Kyivs'koi oblasti, Kyiv (DAKO)
Fond 1   Kievskoe gubernskoe pravlenie
Fond 2   Kantseliariia Kievskogo gubernatora
Fond 864   Kievskii okruzhnoi sud

Deutsches Museum, Archiv, Munich
FA Maschinenfarbrik Esslingen

Dom-muzei P.L. Kapitsy pri Institute Fizicheskikh Problem imeni P.L. Kapitsy, Moscow
Lichnyi fond P.L. Kapitsy

Friedrich Krupp GmbH, Historisches Archiv, Essen
Werksarchiv IV 1361 Niederschrift von Hagemann: Russisches Lokomotivgeschäft, 1920–1937

Gosudarstvennyi arkhiv Orenburgskoi oblasti, Orenburg (GAOO)
Fond 142   Upravlenie Tashkentskoi zheleznoi dorogi
Fond r-316   Upravlenie Tashkentskoi zheleznoi dorogi

Gosudarstvennyi arkhiv Rossiiskoi Federatsii, Moscow (GARF)
Fond 102   Departament politsii Ministerstva vnutrennykh del
Fond 518   Soiuz soiuzov
Fond 601   Nikolai II, Imperator
Fond 640   Aleksandra Fedorovna, Imperatritsa
Fond 1834   Kollektsiia pechatnykh legal'nykh izdanii (listovki, broshiury, ob"iavleniia, instruktsii), otlozhivshikhsia v politseiskikh i sudebnykh organakh dorevoliutsionnoi Rossii
Fond r-130   Sovet narodnykh komissarov RSFSR
Fond r-5674   Sovet truda i oborony pri Sovete narodnykh komissarov SSSR
Fond r-7523   Verkhovnyi sovet SSSR

Hoover Institution Archives, Stanford, CA
Russia: Posol'stvo (U.S.)

Institution of Mechanical Engineers, London
D.R. Carling book collection

Landsarkivet, Gothenburg
Nydqvist & Holm AB papers

Leeds Russian Archive, University of Leeds, Leeds (LRA)
MS 716 G.V. Lomonossoff Collection
MS 717 R.N. Lomonossoff Collection
MS 718 G.G. Lomonossoff Collection

National Archives, London
FO 371   Foreign Office, General Political Correspondence
GFM 34   German Foreign Ministry papers

National Railway Museum, York
Locomotive Test Data, Rugby Locomotive Testing Station

Papers relating to Sir W.G. Armstrong Whitworth & Company Ltd
R.C. Bond Collection

New York State Archives, State Education Department, Albany, NY
Joint Legislative Committee to Investigate Seditious Activities (Lusk Committee)

Oxford University Press (Company Archive), Oxford
Papers relating to G. Lomonossoff, *Introduction to Railway Mechanics*

Pennsylvania State Archives, Harrisburg, PA
MG427   Baldwin Locomotive Works

Rossiiskii gosudarstvennyi arkhiv ekonomiki, Moscow (RGAE)
Fond 413   Narodnyi komissariat vneshnei torgovli
Fond 1637   GOMZA
Fond 1884   Narodnyi komissariat putei soobshcheniia
Fond 2259   Komgosor VSNKh
Fond 3429   Vysshii sovet narodnogo khoziaistva
Fond 4038   Rossiiskaia zheleznodorozhnaia missiia zagranitsei
Fond 4086   Glavmetall
Fond 4372   Gosplan

Rossiiskii gosudarstvennyi arkhiv sotsial'no-politicheskoi istorii, Moscow (RGASPI)
Fond 2   V.I. Lenin
Fond 5   Sekretariat V.I. Lenina
Fond 17   Tsentral'nyi komitet KPSS
Fond 19   Sovnarkom
Fond 74   K.E. Voroshilov
Fond 76   F.E. Dzerzhinskii
Fond 558   I.V. Stalin

Rossiiskii gosudarstvennyi istoricheskii arkhiv, St Petersburg (RGIA)
Fond 25   Uchebnyi otdel Ministerstva torgovli i promyshlennosti
Fond 90   Russkoe tekhnicheskoe obshchestvo
Fond 229   Kantseliariia Ministra putei soobshcheniia
Fond 240   Inzhenernyi sovet Ministerstva putei soobshcheniia
Fond 268   Departament zheleznodorozhnykh del Ministerstva finansov
Fond 273   Upravlenie zheleznykh dorog Ministerstva putei soobshcheniia
Fond 274   Upravlenie po sooruzheniiu zheleznykh dorog Ministerstva putei soobshcheniia
Fond 275   Biuro soveshchatel'nykh s"ezdov predstavitelei russkikh zheleznykh dorog
Fond 560   Obshchaia kantseliariia Ministra finansov

Fond 577   Glavnoe vykupnoe uchrezhdenie
Fond 593   Gosudarstvennyi dvorianskii zemel'nyi bank
Fond 1343   Departament gerol'dii
Fond 1576   Rukhlov, Sergei Vasil'evich

Rossiiskii gosudarstvennyi voenno-istoricheskii arkhiv, Moscow (RGVIA)
Fond 369   Osoboe Soveshchanie dlia obsuzhdeniia i ob"edineniia
                  meropriatii po oborone gosudarstva
Fond 13251   Tsentral'nyi voenno-promyshlennyi komitet

Svenska Riksarkivet, Stockholm
Utrikesdepartamentets arkiv

Swarthmore College Peace Collection, Swarthmore College, PA
Jane Addams Collection, series 1: Correspondence, 1870–1937

Tsentral'nyi derzhavnii istorichnii arkhiv Ukraini, Kyiv (TsDIAU)
Fond 274   Kievskoe gubernskoe zhandarmskoe upravlenie
Fond 275   Kievskoe okhrannoe otdelenie
Fond 442   Kantseliariia kievskogo, podol'skogo i volynskogo General-
                  gubernatora
Fond 693   Upravlenie Iugo–Zapadnoi zheleznoi dorogi
Fond 1691   Upravlenie Khar'kovo–Nikolaevskoi zheleznoi dorogi

Tsentral'nyi gosudarstvennyi arkhiv kinofotofonodokumentov Sankt-Peterburga,
St Petersburg (TsGAKFFD SPb)
Photograph collection

Tsentral'nyi gosudarstvennyi istoricheskii arkhiv Sankt-Peterburga, St Petersburg
(TsGIA SPb)
Fond 381   Institut putei soobshcheniia
Fond 1480   Upravlenie Nikolaevskoi zheleznoi dorogi

Tyne and Wear Archives Service, Newcastle upon Tyne
Papers of Sir W.G. Armstrong Whitworth & Company Ltd

United States National Archives, Washington DC (NA)
RG 39   Bureau of Accounts (Treasury)
RG 43   US Participation in International Conferences, Commissions
                  and Expositions
RG 59   Department of State
RG 62   Council of National Defense
RG 241   Patent and Trademark Office
RG 261   Former Russian Agencies

Cited interviews

Browning, G.M., 25–26 September 1995 and 6 April 2002.
Kuprienko, O.G., 31 March 2003 (informal conversation).

Cited correspondence, research papers and research note

Davidova, L.A. (October Railway Museum), letter to author, 15 August 2002.
Davies, R.D. (Leeds Russian Archive), letter to author, 31 October 2002.
Dike, K. (Institution of Mechanical Engineers), letter to author, 7 July 2008.
Divall, C., 'Technical Change and Railway Systems', www.york.ac.uk/inst/irs/irshome/papers/hulltext.htm (accessed 18 May 2004).
Genis, V.L., letters to author, 17 May 2004, 2 August 2007.
Hellewell, D.S., 'Lomonosov and British Railways', unpublished research note, 30 January 2001.
Knowles, J.W., letters to author, 24 October 2000, 25 April 2001.
Mikheev, D.P. (Consular Department, Embassy of the Russian Federation, London), letter to author, 13 August 2002.
Mozzhukhina, N.P. (Director, Archive of the Foreign Policy of the Russian Federation), letter to author, 6 November 2002.
Nagatsuna, K., 'A Utopian Ideologue in Soviet Industrialisation: S.A. Bessonov and Transport Reconstruction Debates, 1928–1930', paper presented to the Soviet Industrialisation Project Seminar, University of Birmingham, January 1989.
Prew, J. et al. (Cambridge City Crematorium), letters to author, 2, 23–4 July 2007.
Skachkova, N.S., untitled family tree for Iu.V. Lomonosov and M.I. Shelkoplasova, 2008.
Steinberg, J.W., 'Cadet Corps Academies', paper presented to the 35th National Convention of the American Association for the Advancement of Slavic Studies, Toronto, November 2003.
Iurkova, V.A. (Director, Derzhavnii arkhiv Dnipropetrovs'koi oblasti), letter to author, 22 March 2000.
Whittaker, R., 'A.M. Wellington and the Idea of Network Effects', paper presented to the Institute of Railway Studies, University of York, 9 February 2000.

Cited PhD dissertations

Allen, B., 'Worker, Trade Unionist, Revolutionary: A Political Biography of Alexander Shliapnikov, 1905–1922' (Indiana University, 2001).
Argenbright, R.T., 'The Russian Railroad System and the Founding of the Communist State, 1917–1922' (University of California at Berkeley, 1990).
Balzer, H.D., 'Educating Engineers: Economic Politics and Technical Training in Tsarist Industry' (University of Pennsylvania, 1980).

Bingham, R.C., 'The Diesel Locomotive: A Study in Innovation' (Northwestern University, 1962).

Jones, R.H., 'Taylorism and the Scientific Organisation of Work in Russia, 1910–1925' (University of Sussex, 1986).

Oppenheim, S.A., 'Aleksei Ivanovich Rykov (1881–1938): A Political Biography' (Indiana University, 1972).

Reitzer, L.F., 'United States–Russian Economic Relations, 1917–1920' (University of Chicago, 1950).

Sanders, J.E., 'The Union of Unions: Economic, Political, and Human Rights Organizations in the 1905 Russian Revolution' (Columbia University, 1985).

Miscellaneous

Technical Passport of Steam Locomotive E^{SH} No. 4444, Class E^{SH} (courtesy of the former Ministry of Ways of Communication of the Russian Federation).

Published sources

Works by Iu. V. Lomonosov (Russian)

Books

(ed.) *Deiatel'nost' Russkoi zheleznodorozhnoi missii zagranitsei* (Berlin, 1921).

Konspekt kursa parovoza, chitannogo v Kievskom i Varshavskom politekhnicheskikh institutakh i. ob. ekstraordinarnogo professora Iu.V. Lomonosovym (Kiev and Warsaw: Litogr. izdanie, 1901–1903).

Nauchnye problemy ekspluatatsii zheleznykh dorog, 3 edns (Odessa, 1912; Odessa, 1915; Berlin, 1922).

Opytnoe issledovanie tovarnykh vos'mikolesnykh parovozov kompaund normal'nogo tipa, proizvedennoe v 1898–1900 gg. na Khar'kovo–Nikolaevskoi zh.d. (Kiev: Izdanie avtora, 1907).

Opyty nad teplovozami: Teplovoz Iu^E No. 001 i ego ispytanie v Germanii (Berlin: Bukwa, 1925).

Opyty nad tipami parovozov: opyty proizvodivshiesia v 1912–1914 gg. na b. Nikolaevskoi (nyne Oktiabr'skoi) zheleznoi doroge, tom 1 (Berlin, 1925).

Opyty 1925 goda nad teplovozom Iu^E No. 001 na zheleznykh dorogakh SSSR (Berlin: Likvidatsionnoe biuro po postroike i ispytaniiu teplovozov Iu^E, Iu^M, Iu^K, 1927).

(ed. with annotations) *Parovaia mashina i drugie teplovye dvigateli*, by J.A. Ewing (Kiev, 1904).

Parovozy 0-5-0 E, E^{SH}, E^G (Berlin: Bukwa, 1924).

(ed.) *Perevozka parovozov E^{SH} i E^G iz za-granitsy* (Berlin: Rossiiskaia zheleznodorozhnaia missiia, 1923).

Predvaritel'nyi otchet Iu.V. Lomonosova o deiatel'nosti Rossiiskoi zheleznodorozhnoi missii zagranitseiu za vse vremia ee syshchestvovaniia (1920–1923) (Berlin, 1923).

(with E.Shveter) *Proekty teplovozov zakazannykh v 1921–1925 gg. v Germanii dlia SSSR* (Berlin: Likvidatsionnoe biuro po postroike i ispytaniiu teplovozov Iu^E, Iu^M, Iu^K, 1927).

Sravnitel'noe issledovanie tovarnykh parovozov bol'shoi moshchnosti, vypusk I: Tsel' issledovaniia i ego metod (Odessa, 1910).

Sravnitel'noe issledovanie tovarnykh parovozov bol'shoi moshchnosti, vypusk II: Opyty nad parovozami 0-4-0 normal'nogo tipa 1901 g. i 1-4-0 izmenennogo kitaiskogo, proizvedennye v 1908 g. na Ekaterininskoi zh.d. (Odessa, 1915).

Sravnitel'noe issledovanie tovarnykh parovozov bol'shoi moshchnosti, vypusk III: Opyty nad parovozami 0-4-0 normal'nogo tipa na nefti, proizvedeny v 1910 g. na Tashkentskoi zh.d. (Odessa, 1916).

Tekhnicheskie perspektivy zheleznodorozhnogo transporta v blizhaishee vremia (Moscow: Transpechat', 1924).

Tiagovye raschety (Berlin: Biuro inostrannoi nauki i tekhniki, 1922).

Tiagovye raschety i prilozhenie k nim graficheskikh metodov, 2 edns (St Petersburg: MPS, 1912; Odessa, 1915) (first published as an article in *Zhurnal Ministerstva putei soobshcheniia*, which is listed separately below).

Tsel' opytov i ikh metod (St Petersburg: Novoe vremia, 1914).

Vospominaniia o martovskoi revoliutsii 1917 goda (Stockholm and Berlin: STU, 1921; an unauthorised edition, edited by A.S. Senin, was published in Moscow by the Rossiiskii Gosudarstvennyi Gumanitarnyi Universitet, 1994).

Articles, book chapters, pamphlets and letters to editors

'25 let opytov nad parovozami (Tekhnicheskaia avtobiografiia)', *Tekhnicheskii zhurnal NKPS*, 6 (1923): 284–92.

'Evoliutsiia lokomotiva za sto let', in Pravosudovich, M.E. and Manos, I.Ia. (eds), *Stoletie zheleznykh dorog (Trudy Nauchno-Tekhnicheskogo Komiteta Narodnogo Komissariata Putei Soobshcheniia, vypusk 20)* (Moscow: Transpechat', 1925), pp. 126–51.

'Glavneishie rezul'taty issledovaniia tovarnykh parovozov 0-4-0 i 1-4-0', *Izvestiia Obshchego biuro soveshchatel'nykh s"ezdov*, 10 (1912): 863–925; 11 (1912): 1058–95; 12 (1912): 1126–66 (also published as pamphlet, St Petersburg, 1913).

'Itogi opytov nad tovarnymi parovozami (1908–1914)', *Zhurnal Ministerstva putei soobshcheniia*, (1915): 10–25 (reprinted as: 'Itogi opytov nad tovarnymi parovozami 1908–1914 (Iz *Zhurnala Ministerstva Putei Soobshcheniia*)', *Zheleznodorozhnoe delo*, 37–38 (1915): 394–9).

'K ispytaniiu teplovozov', *Tekhnika i ekonomika putei soobshcheniia*, 2 (1924): 190–199.

(with A.O. Chechott) 'K issledovaniiu parovoznykh peregrevatelei', *Inzhener*, 10 (1911): 304–6.

'K teorii dizel' lokomotivov', *Tekhnika i ekonomika putei soobshcheniia*, 2/12 (1923): 705–20.

'K teorii ergometra Duaena', *Zheleznodorozhnoe delo*, 20 (1911): 150–151.

'K voprosu ob ekvivalente dombrovskikh uglei po sravneniiu s donetskimi', *Izvestiia Sobraniia inzhenerov putei soobshcheniia*, 11 (1914): 156.

'K voprosu ob otoplenii parovozov antratsitom', *Vestnik inzhenerov*, 3/16 (1917): 347–54.

'K voprosu o regulirovanii raboty parovoza', *Tekhnika i ekonomika putei soobshcheniia*, 2/10 (1923): 475–6.

'Nabliudeniia nad stepen'iu sukhosti para v reguliatornoi trube parovoza normal'nogo tipa', *Inzhener*, 5 (1905): 135–40; 6 (1905): 181–6 (also published as pamphlet, Kiev, 1905).

'Naivygodneishii sostav tovarnykh poezdov', *Inzhener*, 2 (1904): 52–5 (also published as pamphlet, Kiev, 1904).

'Ob opytakh nad parovozami normal'nogo tipa', *Zheleznodorozhnoe delo*, 3–4 (1911): 1D–25D.

'Opredelenie sostavov tovarnykh poezdov po vesu', *Izvestiia Obshchego biuro soveshchatel'nykh s"ezdov*, 7 (1912): 583–618 (also published as pamphlet, St Petersburg, 1912).

'O predel'nykh pod"emakh magistralei', *Tekhnika i ekonomika putei soobshcheniia*, 8–9 (1922): 310–312.

'Opyty s torfianym otopleniem parovozov: Predvaritel'noe soobshchenie', *Vestnik inzhenerov*, 2/22 (1916): 721–7.

'Organizatsiia opytov nad tipami parovozov na russkikh zheleznykh dorogakh', *Vestnik inzhenerov*, 1/8 (1915): 335–7.

'O teplovozakh', *Krasnyi transportnik*, 1 (5 August 1923): 19–22.

'O teplovozakh, zakazannykh za granitseiu', *Tekhnika i ekonomika putei soobshcheniia*, 2/9 (1923): 321–4.

'Otkrytoe pis'mo inzh. V.N. Shcheglovitovu', *Izvestiia Sobraniia inzhenerov putei soobshcheniia*, 9 (1914): 124–5.

'Otvet V.N. Shcheglovitovu o sebestoimosti', *Izvestiia Sobraniia inzhenerov putei soobshcheniia*, 13 (1915): 283.

(with A.I. Lipets) *Poiasnitel'naia zapiska k proektu kotla Brotana s peregrevatelem Piloka–Slutskogo dlia tovarnogo parovoza kompaund normal'nogo tipa* (Kiev: Obshchestvo Iugo-Vostochnykh zheleznykh dorog, 1906).

'Po povodu broshiury V.N. Shcheglovitova "K voprosu ob opredelenii sebestoimosti"', *Izvestiia Sobraniia inzhenerov putei soobshcheniia*, 8 (1915): 158–60.

'Po povodu stat'i g. Shelesta "Issledovanie raboty teplovoza br. Zultser"', *Vestnik inzhenerov*, 3/7 (1917): 206–7.

'Predislovie', *Tekhnika i ekonomika putei soobshcheniia*, 1/5 (1923): 284 (preface for Kollontai, M., 'Dvigatel' Stilla', in ibid., pp. 284–9).

'Predislovie rukovoditelia', *Izvestiia Kievskogo politekhnicheskogo instituta, Otdel mekhanicheskii i inzhenernyi*, 7/4 (1907): III–IV (preface for Turchaninov,

A.N., 'Poiasnitel'naia zapiska k proektu dachnogo parovoza dlia Kh.-N. zh.d.', in ibid., pp. I–IV, 175–252).

'Sovremennye zadachi passazhirskogo dvizheniia na russkoi seti s tochki zreniia parovoznoi sluzhby', *Inzhener*, 10 (1903): 332–41; 12 (1903): 404–11; 1 (1904): 15–21 (also published as pamphlet, Kiev, 1904).

'Teplovoz Iu^M No. 005', *Zheleznodorozhnoe delo: Obshchii otdel*, 6 (1926): 2–6.

'Tiagovye raschety i prilozhenie k nim graficheskikh metodov', *Zhurnal Ministerstva putei soobshcheniia*, 3 (1912): 1–100 (subsequently published as book, St Petersburg, 1912).

'Tochnyi vyvod uravneniia dvizheniia poezda', *Izvestiia Kievskogo politekhnicheskogo instituta, Otdel mekhanicheskii i inzhenernyi*, 5/3 (1905): 1–6 (also published as pamphlet, Kiev, 1905).

'Tovarnye poezda vesom v 200 000 pudov i bolee', *Vestnik inzhenerov*, 2/19 (1916): 618–20.

['V.A. Pavlovskii', *Vestnik inzhenerov*, 2/19 (1916): 617–18.]

'V kakom vide dolzhny byt' vosstanovleny russkie zheleznye dorogi?', *Ekonomicheskaia zhizn'*, 3 July 1920, pp. 1–2.

'V Narodnom Komissariate Putei Soobshcheniia, noiabr' 1919–ianvar' 1920 (Otryvok iz vospominanii i neizdannaia korrespondentsiia V.I. Lenina)', edited by Aplin, H.A., *Minuvshee: Istoricheskii al'manakh*, 10 (1990): 7–63 (also: extracts reprinted without permission by A. Senin in 12 parts as 'Lozung momenta: Iz vospominanii Iu.V. Lomonosova', *Gudok*, 1–19 August 1992).

'V redaktsiiu "Tekhniki i ekonomiki putei soobshcheniia"', *Tekhnika i ekonomika putei soobshcheniia*, 2/8 (1923): 238.

Experiment instructions, locomotive 'passports', 'passport booklets', raw data and accounts (series: Opyty nad teplovozami; Opyty nad tipami parovozov)

Dopolnenie k pasportnym knizhkam (Petrograd, 1916).

Glavneishie rezul'taty opytov, proizvodivshikhsia v 1908 g. na Ekaterininskoi zh.d. nad parovozami tipa 1-4-0 Shch, 2nd edn (Petrograd, 1915).

Glavneishie rezul'taty opytov, proizvodivshikhsia v 1908 g. na Ek., v 1910 g. na Tashk., v 1913 g. na Nik. i Perm., v 1916 g. na M.–Nizhegorodskoi i v 1918 na M.–Kurskoi zhzh dd. nad parovozami tipa 0-4-0 O^y, 3rd edn (Berlin: Nauchno-Tekhnicheskii Komitet NKPS, 1924).

Glavneishie rezul'taty opytov, proizvodivshikhsia v 1908 g. na Ekaterininskoi zh.d., v 1910 g. na Tashkentskoi zh.d. i v 1913 g. na Nikolaevskoi, Severo–Zapadnykh i Permskoi zh.d. nad parovozami tipa 0-4-0 O^y, 2nd edn (Petrograd, 1915).

Glavneishie rezul'taty opytov, proizvodivshikhsia v 1910 g. na Tashkentskoi zh.d, v 1912 g. na Nikolaevskoi i M–Kurskoi zh.d. i v 1913 g. na Nikolaevskoi zh.d. nad parovozom tipa 2-3-0 B (Petrograd, 1915).

Glavneishie rezul'taty opytov, proizvodivshikhsia v 1911 g. na Nikolaevskoi zh.d. nad parovozom tipa 1-4-0 Shch^P (Petrograd, 1915).

Glavneishie rezul'taty opytov, proizvodivshikhsia v 1912 g. na Nikolaevskoi zh.d. nad parovozom tipa 1-3-0 N^v, 2nd edn (Petrograd, 1915).

Glavneishie rezul'taty opytov, proizvodivshikhsia v 1913 g. na Nikolaevskoi i M–Kurskoi i v 1915 g. na M–Kazanskoi i R.–Ural'skoi zhzh.dd. nad parovozom tipa 2-3-0 K, 2nd edn (Petrograd, 1915).

Glavneishie rezul'taty opytov, proizvodivshikhsia v 1913 g. na Nikolaevskoi i M–Kurskoi i v 1915 g. na M–Kazanskoi i R.–Ural'skoi zhzh.dd. nad parovozom tipa 2-3-0 K^v, 2nd edn (Petrograd, 1915).

Glavneishie rezul'taty opytov, proizvodivshikhsia v 1913 g. na Nikolaevskoi i M–Kurskoi zh.d. nad parovozami tipa 1-3-0 N^{sh} (Petrograd, 1915).

Glavneishie rezul'taty opytov, proizvodivshikhsia v 1913 g. na Nikolaevskoi i M–Kurskoi zh.d. nad parovozami tipa 1-3-0 N^v (Petrograd, 1915).

Glavneishie rezul'taty opytov, proizvodivshikhsia v 1913 g. na Nikolaevskoi i M–Kurskoi zh.d. nad parovozom tipa 2-3-0 U^v ([Petrograd?], [1915?]).

Glavneishie rezul'taty opytov, proizvodivshikhsia v 1913 g. na Nikolaevskoi i M–Kurskoi zh.d. nad parovozom tipa 2-3-0 U^v, 2nd edn (Petrograd, 1915).

Glavneishie rezul'taty opytov, proizvodivshikhsia v 1913 g. na Nikolaevskoi zh.d. nad parovozom tipa 0-4-0 O^o nekompaund (Petrograd, 1915).

Glavneishie rezul'taty opytov, proizvodivshikhsia v 1913 i 1914 g. na Nikolaevskoi zh.d. nad parovozom tipa 0-4-0 Y^{ch}, 2nd edn (Petrograd, 1915).

Glavneishie rezul'taty opytov, proizvodivshikhsia v 1914 g. na Iuzhnykh zh.d. nad parovozom tipa 1-3-0 N^v (Petrograd, 1915).

Glavneishie rezul'taty opytov, proizvodivshikhsia v 1915 g. na Cebepo–Donetskoi zh.d. nad parovozom tipa 0-5-0 E (Petrograd, 1916).

[*Glavneishie rezul'taty opytov, proizvodivshikhsia v 1915 g. na Cebepo–Donetskoi zh.d. i v 1916 g. na Samaro–Zlatoustovskoi i Iuzhnykh zh.d. nad parovozom tipa 0-5-0 E*, 2nd edn (Moscow: Nauchno-eksperimental'nyi institut putei soobshcheniia, 1918).]

[*Glavneishie rezul'taty opytov, proizvodivshikhsia v 1916 g. na Cebepo–Donetskoi zh.d. i Iuzhnykh zhzh.dd. nad parovozom tipa 0-3-0+0-3-0 Fita^{ch}* (Moscow: Nauchno-eksperimental'nyi institut putei soobshcheniia, 1918).]

Glavneishie rezul'taty opytov, v primenenii k parovozu tipa 1-3-0 N^v (Petrograd, 1915).

Neposredstvennye dannye opytov I tsikla s parovozom M.–Kur. zh.d. 2-3-0 B17, proizvodivshikhsia v 1913 g. na Nikolaevskoi zh.d. (Odessa, 1915).

Neposredstvennye dannye opytov I tsikla s parovozom Nikolaevskoi zh.d. 1-3-0 N^v142, proizvodivshikhsia v 1914 g. na Nikolaevskoi zh.d. (Odessa, 1916).

Neposredstvennye dannye opytov I tsikla s parovozom Nikolaevskoi zh.d. 1-3-0 N^v84, proizvodivshikhsia v 1913 g. na Nikolaevskoi zh.d. (Odessa, 1916).

Neposredstvennye dannye opytov I tsikla s parovozom Severo–Zapadnykh zh.d. 1-3-1 S20, proizvodivshikhsia v 1913 g. na Nikolaevskoi zh.d. (Odessa, 1915).

Pamiatnaia knizhka na 1914 g. (St Petersburg, 1914).

Pamiatnaia knizhka na 1924 g. (Stuttgart, 1924).

Pamiatnaia knizhka na 1925 g. (Berlin, 1925).

Tekhnicheskii i denezhnii otchet za 1915 g. (Petrograd, 1916).

Note: similar pamphlets are thought to have been produced for at least the following locomotive classes: K, S, NV, N^Y, N^P, N^{CH}, N^{SHP}, E^F, O^P.

Congress and conference papers

Doklad inzh. Iu.V. Lomonosova ob ispytanii tovarnogo parovoza izmenennogo tipa 1-4-0 Kitaiskoi Vostochnoi dorogi na Ekaterininskoi zheleznoi doroge: Stenogramma zasedanii komissii 28 i 29 maia 1909 g. (St Petersburg: Komissiia podvizhnogo sostava i tiagi, 1912).

'Istoricheskii ocherk zakaza parovozov serii E', in *XXXIII soveshchatel'nyi s"ezd Inzhenerov Podvizhnogo Sostava i Tiagi v Moskve, 18 iiunia 1923 goda* (Moscow, 1924), pp. 676–7.

'Normy remonta parovozov, vvedennye na Nikolaevskoi zhel. dor.', in *XXIX soveshchatel'nyi s"ezd inzhenerov podvizhnogo sostava i tiagi v Rostove, 24 maia–10 iiunia 1912 g.* (St Petersburg, 1913), pp. 159–219.

'O nedostatkakh, zamechennykh v parovozakh normal'nogo tipa, zakaza 1897g.', in *Protokoly zasedanii XXIII soveshchatel'nogo s"ezda inzhenerov sluzhby podvizhnogo sostava i tiagi russkikh zheleznykh dorog* (St Petersburg, 1902), pp. 439–50.

'O novoi organizatsii schetovodstva', in *Protokoly zasedanii XXVIII soveshchatel'nogo s"ezda inzhenerov sluzhby podvizhnogo sostava i tiagi russkikh zheleznykh dorog v Rige, 27 maia–8 iiunia 1911 g.* (St Petersburg, [1912]), pp. 402–22.

'Opredelenie predel'nykh skorostei dvizheniia, v zavisimosti ot konstruktsii puti i parovoza', in *Protokoly zasedanii XXV soveshchatel'nogo s"ezda inzhenerov sluzhby podvizhnogo sostava i tiagi russkikh zheleznykh dorog*, vol. 1 (St Petersburg, 1908), pp. 227–88.

'Opredelenie predel'nykh skorostei dvizheniia, v zavisimosti ot konstruktsii puti i parovoza', in *Protokoly zasedanii XXVII soveshchatel'nogo s"ezda inzhenerov sluzhby podvizhnogo sostava i tiagi russkikh zheleznykh dorog, v Varshave, v avguste 1909 g.* (St Petersburg, 1910), pp. 57–84.

'Opredelenie predel'nykh skorostei dvizheniia, v zavisimosti ot konstruktsii puti i parovoza', in *Protokoly zasedanii XXVIII soveshchatel'nogo s"ezda inzhenerov sluzhby podvizhnogo sostava i tiagi russkikh zheleznykh dorog v Rige, 27 maia–8 iiunia 1911 g.* (St Petersburg, [1912]), p. 76.

'Opredelenie predel'nykh skorostei dvizheniia, v zavisimosti ot konstruktsii puti i parovoza', in *XXIX soveshchatel'nyi s"ezd inzhenerov podvizhnogo sostava i tiagi v Rostove, 24 maia–10 iiunia 1912 g.* (St Petersburg, 1913), pp. 289–321.

'Opredelenie sostavov tovarnykh poezdov po vesu', in *XXIX soveshchatel'nyi s"ezd inzhenerov podvizhnogo sostava i tiagi v Rostove, 24 maia–10 iiunia 1912 g.* (St Petersburg, 1913), pp. 349–85.

'O stroiashchikhsia zagranitsei teplovozakh', in *XXXIII soveshchatel'nyi s"ezd inzhenerov podvizhnogo sostava i tiagi v Moskve, 18 iiunia 1923 goda* (Moscow, 1924), pp. 21–6.

'O teplovozakh', in *Trudy XXXIV soveshchatel'nogo s"ezda inzhenerov podvizhnogo sostava i tiagi v Moskve s 1-go po 9-oe aprelia 1925 goda* (Moscow: NKPS, 1926), pp. 22–9.

'Pereraspredelenie uchastkov tiagi na Nikolaevskoi zh.d.', in *XXIX soveshchatel'nyi s"ezd inzhenerov podvizhnogo sostava i tiagi v Rostove, 24 maia–10 iiunia 1912 g.* (St Petersburg, 1913), pp. 430–509.

'Rezul'taty sravnitel'nykh opytov nad parovozami 4/4 normal'nogo tipa i 4/5 izmenennogo tipa Vostochno–Kitaiskoi zhel. dor.', in *Protokoly zasedanii XXVII soveshchatel'nogo s"ezda inzhenerov sluzhby podvizhnogo sostava i tiagi russkikh zheleznykh dorog, v Varshave, v avguste 1909 g.* (St Petersburg, 1910), p. 355.

Student notes of lectures by Iu.V. Lomonosov

'Parovozy' (Warsaw: [Litogr. izdanie, n.d.]) ('Zapisi stud. Kozlovskogo po lektsiiam prof. Iu.V. Lomonosova, izdano bez razresheniia i vedoma Iu.V. Lomonosova').

'Zapiski studentov po kursu ekspluatatsii zheleznykh dorog, chitannomu v Kievskom politekhnicheskom institute prof. Iu.V. Lomonosovym' (Kiev: Litogr. izdanie, 1903/4).

Books, pamphlets, articles, research papers, discussion contributions, communications and letters to editors by Iu. V. Lomonosov (English)

Contribution to 'Communications on the mobile locomotive testing plant of the LMSR', *Proceedings of the Institution of Mechanical Engineers*, 158 (1948): 471–2.

Contribution to 'Correspondence on Hammer-Blow in Locomotives', *Journal of the Institution of Civil Engineers*, 18/8 Supplement (October 1942): 481–6.

Contribution to 'Discussion on Air Resistance of Passenger Trains', *Proceedings of the Institution of Mechanical Engineers*, 134 (1936): 171–2.

Contribution to 'Discussion on Air Swirl in Engines', *Proceedings of the Institution of Mechanical Engineers*, 128 (1934): 166–7.

Contribution to 'Discussion on Combustion in the Compression-ignition Engine', *Proceedings of the Institution of Mechanical Engineers*, 138 (1938): 459–61.

Contribution to 'Discussion on Combustion of Oil in a Cylinder', *Proceedings of the Institution of Mechanical Engineers*, 121 (1931): 451–5.

Contribution to 'Discussion on Compression-ignition Engine and British Railways', *Proceedings of the Institution of Mechanical Engineers*, 124 (1933): 40–43.

Contribution to 'Discussion on Control of Diesel Railcars', *Proceedings of the Institution of Mechanical Engineers*, 140 (1938): 106–9.

Contribution to 'Discussion on the Development of Locomotive Power at Speed', *Proceedings of the Institution of Mechanical Engineers*, 156 (1947): 422–3.

Contribution to 'Discussion on Diesel Traction on Railways', *Proceedings of the Institution of Mechanical Engineers*, 137 (1937): 156–9.

Contribution to 'Discussion on Energy Balance of Internal Combustion Engine', *Proceedings of the Institution of Mechanical Engineers*, 141 (1939): 329–30.

Contribution to 'Discussion on Exhaust Systems of Two-stroke Engines', *Proceedings of the Institution of Mechanical Engineers*, 138 (1938): 393–4.

Contribution to 'Discussion on the First Gas Turbine Locomotive', *Proceedings of the Institution of Mechanical Engineers*, 151 (1944): 182.

Contribution to 'Discussion on Fuel Injection in Oil Engines', *Proceedings of the Institution of Mechanical Engineers*, 144 (1940): 22.

Contribution to 'Discussion on Heat Liberation in Steam Plants', *Proceedings of the Institution of Mechanical Engineers*, 125 (1933): 263–6.

Contribution to 'Discussion on Hydraulic Couplings', *Proceedings of the Institution of Mechanical Engineers*, 130 (1935): 161–2.

Contribution to 'Discussion on Locomotive Experimental Stations', *Proceedings of the Institution of Mechanical Engineers*, 121 (1931): 40–42.

Contribution to 'Discussion on the Loop Scavenge Diesel Engine', *Proceedings of the Institution of Mechanical Engineers*, 154 (1946): 399–400.

Contribution to 'Discussion on Mechanism of Electric Locomotives', *Proceedings of the Institution of Mechanical Engineers*, 122 (1932): 114–20.

Contribution to 'Discussion on Piston Temperatures', *Proceedings of the Institution of Mechanical Engineers*, 135 (1937): 65.

Contribution to 'Discussion on Piston Temperatures and Supercharging', *Proceedings of the Institution of Mechanical Engineers*, 123 (1932): 445–6.

Contribution to 'Discussion on Problems of Fluid Couplings', *Proceedings of the Institution of Mechanical Engineers*, 139 (1938): 162–5.

Contribution to 'Discussion on Steel Railway Sleepers', *Proceedings of the Institution of Mechanical Engineers*, 121 (1931): 357–60.

Contribution to 'Discussion on Stress Waves in the Tyres of Locomotives', *Proceedings of the Institution of Mechanical Engineers*, 131 (1935): 508–10.

Contribution to 'Discussion on a Three-cylinder Compound Locomotive', *Proceedings of the Institution of Mechanical Engineers*, 2 (1927): 974–6.

Contribution to discussion on C.H. Gibbons, 'Commercial Applications of High-Speed Oil Engines', *Transactions of the American Society of Mechanical Engineers*, 51 (1929), OGP-51-5: 49–51.

Contribution to discussion on L.H. Fry, 'The Locomotive Testing Plant and its Influence on Steam Locomotive Design', *Transactions of the American Society of Mechanical Engineers*, 47 (1925): 1291–2.

(with G. Lomonossoff) 'Condensing Locomotives', *Proceedings of the Institution of Mechanical Engineers*, 152 (1945): 275–88 (with discussion and communications, pp. 289–303).

(with G. Lomonossoff) 'Condensing Locomotives', *The Engineer*, 178/4617 (1944): 5–7; 178/4618: 23–6; 178/4619: 43–6; 178/4620: 62–5; 178/4621: 78–80; 178/4622: 98–100; 178/4625: 156–8; 178/4626: 176–7; 178/4627: 196–9; 178/4628: 218–20.

'Condensing Locomotives', *Railway Gazette*, 80/22 (1944): 569–70, 581.

'Diesel Locomotive Operation in Russia', *Railway Review*, 79/13 (25 September 1926): 457–9.

'Diesel Traction', *Proceedings of the Institution of Mechanical Engineers*, 125 (1933): 537–613.

'Dynamic Loading on Locomotive Wheels', *The Engineer*, 146/3784 (1928): 58–9; 146/3785: 83–5; 146/3786: 108–9.

Introduction to Railway Mechanics (Oxford: Oxford University Press, 1933).

'Locomotives and Bridges', *The Engineer*, 158/4106 (1934): 282–3.

'Locomotive Testing Plants', *The Engineer*, 155/4025 (1933): 225 (letter to editor).

Memoirs of the Russian Revolution (New York: Rand School of Social Science, 1919).

'New European Diesel Locomotives lead American Development', *Diesel Power*, 9/2 (February 1931): 82–5.

On Intervention: Remarks of George V. Lomonossoff, head of Russian Railway Mission in the United States, made at a conference of the Women's International League, held at Rye, NY, July, 20, 1918 ([n.p.], 1918).

Principal Results of Tests of a Prairie Type Locomotive 1-3-1 S: Made in 1913 on the Nicolai Railway and in 1914 on the Southern Railway in Russia (Petrograd, 1916).

'Problems of Railway Mechanics', *Proceedings of the Institution of Mechanical Engineers*, 120 (1931): 643–59.

'Russia at the Cross-Roads', *The Nation*, CVIII/2800 (1 March 1919): 321–2.

(with C.A.J. Elphinston) 'Superheating on Locomotives', *Railway Gazette*, 62/16 (19 April 1935): 728–9.

'A Voice Out of Russia', *The Dial*, LXVI/782 (25 January 1919): 61–6 (also published as a brochure by the Dial Publishing Company).

Books and articles by Iu. V. Lomonosov (German)

Die Diesel-elektrische Lokomotive (Berlin: VDI-Verlag, 1924).

'Diesel-Getriebelokomotive 2-E-1 für die Staatsbahnen der USSR', *Organ für die Fortschritte des Eisenbahnwesens*, 81/11 (1926): 193–8.

Diesellokomotiven (Berlin: VDI-Verlag, 1929; reprinted Düsseldorg: VDI-Verlag, 1985, in the series Klassiker der Technik, with introduction by Wolfgang Messerschmidt).

'Fahrtergebnisse der dieselelektrischen Lokomotive in Russland', *Zeitschrift des Vereines Deutscher Ingenieure*, 69/44 (1925): 1387–9.

'Der gegenwärtige Stand des Diesellokomotivbaues', *Zeitschrift des Vereines Deutscher Ingenieure*, 71/30 (1927): 1046–8.

'Der hundertjährige Werdegang der Lokomotive', *Organ für die Fortschritte des Eisenbahnwesens*, 81/17 (1926): 347–54; 81/18 (1926): 365–71.

'Das Lokomotivstufengetriebe', *Organ für die Fortschritte des Eisenbahnwesens*, 83/19 (1928): 416–22.

Lokomotivversuche in Russland (Berlin: VDI-Verlag, 1926).

'Die mechanisch angetriebene Diesellokomotive mit fester Übersetzung und mehreren, einzeln kuppelbaren Motoren', *Organ für die Fortschritte des Eisenbahnwesens*, 82/15 (1927): 283–4.

'Der russische Lokomotivprüfstand in Esslingen', *Organ für die Fortschritte des Eisenbahnwesens*, 79/8 (1924): 166–70.

(with A. Krukovskii) 'Die Temperaturmessungen im Feuerraum der Dampflokomotive während der Fahrt', *Zeitschrift des Vereines Deutscher Ingenieure*, 53/9 (1909): 345–6.

'Die Thermolokomotive unter der Rubrick sonderantriebe in Eisenbahnwesen', *Die eisenbahntechnische Jagung und ihre Acessteltlungen 1924* (Berlin: VDI Verlag, 1925).

'Ueber den dynamischen Druck der Lokomotivrädar', *Glassers annalen*, 105/5 (1929): 80–84; 105/6 (1929): 92–8.

'Widerstand und Trägheit der diesel-elektrischen Lokomotive', *Organ für die Fortschritte des Eisenbahnwesens*, 83/7 (1928): 133–6.

(with A.O. Czeczott) 'Zur Erforschung der Lokomotivüberhitzer', *Zeitschrift des Vereines Deutscher Ingenieure*, 56/5 (1912): 184–5.

'Zur Theorie der Diesellokomotive', *Zeitschrift des Vereines Deutscher Ingenieure*, 68/9 (1924): 198–202.

'Zur Theorie der Gasübertragung bei Diesellokomotiven', *Zeitschrift des Vereines Deutscher Ingenieure*, 71/38 (1927): 1329–32.

'Zur Untersuchung von Thermolokomotiven', *Zeitschrift des Vereines Deutscher Ingenieure*, 68/33 (1924): 849–52.

Contemporary newspapers

Birzhevye izvestiia
Birzhevye vedomosti
Dneprovskii vestnik
Ekonomicheskaia zhizn'
The Gazette (Montreal)
Gudok
Izvestiia
Krasnaia gazeta
Moskovskie vedomosti
New York Times
New York Tribune
Novoe vremia
Orenburgskaia gazeta
Peterburgskaia gazeta
Peterburgskii listok
Poslednie novosti
Pravda
Pravitel'stvennyi vestnik

Rech'
Russkie vedomosti
Russkoe slovo
Russkoe znamia
Sankt-Peterburgskie vedomosti
The Times
The World

Contemporary periodicals

Armstrong Whitworth Record
Baldwin Locomotives
Biulleten' Narodnogo komissariata putei soobshcheniia
Biulleten' Tekhnicheskogo obshchestva
Biulleten' Tsentral'nogo biuro pechati NKPS
The Dial
Diesel Power
Diesel Railway Traction
The Economist
Elektrichestvo
The Engineer
Engineering
Glassers Annalen
Izvestiia Kievskogo politekhnicheskogo instituta
Izvestiia Obshchego biuro soveshchatel'nykh s"ezdov
Izvestiia Sobraniia inzhenerov putei soobshcheniia
Izvestiia Tsentral'nogo voenno-promyshlennogo komiteta
Journal of the Institution of Civil Engineers
Journal of the Institution of Locomotive Engineers
Katorga i ssylka
Krasnaia letopis'
Krasnyi arkhiv
Krasnyi put' zheleznodorozhnika
Krasnyi transportnik
Inzhener
Letopis' revolutsii
Locomotive, Railway Carriage and Wagon Review (Locomotive Magazine)
Mechanical Engineering
Monthly Bulletin of the International Railway Congress Association (English edition)
The Nation
Oil Engine Power
Organ für die Fortschritte des Eisenbahnwesens
Parovoznoe khoziaistvo

Proceedings of the Institution of Mechanical Engineers
Proletarskaia revoliutsiia
Puti revoliutsii
Puti soobshcheniia
Railway Engineer
Railway Gazette
Railway Mechanical Engineer
Railway Review
Sotsialisticheskii transport (formerly *Transport i khoziaistvo*)
The Spanner
Tekhnicheskii zhurnal NKPS
Tekhnika i ekonomika putei soobshcheniia
Tekhnika i zhizn'
Transactions of the American Society of Mechanical Engineers
Transport i khoziaistvo
Vestnik Ekaterininskoi zheleznoi dorogi
Vestnik finansov, promyshlennosti i torgovli
Vestnik inzhenerov
Vestnik Obshchestva tekhnologov
Vestnik putei soobshcheniia
Der Waggon- und Lokomotiv-Bau
Zapiski Ekaterinoslavskogo otdeleniia Imperatorskogo russkogo tekhnicheskogo obshchestva
Zapiski Imperatorskogo russkogo tekhnicheskogo obshchestva
Zheleznodorozhnaia tekhnika
Zheleznodorozhnaia tekhnika i ekonomika
Zheleznodorozhnoe delo
Zhurnal Ministerstva putei soobshcheniia
Zeitschrift des Vereines Deutscher Ingenieure

Documents, reports, contemporary works and memoirs

A.B.V., 'Iu. Lomonosov. Tiagovye raschety. 2-oe izdanie ispravlennoe i dopolennoe. Odessa 1915. III+295 str. l. chertezhei', *Izvestiia Sobraniia inzhenerov putei sobshcheniia*, 6 (1915): 115–16.

A.C., 'Diamond (E.L.) […], The Horse-power of Locomotives: its Calculation and Measurement – a Pamphlet […] 1935, London, *The Railway Gazette* […]', *Monthly Bulletin of the International Railway Congress Association (English Edition)*, 18/6 (June 1936): 664–5.

Adamovich, E., 'Ianvar' 1905 goda na Ukraine: ianvarskie sobytiia v Kieve', *Letopis' revoliutsii*, 1 (1925): 188–93.

Adibekov, G.M. et al. (eds), *Politburo TsK RKP (b)–VKP (b): Povestki dnia zasedanii, tom 1: 1919–1929: Katalog* (Moscow, 2000).

'A Diesel Electric Locomotive', *The Engineer*, 138 (19 September 1924): 322, 324; and 138 (14 November 1924): 552–4.

'[A.I. Lipetz]', *Railway Mechanical Engineer* (June 1936): 276.

Akhun, M. and Petrov, V., 'Voennaia organizatsiia pri Peterburgskom komitete RS-DRP v 1906 godu', *Krasnaia letopis'*, 1 (1926): 131–61.

Akhun, M. and Petrov, V., 'Voennaia organizatsiia pri Peterburgskom komitete RS-DRP v 1907 godu', *Krasnaia letopis'*, 3 (1926): 143–71.

Akhun, M. and Petrov, V., 'Voennaia organizatsiia pri Peterburgskom komitete RS-DRP v 1907–8 godu', *Krasnaia letopis'*, 4 (1926): 124–37.

Akhun, M. and Petrov, V., 'Vosstanie inzhenernykh voisk v Kieve (iz istorii revoliutsionnogo dvizhenii v armii v 1905 g.)', *Krasnaia letopis'*, 3 (1925): 126–48.

American Society of Mechanical Engineers, *Membership List, Including Society Records, Part 1, 1940 and Professional Service Index* (New York: ASME, 1940).

Amiantov, Iu.N., Akhapkin, Iu.A. and Loginov, V.T. (eds), *V.I. Lenin: Neizvestnye dokumenty, 1891–1922* (Moscow: Rosspen, 1999).

Andreev, V., 'K voprosu o sokrashchenii sroka prokhozhdeniia studentami kursa v Institute Inzhenerov Putei Soobshcheniia', *Izvestiia sobraniia inzhenerov putei soobshcheniia*, 15 (1916): 312–13.

'Annual Report of the Council for the Year 1931', *Proceedings of the Institution of Mechanical Engineers*, 122 (1932): 163–4.

'Annual Report of the Council for the Year 1944', *Proceedings of the Institution of Mechanical Engineers*, 152 (1945): 117.

'Annual Report of the Council for the Year 1952', *Journal of the Institution of Locomotive Engineers*, 43/232 (1953): 160–161.

Antonovich, A.I., *Kak stroit' Moskovskuiu Okruzhnuiu Dorogu s shirokim rel'sovym kol'tsom ili kol'tsom priblizhennym k gorodu* (Moscow, 1897).

Apushkin, V.A., *General ot porazhenii V.A. Sukhomlinov* (Leningrad, 1925).

Babichkov, A., 'Dvadtsat' let tiagi poezdov na dorogakh SSSR', *Sotsialisticheskii transport*, 11–12 (1937): 41–51.

Babichkov, A., 'O novykh pravilakh tiagovykh raschetov', *Zheleznodorozhnaia tekhnika*, 3 (1937): 8–21.

Baedeker, K., *Russia, with Teheran, Port Arthur and Peking: Handbook for Travellers* (London, 1914, facsimile reprint London and Newton Abbot, 1971).

'Baldwin 1000 HP Diesel-electric Locomotive', *Monthly Bulletin of the International Railway Congress Association (English Edition)*, 8/2 (February 1926): 155–9.

Bartenev, N., 'K voprosu o pereustroistve Rostovskogo uzla', *Zheleznodorozhnoe delo: obshchii otdel*, 3 (1928): 17–19.

Beitler, A., 'Reorganizatsiia khoziaistva uchastka tiagi (obraztsovyi uchastok tiagi)', *Zheleznodorozhnoe delo*, 28 (1915): 275–8.

Bekhterev, V., 'Vzryv kotla parovoza 1-5-0 "Dekapod" cep. E^r No. 62 na Permskoi zh.d.', *Zheleznodorozhnoe delo: Podvizhnoi sostav, tiaga i masterskie*, 2–3 (1926): 23–6.

Beleliubskii, V., 'Podschet neobkhodimogo kolichestva tovarnykh vagonov na zheleznykh dorogakh: Kontrol' raboty uzlovykh stantsii', *Zhurnal Ministerstva putei soobshcheniia*, 2 (1910): 35–46.

Beliakov, I., 'O transporte', *Ekonomicheskaia zhizn'*, 22 February 1920, p. 1.

Berezhlivyi, 'Eshche o potrebnosti v tovarnykh vagonakh', *Izvestiia sobraniia inzhenerov putei soobshcheniia*, 15 (1913): 242.

Bernatskii, L.N., 'Naivygodneishii uklon', *Tekhnika i ekonomika putei soobshcheniia*, 2/12 (1923): 677–84.

Bernatskii, L.N., 'Po povodu stat'i prof. Iu. Lomonosova "O predel'nykh pod"emakh magistralei"', *Tekhnika i ekonomika putei soobshcheniia*, 1/1 (1923): 3–4.

'Beseda po povodu nedostatka vagonov', *Zheleznodorozhnoe delo*, 32 (1915): 317–32.

Beskrovnyi, L.G., Voronin, E.P., Korablev, Iu.I. et al. (eds), *Zhurnaly Osobogo soveshchaniia dlia obsuzhdeniia i ob"edineniia meropriatii po obornoe gosudarstva (Osoboe soveshchanie po obornoe gosudarstva), 1915–1918 g.g.: Publikatsiia* (Moscow: Institut Istorii SSSR AN SSSR, 1975).

Bezrukikh, P., 'O samokritike', *Transport i khoziaistvo*, 6 (1928): 3–6.

Bilek, J., 'Traction Diagrams Based on the Ratio of the Power to the Weight Hauled', *Monthly Bulletin of the International Railway Congress Association (English Edition)*, 18/1 (January 1936): 41–50.

Bolonov, V., 'Parovozoispitatel'naia laboratoriia', *Krasnyi transportnik*, 3 (1923): 13–15.

Bolonov, V., 'Sovremennye bolezni parovozov', *Krasnyi transportnik*, 4 (1923): 20–24.

Bond, R.C., *A Lifetime with Locomotives* (Cambridge: Goose & Son, 1975).

'Boris A. Bakhmeteff, 1880–1951', *Russian Review*, 10/4 (1951): 311–12.

Borodin, A., 'Experiments on the Steam-Jacketing and Compounding of Locomotives in Russia', *Proceedings of the Institution of Mechanical Engineers* (1886): 297–354, 363–409.

Brandt, A.A., *Kurs parovykh mashin: Lektsii, chitannye v Institute inzhenerov putei soobshcheniia Imperatora Aleksandra I*, 2nd edn (St Petersburg, 1896).

Brandt, A.A., *List'ia pozheltelye: Peredumannoe i perezhitoe* (Belgrade, 1930).

Brockway, F., *Inside the Left: Thirty Years of Platform, Press, Prison and Parliament* (London: George Allen and Unwin, 1942).

Brockway, F., *Outside the Right: A Sequel to 'Inside the Left', with a Lost Play by G. Bernard Shaw* (London: George Allen and Unwin, 1963).

Browder, R.P. and Kerensky, A.F. (comps and eds), *The Russian Provisional Government 1917: Documents*, vol. 1 (Stanford: Stanford University Press, 1961).

[Bublikov, A.A.,] 'Plan zheleznodorozhnoi seti i ego finansirovanie', *Izvestiia Sobraniia inzhenerov putei soobshcheniia*, 1 (1917): 20–27.

Bublikov, A.A., *Russkaia revoliutsiia (ee nachalo, arest tsaria, perspektivy): vpechatleniia i mysli ochevidtsa i uchastnika* (New York, 1918).

Budnitskii, O.V. (ed.), *'Sovershenno lichno i doveritel'no!': B.A. Bakhmetev–V.A. Maklakov, perepiska, 1919–1951, tom 1: avgust 1919–sentiabr' 1921* (Moscow and Stanford: ROSSPEN and Hoover Institution Press, 2001).

Buharin, N. and Preobrazhensky, E., *The ABC of Communism: A Popular Explanation of the Program of the Communist Party of Russia* ([n.p.], 1922).

'Builder of First Diesel Meets "Student" Here', *The Gazette* (Montreal), 31 May 1952, p. 7.

'Built First Diesel Locomotive, Dr G.V. Lomonossoff Dies Here', *The Gazette* (Montreal), 21 November 1952, p. 27.

Carling, D.R., 'The Late Dr G.V. Lomonossoff', *Railway Gazette*, 97/23 (5 December 1952): 622.

Carling, D.R., 'Locomotive Testing on British Railways', *Journal of the Institution of Locomotive Engineers*, 40/217 (September–October 1950): 496–530 and discussions 530–591.

Census of England and Wales, 1921: General Report with Appendices (London: HMSO, 1927).

Census of England and Wales, 1921: General Tables (London: HMSO, 1925).

Chapelon, A., *La Locomotive à Vapeur*, 2nd edn, English translation (Rode: Camden Miniature Steam Services, 2000).

Chechott, A.O. – see also Czeczott, A.O.

Chechott, A.O., 'Doklad inzhenera A.O. Chechotta v komissii po "Ekonomike izyskanii"', *Izvestiia Sobraniia inzhenerov putei soobshcheniia*, 18 (1916): 371–4.

Chechott, A.O., 'Inzhenery putei soobshcheniia i prikladnaia mekhanika (K voprosu o reformakh prepodavaniia v Institute)', *Izvestiia Sobraniia inzhenerov putei soobshcheniia*, 18 (1914): 266–72.

Chechott, A.O., 'K voprosu opytnogo issledovaniia parovozov', *Inzhener*, 2 (1904): 48–52.

Chechott, A.O., 'Nekotorye soobrazheniia o tipakh parovozov dlia proektiruemykh i vnov' stroiashchikhsia dorog', *Izvestiia Sobraniia inzhenerov putei sobshcheniia*, 9 (1915): 185–8.

Chechott, A.O., 'Nekotorye zamechaniia po povodu tablits tiagovykh raschetov inzhenera Godytskogo-Tsvirko', *Zheleznodorozhnoe delo*, 27 (1915): 257–61.

Chechott, A.O., 'O nenormal'nom polozhenii nashei professury', *Izvestiia Sobraniia inzhenerov putei sobshcheniia*, 25 (1916): 518–21.

Chechott, A.O., 'O programme opytov dlia opredeleniia novykh norm tormazheniia tovarnykh poezdov ne po chislu vagonov, a po vesu poezdov', in *XXX soveshchatel'nyi s"ezd inzhenerov podvizhnogo sostava i tiagi v Khar'kove, 25 avgusta 1913 g.* (St Petersburg, 1914), pp. 431–45.

Churilov, G., 'Diletanty o transporte', *Vestnik putei soobshcheniia*, 27 (1925): 1–3.

Coller, P., 'Question V', *Monthly Bulletin of the International Railway Congress Association (English Edition)*, 12/5 (May 1930): 1383–90.

Coller, P., 'Report No.3 (all countries excluding America, the British Empire, China, Japan, Belgium, France, Italy, Portugal, Spain and their Colonies) on the Question of Locomotives of New Types', *Monthly Bulletin of the International Railway Congress Association (English Edition)*, 12/4 (April 1930): 1259–307.

Cox, E.S., *Chronicles of Steam* (London: Ian Allan, 1967).

Cox, E.S., *Locomotive Panorama*, vol. 1 (London: Ian Allan, 1965).

Cox, E.S., *Speaking of Steam* (London: Ian Allan, 1971).

Czeczott, A. – see also Chechott, A.O.

Czeczott, A., 'A Description of the Method of Carrying Out Locomotive Tests on the Polish State Railways', *Monthly Bulletin of the International Railway Congress Association (English Edition)*, 13/7 (July 1931): 575–605.

Daley, W.E., 'Characteristic Energy Diagrams for Steam Locomotives', *Engineering* (19 August 1910): 255–9; and (26 August 1910): 289–90.

Degterev, N.N., 'Integrirovanie uravneniia dvizheniia poezda', *Tekhnika i ekonomika putei soobshcheniia*, 8 (1924): 897–908.

Degterev, N.N., 'Naivygodneishii pod"em i razvitie linii', *Tekhnika i ekonomika putei soobshcheniia*, 2/8 (1923): 156–9.

Degterev, N., 'Otkrytoe pis'mo gg. professoram i prepodavateliam Instituta inzhenerov putei soobshcheniia', *Izvestiia Obshchego biuro soveshchatel'nykh s"ezdov*, 40 (1911): 6.

Dement'ev, K.G., *Stranitsa iz istorii Kievskogo politekhnicheskogo instituta* (Kiev, 1911).

Departament Zheleznodorozhnykh Del, *Kratkii otchet o deiatel'nosti tarifnykh uchrezhdenii i Departamenta zheleznodorozhnykh del za 1889–1913 g.g.* (St Petersburg, 1914).

Diamond, E.L., 'The Horse-power of Locomotives: Its Calculation and Measurement', *Monthly Bulletin of the International Railway Congress Association (English Edition)*, 18/2 (February 1936): 150–89.

'Diesel Electric Locomotive Surpasses Steam on Test', *Oil Engine Power*, 2/12 (1924): 633–7.

'The Diesel Locomotive in Russia', *Railway Gazette*, 44/4 (22 January 1926): 113–14.

'Diesel Locomotives for Main Line Traffic', *Railway Engineer*, 47/561 (October 1926): 357–63, 368.

'Diesel Traction: Summary of a Paper Read Before the Institution of Mechanical Engineers at Manchester Last Night by Professor G. Lomonossoff, Dr. Ing., MIME', *Diesel Railway Traction Supplement to the Railway Gazette* (6 October 1933): 501–3.

Dmitrievskii, S., *Sud'ba Rossii (Pis'ma k druz'iam)* ([Berlin?]: Strela, 1930).

Dmokhovskii, K., 'N.A. Dobrovol'skii – Teplovoz Iu^M 005 i ego ispytaniia v Germanii', *Transport i khoziaistvo*, 2/6 (1927): 143–4.

Dmokhovskii, V., 'Prof. Iu. Lomonosov, "Tekhnicheskie perspektivy zheleznodorozhnogo transporta v blizhaishee vremia"', *Zheleznodorozhnoe delo*, 6 (1924): 120.

Dobrovol'skii, N.A., 'Die dieselelektrische Lokomotive auf den Bahnen der UdSSR', *Organ für die Fortschritte des Eisenbahnwesens*, 82/15 (1927): 286.

Dobrovol'skii, N.A., 'Opyt szhiganiia mazuta v tsilindrakh dvigatelia dizelia teplovoza Iu^E No. 001', *Zheleznodorozhnoe delo: Podvizhnoi sostav, tiaga i masterskie*, 4–5 (1926): 10–11.

Dobrovol'skii, N.A., 'Po povodu statei o dinamometricheskom vagone Iugo–Vost. zh.d. i znachenii dinamometricheskikh lent', *Vestnik Obshchestva tekhnologov*, 16 (1914): 628–9.

Dobrovol'skii, N.A., 'Prof. Iu. Lomonosov: Tiagovye raschety', *Vestnik inzhenerov*, 12 (1915): 564–5.

Dobrovol'skii, N.A., 'Sravnitel'noe ispytanie teplovozov E-MKh-3 i E-EL-2 v 1927 godu', *Zheleznodorozhnoe delo: Podvizhnoi sostav, tiaga i masterskie*, 9–10 (1927): 3–5.

Dobrovol'skii, N.A., *Teplovoz Iu^M 005 i ego ispytanie v Germanii* (Moscow: Transpechat', 1927).

Dodonov, B.F. (ed.), *Zhurnaly zasedanii Vremennogo pravitel'stva, tom 1: Mart–aprel' 1917 goda* (V.A. Kozlov and S.V. Mironenko (eds), *Arkhiv noveishei istorii Rossii, seriia 'Publikatsii', tom VII*) (Moscow: Rosspen, 2001).

'Dokumenty k "Vospominaniiam" gen. A. Lukomskogo', in Gessen, I.V. (ed.), *Arkhiv russkoi revoliutsii*, vol. 3 (reprinted edition Moscow: Terra, 1991), pp. 247–70.

Dombrovskii, A., 'Opisanie teplovoza Iu^E tipa 1-5-1', *Zheleznodorozhnoe delo: Podvizhnoi sostav, tiaga i masterskie*, 4–5 (1926): 9–10.

'Dopros Gen. N.I. Ivanova, 28 iiunia 1917 goda', in Shchegolev, P.E. (ed.), *Padenie tsarskogo rezhima: stenograficheskie otchety doprosov i pokazanii, dannykh v 1917 g. v Chrezvychainoi Sledstvennoi Komissii Vremennogo Pravitel'stva*, vol. 5 (Moscow and Leningrad: Gosudarstvennoe izdatel'stvo, 1926), pp. 313–35.

Dubelir, G.D., 'Blizhaishie zadachi elektrifikatsii zh.d.', *Tekhnika i ekonomika putei soobshcheniia*, 2/11 (1923): 561–4.

Dubelir, G.D., 'Nauka i transport', *Krasnyi transportnik*, 6 (1923): 3–4.

Dul'nev, A.B., 'K voprosu o naivygodneishem sostave tovarnykh poezdov', *Zheleznodorozhnoe delo*, 7–8 (1931): 38–40.

Dva dostizheniia v oblasti sovetskoi zheleznodorozhnoi tekhniki: Doklady VI Plenumu Biuro pravlenii zheleznykh dorog (Moscow, 1925).

Dzerzhinskii, F.E., *Izbrannye proizvedeniia v dvukh tomakh*, vol. 1 (Moscow: Izdatel'stvo politicheskoi literatury, 1977).

Egorchenko, V.F., 'Forsirovka parovozov', *Parovoznoe khoziaistvo*, 8 (1935): 10–11.

Egorchenko, V.F., 'G.S. Kizilov. "Osnovy tiagovykh raschetov". Odobreno TsOP-KADROM NKPS v kachestve uchebnogo posobiia dlia shkol parovoznykh mashinistov, Transzheldorizdat NKPS 1934 g.', *Parovoznoe khoziaistvo*, 4 (1935): 38–9.

Egorchenko, V.F., 'K predstoiashchim ispytaniiam tormazov Kazantseva i Kuntse-Knorra v SSSR v sviazi s vyborom sistemy tormaza dlia nashego tovarnogo parka', *Zheleznodorozhnoe delo*, 10 (1925): 29–41.

Egorchenko, V.F. et al., 'Khorosha li vvodimaia sistema numeratsii parovozov', *Zheleznodorozhnoe delo*, 2 (1929): 66–7.

'Ekonomika izyskanii zheleznykh dorog: Otchet o deiatel'nosti komissii, obrazovannoi pri Sobranii Inzhenerov Putei Soobshcheniia dlia osveshcheniia voprosov, zatronutykh v trude K.N. Kashkina', *Izvestiia Sobraniia inzhenerov putei soobshcheniia*, 17 (1916): 350–354; 18 (1916): 367–71; and 19 (1916): 393-6.

El'kin, L., 'Narekaniia na kazennoe khoziaistvo zheleznykh dorog', *Izvestiia Sobraniia inzhenerov putei soobshcheniia*, 9 (1911): 1–5.

Ell, S.O., 'Developments in Locomotive Testing', *Journal of the Institution of Locomotive Engineers*, 43/235 (1953): 561–633.

Ell, S.O., 'The Mechanics of the Train in the Service of Railway Operation', *Journal of the Institute of Locomotive Engineers*, 48/5 (1958/1959): 528–61 and discussion pp. 561–91.

Enaes, [engineer], 'K voprosu o snabzhenii zheleznykh dorog zheleznodorozhnymi prinadlezhnostiami v sviazi s voinoi', *Zheleznodorozhnoe delo*, 12 (1915): 86–7.

Ermolaev, N.P., 'Pasportnye krivye parovoza serii Ts', *Zheleznodorozhnoe delo*, 2 (1931): 22–6.

Erofeev, N.S., 'K voprosu o proektirovaniia novykh parovozov', *Vestnik inzhenerov*, 3/17–18 (1917): 368–73.

Eshchenko, V.D., 'Umen'shenie tiazhelykh sostavov i uvelichenie skorosti dvizheniia – vykhod iz polozheniia (K razrabotke piatiletnego plana)', *Zheleznodorozhnoe delo: obshchii otdel*, 1 (1928): 1–4.

'Evoliutsiia lokomotiva za 100 let (Doklad professora Iu.V. Lomonosova)', *Vestnik putei soobshcheniia*, 47 (1925): 16.

Ewing, A.W., *The Man of Room 40: The Life of Sir Alfred Ewing* (London: Hutchinson, 1939).

Ewing, J.A., *An Engineer's Outlook* (London: Methuen, 1933).

Ewing, J.A., *The Steam-Engine and Other Heat-Engines*, 2nd edn (Cambridge: Cambridge University Press, 1897).

F.K., 'Neskol'ko slov zashchitniku kazennogo zheleznodorozhnogo khoziaistva (Po povodu stat'i inzh. El'kina: "Narekaniia na kazennoe khoziaistvo zheleznykh dorog")', *Izvestiia Sobraniia inzhenerov putei soobshcheniia*, 19 (1911): 2–3.

F.K., 'O parovoznom khoziaistve kazennykh zh.d. (Zametki po povodu truda inzhenerov S.N. Kul'zhinskogo i L.L. El'kina "Parovoznyi park kazennykh zheleznykh dorog")', *Izvestiia Sobraniia inzhenerov putei soobshcheniia*, 17 (1911): 4–8.

Fedorov, M., 'Pamiati Mikhaila Aleksandrovicha Izrail'sona', *Zheleznodorozhnoe delo*, 4 (1927): 18.

Fel'dt, V., 'K stat'e inzhenera Nagrodskogo "Po voprosu o prepodavaniia eksploatatsii zheleznykh dorog"', *Izvestiia Sobraniia inzhenerov putei soobshcheniia*, 2 (1917): 33–4.

Fel'dt, V., '"Prezhde i teper"': Voenno-zheleznodorozhnye voprosy v sovremennoi Velikoi voine', *Zheleznodorozhnoe delo*, 21–2 (1915): 189–97; 13–14 (1916): 113–22; 23–4 (1916): 209–11; 29–30 (1916): 259–63; 33–4 (1916): 293–6; 39–40 (1916): 356–61; and 41–2 (1916): 379–84.

Fowler, H., 'Development of the Locomotive', *Monthly Bulletin of the International Railway Congress Association (English Edition)*, 5/12 (December 1923): 1097–100.

Franco, I. and Labryn, P., *Internal Combustion Locomotives and Motor Coaches* (The Hague: Martinus Nijhoff, 1931).

Frolov, A.N., 'Deiatel'nost' Osobogo Soveshchaniia po perevozkam', *Zheleznodorozhnoe delo*, 12 (1916): 105–6 and 15 (1916): 125–30.

Frolov, A., 'K voprosu o kazennom i chastnom khoziaistvakh na zheleznykh dorogakh', *Izvestiia Sobraniia inzhenerov putei soobshcheniia*, 16 (1913): 253–6.

Frolov, A.N., 'O metodakh rascheta vygodnosti podtalkivaniia', *Zheleznodorozhnoe delo*, 8–9 (1930): 10.

Frolov, A.N., 'O sposobakh regulirovki perevozki gruzov po zheleznym dorogam: Stenograficheskii otchet o besede v VIII-m Otdele I.R. Tekhnicheskogo [Obshchestva] 4 fevralia 1916 goda pod predsedatel'stvom A. Frolova', *Zheleznodorozhnoe delo*, 19–20 (1916): 169–73; 21–2 (1916): 187–9; 23–4 (1916): 205–9; 25–6 (1916): 223–7; and 27–8 (1916): 241–5.

Frolov, A.N., 'O zheleznodorozhnykh zatrudneniiakh (Doklad A.N. Frolova na Vserossiiskom Soveshchanii po bor'be s dorogoviznoi 11–13 iiulia 1915 g. v Moskve)', *Zheleznodorozhnoe delo*, 28 (1915): 269–75.

Fry, L.H., 'The Locomotive Testing Plant and its Influence on Steam Locomotive Design', *Transactions of the American Society of Mechanical Engineers*, 47 (1925): 1267–85 and discussion 1285–93.

Fuhrmann, J.T. (ed.), *The Complete Wartime Correspondence of Tsar Nicholas II and the Empress Alexandra: April 1914–March 1917* (Westport: Greenwood, 1999).

G.N., 'Pervyi teplovoz – v Moskve', *Tekhnika i zhizn'*, 2 (1925): 6.

Gakkel', Ia.M., 'O teplovoze G^{E} 001', in *Trudy XXXIV soveshchatel'nogo s"ezda inzhenerov podvizhnogo sostava i tiagi v Moskve s 1-go po 9-oe aprelia 1925 goda* (Moscow, 1926), pp. 29–52.

Ganitskii, I. (ed.), *Illiustrirovannyi sbornik materialov k istorii vozniknoveniia Kievskogo Politekhnicheskogo Instituta: Pamiati Viktora L'vovicha Kirpicheva* (Kiev, 1914).

Ge, I., 'Ekspeditsiia gen. Ivanova na Petrograd', *Krasnyi arkhiv*, 4 (1926): 225–32.

Gel'zin, S.L., 'Iuzhnoe Voenno-Tekhnicheskoe Biuro pri TsK RS-DRP', *Katorga i ssylka*, 12 (1929): 28–42.

Genkin, I., 'Revoliutsioner-podvizhnik – B.P. Zhadanovskii', *Katorga i ssylka*, 3 (1925): 200–211.

Gerasimov, A.V., *Na lezvii s terroristami* (Paris: YMCA, 1985).

German State Railways General Management, 'The Horse-power of Locomotives: Its Calculation and Measurement, by E.L. Diamond', *Monthly Bulletin of the International Railway Congress Association (English Edition)*, 19/8 (August 1937): 1796–7.

Gersevanov, M.N., *Institut Inzhenerov putei soobshcheniia Imperatora Aleksandra I v period 1890–1896 gg.* (St Petersburg: Izdanie Sobraniia inzhenerov putei soobshcheniia, 1896).

Gessen, I.V. (ed.), *Arkhiv russkoi revoliutsii*, 22 vols (reprinted edition Moscow: Terra, 1991).

Gibbons, C.H., 'Commercial Applications of High-Speed Oil Engines', *Transactions of the American Society of Mechanical Engineers*, 51 (1929), OGP-51-5: 41–8 and discussion 48–52.

Glavneishie rezul'taty opytov nad parovozom tipa 0-5-0, serii E^v, 2nd edn (Moscow: Transzheldorizdat, 1935).

Glazenap, V.E., 'Teplovoz sistemy prof. Iu.V. Lomonosova', *Elektrichestvo*, 2 (1925): 105–7.

'Godovshchina stakhanovsko-krivonosovskogo dvizheniia', *Sotsialisticheskii transport*, 6 (1936): 10–24.

Godytskii-Tsvirko, A.M., 'Neskol'ko slov na "Zamechaniia" A.O. Chechotta po povodu "Tablits tiagovykh raschetov"', *Zheleznodorozhnoe delo*, 39 (1915): 410–11.

Godytskii-Tsvirko, A., 'Ratsional'noe proektirovanie zheleznodorozhnoi linii', *Izvestiia Sobraniia inzhenerov putei soobshcheniia*, 4 (1918): 65–72.

Godytskii-Tsvirko, A.M., 'Tablitsy tiagovykh raschetov', *Zheleznodorozhnoe delo*, 3 (1915): 12–16 and 4 (1915): 21–6.

Goldstein, H., *Classical Mechanics*, 2nd edn (Reading, MA: Addison-Wesley, 1980).

Golikov, G.N., Antoniuk, D.I., Mchedlov, M.P. et al. (eds), *Vladimir Il'ich Lenin: Biograficheskaia khronika*, 12 vols (Moscow: Izdatel'stvo politicheskoi literatury, 1970–1982).

Gololobov, M.V., 'Ob integrirovaniia uravnenii dvizheniia zheleznodorozhnogo poezda', *Vestnik inzhenerov*, 2/10 (1916): 383–5.

Gololobov, M.V., *Opytnaia parovoznaia stantsiia Putilovskogo zavoda* (St Petersburg, 1907).

Gololobov, M.V., 'Opytnaia parovoznaia stantsiia Putilovskogo zavoda, zameniaiushchaia obkatku parovozov', *Zheleznodorozhnoe delo*, 20 (1905): 221–33.

Gololobov, M.V., 'O ratsional'nom vide zavisimosti N/H ot skorosti', *Izvestiia Sobraniia inzhenerov putei soobshcheniia*, 3 (1918): 47–53; and 4 (1918): 73–9.

Gololobov, M.V., 'O sluzhbe parovozov, snabzhennykh peregrevateliami', in *Protokoly zasedanii XXVIII soveshchatel'nogo s''ezda inzhenerov sluzhby podvizhnogo sostava i tiagi russkikh zheleznykh dorog v Rige, 27 maia–8 iiunia 1911 g.* (St Petersburg, [1912]), pp. 22–32.

Gololobov, M.V., 'Otvet inzh. G.V. Lebedevu po povodu zametki "K voprosu ob integrirovaniia i t.d."', *Vestnik inzhenerov*, 3/1 (1917): 28.

Gololobov, M.V., 'Proekt parovoznoi laboratorii dlia russkikh zheleznykh dorog', in *Protokoly zasedanii XXVII soveshchatel'nogo s''ezda inzhenerov sluzhby podvizhnogo sostava i tiagi russkikh zheleznykh dorog, v Varshave, v avguste 1909 g.* (St Petersburg, 1910), pp. 251–90.

Gololobov, M.V., 'Sravnenie dvukh sposobov issledovaniia rabotosposobnosti parovozov na spetsial'nykh stantsiiakh i probnykh poezdkakh i organizatsiia pervoi takoi stantsii na russkikh zhel. dorogakh', in *XXIX soveshchatel'nyi s''ezd inzhenerov podvizhnogo sostava i tiagi v Rostove, 24 maia–10 iiunia 1912 g.* (St Petersburg, 1913), pp. 85–7.

Gololobov, M.V., 'Sravnenie dvukh sposobov issledovaniia rabotosposobnosti parovozov – na spetsial'nykh stantsiiakh i probnykh poezdkakh, i organizatsiia pervoi takoi stantsii na russkikh zhel. dorogakh', in *XXX soveshchatel'nyi s''ezd inzhenerov podvizhnogo sostava i tiagi v Khar'kove, 25 avgusta 1913 g.* (St Petersburg, 1914), pp. 64–77.

Gololobov, M.V. and Solodovnikov, A.A., 'K voprosu o soprotivlenii vozdukha dvizheniiu poezdov: Po povodu otveta inzh. N.A. Rynina', *Izvestiia Obshchego biuro soveshchatel'nykh s''ezdov*, 5 (1914): 481–96.

Gololobov, M.V. and Solodovnikov, A.A., 'K voprosu o soprotivlenii vozdukha dvizheniiu poezdov (Po povodu stat'i Inzh. N.A. Rynina pod tem zhe zaglaviem)', *Izvestiia Obshchego biuro soveshchatel'nykh s''ezdov*, 1 (1914): 34–63.

Goraiskii, K., 'Po povodu st. L.A. "Novye inzhenery putei soobshcheniia", pomeshch. v No.40 "Izvestii Sobr. I.P.S." za 1913 god', *Izvestiia Sobraniia inzhenerov putei soobshcheniia*, 5 (1914): 63–4.

Gordeenko, Ia.N., *Kurs zheleznykh dorog po programme utverzhdennoi g. Ministrom putei soobshcheniia 4-go aprelia 1888 goda dlia ekzamena na zvanie tekhnika putei soobshcheniia* (St Petersburg, 1898).

Gosudarstvennaia Duma, Sozyv IV, Sessiia 1-ia (1912–1913): Zakonproekty Ministerstva putei soobshcheniia, 3 vols ([St Petersburg, 1913]).

Gosudarstvennaia Duma, Sozyv IV, Sessiia 2-ia (1913–1914): Zakonproekty Ministerstva putei soobshcheniia, 2 vols ([St Petersburg, 1914]).

Gosudarstvennaia Duma, Sozyv IV, Sessiia 3-e (1914–1915): Zakonproekty Ministerstva putei soobshcheniia ([St Petersburg, 1915]).

Gosudarstvennaia Duma, Sozyv IV, Sessiia 4-ia (1915–1916): Zakonproekty Ministerstva putei soobshcheniia ([St Petersburg, 1916]).

Gosudarstvennaia Kantseliariia (comp.), *Obzor deiatel'nosti Vtorogo Departamenta Gosudarstvennogo Soveta po rassmotreniiu del o chastnykh zheleznykh dorogakh za vremia s 1906 po 1913 god* (St Petersburg, 1914).

Gresley, H.N., 'Locomotive Experimental Stations', *Monthly Bulletin of the International Railway Congress Association (English Edition)*, 14/1 (January 1932): 12–24.

Gresley, H.N., 'Question V: Recent Improvements in Steam Locomotives of the Usual Type and Test of New Designs (High-pressure Reciprocating and Turbine Locomotives) as Regards Construction, Quality of Materials Used, Efficiency, Working Conditions, Maintenance and Financial Results; Testing Locomotives at Locomotive Experimental Stations, and in Service with Dynamometer Cars and Brake Locomotives', *Monthly Bulletin of the International Railway Congress Association (English Edition)*, 19/6 (June 1937): 1521–7.

Gresley, H.N., 'Report: Great Britain, Dominions and Colonies, America, China and Japan', *Monthly Bulletin of the International Railway Congress Association (English Edition)*, 18/12 (December 1936): 1291–445.

Grigor'ev, I., 'Pora perestroit' nauchno-issledovatel'skuiu rabotu na transporte', *Sotsialisticheskii transport*, 7 (1938): 84–7.

Grinenko, R. and Isaakian, O., 'Neuere Versuche mit russischen Dampflokomotiven', *Zeitschrift des Vereines Deutscher Ingenieure*, 73/10 (1929): 339–44.

Grinenko, R.P., 'O szhiganii antratsita v parovoznoi topke', *Tekhnika i ekonomika putei soobshcheniia*, 2/11 (1923): 588–611.

Grinevetskii, V.I., *Poslevoennye perspektivy russkoi promyshlennosti* (Moscow, 1922 edn).

Grinevetskii, V.I., *Problema teplovoza i ego znachenie dlia Rossii* (Moscow: Izdatel'stvo Teplotekhnicheskogo instituta, 1923).

'G.V. Lomonossoff, Engineer, was 76', *New York Times*, 21 November 1952, p. 25:4.

Harcave, S. (transl. and ed.), *The Memoirs of Count Witte* (Armonk: M.E. Sharpe, 1990).

Harcavi, G., 'The Co-ordination of the Duties of the Operating and Locomotive Running Departments and their Organic Reunion', *Monthly Bulletin of the International Railway Congress Association (English Edition)*, 11/1 (January 1929): 44–58.

I., '25 let opytnogo issledovaniia parovozov na Rossiiskoi zheleznodorozhnoi seti', *Tekhnika i ekonomika putei soobshcheniia*, 2/12 (1923): 760–762.

I.T., 'Predstoiashchee razvitie seti zheleznykh dorog v Rossii', *Izvestiia Sobraniia inzhenerov putei soobshcheniia*, 26 (1912): 3–5.

Iakovlev, A.N. et al. (eds), *Rossiia i SShA: diplomaticheskie otnosheniia, 1900–1917* (Moscow: Mezhdunarodnyi Fond 'Demokratiia', 1999).

Iakovlev, A.N. et al. (eds), *Sovetsko–amerikanskie otnosheniia: Gody nepriznaniia, 1918–1926* (Moscow: Mezhdunarodnyi Fond 'Demokratiia', 2002).

Ianitskii, N.P., 'Metod proizvodstva opytnykh poezdok', *Parovoznoe khoziaistvo*, 5–6 (1935): 23–7.

Ianushevskii, P.S., 'Ergometr inertsii', *Zheleznodorozhnoe delo*, 2 (1911): 14–19.

Ianushevskii, P.S., 'K voprosu ob uvelichenii pod"emnoi sily tovarnykh vagonov', *Izvestiia Obshchego biuro soveshchatel'nykh s"ezdov*, 6 (1912): 522–7.

Ianushevskii, P.S., 'Teplovozy', *Zheleznodorozhnoe delo*, 8 (1924): 17–24.

Iaroslavskii, E.M. (ed.), *Protokoly s"ezdov i konferentsii Vsesoiuznoi kommunisticheskoi partii (b): Pervaia konferentsiia voennykh i boevykh organizatsii RSDRP, noiabr' 1906 god* (Moscow: Partizdat, 1932).

Iastrebov, F., *1905 rik u Kievi* (Khar'kov-Kiev: Proletar', 1930).

Iatsina, V., 'Otyskanie vygodneishego napravleniia proektiruemykh zhelez-nodorozhnykh linii po naimen'shei summe stroitel'nykh i eksploatatsionykh raskhodov', *Zhurnal Ministerstva putei soobshcheniia*, 8 (1907): 3–47; and 9 (1907): 82–127.

Idel'son, V.G. and Korbelov, Sh.V., 'Vzryv kotla parovoza serii E^G No.5213 na Zakavkazskikh zhel. dorogakh', *Zheleznodorozhnoe delo: Podvizhnoi sostav, tiaga i masterskie*, 3–4 (1927): 3–4.

Ignat'ev, A.A., *Piat'desiat' let v stroiu* (Stalingrad, 1941).

'I.N. Borisov (1860–1928g.)', *Transport i khoziaistvo*, 5 (1928): 3–5.

'Intensifikatsiia Moskovsko–Kurskoi zhel. dor.', *Izvestiia Sobraniia inzhenerov putei soobshcheniia*, 10 (1916): 186–9.

'The Internal-Combustion Locomotive', *Engineering* (5 September 1913): 326.

'Introduction to Railway Mechanics', *The Railway Engineer*, 54/638 (April 1933): 123.

'Introduction to Railway Mechanics', *Railway Gazette*, 58/12 (24 March 1933): 412.

Inzhener, 'Kazennye i chastnye zheleznye dorogi', *Izvestiia Sobraniia inzhenerov putei soobshcheniia*, 15 (1912): 2–4.

Inzhener, 'Omolozhenie', *Izvestiia Obshchego biuro soveshchatel'nykh s"ezdov*, 3 (1912): 10–11.

Inzhener, '1000 pudov', *Izvestiia Sobraniia inzhenerov putei soobshcheniia*, 19 (1911): 3–4.

Inzhenernyi sovet MPS, *Zhurnaly komissii podvizhnogo sostava, tiagi i masterskikh* [from 1911: *Zhurnaly komissii podvizhnogo sostava i tiagi*] (St Petersburg, 1907–1917).

Iofe, A.F. et al., *Nauka i tekhnika SSSR, 1917–1927*, vol. 3 (Moscow: Rabotnik prosveshcheniia, 1928).

Isaakian, O., 'Proekty teplovozov, razrabotannye Teplovoznym Biuro "GOMZY"', *Tekhnika i ekonomika putei soobshcheniia*, 9–10 (1924): 1114–16.

'Ispytaniia tormozov Kazantseva i Kuntse-Knorre', *Vestnik putei soobshcheniia*, 45 (1925): 23.

'Ispytatel'naia stantsiia dlia issledovaniia podvizhnogo sostava i materialov otopleniia i smazki i o vybore mesta dlia ee ustroistva: Doklad Obshchego biuro', *Izvestiia Obshchego biuro soveshchatel'nykh s"ezdov*, 2 (1910): 145–9.

Ivanov, P.G., 'Po povodu stat'i "Uvelichenie parka tovarnykh vagonov"', *Izvestiia Sobraniia inzhenerov putei sobshcheniia*, 6 (1915): 107–9.

Ivanovskii, [I.K.], 'Pis'm[o] v redaktsiiu', *Izvestiia Sobraniia inzhenerov putei soobshcheniia*, 21 (1915): 448–9.

'Iz pisem v redaktsiiu', *Izvestiia Sobraniia inzhenerov putei soobshcheniia*, 12 (1911): 12–13.

Izvlechenie iz zhurnala inzhenernogo soveta, No.49 – 1909 g. Po voprosu o rezul'tatakh sravnitel'nykh ispytanii, proizvedennykh v 1907–1908 g.g., parovozov normal'nogo tipa o 4/4 osiakh i izmenennogo tipa Vostochnoi Kitaiskoi zheleznoi dorogi o 4/5 osiakh i nabliudenii nad deformatsiei puti i mostovykh (St Petersburg, 1910).

Johnson, R.P., 'Dynamometer Cars', *Baldwin Locomotives*, 5/3 (January 1927): 35–43.

K., 'Vosstanovlenie nashikh zheleznykh dorog', *Ekonomicheskaia zhizn'*, 11 July 1920, p. 1.

K.O., 'Po povodu stat'i inzhenera Larionova o sravnitel'noi khoziaistvennoi eksploatatsii kazennykh i chastnykh zheleznykh dorog', *Izvestiia Sobraniia inzhenerov putei soobshcheniia*, 15 (1913): 234–40.

Kaganovich, L.M., 'God pod''ema i blizhaishie zadachi zheleznodorozhnogo transporta', *Zheleznodorozhnaia tekhnika*, 3–4 (1936): 3–4.

[Kaganovich, L.M.,] 'Ob uluchshenii ispol'zovaniia parovozov i organizatsii dvizheniia poezdov: prikaz Narodnogo komissara putei soobshcheniia No.183/Ts, 7 avgusta 1935 g.', *Parovoznoe khoziaistvo*, 8 (1935): 4–7

Kaganovich, L.M., *Voprosy zheleznodorozhnogo transporta v sviazi s stakhanovskim dvizheniem: Doklad na plenume TsK VKP(b) 22 dekabria 1935 g.* (Moscow: Partizdat TsK VKP(b), 1936).

Kandaurov, P., 'O tekhnicheskikh meropriiatiiakh, neobkhodimykh dlia vozmozhnosti osushchestvleniia intensivnoi postroiki zheleznykh dorog posle voiny', *Zheleznodorozhnoe delo*, 37–38 (1916): 331–9.

Kapitsa, E.L. and Rubinin, P.E., *Petr Leonidovich Kapitsa: Vospominaniia, pis'ma, dokumenty* (Moscow: Nauka, 1994).

Kareisha, S.D., *Istoriia Instituta s 1905–1934 g. (posmertnyi trud prof. S.D. Kareishi po zakazu V.I. Romanova)* (Leningrad, 1933–1934).

Kartashov, N.I., 'Noveishie dopolneniia k tiagovym raschetam', *Zheleznodorozhnaia tekhnika*, 9 (1936): 7–13.

Kartashov, N.I., *Opytnoe issledovanie parovozov* (Khar'kov, 1902)

Kartashov, N.I., 'Parovozy mogut rabotat' na liubom toplive', *Zheleznodorozhnaia tekhnika*, 1 (1937): 16–22.

Kartashov, N.I., 'Podnimem paroobrazovanie v parovoznykh kotlakh na dolzhnuiu vysotu', *Zheleznodorozhnaia tekhnika*, 14 (1936): 5–17.

Kartashov, N.I., 'Regulirovanie parovoznoi mashiny', *Zheleznodorozhnaia tekhnika*, 16 (1936): 21–8.

Kashkin, K.N., 'Ekonomika izyskanii zheleznykh dorog (Vstuplenie k besede, sostoiavsheisia v Sobranii I.P.S. 28 noiabria 1914 g.', *Izvestiia Sobraniia inzhenerov putei soobshcheniia*, 35 (1914): 566–7.

Kashkin, K.N., 'O podschetakh inzhenera Slovikovskogo vygodnosti uvelichenii ili umen'shenii sostavov poezdov: Prilozhenie No.3 k "Otchetu o deiatel'nosti komissii po ekonomike izyskanii": Doklad K.N. Kashkina', *Izvestiia Sobraniia inzhenerov putei soobshcheniia*, 19 (1916): 396–7.

'K diskussii po predlozheniiu V.D. Eshchenko: "Umen'shenie tiazhelykh sostavov i uvelichenie skorosti dvizheniia – vykhod iz polozheniia"', *Zheleznodorozhnoe delo: obshchii otdel*, 5 (1928): 10.

Kemmer, N.P., 'O predel'nykh skorostiakh podkhoda tovarnykh poezdov k signalam (Raschety tormozheniia)', *Izvestiia Obshchego biuro soveshchatel'nykh s"ezdov*, 4 (1915): 292–323.

Kerensky, A.F., *The Catastrophe: Kerensky's Own Story of the Russian Revolution* (New York: Appleton, 1927).

Kerensky, A., *The Crucifixion of Liberty* (London: Arthur Barker, 1934).

Kharitonov, I., 'Gidravlicheskaia peredacha v teplovozakh', *Tekhnika i ekonomika putei soobshcheniia*, 3–4 (1924): 416–21.

Khlevniuk, O.V., Davies, R.W., Kosheleva, L.P. et al. (comps), *Stalin i Kaganovich: Perepiska, 1931–1936 gg.* (Moscow: Rosspen, 2001).

Khodanovich, V.D., 'O naivygodneishem vese tovarnykh poezdov', *Zheleznodorozhnoe delo: obshchii otdel*, 4 (1928): 1–5.

Khomiakov, N., 'Pamiati Vladimira Grigor'evicha Lomonosova', *Dneprovskii vestnik*, 197 (1905).

Klenov, D., 'Rabota Nauchno-Tekhnicheskogo komiteta v 1922–1923 godu (Po materialam nauchnogo sekretariata NTK)', *Tekhnika i ekonomika putei soobshcheniia*, 1 (1924): 83–90.

Koestler, A., *Darkness at Noon* (London: Cape, 1940).

Kokoshko, S., *Bil'shoviky u Kyievi naperedodni i za revoliutsii 1905–1907 rr.* (Kharkiv and Kiev: Proletar', 1930).

Kollontai, A.M., *Diplomaticheskie dnevniki, 1922–1940*, 2 vols (Moscow: Academia, 2001).

Kollontai, M., 'Dvigatel' Stilla', *Tekhnika i ekonomika putei soobshcheniia*, 1/5 (1923): 284–9 (including 'Predislovie' by Iu.V. Lomonosov, p. 284).

Komarov, N., 'Ocherki po istorii mestnykh i oblastnykh boevykh organizatsii partii sots.-revoliutsionerov 1905–1909 gg. Ocherk 1: Ukrainskaia boevaia gruppa p.s.-r. (1905–1908 gg.)', *Katorga i ssylka*, 4 (1926): 56–81.

Konshin, S.I., 'Uproshchenie tiagovykh raschetov pomoshch'iu spetsial'nykh treugol'nika i lineiki', *Zheleznodorozhnoe delo: Podvizhnoi sostav, tiaga i masterskie*, 3–4 (1927): 14–15.

Koriukin, E., 'Ob opytakh s parovozami Malleta i Dekapod', *Zheleznodorozhnoe delo*, 37–8 (1916): 353–4.

Korsh, E.V., 'Delo perveishei vazhnosti', *Izvestiia Obshchego biuro soveshchatel'nykh s"ezdov*, 42 (1911): 1–3.

Korsh, E.V., 'Razvitie russkoi zheleznodorozhnoi seti', *Izvestiia Sobraniia inzhenerov putei soobshcheniia*, 34 (1912): 2–9.

Kosiakin, L.D., 'K uvelicheniiu provozosposobnosti Nikolaevskoi zheleznoi dorogi', *Zheleznodorozhnoe delo*, 12 (1916): 103–5.

Kosiakin, L.D., 'Neskol'ko slov o primenenii metodov uplotneniia na Moskovsko–Kurskoi doroge', *Zheleznodorozhnoe delo*, 25–6 (1916): 227–30.

Kosiakin, L., 'Uplotnenie i parallel'nost' rabot v primenenii k tekhnicheskomu osmotru i tekushchemu remontu tovarnykh vagonov pri sledovanii ikh komplektami', *Zheleznodorozhnoe delo*, 24 (1915): 225–31.

Kovalev, [inzhener] and Makulov, [inzhener] kniaz', 'Rezul'taty primeneniia smennoi i podsmennoi ezdy na Iuzhnykh dorogakh za 1915 god', *Zheleznodorozhnoe delo*, 27–8 (1916): 245–8.

Kozhevnikov, V., 'Ekonomkia predel'nykh pod"emov', *Izvestiia Sobraniia inzhenerov putei soobshcheniia*, 13 (1916): 262–7.

Krasin, L.B., *Bol'sheviki v podpol'e* (Moscow: Molodaia gvardiia, 1932).

Krasin, L.B., 'Bol'shevistskaia partiinaia tekhnika', in *Tekhnika bol'shevistskogo podpol'ia: sbornik statei i vospominanii*, vyp. 1 (Moscow and Leningrad: Gosudarstvennoe Izdatel'stvo, 1924): 7–15.

Krasin, L.B., *Delo davno minuvshikh dnei* (Moscow: Molodaia gvardiia, 1931).

Krasovskii, P.I., 'Amerikanskie teplovozy (Iz zagranichnykh zhurnalov)', *Zheleznodorozhnoe delo: Podvizhnoi sostav, tiaga i masterskie*, 1 (1926): 9–11.

Krasovskii, P.I., 'O sravnenii vygodnosti eksploatatsii moshchnykh i slabykh parovozov', *Tekhnika i ekonomika putei soobshcheniia*, 2 (1924): 160–164.

Krasovskii, P.I., 'Podvizhnoi sostav i tiaga za desiatiletie 1917–1927 g.g.', *Zheleznodorozhnoe delo*, 10–11 (1927): 23–6.

Krasovskii, P.I., 'Znachenie opytov nad tipami russkikh parovozov dlia otsenki postroennykh parovozov, vybora tipa i opredeleniia razmerov proektiruemykh parovozov: K XXV-letiiu pervoi opytnoi poezdki Iu.V. Lomonosova', *Zheleznodorozhnoe delo*, 1 (1924): 7–10.

Kratkii istoricheskii ocherk uchebnykh zavedenii vedomstva putei soobshcheniia (St Petersburg, 1900).

Kratkii ocherk deiatel'nosti Russkikh zheleznykh dorog vo vtoruiu otechestvennuiu voinu, chast' 1-aia (S nachala voiny po 1 ianvaria 1915 goda) (Petrograd: Upravlenie Zheleznykh Dorog, 1916).

Kratkii ocherk deiatel'nosti Russkikh zheleznykh dorog vo vtoruiu otechestvennuiu voinu, chast' 2-aia (Pervoe polugodie 1915 goda) (Petrograd: Upravlenie Zheleznykh Dorog, 1916).

Kravets, F.P., 'Eksploatatsiia zheleznykh dorog za 10 let', *Zheleznodorozhnoe delo*, 10–11 (1927): 16–23.

Kriger-Voinovskii, E.B., *Doklad Sovetu ministrov: o putiakh soobshcheniia i usloviiakh perevozok v tretii god voiny* ([Petrograd, 1917]).

Kriger-Voinovskii, E.B., 'O premirovanii rezul'tatov eksploatatsii po sluzhbe podvizhnogo sostava i tiagi', in *Protokoly zasedanii XXIV soveshchatel'nogo s"ezda inzhenerov sluzhby podvizhnogo sostava i tiagi russkikh zheleznykh dorog, sozvannogo v Varshave na 23 noiabria 1902 g.* (St Petersburg, 1903), pp. 91–193.

Kriger-Voinovskii, E.B., *Zapiski inzhenera* (Moscow: Russkii put', 1999).

Krivonos, P.F., 'Prizyv Petra Krivonosa', *Parovoznoe khoziaistvo*, 9 (1935): 4–6.

Krushinskaia, A.A., 'Nauchno-issledovatel'skaia rabota v oblasti zheleznodorozhnogo transporta v Soed. Shtatakh Severnoi Ameriki', *Zheleznodorozhnoe delo: obshchii otdel*, 11–12 (1928): 39–40.

Krylov, A.N., *Moi vospominaniia* (Moscow: AN SSSR, 1945).

Krzhichkovskii, K., 'Po povodu doklada inzhenera N.M. Khlebnikova "O meropriiatiiakh dlia ponizheniia protsenta bol'nykh vagonov na zheleznykh dorogakh", chitannogo v Petrograde, Moskve i Khar'kove', *Zheleznodorozhnoe delo*, 40 (1915): 419–20.

Kucherenko, S.M., 'Povysim konstruktivnye skorosti', *Zheleznodorozhnaia tekhnika*, 1 (1936): 7–11.

Kucherenko, S., 'Problema dinamiki lokomotivov', *Zheleznodorozhnaia tekhnika*, 2 (1937): 7–27.

Kul'zhinskii, S., 'Ob uluchshenii Moskovskogo zheleznodorozhnogo uzla', *Izvestiia Sobraniia inzhenerov putei soobshcheniia*, 9 (1913): 132–4.

Kunitskii, S.K. (ed.), *Kratkii istoricheskii ocherk deiatel'nosti Inzhenernogo Soveta za XXV let, s 1892 g. po 1917 g.* (Petrograd, 1917).

'K voprosu o vosstanovlenie nashikh zheleznykh dorog', *Ekonomicheskaia zhizn'*, 25 July 1920, p. 2.

L.A., 'Neudachnaia kritika', *Izvestiia Sobraniia inzhenerov putei sobshcheniia*, 5 (1915): 81–2.

L.A., 'O zheleznodorozhnykh zatrudneniiakh', *Izvestiia Sobraniia inzhenerov putei soobshcheniia*, 5 (1916): 73–4.

L.A., 'Stroitel'nye grekhi', *Izvestiia Sobraniia inzhenerov putei soobshcheniia*, 15 (1915): 314–16.

Larionov, A.M., *Istoriia Instituta inzhenerov putei soobshcheniia imperatora Aleksandra I-go za pervoe stoletie ego sushchestvovaniia, 1810–1910* (St Petersburg, 1910).

Larionov, A.M., 'K voprosu o sravnitel'noi khoziaistvennosti eksploatatsii kazennykh i chastnykh zheleznykh dorog (Po povodu stat'i L.F. Shukhtana "O rezul'tatakh eksploatatsii chastnykh zhel. dor. v 1911 g")', *Izvestiia Sobraniia inzhenerov putei soobshcheniia*, 3 (1913): 30–32; 4 (1913): 45–52; and 5 (1913): 63–72.

Larionov, A.M., *O sravnenii khoziaistvennosti ekspluatatsii kazennykh i chastnykh zheleznykh dorog* (St Petersburg, 1913).

Larsons, M.Ia. [M.Ia. Lazerson], *Na sovetskoi sluzhbe: Zapiski spetsa* (Paris: Rodnik, 1930).

Larsons, M.Ia. [M.Ia. Lazerson], *V sovetskom labirinte: Epizody i siluety* (Paris: Strela, 1932).

Laserson, M. [see also Larsons, M.Ia.], *From Russian Skies to Southern Cross: Diary of an Intellectual* (Sydney: Hebrew Standard, 1950).

Lavrov, A., 'O korennykh nedostatkakh v organizatsii zheleznodorozhnogo dela v Rossii', *Izvestiia Sobraniia inzhenerov putei soobshcheniia*, 1 (1917): 4–14.

Lavrov, A., 'Otkrytoe pis'mo k chitateliam "Izvestii Sobraniia I.P.S."', *Izvestiia Sobraniia inzhenerov putei soobshcheniia*, 21 (1914): 327–8.

Lavrov, A., 'Po voprosu o vvedenii v Institute Inzhenerov Putei Soobshcheniia spetsializatsii', *Izvestiia Sobraniia inzhenerov putei soobshcheniia*, 35 (1914): 570–75.

Lavrov, L., 'K voprosu o merakh umen'sheniia bespoleznogo prostoia tovarnykh vagonov', *Izvestiia Sobraniia inzhenerov putei soobshcheniia*, 11 (1915): 227–8.

Lavrov, V.S., 'Dve politiki', *Izvestiia Sobraniia inzhenerov putei sobshcheniia*, 4 (1915): 71–3.

Lavrov, V.S., 'K voprosu o zheleznodorozhnom stroitel'stve tekushchego momenta', *Zheleznodorozhnoe delo*, 9–10 (1914): 77–8.

Lazarev, M., 'Protiv "predela", za perevooruzhenie fronta proektirovaniia', *Sotsialisticheskii transport*, 2 (1936): 36–48.

Lebedev, G.V., 'K voprosu ob integrirovaniia uravneniia dvizheniia poezda', *Vestnik inzhenerov*, 3/1 (1 January 1917): 25–7.

Lebedev, G.V., 'Nekotorye vyvody iz opytnykh dannykh o sile tiagi i paroproizvoditel'nosti nefti dlia passazhirskikh parovozov s peregrevateliami', *Izvestiia Obshchego biuro soveshchatel'nykh s"ezdov*, 9 (1916): 729–81.

Lebedev, G.V., 'Opyty 1912–1913 gg. nad soprotivleniem passazhirskikh parovozov i vagonov russkoi seti', *Izvestiia Obshchego biuro soveshchatel'nykh s"ezdov*, 3 (1914): 192–214.

Lebedev, S., 'O reforme vysshego tekhnicheskogo obrazovaniia v Rossii', *Izvestiia Sobraniia inzhenerov putei soobshcheniia*, 10 (1916): 184–6.

'Lektsii po parovoznomu khoziaistvu', *Vestnik Obshchestva tekhnologov*, 2 (1913): 62.

Lenin, V.I., *Pol'noe sobranie sochinenii*, 5th edn, 55 vols (Moscow: Izdatel'stvo politicheskoi literatury, 1960–1965).

Leninskii sbornik, 40 vols (Moscow: Gosudarstvennoe izdatel'stvo/Izdatel'stvo politicheskoi literatury, 1925–85).

Levi, L.M., 'Eshche po povodu zheleznodorozhnoi opytnoi stantsii imeni A.P. Borodina', *Inzhener*, 8 (1911): 227–39.

Levi, L.M., 'K voprosu o rasstroistve transporta: Doklad sdelannyi L.M. Levi v Sobranii I. P. S. 27 oktiabria 1915 goda', *Izvestiia Sobraniia inzhenerov putei soobshcheniia*, 25 (1915): 514–17; and 26 (1915): 535–9 (discussion).

Levi, L.M., 'O prakticheskom opredelenii velichiny sostavov poezdov v zavisimosti ot razmerov parovozov i uslovii ikh sluzhby', *Inzhener*, 4 (1902): 127–32; and 5 (1902): 180–84.

Levi, L.M., 'Osnovaniia dlia ustanovleniia odnoobraznoi programmy ispytaniia novykh parovozov na russkikh zheleznykh dorogakh', *Inzhener*, 2 (1913): 33–7; and 3 (1913): 79–85; earlier version in *XXIX soveshchatel'nyi s"ezd inzhenerov podvizhnogo sostava i tiagi v Rostove, 24 maia–10 iiunia 1912 g.* (St Petersburg, 1913), pp. 33–71.

Levi, L.M., 'O zheleznodorozhnoi opytnoi stantsii imeni A.P. Borodina', in *Protokoly zasedanii XXVIII soveshchatel'nogo s"ezda inzhenerov sluzhby podvizhnogo sostava i tiagi russkikh zheleznykh dorog v Rige, 27 maia–8 iiunia 1911 g.* (St Petersburg, [1912]), pp. 117–38.

Levi, L.M., 'Rasstroistvo zheleznodorozhnogo transporta, ego neposredstvennye prichiny i prosteishie mery protivodeistviia', *Inzhener*, 2–3 (1916): 33–9.

Levi, L.M., 'Torzhestvo idei', *Inzhener*, 2 (1913): 63–4.

Levi, L.M., 'Zadacha snabzheniia zheleznodorozhnogo podvizhnogo sostava avtomaticheskimi stsepnymi priborami i ee prosteishee reshenie v Rossii (Doklad inzhenera L.M. Levi)', *Izvestiia Sobraniia inzhenerov putei soobshcheniia*, 7 (1915): 122–7.

Levi, L.M. et al., 'Programma proizvodstva opytov opredeleniia velichiny sostavov poezdov', in *Protokoly zasedanii XXIV soveshchatel'nogo s"ezda inzhenerov sluzhby podvizhnogo sostava i tiagi russkikh zheleznykh dorog, sozvannogo v Varshave na 23 noiabria 1902 g.* (St Petersburg, 1903), pp. 411–18.

Liadov, M.N. and Pozner, S.M. (eds), *Leonid Borisovich Krasin ('Nikitich'): gody podpol'ia: sbornik vospominanii, statei i dokumentov* (Moscow and Leningrad: Gosudarstvennoe izdatel'stvo, 1928).

Liberman, S.I., *Building Lenin's Russia* (Chicago: University of Chicago Press, 1945).

Lipets, A.I., 'O sluzhbe parovozov s topkami brotana na russkikh zheleznykh dorogakh', in *XXX soveshchatel'nyi s"ezd inzhenerov podvizhnogo sostava i tiagi v Khar'kove, 25 avgusta 1913 g.* (St Petersburg, 1914), pp. 83–103.

Lipets, A.I., 'Po povodu stat'i inzhenera A.P. Kuchkina: "K voprosu o naibol'shikh dopuskaemykh verkhnim stroeniiem skorostiakh dvizheniia poezdov"', *Inzhener*, 5 (1910): 167–70.

Lipets, A.I., 'Rezul'taty opytov s parovozami Malleta na Tashkentskoi zheleznoi doroge', in *XXX soveshchatel'nyi s"ezd inzhenerov podvizhnogo sostava i tiagi v Khar'kove, 25 avgusta 1913 g.* (St Petersburg, 1914), pp. 78–82.

Lipets, A.I., 'Uproshchennye priemy rascheta vremen khoda poezdov', in *XXIX soveshchatel'nyi s"ezd inzhenerov podvizhnogo sostava i tiagi v Rostove, 24 maia–10 iiunia 1912 g.* (St Petersburg, 1913), pp. 322–48.

Lipets, A.I., 'Uproshchennyi sposob graficheskogo rascheta khoda poezdov', *Izvestiia Kievskogo politekhnicheskogo instituta, Otdel mekhanicheskii i inzhenernyi*, 7/3 (1907): 163–73.

Lipets, A.I., 'Usilenie i usovershenstvovanie parovoza i ikh vliianiia na zheleznodorozhnoe khoziaistvo', *Inzhener*, 9 (1908): 261–7; 10 (1908): 308–12; and 12 (1908): 376–81.

Lipetz, A.I., 'Attempts to Increase Steam Locomotive Efficiency: A Review of the Progress Attained up to the Present Time, and of Recent Developments', *Railway Mechanical Engineer*, 102/7–8 (1928): 384–91 and 437–42.

Lipetz, A.I., 'Report No.4 (America) on the Question of Locomotives of New Types', *Monthly Bulletin of the International Railway Congress Association (English Edition)*, 12/3 (March 1930): 863–950.

'Lipetz, 68, Retired Alco Engineer, Dies', *Schenectady Gazette*, 18 April 1950, p. 18.

Liubitskii, A., 'K voprosu o nashei korporatsii i novykh inzhenerakh putei soobshcheniia', *Izvestiia Sobraniia inzhenerov putei soobshcheniia*, 3 (1914): 38–40.

'Locomotive Testing Plant at Vitry-sur-Seine, France', *The Locomotive*, 39/496 (15 December 1933): 355–8.

'Locomotive Transmission Methods', *Railway Gazette*, 48/13 (30 March 1928): 452–3.

'Lokomotivversuche in Russland', *The Locomotive*, 32/404 (15 April 1926): 136.

Lomonosov, V.Iu., *Elektrotekhnika* (Moscow, 1947; final revised edn with V.M. Polivanov and O.P. Mikhailov – Moscow: Energoatomizdat, 1990).

Lomonossoff, S., 'Ueber die Beeinflussung der Wirkung narkotischer Medikamente durch Antipyretica', Diss. med. (Bern, 1910/1911).

Lontkevich, E.E., 'Teplovoz', in *Stoletie zheleznykh dorog (Trudy Nauchno-Tekhnicheskogo Komiteta Narodnogo Komissariata Putei Soobshcheniia, vypusk 20)* (Moscow: Transpechat', 1925), pp. 164–74.

Lopushinskii, P.I., 'Programma proizvodstva parallel'nykh ispytanii parovozov v poezdakh', in *Protokoly zasedanii XXIV soveshchatel'nogo s"ezda inzhenerov sluzhby podvizhnogo sostava i tiagi russkikh zheleznykh dorog, sozvannogo v Varshave na 23 noiabria 1902 g.* (St Petersburg, 1903), pp. 419–48.

Lorenz, R., 'Diesellokomotiven. Von G. Lomonossoff. Übersetzt von E. Mrongrovius, durchgesehen von F. Meineke. Berlin 1929, VDI-Verlag', *Zeitschrift des Vereines Deutscher Ingenieure*, 73/29 (1929): 1039–40.

Los', F.E., Oleinik, I.P. and Sheludchenko, V.I. (eds), *Revoliutsiia 1905–1907 gg. na Ukraine: sbornik dokumentov i materialov v dvykh tomakh*, 2 vols (Kiev: Gospolitizdat, 1955).

Los', F.E., Plygunov, A.S. and Chernenko, N.V., *Iz istorii Kievskogo politekhnicheskogo instituta: Sbornik dokumentov i materialov, tom 1 (1898–1917 gg.)* (Kiev: Izdatel'stvo Kievskogo universiteta, 1961).

M.K., 'Bol'sheviki v 1905 g. v Kieve (Po materialam vechera vospominanii)', *Letopis' revoliutsii*, 4 (1925): 45–53.

M.Sh., 'A.N. Shelest, "Problemy ekonomichnykh lokomotivov"', *Zheleznodorozhnoe delo*, 6 (1924): 122–3.

Maiorov, M., *Bol'shevizm i revoliutsiia 1905 g. na Ukraine* (Tashkent: Saogiz, 1934).

Manilov, V., 'Kievskaia voennaia organizatsiia RSDRP i vosstanie saper v noiabre 1905 g.', *Letopis' revoliutsii*, 5–6 (1925): 176–225.

Manilov, V., *Kievskii Sovet rabochikh deputatov v 1905 g.* (Kiev: Gosudarstvennoe izdatel'stvo Ukrainy, 1926).

Manilov, V., *Vooruzhennoe vosstanie v chastiakh Kievskogo garnizona* (Gosudarstvennoe izdatel'stvo Ukrainy, 1926).

Manilov, V. and Marenko, G., (eds), *1905 rik u Kiivi ta na Kiivshchini: zbirnik stattiv ta spogadiv* (Kiev: Derzhavne Vidavnishchtvo Ukraini, 1926).

Markovich, V.S., 'Sila tiagi parovozov serii E', *Zheleznodorozhnoe delo*, 9 (1929): 33–5.

Markovich, V.S., 'Uvelichenie moshchnosti parovozov serii E', *Zheleznodorozhnoe delo*, 4–5 (1930): 15–22; and 6 (1930): 27–8.

Mascini, A. and Corbellini, G., 'Methods Employed by the Italian State Railways for Testing Locomotive Performances', *Monthly Bulletin of the International Railway Congress Association (English Edition)*, 7/13 (December 1925): 2523–55.

Materialy dlia otsenki zemel' Smolenskoi gubernii, tom 2: Sychevskii uezd: vypusk I: Estestvenno-istoricheskaia chast' (Smolensk: Izdanie Smolenskogo Gubernskogo Zemstva, 1904).

Matveenko, V., 'K voprosu o nedostatke u nas tovarnykh vagonov', *Izvestiia Sobraniia inzhenerov putei soobshcheniia*, 7 (1915): 127–9; 20 (1916): 409–12; and 21 (1916): 432–7.

Mebel', M.I. (ed.), *1905 god na Ukraine: Khronika i materialy, t.1: ianv.-sent. 1905* (Khar'kov: Proletarii, 1926).

Medel', [inzhener], 'Ob eksploatatsionnykh pasportakh', *Tekhnika i ekonomika putei soobshcheniia*, 2/10 (1923): 477–80.

Meijer, J.M. (ed.), *The Trotsky Papers* (The Hague: Mouton, 1964–1971).

Meineke, F., 'Betriebs- und Versuchsergebnisse der russischen dieselelektrischen Lokomotive', *Zeitschrift des Vereines Deutscher Ingenieure*, 69/42 (1925): 1321–3

Meineke, F., 'Neue Wege im Lokomotivbau', *Zeitschrift des Vereines Deutscher Ingenieure*, 68/37 (1924): 937–42.

Meineke, F., 'Professor Lomonossff †', *Glasers Annalen* (January 1953): 24.

Meineke, F., 'Professor G. Lomonossoff: Lokomotiv-Versuche in Russland. VDI-Verlag GmbH, Berlin 1926', *Organ für die Fortschritte des Eisenbahnwesens*, 81/7 (1926): 140.

Meineke, F., 'Vergleichversuche zwischen Diesel- und Dampflokomotive', *Zeitschrift des Vereines Deutscher Ingenieure*, 69/10 (1925): 321–2.

Mekk, N.K., 'Ekonomika zheleznykh dorog i ee evoliutsiia za stoletie', in *Stoletie zheleznykh dorog (Trudy Nauchno-Tekhnicheskogo Komiteta Narodnogo Komissariata Putei Soobshcheniia, vypusk 20)* (Moscow: Transpechat', 1925), pp. 236–53.

Mel'gunov, S.P., *Martovskie dni 1917 goda* (Moscow: Veche, 2006).

Meshcheriakov, N., 'Iu.V. Lomonosov: "Vospominaniia o martovskoi revoliutsii, 1917g.", Stokgol'm-Berlin 1921 god, 86 str.', *Proletarskaia revoliutsiia*, 5 (1922): 275–9.

Messerschmidt, W., 'The Esslingen Roller Test Plant', *Diesel Railway Traction* (December 1956), pp. 477–9.

Mikhailov, G., 'Iz spogadiv pro pratsiu v Kiivs'kii bil'shovits'kii grupi', in Manilov, V. and Marenko, G. (eds), *1905 rik u Kiivi ta na Kiivshchini: zbirnik stattiv ta spogadiv* (Kiiv, 1926), pp. 308–22.

Mikhailov, V.T., 'Primenenie dvigatelia vnutrennego sgoraniia na zheleznodoro-zhnykh lokomotivakh', *Vestnik Obshchestva tekhnologov*, 9 (1913): 284–90; 10 (1913): 318–24; 11 (1913): 349–55; and 12 (1913): 384–9.

Mikhin, S.I., 'Opredelenie graficheskim sposobom vremeni khoda poezdov', *Zheleznodorozhnoe delo: Podvizhnoi sostav, tiaga i masterskie*, 1–2 (1927): 14–19.

Miliukov, P., *Political Memoirs, 1905–1917*, ed. by A.P. Mendel, transl. by C. Goldberg (Ann Arbor: University of Michigan Press, 1967).

Miliukov, P., *Rossiia na perelome: Bol'shevistskii period russkoi revoliutsii, tom 1: Proiskhozhdenie i ukreplenie bol'shevistskoi diktatury* (Paris, 1927).

Miliukov, P.N., *The Russian Revolution, Vol. 1: The Revolution Divided: Spring 1917*, ed. by R. Stites, transl. by T. and R. Stites (Gulf Breeze: Academic International Press, 1978).

Milonov, Iu., 'Partiia i professional'nye soiuzy v 1905 g.', *Proletarskaia revolutsiia*, 1 (1926): 93–118.

Ministerstvo Putei Soobshcheniia, *Smeta dokhodov i raskhodov Upravleniia zheleznykh dorog na 1914 god* (St Petersburg, 1913).

'Ministr Putei Soobshcheniia S.V. Rukhlov (1909–1915 gg)', *Izvestiia Sobraniia inzhenerov putei soobshcheniia*, 1 (1916): 2–7; and 2 (1916): 18–22.

Mogilenskii, E.A., 'O proektiruemykh merakh k otvlecheniiu chastnykh kapitalov ot zheleznodorozhnogo stroitel'stva v Rossii', *Izvestiia Sobraniia inzhenerov putei soobshcheniia*, 19 (1915): 407–11.

Mokrshitskii, E.I., 'Perestroika ili postroika uglevoznoi magistrali', *Transport i khoziaistvo*, 7 (1928): 74–85.

Morozov, N., 'O predel'shchikakh–parovoznikakh i ikh "ideologakh"', *Sotsialis-ticheskii transport*, 1 (1936): 25–37.

Morrison, L.H., *American Diesel Engines*, 2nd edn (New York and London: McGraw Hill, 1939).

Mrongovius, E., 'Polozhenie teplovoznoi problemy v Zap. Evrope i Amerike', *Vestnik putei soobshcheniia*, 29 (1925): 1–3.

Mrongovius, 'Teplovozy za granitsei', *Zheleznodorozhnoe delo*, 7 (1924): 35–9.

Myshenkov, K.S., 'K voprosu o vygodnosti elektrifikatsii zh.d. so srednem razmerom dvizheniia, nakhodiashchikhsia v raione sushchestvuiushchikh i stroiashchikhsia moshchnykh silovykh stantsii', *Tekhnika i ekonomika putei soobshcheniia*, 2/11 (1923): 565–73.

N., 'Doklad o rabotakh po Moskovsko–Kurskoi zh.d. i Moskovskomu uzlu', *Zheleznodorozhnoe delo*, 19–20 (1916): 173–6.

N., 'Novaia oblast' prilozheniia truda inzhenerovov putei soobshcheniia', *Izvestiia Sobraniia inzhenerov putei soobshcheniia*, 5 (1912): 3–4.

N., 'Obsuzhdenie proekta zakona o zheleznodorozhnom stroitel'stve v komissii pri Sovete s"ezdov predstavitelei promyshlennosti i torgovli', *Izvestiia Sobraniia inzhenerov putei soobshcheniia*, 5 (1911): 12–15.

N.A., 'Eshche o "Poezde vesom 100,000 pudov"', *Izvestiia Sobraniia inzhenerov putei soobshcheniia*, 26 (1915): 534–5.

N-e, 'Iubileinoe chestvovanie inzhener-generala Nikolaia Pavlovicha Petrova, pochetnogo chlena Sobraniia inzhenerov putei soobshcheniia', *Izvestiia Sobraniia inzhenerov putei soobshcheniia*, 17 (1911): 2–3.

N-skii, 'Malen'kaia knizhka bol'shogo znacheniia (Iu. Lomonosov, "Nauchnye problemy eksploatatsii zheleznykh dorog")', *Izvestiia Sobraniia inzhenerov putei soobshcheniia*, 12 (1912): 12–14.

N-skii, V., 'Kazennyi dukh', *Izvestiia Sobraniia inzhenerov putei soobshcheniia*, 26 (1910): 95–6.

N.V., 'V edinenii – sila', *Izvestiia Sobraniia inzhenerov putei soobshcheniia*, 42 (1910): 183–4.

Nabokov, V., '"Fil'm" prof. Lomonosova', *Rul'*, 5 August 1921.

Nabokov, V.D., 'Vremennoe Pravitel'stvo', in Gessen, I.V. (ed.), *Arkhiv russkoi revoliutsii*, vol. 1 (reprinted edition Moscow: Terra, 1991), pp. 9–96.

'Nad chem rabotaiut nauchno-issledovatel'skie instituty', *Zheleznodorozhnaia tekhnika*, 1 (1936): 58–60.

Nagel', R., 'Neskol'ko slov o tiagovykh raschetakh pri postroike zheleznykh dorog', *Izvestiia Sobraniia inzhenerov putei soobshcheniia*, 5–6 (1918): 88–92.

Nagrodskii, V.A., *Chego Institut Inzhenerov putei soobshcheniia ne daet svoim pitomtsam?* (St Petersburg, 1909).

Nagrodskii, [V.A.], 'K voprosu o prepodavanii eksploatatsii zheleznykh dorog', *Izvestiia Sobraniia inzhenerov putei soobshcheniia*, 23 (1916): 472–3.

Naporko, A. (comp.), *Zheleznodorozhnyi transport SSSR v dokumentakh Kommunisticheskoi partii i Sovetskogo pravitel'stva* (Moscow: Gosudarstvennoe transportnoe zheleznodorozhnoe izdatel'stvo, 1957).

'Nauchno-mekhanicheskii kruzhok', *Vestnik Obshchestva tekhnologov*, 19 (1913): 649–50.

'Nauchno-mekhanicheskii kruzhok', *Vestnik Obshchestva tekhnologov*, 22 (1913): 764–5.

Nauchno-tekhnicheskii komitet NKPS, 'Nekrolog: Professor A.S. Raevskii', *Tekhnika i ekonomika putei soobshcheniia*, 7 (1924): 801–4.

Ne-inzhener, 'Porcha kazennykh koeffitientov', *Izvestiia Sobraniia inzhenerov putei soobshcheniia*, 7 (1911): 2–5.

Nekrasov, E., 'Pereustroistvo petrogradskogo zheleznodorozhnogo uzla', *Izvestiia Sobraniia inzhenerov putei soobshcheniia*, 21 (1916): 426–9.

Nekrasov, V.P., 'Osnovy v metodakh prepodavaniia (K voprosu o reformakh prepodavaniia v Institute Imperatora Aleksandra I)', *Izvestiia Sobraniia inzhenerov putei soobshcheniia*, 25 (1914): 394–6.

Nemenov, M., 'Kievskaia i ekaterinoslavskaia voennye organizatsii v 1905 g.', *Proletarskaia revoliutsiia*, 4 (1926): 193–215.

Neopikhanov, A.A., 'Transport', in Iofe, A.F. et al., *Nauka i tekhnika SSSR, 1917–1927*, vol. 3 (Moscow: Rabotnik prosveshcheniia, 1928), pp. 455–91.

Ne professor, 'Institutskii krizis', *Izvestiia Sobraniia inzhenerov putei soobshcheniia*, 5 (1912): 2–3.

Nevskii, V., *Materialy dlia biograficheskogo slovaria sotsial-demokratov, vstupivshikh v rossiiskoe rabochee dvizhenie za period ot 1880 do 1905 g., vypusk 1: A-D* (Moscow and Petrograd: Gosudarstvennoe izdatel'stvo, 1923).

Nevskii, V., 'Otvet moemu kritiku', *Proletarskaia revoliutsiia*, 9 (1922): 70–76.

Nikolaev, V., 'Tsennyi vklad v nauku o tiagovykh raschetakh', *Zheleznodorozhnaia tekhnika*, 3–4 (1936): 116–17.

Nikol'skii, I. and Romanov, V., 'Za planovoe sochetanie vsekh vidov moshchnykh dvigatelei na transporte', *Sotsialisticheskii transport*, 1–2 (1932): 139–41.

Nikol'skii, L.V., 'Protov liberalizma k "predel'cheskim retsidivam"', *Parovoznoe khoziaistvo*, 9 (1935): 3.

Nordmann, H., 'Lokomotivversuche in Russland. Von Prof. G. Lomonossoff. Übersetzt aus dem Russischen von Dr.-Ing. E. Mrongrovius. Berlin 1926, VDI-Verlag', *Zeitschrift des Vereines Deutscher Ingenieure*, 70/52 (1926): 1772.

Nordmann, H., 'Report No. 5 (Germany) on the Question of Locomotives of New Types', *Monthly Bulletin of the International Railway Congress Association (English Edition)*, 12/1 (January 1930): 259–307.

Novopolin, G., 'Iz istorii ekaterinoslavskogo "Soiuza Soiuzov"', *Puti revoliutsii*, 1 (1926): 67–75.

'Novye pravila tekhnicheskoi ekspluatatsii', *Zheleznodorozhnaia tekhnika*, 2 (1936): 3–6.

'Novye tipy parovozov Kolomenskogo zavoda', *Izvestiia Obshchego biuro soveshchatel'nykh s"ezdov*, 1 (1910): 24–6.

'Novyi Nakaz Inzhenernomu Sovetu Ministerstva Putei Soobshcheniia', *Izvestiia Sobraniia inzhenerov putei soobshcheniia*, 1 (1917): 4.

Nyrkov, 'Nuzhen-li plan zheleznodorozhnogo stroitel'stva', *Izvestiia Sobraniia inzhenerov putei soobshcheniia*, 23 (1916): 474–5.

Nyrkov, 'Ob instituskom krizise', *Izvestiia Sobraniia inzhenerov putei soobshcheniia*, 12 (1912): 4.

Nyrkov, 'O plane zheleznodorozhnogo stroitel'stva', *Izvestiia Sobraniia inzhenerov putei soobshcheniia*, 22 (1916): 446–51.

Nyrkov, I.N., 'O plane zheleznodorozhnogo stroitel'stva', *Zheleznodorozhnoe delo*, 35–6 (1916): 313–18.

Nyrkov, N., 'Po povodu stat'i inzh. A.S. Lavrova "O vvedenii v institut Inzhenerov Putei Soobshcheniia spetsializatsii', *Izvestiia Sobraniia inzhenerov putei sobshcheniia*, 3 (1915): 53–4

'Ob ekonomichnosti teplovoza Iu^E', *Vestnik putei soobshcheniia*, 31 (1925): 23.

'Obituary: Dr G.V. Lomonossoff', *The Engineer* (5 December 1952): 763.

'Obituary: Dr G.V. Lomonossoff', *Engineering*, 174 (1952/2): 733.

'Obituary: Mr A.E.L. Chorlton, CBE', *Engineering*, 162 (1946/2): 354.

Obraztsov, V.N., 'Zadachi i perspektivy nauchno-issledovatel'skoi raboty na transporte', *Sotsialisticheskii transport*, 6 (1938): 30–44.

Obraztsov, V.N., Oppengeim, K.A., Syromiatnikov, S.P. et al., 'Ko vsem professoram, dotssentam, assistentam i nauchnym rabotnikam SSSR', *Zheleznodorozhnaia tekhnika*, 3–4 (1936): 59–61.

Obshchestvo Iugo–Vostochnykh Zheleznykh Dorog, *Otchet po ekspluatatsii dorog za 1905 g.* (St Petersburg, 1906).

Obshchestvo Iugo–Vostochnykh Zheleznykh Dorog, *Otchet po ekspluatatsii dorog za 1905 g.: Prilozheniia k otchetu, No. 17–48* (St Petersburg, 1906).

'Oktiabr'skie sobytiia 1905 goda v Kieve v izobrazhenii okhrannogo otdeleniia', *Letopis' revoliutsii*, 2 (1925): 112–24.

'O nedostatkakh parovozov 4/5 izmenennogo tipa Kitaiskoi Vostochnoi zh.d.', in *XXX soveshchatel'nyi s"ezd inzhenerov podvizhnogo sostava i tiagi v Khar'kove, 25 avgusta 1913 g.* (St Petersburg, 1914), pp. 31–63.

'O nedostatkakh, zamechennykh v parovozakh normal'nogo tipa, zakaza 1897 goda', in *Protokoly zasedanii XXIV soveshchatel'nogo s"ezda inzhenerov sluzhby podvizhnogo sostava i tiagi russkikh zheleznykh dorog, sozvannogo v Varshave na 23 noiabria 1902 g.* (St Petersburg, 1903), pp. 265–93.

'O nedostatkakh, zamechennykh v parovozakh normal'nogo tipa, zakaza 1897 goda; rassmotrenie doklada inzh. Iu.V. Lomonosova po tomu zhe predmetu', in *Protokoly zasedanii XXIII soveshchatel'nogo s"ezda inzhenerov sluzhby podvizhnogo sostava i tiagi russkikh zheleznykh dorog, sozvannogo v Moskve na 23 noiabria 1901 g.* (St Petersburg, 1902), pp. 439–62.

'O novom tipe passazhirskikh parovozov', *Tekhnika i ekonomika putei soobshcheniia*, 1/5 (1923): 274–7.

Oppengeim, K., 'Nasushchnaia potrebnost' (Pis'mo v redaktsiiu)', *Izvestiia Sobraniia inzhenerov putei sobshcheniia*, 6 (1915): 110.

'Opredelenie norm godovykh probegov inventarnykh edinits podvizhnogo sostava na otdel'nykh dorogakh i potrebnogo na 1912 g. inventarnogo kolichestva podvizhnogo sostava', in *Protokoly zasedanii XXVI vneocherednego soveshchatel'nogo s"ezda inzhenerov sluzhby podvizhnogo sostava i tiagi russkikh zheleznykh dorog* (St Petersburg, 1908), pp. 37–53.

'Opredelenie priemov rascheta ozhidaemogo na dorogakh v 1912 g. probega poezdov i podvizhnogo sostava i opredelenie razmerov etogo probega', in *Protokoly zasedanii XXVI vneocherednego soveshchatel'nogo s"ezda inzhenerov sluzhby podvizhnogo sostava i tiagi russkikh zheleznykh dorog* (St Petersburg, 1908), pp. 27–36.

O rabote zheleznodorozhnoi seti v usloviiakh nastoiashchei voiny (Petrograd, 1916).

'O rezul'tatakh ispytaniia probnymi poezdkami s predel'nymi sostavami na predel'nykh pod"emakh tovarnykh parovozov noveishikh konstruktsii, a imenno parovozov: 4/4 normal'nykh s peregrevatelem; Kompaund bez peregrevatelia; 4/5 Tandem Kompaund tipa M–V–Rybinskoi zh.d. i 4/5 Kompaund tipa Kitaiskoi–Vostochnoi zh.d.', in *Protokoly zasedanii XXVI vneocherednego soveshchatel'nogo s"ezda inzhenerov sluzhby podvizhnogo sostava i tiagi russkikh zheleznykh dorog* (St Petersburg, 1908), pp. 268–86.

'O rezul'tatakh sluzhby parovozov normal'nogo tipa, peredelannykh soglasno ukazaniiam, pomeshchannym v zakliuchenii XXIII soveshchatel'nogo s"ezda po voprosu 19, punkt b', in *Protokoly zasedanii XXIV soveshchatel'nogo s"ezda inzhenerov sluzhby podvizhnogo sostava i tiagi russkikh zheleznykh dorog, sozvannogo v Varshave na 23 noiabria 1902 g.* (St Petersburg, 1903), pp. 294–300.

Orlov, V., 'Protiv "teorii" predela za razrabotku novoi metodologii sebestoimosti perevozok', *Sotsialisticheskii transport*, 3 (1936): 65–76.

O.S. and B.R., 'The Early History of Electric Transmission Control', *Diesel Railway Traction Supplement to the Railway Gazette* (October 1952): 254–6.

'Osnovnye raboty IRT za mai–dekabr' 1934 goda: Parovozo-teplotekhnicheskii sektor', *Parovoznoe khoziaistvo*, 2 (1935): 43.

'O sposobakh k ustraneniiu nedostatkov, zamechennykh v tovarnom parovoze normal'nogo tipa, zakaza 1897 goda', in *Protokoly zasedanii XXII soveshchatel'nogo s"ezda inzhenerov sluzhby podvizhnogo sostava i tiagi russkikh zheleznykh dorog, sozvannogo v Kieve na 25 noiabria 1900 g.* (St Petersburg, 1901), pp. 157–201.

Ostapenko, S.S., *Transport i voina* (Ekaterinoslav, 1916).

'Otchet Ekaterinoslavskogo Otdeleniia za 1907 god', *Zapiski Imperatorskogo russkogo tekhnicheskogo obshchestva*, 9–10 (1908): 362–6.

'Otchet Kievskogo Otdeleniia za 1907 god', *Zapiski Imperatorskogo russkogo tekhnicheskogo obshchestva*, 9–10 (1908): 366–7.

'Otchet o deiatel'nosti Imperatorskogo Russkogo Tekhnicheskogo Obshchestva v 1908 g.', *Zapiski Imperatorskogo russkogo tekhnicheskogo obshchestva*, 6–7 (1909): 189–217.

Otchet o deiatel'nosti Osobogo soveshchaniia dlia obsuzhdeniia i ob"edineniia meropriatii po perevozke topliva i prodovol'stvennykh i voennykh gruzov: Za period sentiabr' 1915 g.–sentiabr' 1916 g. (Petrograd, 1916).

Otchet o sostoianii Instituta inzhenerov putei soobshcheniia Imperatora Aleksandra I, s 1-go iiunia 1893 goda po 1-e iiune 1894 goda (St Petersburg, 1895).

Otchet o sostoianii Instituta inzhenerov putei soobshcheniia Imperatora Aleksandra I, s 1-go iiunia 1897 goda po 1-e iiune 1898 goda (St Petersburg, 1899).

Otchet po ekspluatatsii Tashkentskoi kazennoi zheleznoi dorogi za 1909 g. (Orenburg, 1910).

Otchet po ekspluatatsii Tashkentskoi kazennoi zheleznoi dorogi za 1910 g. (Orenburg, 1911).

'Ot redaktsii', *Sotsialisticheskii transport*, 6 (1936): 94–5.

'Ot Vremennogo Tsentral'nogo Soveta Soiuza Inzhenerov i tekhnikov, rabotaiushchikh po putiam soobshcheniia', *Izvestiia Sobraniia inzhenerov putei soobshcheniia*, 4 (1917): 93–4.

'O vozobnovlenii zheleznodorozhnogo stroitel'stva', *Izvestiia Sobraniia inzhenerov putei soobshcheniia*, 4 (1916): 58–60.

'O vybore predstavitelei dorog v stroitel'nyi komitet parovoznoi, imeni A.P. Borodina, opytnoi stantsii russkikh zheleznykh dorog', in *XXX soveshchatel'nyi s"ezd inzhenerov podvizhnogo sostava i tiagi v Khar'kove, 25 avgusta 1913 g.* (St Petersburg, 1914), pp. 496–8.

Ozols, K.V., *Memuary poslannika* (Paris: Dom knigi, 1938).

P.Ia., 'Teplovozy dlia russkikh zheleznykh dorog', *Zheleznodorozhnoe delo*, 5 (1924): 24–8.

Pamiati Nikolaia Leonidovicha Shchukina, zasluzhennogo professora, chlena Soveta Nauchno-Tekhnicheskogo Komiteta (Moscow: Transpechat', 1925).

Pamiatnaia knizhka ([n.p.]: RSFSR Zheleznodorozhnaia missiia za-granitsei, [1922?]).

Papers Relating to the Foreign Relations of the United States (Washington, DC, 1864–).

Paton, E.O., *Vospominaniia: Literaturnaia zapis' Iuriia Buriakovskogo* (Kiev: Derzhavne Vidavnitstvo Khudozhn'oi Literaturi, 1962).

Pavlov, B., 'Al'bom skhem parovozov i pasportov: Izdanie Tsentral'nogo upravleniia parovoznogo khoziaistva 1935 g.', *Zheleznodorozhnaia tekhnika*, 11 (1936): 62–3.

Pavlov, B., 'K voprosu o sozdanii plana zheleznodorozhnogo stroitel'stva (Po povodu dvadtsatiletiia Komissii o novykh zheleznykh dorogakh)', *Zheleznodorozhnoe delo*, 5–6 (1916): 34–8.

Pavlovskii, A.A., *Sovremennoe russkoe parovoznoe khoziaistvo*, 2 vols (St Petersburg, 1908).

Pavlovskii, V.A., *Ispytaniia donetskikh, dombrovskikh i angliiskikh uglei proizvedennye na Severo–Zapadnykh zhel. dor. v 1913 g.* (Berlin: Rossiiskaia zheleznodorozhnaia missiia, 1923).

Perekrestov, P., 'Problema korennogo pereustroistva zheleznodorozhnykh vykhodov iz Donbassa na sever', *Transport i khoziaistvo*, 7 (1928): 64–74.

'Pereotsenka odnoi tsennosti', *Zheleznodorozhnoe delo*, 16 (1915): 137.

Perminov, V.V., 'Vzaimnoe soedinenie putei na stantsiiakh zheleznykh dorog, ego konstruktivnye osobennosti, skhematicheskoe izobrazhenie i tochnyi raschet: Chast' I', *Izvestiia Kievskogo politekhnicheskogo instituta, Otdel mekhanicheskii i inzhenernyi*, 3/1 (1903): 1–96; and 2 (1903): 97–168.

Persivanov, A.A., 'Institut inzhenerov eksploatatsii zh.d.', *Zheleznodorozhnoe delo*, 9–10 (1917): 88–92.

'Pervyi teplovoz v SSSR', *Ekonomicheskaia zhizn'*, 27 November 1924, p. 3.

Petrov, N., *Ekonomicheskoe znachenie russkikh zheleznykh dorog* (St Petersburg, 1910). (Vysochaishe uchrezdennaia Osobaia Vysshaia Komissia Vsestoronnego Issledovaniia zheleznodorozhnogo dela v Rossii, vypusk XXVIII).

Petrov, N.P., 'Nuzhda v zheleznykh dorogakh', *Izvestiia Sobraniia inzhenerov putei soobshcheniia*, 19 (1913): 302–5.

Petrov, N., *Soprotivlenie poezda na zheleznoi doroge: Kurs Prakticheskogo Tekhnicheskogo Instituta* (St Petersburg, 1889).

Phillipson, E.A., 'Notes on Locomotive Running Trials', *Journal of the Institution of Locomotive Engineers*, 14/65 (May–June 1924): 363–86.

'Piatoe prisuzhdenie premii imeni Aleksandra Parfen'evicha Borodina: Trud Iu.V. Lomonosova – "Opytnye issledovaniia tovarnykh vos'mikolesnykh parovozov kompaund normal'nogo tipa"', *Zheleznodorozhnoe delo*, 13–14 (1912): 31Д–32Д.

'Plan zheleznodorozhnogo stroitel'stva', *Izvestiia Sobraniia inzhenerov putei soobshcheniia*, 24 (1916): 506–10.

'Plan zheleznodorozhnogo stroitel'stva na piatiletie 1917–1922 gg.', *Izvestiia Sobraniia inzhenerov putei soobshcheniia*, 18 (1916): 374–7.

'Plan zheleznodorozhnogo stroitel'stva, sostavlennyi Soveshchaniem pod predsedatel'stvom Tovarishcha Ministra putei soobshcheniia I.N. Borisova (Kratkoe reziume)', *Zheleznodorozhnoe delo*, 35–6 (1916): 318–23.

'Plan zheleznodorozhnogo stroitel'stva v Rossii', *Inzhener*, 11 (1914): 352.

'Po-bol'shevistski perestroim parovoznoe khoziaistvo', *Parovoznoe khoziaistvo*, 8 (1935): 8–9.

'Poezd vesom 100,000 pudov: Beglye vpechatleniia po opytam prof. Iu.V. Lomonosova', *Izvestiia Sobraniia inzhenerov putei soobshcheniia*, 23 (1915): 478–9.

Pokrovskii, I., 'Zamechanie po povodu stat'i Vl. Sventsitskogo "O prichinakh mnogoletnego zasizhivaniia studentov v spetsial'nykh vysshikh uchebnykh zavedeniiakh i nekotorykh merakh k ustraneniiu etogo nenormal'nogo iavleniia"', *Izvestiia Sobraniia inzhenerov putei soobshcheniia*, 13 (1916): 273–4.

Polovtsov, A.V., 'Rekvizitsiia kamennogo uglia i ee vozmozhnoe vliianie na zheleznodorozhnye perevozki: Doklad, sdelannyi 22 oktiabria s.g. v Sobranii Inzhenerov Putei Soobshcheniia', *Izvestiia Sobraniia inzhenerov putei soobshcheniia*, 22 (1915): 456–9; and 24 (1915): 494–9.

'Po perestroike seti zhurnalov na zheleznodorozhnom transporte (Postanovlenie kollegii NKPS)', *Zheleznodorozhnoe delo*, 12 (1931): 73.

'Postanovlenie VIII otdela Imperatorskogo russkogo tekhnicheskogo obshchestva po voprosu o nedostatke vagonov', *Zheleznodorozhnoe delo*, 18–19 (1915): 177.

Pozner, S.M. (comp.), *1905: Boevaia gruppa pri TsK RSDRP(b) (1905–1907 gg.): stat'i i vospominaniia* (Moscow and Leningrad: Gosudarstvennoe izdatel'stvo, 1927).

Pozner, S.M. (comp.), *Pervaia boevaia organizatsiia bol'shevikov, 1905–1907 gg.: stat'i, vospominaniia i dokumenty* (Moscow: Staryi Bol'shevik, 1934).

Pozner, S.M., 'Pervaia organizatsiia bol'shevikov po podgotovke k vooruzhennomu vosstaniiu', *Krasnaia letopis'*, 5 (1922): 329–34.

Pozner, S.M. (comp.), 'Rabota boevykh bol'shevistskikh organizatsii 1905–1907 gg.', *Proletarskaia revoliutsiia*, 7 (1925): 82–111.

Pravdin, V. [Vakar, V.V.], 'Epokha "neslykhannoi smuty" na Kievshchine: nabroski', *Letopis' revoliutsii*, 2 (1925): 91–111.

Pravosudovich, M.E., 'Novyi metod sravnitel'nogo opredeleniia dinamicheskikh svoistv lokomotivov opytnym putem', *Tekhnika i ekonomika putei soobshcheniia*, 2/8 (1923): 186–7.

Pravosudovich, M.E., 'Popovodu stat'i A.S. Lavrova po voprosu o vvedenii v institutakh Inzhenerov Putei Soobshcheniia spetsializatsii', *Izvestiia Sobraniia inzhenerov putei sobshcheniia*, 7 (1915): 130–131.

Pravosudovich, M.E. and Manos, I.Ia. (eds), *Stoletie zheleznykh dorog (Trudy Nauchno-Tekhnicheskogo Komiteta Narodnogo Komissariata Putei Soobshcheniia, vypusk 20)* (Moscow: Transpechat', 1925).

'Pribytie pervogo russkogo teplovoza', *Ekonomicheskaia zhizn'*, 17 January 1925, p. 3.

'Prof. Dr.-Ing. Meineke †', *Glasers Annalen*, 79 (1955): 254–6.

'Professor G.V. Lomonossoff', *Railway Gazette*, 97/22 (28 November 1952): 594–5.

Prostnev, I., 'O zheleznodorozhnom stroitel'stve posle voiny', *Izvestiia Sobraniia inzhenerov putei soobshcheniia*, 23 (1916): 475–6.

'Protokol No.35 zasedaniia Biuro Soveshchatel'nykh s"ezdov inzhenerov sluzhby Podvizhnogo sostava i tiagi', *Izvestiia Obshchego biuro soveshchatel'nykh s"ezdov*, 4 (1912): 303–5.

'Protokol No.47 ekstrennogo zasedaniia Biuro soveshchatel'nogo s"ezda inzhenerov sluzhby podvizhnogo sostava i tiagi', *Izvestiia Obshchego biuro soveshchatel'nykh s"ezdov*, 8–9 (1917): 441–6.

'Protokol No.55 zasedaniia Obshchego biuro Tekhnicheskikh soveshchatel'nykh s"ezdov', *Izvestiia Obshchego biuro soveshchatel'nykh s"ezdov*, 5 (1912): 382–95.

'Protokol No.56 zasedaniia Obshchego biuro Tekhnicheskikh soveshchatel'nykh s"ezdov', *Izvestiia Obshchego biuro soveshchatel'nykh s"ezdov*, 8 (1912): 661–9.

'Protokol No.62 zasedaniia Obshchego biuro Tekhnicheskikh soveshchatel'nykh s"ezdov', *Izvestiia Obshchego biuro soveshchatel'nykh s"ezdov*, 6 (1914): 552–79.

'Protokol No.64 zasedaniia Obshchego biuro tekhnicheskikh soveshchatel'nykh s"ezdov', *Izvestiia Obshchego biuro soveshchatel'nykh s"ezdov*, 9 (1914): 772–883.

Protokoly zasedanii XXII soveshchatel'nogo s"ezda inzhenerov sluzhby podvizhnogo sostava i tiagi russkikh zheleznykh dorog, sozvannogo v Kieve na 25 noiabria 1900 g. (St Petersburg, 1901).

Protokoly zasedanii XXIII soveshchatel'nogo s"ezda inzhenerov sluzhby podvizhnogo sostava i tiagi russkikh zheleznykh dorog, sozvannogo v Moskve na 23 noiabria 1901 g. (St Petersburg, 1902).

Protokoly zasedanii XXIV soveshchatel'nogo s"ezda inzhenerov sluzhby podvizhnogo sostava i tiagi russkikh zheleznykh dorog, sozvannogo v Varshave na 23 noiabria 1902 g. (St Petersburg, 1903).

Protokoly zasedanii XXV soveshchatel'nogo s"ezda inzhenerov sluzhby podvizhnogo sostava i tiagi russkikh zheleznykh dorog, vol. 1 (St Petersburg, 1908).

Protokoly zasedanii XXVI vneocherednego soveshchatel'nogo s"ezda inzhenerov sluzhby podvizhnogo sostava i tiagi russkikh zheleznykh dorog (St Petersburg, 1908).

Protokoly zasedanii XXVII soveshchatel'nogo s"ezda inzhenerov sluzhby podvizhnogo sostava i tiagi russkikh zheleznykh dorog, v Varshave, v avguste 1909 g. (St Petersburg, 1910).

Protokoly zasedanii XXVIII soveshchatel'nogo s"ezda inzhenerov sluzhby podvizhnogo sostava i tiagi russkikh zheleznykh dorog v Rige, 27 maia–8 iiunia 1911 g. (St Petersburg, [1912]).

Protokoly zasedanii XXIX soveshchatel'nyi s"ezd inzhenerov podvizhnogo sostava i tiagi v Rostove, 24 maia–10 iiunia 1912 g. (St Petersburg, 1913).

Protokoly zasedani XXX soveshchatel'nyi s"ezd inzhenerov podvizhnogo sostava i tiagi v Khar'kove, 25 avgusta 1913 g. (St Petersburg, 1914).

Protokoly zasedani XXXIII soveshchatel'nyi s"ezd inzhenerov podvizhnogo sostava i tiagi v Moskve, 18 iiunia 1923 g. (Moscow, 1924).

Putiat, N.E., 'Ob ustroistve ispytatel'noi stantsii dlia opredeleniia kachestva vody, topliva i parovykh kotlov', in *Protokoly zasedanii XXII soveshchatel'nogo s"ezda inzhenerov sluzhby podvizhnogo sostava i tiagi russkikh zheleznykh dorog, sozvannogo v Kieve na 25 noiabria 1900 g.* (St Petersburg, 1901), p. 92.

'Question V – Locomotives of New Types; in Particular Turbine Locomotives and Internal Combustion Motor Locomotives', *Monthly Bulletin of the International Railway Congress Association (English Edition)*, 13/2 (February 1931): 87–111.

R., 'Razrukha zheleznodorozhnogo transporta', *Izvestiia Sobraniia inzhenerov putei sobshcheniia*, 5 (1915): 82–3 (reprinted from *Promyshlennost' i torgovlia*, 3 (1915)).

Raaben, E., 'Za bol'shevistskuiu nauku o sebestoimosti perevozok', *Sotsialisticheskii transport*, 5 (1936): 38–44.

Radtsig, A.A., 'Matematicheskaia teoriia obmena tepla v tsilindrakh parovykh mashin', *Izvestiia Kievskogo politekhnicheskogo instituta, Otdel mekhanicheskii i inzhenernyi*, 3/4 (1903): 1–158.

'Railway-Locomotive Development', *Engineering*, 142/3688 (18 September 1936): 315–16.

'Railway Traction', *The Engineer*, 154/4001 (16 September 1932): 274–5.

'Razlichnye izvestiia', *Izvestiia Sobraniia inzhenerov putei soobshcheniia*, 14 (1915): 309.

'Razvitie zheleznodorozhnoi seti', *Izvestiia Sobraniia inzhenerov putei soobshcheniia*, 31 (1913): 546–7.

Raymond, W.G., Riggs, H.F. and Sadler, W.C., *Elements of Railroad Engineering*, 6th edn (New York: John Wiley and Sons, 1947).

'Reaktsionery parovoznogo khoziaistva prigvozhdeny', *Gudok*, 9 January 1936, p. 1.

Reed, B., 'Development of Diesel Traction', *Diesel Railway Traction Supplement to the Railway Gazette* (21 April 1933): 2–5.

Reed, B., 'Development of Diesel Traction VI: Locomotives in North America', *Diesel Railway Traction Supplement to the Railway Gazette* (3 November 1933): 678–81.

Reed, B., 'Diesel Traction in Extra-European Countries', *Diesel Railway Traction Supplement to the Railway Gazette* (24 February 1933): 2–4.

Revoliutsiia 1905–1907 gg. v Rossii: Dokumenty i materialy (Moscow: AN SSSR, 1955–1965).

'Rezul'taty poezdki teplovoza Iu^E No. 001 na Kavkaze (Doklad prof. Iu.V. Lomonosova Kollegii NKPS'a)', *Vestnik putei soobshcheniia*, 28 (1925): 18–19.

Rich, F., 'You'll Go to Jail, Young Man', *Steam World*, 215 (May 2005): 8–14.

Rikhter, I., 'Prinuditel'nyi otdykh i rasstristvo tyla', *Zheleznodorozhnoe delo*, 47 (1915): 483–7.

Rimmer, A., *Testing Times at Derby: A "Privileged" View of Steam* (Usk: Oakwood Press, 2004).

Rodzianko, M.V., 'Gosudarstvennaia Duma i fevral'skaia 1917 goda revoliutsiia', in Gessen, I.V. (ed.), *Arkhiv russkoi revoliutsii*, vol. 6 (reprinted edition Moscow: Terra, 1991), pp. 5–80.

Rogovskii, N.I. et al. (eds), *Protokoly Prezidiuma Gosplana za 1922 god*, 2 vols (Moscow: Ekonomika, 1979).

Romanov, A.D., *Parovozy: Kurs, prepodavaemyi v Institute Inzhenerov Putei Soobshcheniia Imperatora Aleksandra I*, 2nd edn (St Petersburg, 1900).

Romanov, V., 'Protiv burzhuaznoi apologetiki v ekonomicheskikh izyskaniiakh', *Sotsialisticheskii transport*, 3 (1932): 177–80.

Rostov, N., 'Iuzhnoe Voenno-Tekhnicheskoe Biuro pri Tsentral'nom Komitete RS-DRP', *Katorga i ssylka*, 1 (1926): 93–108.

Rostov, N., 'Samoubiistvo M.F.Vetrovoi i studencheskie besporiadki 1897 g.', *Katorga i ssylka*, 2 (1926): 50–66.

Rozenblium, K.I., *Voennye organizatsii Bol'shevikov 1905–1907 gg.* (Moscow: Gos.-Sotsial'no-Ekonomicheskoe Izdatel'stvo, 1931).

Rud-v, N., 'K voprosu o napravlenii politiki zheleznodorozhnogo stroitel'stva', *Izvestiia Sobraniia inzhenerov putei sobshcheniia*, 1 (1915): 4–8.

Rudyi, Iu.V., 'Ivan Nikolaevich Borisov', *Zheleznodorozhnoe delo: obshchii otdel*, 6 (1928): 2.

Rudyi, Iu.V., 'Zadachi zheleznodorozhnogo transporta v deviatuiu godovshchinu Oktiabria', *Vestnik putei soobshcheniia*, 46 (1925): 3.

'Russkii zheleznodorozhnyi otdel na Vsemirnoi Vystavke v Parizhe', *Inzhener*, 8–9 (1900): 348–50.

Rynin, N.A., 'K voprosu o soprotivlenii vozdukha dvizheniiu poezdov', *Vestnik Obshchestva tekhnologov*, 9 (1913): 279–83; and 10 (1913): 314–17.

Rynin, N.A., 'Po povodu statei inzhenera M.V. Gololobova i inzhenera A.A. Solodovnikova', *Izvestiia Obshchego biuro soveshchatel'nykh s"ezdov*, 5 (1914): 496–500.

Rynin, N.A., 'Po povodu statei "K voprosu o soprotivlenii vozdukha dvizheniiu poezdov" (Otvet inzheneram M.V. Gololobovu i A.A. Solodovnikovu)', *Izvestiia Obshchego biuro soveshchatel'nykh s"ezdov*, 1 (1914): 64–77.

S., 'Po povodu rabot A.N. Shelesta', *Tekhnika i ekonomika putei soobshcheniia*, 2/12 (1923): 777–8.

S., 'Teplovozy', *Tekhnika i ekonomika putei soobshcheniia*, 3–4 (1924): 378–9.

S.M.E., 'Pis'mo v redaktsiiu', *Izvestiia Sobraniia inzhenerov putei soobshcheniia*, 25 (1915): 517–18.

S.P., 'O printsipe "uplotneniia rabot", predlagaemom inzhenerom Voskresenskim', *Izvestiia Sobraniia inzhenerov putei soobshcheniia*, 15 (1915): 316–18.

Sakharov, V., 'Teplovoz na linii Orsk–Astrakhan'–Manych–Torgovaia', *Vestnik putei soobshcheniia*, 51 (1925): 9–10.

Sakharov, V., 'Teplovozy i zheleznodorozhnoe stroitel'stvo', *Vestnik putei soobshcheniia*, 34 (1925): 4–5.

Sbornik statisticheskikh svedenii po Smolenskoi gubernii, tom 2: Sychevskii uezd (Smolensk, 1885).

Schulz, K., 'Die dieselelektrische Lokomotive. Von Prof. G. Lomonossoff. Aus dem Russischen übersetzt von Dr.-Ing. Erich Mrongrovius. Berlin 1924, VDI-Verlag', *Zeitschrift des Vereines Deutscher Ingenieure*, 68/38 (1924): 1008.

Sergeev, B.N., 'Novye pravila rascheta zheleznodorozhnogo puti', *Zheleznodorozhnoe delo*, 7–8 (1931): 56–61.

Shapiro, A., 'Iz vospominanii o B.P. Zhadanovskom', *Katorga i ssylka*, 7 (1927): 190–194.

Shargin, V., 'Iu.Lomonosov: Nauchnye problemy eksploatatsii zheleznykh dorog', *Vestnik inzhenerov*, 11 (1915): 509.

Shatunovskii, Ia., *Vosstanovlenie transporta: Puti soobshcheniia i puti revoliutsii* (Moscow and Petrograd, 1920).

Shcheglovitov, V.N., 'Kratkii otvet prof. Iu.V. Lomonosovu po povodu ego zametki o moei broshiure "K voprosu ob opredelenii sebestoimosti pudoversty"', *Izvestiia Sobraniia inzhenerov putei soobshcheniia*, 10 (1915): 212–13; and 11 (1915): 228–30.

Shcheglovitov, V., 'K voprosu ob usilenii propusknoi sposobnosti odnoputnykh linii', *Zheleznodorozhnoe delo*, 41–2 (1916): 384–8; 43–4 (1916): 403–6; and 45–6 (1916): 427–9.

Shcheglovitov, V., 'K voprosu ob uvelichenii skorosti peredvizheniia tovarnykh poezdov putem umen'sheniia sostavov, v tseliakh uluchsheniia oborota vagonov, i uvelicheniia propusknoi sposobnosti', *Zheleznodorozhnoe delo*, 29–30 (1916): 263–5.

Shcheglovitov, V.N., *K voprosu o naivygodneishikh sostavakh tovarnykh poezdov* (Warsaw, 1910).

Shcheglovitov, V.N., 'Pis'mo inzh. V.N. Shcheglovitova inzh. Iu.V. Lomonosovu', *Izvestiia Sobraniia inzhenerov putei soobshcheniia*, 9 (1914): 125.

Shchegolev, P.E. (ed.), *Padenie tsarskogo rezhima: stenograficheskie otchety doprosov i pokazanii, dannykh v 1917 g. v Chrezvychainoi Sledstvennoi*

Komissii Vremennogo Pravitel'stva, vols 5–7 (Moscow and Leningrad: Gosudarstvennoe izdatel'stvo, 1926–1927).

Shchtol'tsman, S., ''Nekotorye soobrazhenoiia o zadache pravitel'stvennykh zheleznodorozhnykh izyskanii', *Izvestiia Sobraniia inzhenerov putei soobshcheniia*, 33 (1913): 580–585.

Shchukin, N.L., *Istoricheskii ocherk deiatel'nosti Komissii podvizhnogo sostava i tiagi pri Inzhenernom sovete* (Petrograd, [1916]).

Shelest, A.N., 'Issledovanie raboty teplovoza Br. Zul'tser v Shveitsarii', *Vestnik inzhenerov*, 3/3 (1917): 71–80.

Shelest, A.N., 'Po povodu zametki prof. Iu.V. Lomonosova o stat'e "Issledovanie raboty teplovoza Br. Zul'tser v Shveitsarii"', *Vestnik inzhenerov*, 3/15 (1917): 336–8.

Shelest, A.N., *Problemy ekonomichnykh lokomotivov* (Moscow, 1923).

Shelest, A.N., 'Teplovoz s pnevmaticheskoi peredachei i ego budushchnost", *Zheleznodorozhnoe delo*, 6 (1924): 22–7.

Sheligovskii, A.K., *Ekspluatatsiia podvizhnogo sostava zheleznykh dorog* (Moscow, 1913).

Sheridan, C., *Russian Portraits* (Cambridge: Ian Faulkner, 1992).

Shlikhter, A.G., *Il'ich, kakim ia ego znal: koe-chto iz vstrech i vospominanii* (Moscow: Politizdat, 1970).

Shlikhter, A.G., 'Kiev v oktiabr'skie dni 1905 goda', *Puti revoliutsii*, 1 (1926): 37–58.

Shlikhter, O., *3 chasiv revoliutsii 1905 roku* (Khar'kov: Proletar, 1931).

Shlikhter, O.G., *Vibrani tvori, t.5* (Kiev: AN URSR, 1959).

Shmen-de-Fernikov, 'Pravaia ruka ne znaet, chto delaet levaia', *Izvestiia Sobraniia inzhenerov putei soobshcheniia*, 38 (1913): 678–9.

Shmukker, M.M., *Ocherki finansov i ekonomiki zhel.-dor. transporta za 1913–1922 gody (v sviazi s obshchimi ekonomicheskimi iavleniiami zhizni strany* (Moscow: Transpechat', 1923).

Shtange, D.A., 'Vybor naivygodneishikh sostavov poezdov i skorosti dvizheniia (Po povodu stat'i V.D. Eshchenko v No. 1 "Zheleznodorozhnogo dela")', *Zheleznodorozhnoe delo: obshchii otdel*, 3 (1928): 5–7.

Shukhtan, L., 'Dolgosrochnyi zakaz vagonov', *Izvestiia Sobraniia inzhenerov putei soobshcheniia*, 21 (1915): 441–2.

Shukhtan, L.F., 'Nasha zheleznodorozhnaia politika', *Izvestiia Sobraniia inzhenerov putei soobshcheniia*, 9 (1914): 118–24; and 12 (1914): 166–72.

Shukhtan, L., 'Ob uluchshenii eksploatatsii kazennykh zheleznykh dorog', *Izvestiia Sobraniia inzhenerov putei soobshcheniia*, 13 (1913): 196–9.

Shulgin, V.V., *Days of the Russian Revolution: Memoirs from the Right*, ed. and transl. by B.F. Adams (Gulf Breeze: Academic International Press, 1990).

Sidorov, A.L. et al. (eds), *Ekonomicheskoe polozhenie Rossii nakanune Velikoi Oktiabr'skoi Sotsialisticheskoi Revoliutsii: Dokumenty i materialy, mart–oktiabr' 1917 g.*, 3 vols (Moscow and Leningrad, 1957).

'Smeta Departamenta zheleznodorozhnykh del na 1914 god', *Izvestiia Sobraniia inzhenerov putei soobshcheniia*, 38 (1913): 675–8.

'Sobranie aktiva Narodnogo komissariata putei soobshcheniia', *Sotsialisticheskii transport*, 3 (1937): 4–7.

'Sobranie aktiva Narodnogo komissariata putei soobshcheniia: Doklad tovarishcha L.M. Kaganovicha ob itogakh Plenuma TsK VKP(b)', *Zheleznodorozhnaia tekhnika*, 3 (1937): 4–7.

'Soiuz inzhenerov (Vsekh spetsial'nostei)', *Izvestiia Sobraniia inzhenerov putei soobshcheniia*, 5 (1917): 129–30.

Sokolov, V.N., 'Transportno-tekhnicheskoe biuro TsK (b)', *Tekhnika bol'shevistskogo podpol'ia: sbornik statei i vospominanii, vyp. 1* (Moscow–Leningrad: Gosudarstvennoe Izdatel'stvo, 1924), pp. 28–176.

Sokovich, V.A., 'K voprosu o sebestoimosti perevozok', *Zheleznodorozhnoe delo: obshchii otdel*, 6 (1928): 8–11.

Solomon, G., *Among the Red Autocrats: My Experience in the Service of the Soviets* (New York, 1935).

Solomon, G., '"Prisoedinivshiisia", ili Istoriia odnoi druzhby', publication by V.L. Genis, *Voprosy istorii*, 2 (2009): 3–39.

'Sorok let na zheleznodorozhnom transporte', *Zheleznodorozhnaia tekhnika*, 7 (1937): 69–70.

Spiridovich, A.I., *Zapiski zhandarma*, 2nd edn (Khar'kov: Proletarii, 1928).

Spisok naselennykh mest po svedeniiam 1859 goda, [vypusk] XL: Smolenskaia guberniia (St Petersburg: Tsentral'nyi statisticheskii komitet MVD, 1868).

Spravochnaia knizhka po lichnomu sostavu sluzhashchikh Tashkentskoi zh.d. 1914 goda (Orenburg, 1914).

'Sravnitel'nye ispytaniia tovarnykh parovozov', *Zhurnal Ministerstva putei soobshcheniia*, 10 (1908): 57–151.

[Stalin, I.V.,] 'O nedostatkakh partiinoi raboty i merakh likvidatsii trotskistskikh i inykh dvurushnikov: Doklad tovarishcha Stalina na Plenume TsK VKP(b) 3 marta 1937 g.', *Zheleznodorozhnaia tekhnika*, 4 (1937): 3–18.

Stanier, W.A., 'The Development and Testing of Locomotives', *Engineering*, 142/3688 (18 September 1936): 305–8.

'The Steam Locomotive', *Engineering*, 142/3690 (2 October 1936): 368–9.

Stempkovskii, S.F., 'Ob organizatsii zheleznodorozhnoi ispytatel'noi stantsii', in *Protokoly zasedanii XXIII soveshchatel'nogo s"ezda inzhenerov sluzhby podvizhnogo sostava i tiagi russkikh zheleznykh dorog, sozvannogo v Moskve na 23 noiabria 1901 g.* (St Petersburg, 1902), pp. 899–940.

Stepanov, P., 'Sravnitel'nye dannye o raskhode para i topliva parovozami serii E postroiki russkikh zavodov i E^G, E^{sh} postroiki zagranichnykh zavodov', *Zheleznodorozhnoe delo*, 8 (1925): 37–40.

Straten, V., 'K istorii rabochego dvizheniia na iuge Rossii', *Proletarskaia revoliutsiia*, 9 (1922): 62–9.

Sukhomlinov, V., *Vospominaniia* (Berlin: Russkoe universal'noe izdatel'stvo, 1924).

'The Sulzer–Diesel Locomotive', *Engineering* (5 September 1913): 317–21.

Sushinskii, B.B., 'Ispitatel'nyi briketnyi zavod', *Izvestiia Obshchego biuro soveshchatel'nykh s"ezdov*, 11 (1913): 958–9.

Sushinskii, B.B., 'Mery k umen'sheniiu kolichestva bol'nykh tovarnykh vagonov', *Izvestiia Obshchego biuro soveshchatel'nykh s"ezdov*, 6 (1915): 612–14.

Sushinskii, B.B., 'Opytnaia parovoznaia stantsiia imeni A.P. Borodina russkikh zheleznykh dorog', *Izvestiia Obshchego biuro soveshchatel'nykh s"ezdov*, 3 (1917): 176–7.

Sventsitskii, V., 'O prichinakh mnogoletnego zasizhivaniia studentov v spetsial'nykh vysshikh uchebnykh zavedeniiakh i nekotorykh merakh k ustraneniiu etogo nenormal'nogo iavleniia', *Izvestiia Sobraniia inzhenerov putei soobshcheniia*, 5 (1916): 75–8; 6 (1916): 91–9; and 7 (1916): 112–18.

Sverchkov, D., 'Soiuz soiuzov', *Krasnaia letopis'*, 3 (1925): 149–62.

Syromiatnikov, I., 'B.D. Voskresenskii Inzh p.s. "Naibol'shee uplotnenie rabot putem primeneniia parallel'nykh i posledovatel'nykh rabot, kak osnovnoi metod eksploatatsii zheleznykh dorog. Usilenie: A) provoznoi sposobnosti uchastkov i uzlov zheleznykh dorog i B) raboty parovoznogo i vagonnykh parkov" 2-e izdanie 19 str. 8°. Tsena 50 kop.', *Izvestiia Sobraniia inzhenerov putei soobshcheniia*, 3 (1916): 48–9.

Syromiatnikov, I., 'Inzhenery B.D. Voskresenskii i V.I. Taranov-Belozerov "Uplotnenie raboty poezdov, parovozov i parovoznykh sposobnostei" 8° str. 78, chert. 4 lista – tsena 2 rub.', *Izvestiia Sobraniia inzhenerov putei soobshcheniia*, 3 (1916): 49–50.

Syromiatnikov, I., 'S.K. Kudrevatov, inzh p.s. "Primenenie metoda uplotneniia i parallel'nosti rabot na peredatochnykh, uzlovykh i sortirovochnykh stantsiiakh dlia dostizheniia naimen'shego prostoia vagonov i gruzov". str. 20, chert. 11 listov. Tsena 1 rub.', *Izvestiia Sobraniia inzhenerov putei soobshcheniia*, 3 (1916): 50–51.

Syromiatnikov, I., 'Zheleznodorozhnoe stroitel'stvo v Rossii', *Izvestiia Obshchego biuro soveshchatel'nykh s"ezdov*, 22 (1911): 6–7.

Syromiatnikov, S.P., *Issledovanie rabochego protsessa parovoznogo kotla i paroperegrevatelia (po opytam prof. Iu.V. Lomonosova i s ego predisloviem)* (Berlin: Rossiiskaia zheleznodorozhnaia missiia, 1923).

Syromiatnikov, S.P., 'K voprosu ob eksploatatsii parovozov na nizkosortnom toplive', *Tekhnika i ekonomika putei soobshcheniia*, 2/9 (1923): 351–69; 2/10 (1923): 482–508; and 2/11 (1923): 574–87.

Syromiatnikov, S.P., 'Novyi metod postroeniia diagramm naivygodneishego regulirovaniia rabot parovozov i ego prakticheskoe primenenie', *Tekhnika i ekonomika putei soobshcheniia*, 1/1 (1923): 29–36.

Syromiatnikov, S.P., 'Osnovnye etapy razvitiia moshchnosti i ekonomichnosti parovoza za 100 let', in *Stoletie zheleznykh dorog (Trudy Nauchno-Tekhnicheskogo Komiteta Narodnogo Komissariata Putei Soobshcheniia, vypusk 20)* (Moscow: Transpechat', 1925), pp. 152–63.

Syromiatnikov, S.P., 'Put' i parovoz (K voprosu o povyshenii moshchnosti nashikh parovozov)', *Krasnyi transportnik*, 5 (1923): 16.

Syromiatnikov, S.P., 'Regulirovanie raboty parovozov (Po povodu statei prof. Iu.V. Lomonosova i inzh. Medel')', *Tekhnika i ekonomika putei soobshcheniia*, 2/10 (1923): 480–482.

'S"ezd predstavitelei kazen. i chastn. zh.d. 17 maia v Khar'kove, dlia vyslushaniia rezul'tatov primeneniia metoda uplotneniia rabot v razlichnykh otrasliakh khoziaistva Iuzhnykh zh.d.', *Zheleznodorozhnoe delo*, 24 (1915): 221–3.

T., 'Iu.V. Lomonosov, Tekhnicheskie perspektivy zheleznodorozhnogo transporta v blizhaishee vremia', *Vestnik putei soobshcheniia*, 14 (1924): 32.

Taranov-Belozerov, [V.I.], 'K voprosu ob intensifikatsii raboty tovarnykh poezdov', *Zheleznodorozhnoe delo*, 24 (1915): 224–5.

Taranov-Belozerov, V.I., 'Vzryv parovoznogo kotla serii E^G na Syzrano–Viazemskoi zh.d. 10 iiunia 1925 g.', *Zheleznodorozhnoe delo: Podvizhnoi sostav, tiaga i masterskie*, 4–5 (1926): 13–16.

Taylor, F.W., *The Principles of Scientific Management* (New York: Harper, 1911).

Tekhnika Bol'shevistskogo podpol'ia: sbornik statei i vospominanii, vyp. 1 (Moscow and Leningrad: Gosudarstvennoe Izdatel'stvo, 1924).

'Temperature tests on Superheater Locomotives', *Engineering*, 94 (26 July 1912): 134.

'Teplovozy sistemy prof. A.N. Shelesta (Doklad prof. A.N. Shelesta v NTK)', *Vestnik putei soobshcheniia*, 34 (1925): 18.

'Tikhii khod', *Izvestiia Sobraniia inzhenerov putei soobshcheniia*, 5 (1916): 74–5.

Timoshenko, S.P., *As I Remember: The Autobiography of Stephen P. Timoshenko* (Princeton: D. Van Nostrand, 1968).

Tolstopiatov, V., 'Ivan Nikolaevich Borisov', *Zheleznodorozhnoe delo: obshchii otdel*, 6 (1928): 1–2.

Trinkler, G., 'K voprosu o naivygodneishem tipe teplovoza', *Zheleznodorozhnoe delo*, 1 (1925): 8–14.

Tritton, S.B., 'Railway Traction by Steam Power', *The Engineer*, 154/4001 (16 September 1932): 279.

Trofimenko, A., 'K istorii Voenno-tekhnicheskogo biuro iuga Rossii v 1905–1906 gg.', *Letopis' revoliutsii*, 5–6 (1925): 99–105.

Troianovskii, A., 'Vosstanie v Kieve v 1907 g.', *Katorga i ssylka*, 3 (1922): 181–8.

'Trotskistskie bandity – predateli rodiny', *Sotsialisticheskii transport*, 2 (1937): 9–18.

Trotsky, L.D., *My Life* (New York: Grosset and Dunlap, 1960).

Trudy XXXII-go soveshatel'nogo s"ezda inzhenerov sluzhby tiagi (Moscow, 1923).

Trudy XXXIV soveshatel'nogo s"ezda inzhenerov podvizhnogo sostava i tiagi v Moskve s 1-go po 9-oe aprelia 1925 goda (Moscow, 1926).

Trudy komissii pod predsedatel'stvom chlena Inzhenernogo soveta d.s.s. Kunitskogo za vremia s 31-go maia po 23-e oktiabria 1912 g., po voprosam:

'*O napriazheniiakh v rel'sakh i v ostal'nykh chastiakh verknego stroeniia zheleznodorozhnogo puti*': *Zhurnaly komissii* (St Petersburg, 1913).

Trutch, C.J.H., 'Armstrong Whitworth Diesel Traction Development', *Armstrong Whitworth Record*, 2/4 (1933): 2–11.

Trutch, C.J.H., 'Armstrong Whitworth Diesel Traction Progress', *Armstrong Whitworth Record*, 2/2 (1932): 2–9.

Trutch, C.J.H., 'Diesel Rail Traction', *Armstrong Whitworth Record*, 1/2 (1930): 13–17.

Tschebotarioff, G.P., *Russia, My Native Land* (New York: McGraw Hill, 1964).

Tuplin, W.A., 'Some Questions about the Steam Locomotive', *Journal of the Institution of Locomotive Engineers*, 43/236 (1953): 637–714.

Turchaninov, A.N., 'Poiasnitel'naia zapiska k proektu dachnogo parovoza dlia Kh.–N. zh.d.', *Izvestiia Kievskogo politekhnicheskogo instituta, Otdel mekhanicheskii i inzhenernyi*, 7/4 (1907): I–IV, 175–252.

Universität Bern, *Behörden, Lehrer und Studierende*, WS 1906/7, Medizinische Fakultät (Bern, [n.d.]).

[Upravlenie Zheleznykh Dorog], *Kratkii ocherk deiatel'nosti russkikh zheleznykh dorog vo vtoruiu otechestvennuiu voinu, chast' 1-aia (S nachala voiny po 1 ianvaria 1915 g.)* (Petrograd, 1916).

[Upravlenie Zheleznykh Dorog], *Kratkii ocherk deiatel'nosti russkikh zheleznykh dorog vo vtoruiu otechestvennuiu voinu, chast' 2-aia (Pervoe polugodie 1915 goda)* (Petrograd, 1916).

'Upravliaiushchii Ministerstvom putei soobshcheniia E.B. Voinovskii-Kriger', *Izvestiia Sobraniia inzhenerov putei soobshcheniia*, 1 (1917): 4.

Urquhart, T., 'Supplementary Paper on the Use of Petroleum Refuse as Fuel in Locomotive Engines', *Proceedings of the Institution of Mechanical Engineers* (1889): 36–84.

'Uvelichenie parka tovarnykh vagonov', *Izvestiia Sobraniia inzhenerov putei sobshcheniia*, 3 (1915): 51–2.

V.F., 'Nekrolog[: P.N. Borshchov]', *Vestnik Ekaterininskoi zheleznoi dorogi*, 56 (1908): 222.

Vl. Sh., 'Iubilei tiagovikov', *Krasnyi transportnik*, 7 (1923): 3–4.

Vl. Sh., 'Nashi izobretateli: mashinist Florentii Pimenovich Kazantsev', *Krasnyi transportnik*, 1 (1923): 46.

Vl. Sh., 'Vozdushnyi avtomaticheskii tormaz sistemy F.P. Kazantseva', *Krasnyi transportnik*, 1 (1923): 26–7.

V.N., 'Na novom meste: Po povodu perekhoda komissii o novykh dorogakh v Ministerstvo putei soobshcheniia', *Izvestiia Sobraniia inzhenerov putei soobshcheniia*, 24 (1916): 497–9.

V.V., 'Postroika, eksploatatsiia i organizatsiia', *Izvestiia Sobraniia inzhenerov putei soobshcheniia*, 20 (1915): 423–6.

Vakar, V.V., *Nakanune 1905 goda v Kieve (Iiul'skaia stachka 1903 goda)* (Khar'kov: Knigospilka, 1925).

Vasil'ev, G.P., 'A.M. Babichkov. "Teoriia tiagi poezdov i tiagovye raschety"', *Parovoznoe khoziaistvo*, 2 (1935): 42.

Vasil'ev, M.I., 'K voprosu o nuzhdaemosti russkikh zheleznykh dorog v tovarnykh vagonakh i merakh k ee udovletvoreniiu: Doklad inzhenera M.I. Vasil'eva v Sobranii 13 marta 1915 goda', *Izvestiia Sobraniia inzhenerov putei sobshcheniia*, 8 (1915): 140–153; 9 (1915): 168–74 (discussion); 14 (1915): 303–7; and 15 (1915): 318–26.

Vasil'ev, M.I., *O sovremennom polozhenii zheleznodorozhnogo transporta i vazhneishikh ego elementakh* (Petrograd, 1916).

Vauclain, S.M., 'Internal Combustion Locomotives and Vehicles', *Baldwin Locomotives*, 5/1 (July 1926): 43–9.

Vauclain, S.M., *Steaming Up!* (San Marino: Golden West Books, 1973).

'Velikaia godovshchina', *Zheleznodorozhnaia tekhnika*, 10 (1936): 3–6.

Verkhovskoi, N.P., 'Gruzy Tashkentskoi zheleznoi dorogi', *Zhurnal Ministerstva putei soobshcheniia*, 1 (1910): 39–136; and 2 (1910): 3–34.

Verkhovskoi, N.P., 'Kolichestvo prodovol'stvennykh gruzov i topliva, dostavlennykh v Petrograd po zheleznym dorogam v 1914 i 1915 godakh', *Zheleznodorozhnoe delo*, 5–6 (1916): 38–9.

Viazemskii, V., 'O tipakh parovozov, preimushchestvenno passazhirskikh', *Izvestiia Sobraniia inzhenerov putei soobshcheniia*, 4 (1916): 60–69.

Viazemskii, V., 'Psikhologicheskie prichiny dolgogo prebyvaniia studentov v vysshikh uchebnykh zavedeniiakh', *Izvestiia Sobraniia inzhenerov putei soobshcheniia*, 5 (1915): 86–9.

Viazemskii, V., 'Spetsializatsiia', *Izvestiia Sobraniia inzhenerov putei soobshcheniia*, 6 (1916): 99.

'Vice-President Meets Scientist Who Designed and Built World's First Diesel-Electric Locomotive', *The Spanner* (July 1952): 8.

Vinogradov, G.Z., '35-yi Soveshchatel'nyi S"ezd Inzhenerov tiagi', *Zheleznodorozhnoe delo: Podvizhnoi sostav, tiaga i masterskie*, 9–10 (1927): 1–3.

'Vladimir Antonovich Shtukenberg (Nekrolog)', *Izvestiia Sobraniia inzhenerov putei soobshcheniia*, 29 (1913): 506–8.

Vladimirov, M., 'Kontsentrat predel'chestva', *Sotsialisticheskii transport*, 7 (1936): 110–115.

Vlast' Sovetov za desiat' let, 1917–27 (Leningrad: Krasnaia gazeta, 1927).

'Vodit' poezda po-krivonosovski! Likvidirovat' okhvost'e predel'shchikov!', *Parovoznoe khoziaistvo*, 10 (1935): 3–4.

Voenno-tipograficheskaia karta Smolenskoi gubernii (St Petersburg, 1858–1871).

Voinovskii-Kriger, E.B. – see Kriger-Voinovskii, E.B.

Voskresenskii, B.D., 'Naibol'shee uplotnenie rabot putem primeneniia parallel'nykh i posledovatel'nykh rabot, kak osnovnoi metod eksploatatsii zheleznykh dorog', *Izvestiia Sobraniia inzhenerov putei sobshcheniia*, 9 (1915): 189–94.

Voskresenskii, B.D., 'Naibol'shee uplotnenie rabot putem primeneniia parallel'nykh i posledovatel'nykh rabot, kak osnovnoi metod eksploatatsii zhel. dorog: Metod inzh. Khlebnikova v primenenii k usileniiu provoznoi sposobnosti uchastkov i uzlov zhel. dorog i raboty parovoznogo i vagonnogo parkov', *Zheleznodorozhnoe delo*, 16 (1915): 137–41.

Voskresenskii, B.D., *Teoriia raboty zheleznodorozhnykh poezdov: Poiasnitel'naia zapiska k polozheniiu ob uchete raboty podvizhnogo sostava na otdeleniiakh Sluzhby dvizheniia Ekaterininskoi zhel. dor.* (Ekaterinoslav, 1903).

'Vsevolod Iur'evich Lomonosov', *Elektrichestvo*, 12 (1962): 88.

'V teplovoznoi komissii', *Vestnik putei soobshcheniia*, 47 (1925): 17.

'Vykup v Kaznu chastnykh zheleznykh dorog: Izvlechenie iz smety Departamenta Zheleznodorozhnykh Del', *Izvestiia Sobraniia inzhenerov putei soobshcheniia*, 24 (1915): 492–3.

Vysochaishe uchrezhdennaia Osobaia vysshaia komissiia dlia vsestoronnego issledovaniia zheleznodorozhnogo dela v Rossii, *Doklady po obsledovaniiu zheleznykh dorog, Doklad No.17: Tashkentskaia zheleznaia doroga* (St Petersburg, 1913).

Vysochaishe uchrezhdennaia Osobaia vysshaia komissiia dlia vsestoronnego issledovaniia zheleznodorozhnogo dela v Rossii, [*Trudy,*] *vypusk 9: Zapiska N.K. Gofmana o glavnykh masterskikh dlia pochinki podvizhnogo sostava; o glavnykh material'nykh skladakh; ob ispitatel'nykh laboratoriiakh i o tekhnicheskikh kontorakh pri sluzhbe podvizhnogo sostava i tiagi na kazennykh zheleznykh dorogakh* (St Petersburg, 1909).

Vysochaishe uchrezhdennaia Osobaia vysshaia komissiia dlia vsestoronnego issledovaniia zheleznodorozhnogo dela v Rossii, [*Trudy,*] *vypusk 42: Materialy k voprosu o reorganizatsii otchetnosti kazennykh zheleznykh dorog: Otvety mestnykh upravlenii kazennykh zheleznykh dorog na voprosy, postavlennye pis'mom g. predsedatelia Osoboi vysshei komissii ot 18 dekabria 1909 g., chast' III: Otvety Upravlenii kazennykh zheleznykh dorog: Sredne–Aziatskoi, Syzrano–Viazemskoi, Severnykh, Severo–Zapadnykh, Tashkentskoi, Iugo–Zapadnykh i Iuzhnykh* (St Petersburg, 1910).

Vysochaishe uchrezhdennaia Osobaia vysshaia komissiia dlia vsestoronnego issledovaniia zheleznodorozhnogo dela v Rossii, [*Trudy,*] *Vypusk 96: Materialy po obsledovaniiu zheleznykh dorog: Tashkentskaia zhel. dor.* (St Petersburg, 1913).

Vysochaishe uchrezhdennaia Osobaia vysshaia komissiia dlia vsestoronnego issledovaniia zheleznodorozhnogo dela v Rossii, *Zhurnal No. 43* (17 December 1912) [St Petersburg, 1912].

Vysochaishe uchrezhdennaia Osobaia vysshaia komissiia dlia vsestoronnego issledovaniia zheleznodorozhnogo dela v Rossii, *Zhurnal No. 51* (7 March 1913) [St Petersburg, 1913].

Wellington, A.M., *The Economic Theory of the Location of Railways: An Analysis of the Conditions Controlling the Laying Out of Railways to Effect the Most Judicious Expenditure of Capital*, 6th edn (New York: Chapman and Hall, 1911).

Willigens, W., 'Russische 1E1 Diesel-elektrische Lokomotive Bauart Professor Lomonossoff–Hohenzollern AG Düsseldorf', *Der Waggon- und Lokomotiv-Bau*, 6 (19 March 1925): 81–6.

Z., 'Neskol'ko slov o podtalkivanii', *Izvestiia Sobraniia inzhenerov putei soobshcheniia*, 13 (1915): 282–3.

'Za bol'shevistskuiu nauku, protiv "teorii" predela', *Sotsialisticheskii transport*, 1 (1936): 1–8.

'Za dal'neishee razvitie nauchnoi mysli na transporte', *Sotsialisticheskii transport*, 9 (1938): 1–4.

Zagliadimov, D.P., Aksenov, I.Ia and Petrov, A.P., 'Grafiki i raspisanie dvizheniia poezdov – zheleznyi zakon raboty transporta', *Zheleznodorozhnaia tekhnika*, 1 (1936): 27–36.

'Zakliucheniia podkomissii pri Osoboi Vysshei Zheleznodorozhnoi Komissii po voprosu o reorganizatsii vagonnogo i parovoznogo khoziaistva na kazennykh zheleznykh dorogakh', *Izvestiia Sobraniia inzhenerov putei soobshcheniia*, 5 (1911): 10–12.

Zamlinskii, V.A. (comp.), *Kievskii universitet: dokumenty i materialy, 1834–1984* (Kiev: Vishcha shkola, 1984).

'Zapiska komissii, obrazovannoi pri Sobranii Inzhenerov Putei Soobshcheniia dlia obsuzhdeniia voprosa o polozhenii uchebnogo dela v Institute Putei Soobshcheniia Imperatora Aleksandra I: Ot Soveta tekhnicheskogo otdela sobraniia inzhenerov putei soobshcheniia', *Izvestiia Sobraniia inzhenerov putei soobshcheniia*, 5 (1916): 78–85.

Zausailov, I., 'K voprosu ob uvelichenii sostava tovarnykh poezdov', *Zheleznodorozhnoe delo*, 40 (1915): 420–422.

Zheleznodorozhnik, 'Kazna i zheleznodorozhnoe stroitel'stvo', *Izvestiia Sobraniia inzhenerov putei soobshcheniia*, 11 (1911): 5–7.

Zhitkov, M.N., 'Ob osnovnykh prichinakh umen'sheniia rabotosposobnosti nashego parovoznogo parka i meropriiatiiakh dlia predel'nogo podniatiia etoi rabotosposobnosti', *Zheleznodorozhnoe delo*, 9 (1931): 24–8.

Zhitkov, M.N., 'O prichinakh nedostatochnoi moshchnosti i ekonomichnosti sovremennykh parovozov v zheleznodorozhnoi eksploatatsii i merakh bor'by s nei', *Vestnik Obshchestva tekhnologov*, 15 (1913): 62.

Zhitkov, S.M., *Biografii inzhenerov putei soobshcheniia*, 3 vols (St Petersburg, 1889–1902).

Zhitkov, S.M., *Institut inzhenerov putei soobshcheniia Imperatora Aleksandra I: istoricheskii ocherk* (St Petersburg, 1899).

Zhurnal komissii dlia rassledovaniia neporiadkov i zloupotreblenii po perevozkam gruzov na Tashkentskoi zh.d. (Orenburg, 1909).

Zhurnaly soveta Instituta inzhenerov putei soobshcheniia Imperatora Aleksandra I (St Petersburg, 1905–1915).

Zhurnaly Vtorogo departamenta Gosudarstvennogo soveta sessii 1913–14 g. ([St Petersburg, 1914]).

Zhurnaly Vtorogo departamenta Gosudarstvennogo soveta sessii 1914–15 g. ([St Petersburg, 1914]).

Zuev, I.V. and Reingardt, V.O., 'O neftianykh dvigateliakh sistemy Dizelia', in *Protokoly zasedanii XXVII soveshchatel'nogo s"ezda inzhenerov sluzhby podvizhnogo sostava i tiagi russkikh zheleznykh dorog, v Varshave, v avguste 1909 g.* (St Petersburg, 1910), pp. 477–515.

Selected secondary works

Afonina, G.M. (comp.), *Kratkie svedeniia o razvitii otechestvennykh zheleznykh dorog s 1838 po 1990 g.* (Moscow: MPS RF, 1996).

Aldcroft, D.H., *British Railways in Transition: The Economic Problems of Britain's Railways Since 1914* (London: Macmillan, 1968).

Alloy, V. (ed.), *Diaspora: Novye materialy*, vol. 1 (Paris and St Petersburg: Athenaeum–Feniks, 2001).

Alston, P., *Education and the State in Tsarist Russia* (Stanford: Stanford University Press, 1969).

Andreeva, T., 'Nash teplovoz, vpered leti', *Gudok*, 15 September 1999, p. 2.

Andrews, J.T., *Science for the Masses: The Bolshevik State, Public Science, and the Popular Imagination, 1917–1934* (College Station: Texas A&M Press, 2003).

Aplin, H.A., 'George Lomonossoff (1908–1954)', in his *Catalogue of the Lomonossoff Collections* (Leeds: University of Leeds Press, 1988), pp. xxvii–xxix.

Aplin, H.A., 'Iurii Vladimirovich Lomonosov (1876–1952)', in his *Catalogue of the Lomonossoff Collections* (Leeds: University of Leeds Press, 1988), pp. vii–xx.

Aplin, H.A., 'Raisa Nikolaevna Lomonosova (1888–1973)', in his *Catalogue of the Lomonossoff Collections* (Leeds: University of Leeds Press, 1988), pp. xxi–xxvi.

Argenbright, R.T., 'Lethal Mobilities: Bodies and Lice on Soviet Railroads, 1918–1922', *The Journal of Transport History*, 29/2 (2008): 259–76.

Argenbright, R.T., 'Marking NEP's Slippery Path: The Krasnoshchekov Show Trial', *The Russian Review*, 61/2 (2002): 249–75.

Armstrong, J., Bouneau, C. and Vidal Olivares, J. (eds), *Railway Management and its Organisational Structure: Its Impact on and Diffusion into the General Economy* (Seville, 1998).

Armytage, W.H.G., *A Social History of Engineering*, 3rd edn (London: Faber & Faber, 1970).

Ascher, A., *The Revolution of 1905*, 2 vols (Stanford: Stanford University Press, 1988–1992).

Atkins, P., 'On Test: Rugby Locomotive Testing Station', *Backtrack*, 3/4 (April 1989): 148–54.

Badash, L., *Kapitza, Rutherford, and the Kremlin* (New Haven: Yale University Press, 1985).

Bailes, K.E., 'Alexei Gastov and the Soviet Controversy over Taylorism, 1918–1924', *Soviet Studies*, 29/3 (1977): 373–94.

Bailes, K.E., 'Natural Scientists and the Soviet System', in Koenker, D.P., Rosenberg, W.G. and Suny, R.G. (eds), *Party, State, and Society in the Russian Civil War: Explorations in Social History* (Bloomington and Indianapolis: Indiana University Press, 1989), pp. 267–95.

Bailes, K.E., *Science and Russian Culture in an Age of Revolutions: V.I. Vernadsky and His Scientific School, 1863–1945* (Bloomington and Indianapolis: Indiana University Press, 1990).

Bailes, K.E., *Technology and Society under Lenin and Stalin: Origins of the Soviet Technical Intelligentsia, 1917–1941* (Princeton: Princeton University Press, 1978).

Baker, J.F., rev. J. Heyman, 'Inglis, Sir Charles Edward (1875–1952)', in *Oxford Dictionary of National Biography*, vol. 29 (Oxford: Oxford University Press, 2004), pp. 258–9.

Balzer, H.D., 'The Engineering Profession in Tsarist Russia', in his (ed.), *Russia's Missing Middle Class: The Professions in Russian History* (Armonk: M.E. Sharpe, 1996), pp. 55–88.

Balzer, H.D., 'Engineers: The Rise and Decline of a Social Myth', in Graham, L.R. (ed.), *Science and the Soviet Social Order* (Cambridge, MA: Harvard University Press, 1990), pp. 141–67.

Balzer, H.D., 'The Problem of Professions in Imperial Russia', in Clowes, E.W., Kassow, S.D. and West, J.L. (eds), *Educated Society and the Quest for Public Identity in Late Imperial Russia* (Princeton: Princeton University Press, 1991), pp. 183–98.

Balzer, H.D. (ed.), *Russia's Missing Middle Class: the Professions in Russian History* (Armonk: M.E. Sharpe, 1996).

Basalla, G., *The Evolution of Technology* (Cambridge: Cambridge University Press, 1988).

Bastrakova, M.S., *Stanovlenie sovetskoi sistemy organizatsii nauki (1917–1922)* (Moscow: Nauka, 1973).

Becker, S., *Nobility and Privilege in Late Imperial Russia* (DeKalb: Northern Illinois University Press, 1985).

Berberova, N.N., *Liudi i lozhi: Russkie masony XX stoletiia* (New York: Russica, 1986).

Bernstein, S., *Histoire du Parti Radical: La recherche de l'age d'or, 1919–1926* (Paris: Presses de la Fondation Nationale des Sciences Politiques, 1980).

Beveridge, rev. Jose Harris, 'Stamp, Josiah Charles, first Baron Stamp (1880–1941)', in *Oxford Dictionary of National Biography* (Oxford: Oxford University Press, 2004), pp. 85–8.

Bijker, W., Hughes, T.P. and Pinch, T.J. (eds), *The Social Construction of Technological Systems: New Directions in the Sociology and History of Technology* (Cambridge, MA: MIT Press, 1989).

Bogdanovich, A.O. et al. (eds), *Ministry i Narkomy Putei Soobshcheniia* (Moscow: Transport, 1995).

Bokhanov, A.N., 'Aleksandr Ivanovich Guchkov', in Tiutiukin, S.V. (ed.), *Istoricheskie siluety* (Moscow: Nauka, 1991), pp. 63–105.

Bokhanov, A., 'Hopeless Symbiosis: Power and Right-Wing Radicalism at the Beginning of the Twentieth Century', in Geifman, A. (ed.), *Russia under the Last Tsar: Opposition and Subversion, 1894–1917* (Oxford: Blackwell, 1999), pp. 199–213.

Bonavia, M.R., *The Four Great Railways* (Newton Abbot: David & Charles, 1980).

Bondarenko, I.N., *Bol'sheviki Kieva v pervoi russkoi revoliutsii (1905–1907 gg.)* (Kiev: Izdatel'stvo Kievskogo universiteta, 1960).

Brovkin, V.N. (ed.), *The Bolsheviks in Russian Society: The Revolution and the Civil Wars* (New Haven: Yale University Press, 1997).

Brown, J., 'Vauclain, Samuel Matthews (18 May 1856–4 Feb 1940)', in *American National Biography*, vol. 22 (New York, 1999), pp. 285–6.

Brown, V.B., 'Addams, Jane (6 Sept 1860–21 May 1935)', in *American National Biography*, vol. 1 (New York, 1999), pp. 139–41.

Bryant, L., 'The Development of the Diesel Engine', *Technology and Culture*, 17 (July 1976): 432–46.

Buchanan, R.A., *The Engineers: A History of the Engineering Profession in Britain, 1750–1914* (London: Jessica Kingsley, 1989).

Budaev, D.I., *Krest'ianskaia reforma 1861 goda v Smolenskoi gubernii (K voprosu o realizatsii 'Polozhenii 19 fevralia')* (Smolensk: Smolenskii Gosudarstvennyi Pedagogicheskii Institut, 1967).

Budaev, D.I., *Smolenskaia derevnia v kontse XIX–nachale XX vv.: K voprosu o tempakh, urovne i stepeni razvitiia agrarnogo kapitalizma* (Smolensk: Smolenskii Gosudarstvennyi Pedagogicheskii Institut, 1972).

Budaev, D.I., *Smolenskaia guberniia v 1861–1917 godakh: Uchebnoe posobie k spetskursu* (Smolensk: Smolenskii gosudarstvennyi pedagogicheskii institut imeni Karla Marksa, 1990).

Bugliarello, G. and Doner, D.B. (eds), *The History and Philosophy of Technology* (Chicago: University of Illinois, 1979).

Burstall, A.F., *A History of Mechanical Engineering* (London: Faber & Faber, 1963).

Cantor, G., *Quakers, Jews, and Science: Religious Responses to Modernity and the Sciences in Britain, 1650–1900* (Oxford: Oxford University Press, 2005).

Cardwell, D.S.L., *Technology, Science and History: A Short Study of the Major Developments in the History of Western Mechanical Technology and their Relationships with Science and Other Forms of Knowledge* (London: Heinemann, 1972).

Carling, D.R., 'The Development of Locomotive Testing', *Railway World*, 37/437 (September 1976): 372–5.

Carling, D.R., 'Locomotive Testing Stations (Part I)', *Transactions of the Newcomen Society*, 45 (1972–1973): 105–44.

Carling, D.R., 'Locomotive Testing Stations (Part II)', *Transactions of the Newcomen Society*, 45 (1972–1973): 145–82.

Carlyle, E.I., rev. W.H. Brock, 'Ewing, Sir (James) Alfred (1855–1935)', in *Oxford Dictionary of National Biography*, vol. 18 (Oxford: Oxford University Press, 2004), pp. 822–3.

Carr, E.H., *Socialism in One Country, 1924–1926*, 3 vols (London: Macmillan, 1958–1964).

Carr, E.H. and Davies, R.W., *Foundations of a Planned Economy, 1926–1929*, 3 vols (London: Macmillan, 1969–1978).

Carvill, J., *Famous Names in Engineering* (London: Butterworths, 1981).

Chacksfield, J.E., *Sir William Stanier: A New Biography* (Usk: Oakwood Press, 2001).

Chester, K.R. (ed.), *Parovozy, Vol. 2: Elements of Locomotive Development in Russia and the USSR* (Skipton: Trackside Publications, [2002]).

Chorlton, F.O.L., 'Alan Ernest Leofric Chorlton, CBE', *Proceedings of the Institution of Mechanical Engineers*, 156 (1947): 245.

Churella, A.J., 'Business Strategies and Diesel Development', *Railroad History, Millenium Special* (2000): 22–37.

Churella, A.J., *From Steam to Diesel: Managerial Customs and Organizational Capabilities in the Twentieth-Century American Locomotive Industry* (Princeton: Princeton University Press, 1998).

Clowes, E.W., Kassow, S.D. and West, J.L. (eds), *Educated Society and the Quest for Public Identity in Late Imperial Russia* (Princeton: Princeton University Press, 1991).

Cohen, S.F., *Bukharin and the Bolshevik Revolution: A Political Biography, 1888–1938* (Oxford: Oxford University Press, 1980).

Collins, D.N., 'The Franco–Russian Alliance and Russian Railways, 1891–1914', *The Historical Journal*, 16/4 (1973): 777–88.

Cookson, G. and Hempstead, C., *A Victorian Scientist and Engineer: Fleeming Jenkin and the Birth of Electrical Engineering* (Aldershot: Ashgate, 2000).

Cooper, J. et al., *Soviet History, 1917–53: Essays in Honour of R.W. Davies* (Basingstoke: Macmillan, 1995).

Coopersmith, J., *The Electrification of Russia, 1880–1926* (Ithaca: Cornell University Press, 1992).

Cossons, N., Patmore, A. and Shorland-Ball, R. (eds), *Perspectives on Railway History and Interpretation: Essays to Celebrate the Work of Dr John Coiley at the National Railway Museum, York* (York: National Railway Museum, 1992).

Cox, E.S., *World Steam in the Twentieth Century* (London: Ian Allan, 1969).

Crosse, J., 'The Scientific Research Department of the LMS', *Backtrack*, 22/4 (April 2008): 236–7.

Cummins, C.L., Jr, *Internal Fire*, revised edn (Warrendale: Society of Automotive Engineers, 1989).

Daly, J.W., *Autocracy under Siege: Security Police and Opposition in Russia, 1866–1905* (DeKalb: Northern Illinois University Press, 1998).

Daly, J.W., *The Watchful State: Security Police and Opposition in Russia, 1906– 1917* (DeKalb: Northern Illinois University Press, 2004).

Davidova, L., 'Gosudar' oschastlivil ego svoim rukopozhatiem: Ivanovskii, Ippolit Konstantinovich', *Oktiabr'skaia magistral'*, 13 January 1993.

Davies, R.W. (ed.), *From Tsarism to the New Economic Policy: Continuity and Change in the Economy of the USSR* (Basingstoke: Macmillan, 1990).

Davies, W.J.K., *Diesel Rail Traction: An Illustrated History of Diesel Locomotives, Rail-cars, and Trains* (London: Altmark, 1973).

Day, R.B., *Leon Trotsky and the Politics of Economic Isolation* (London: Cambridge University Press, 1973).

Day-Lewis, S., *Bulleid: Last Giant of Steam* (London: George Allen and Unwin, 1964).

Debo, R., *Revolution and Survival: The Foreign Policy of Soviet Russia, 1917– 1918* (Liverpool: Liverpool University Press, 1979).

Debo, R., *Survival and Consolidation: The Foreign Policy of Soviet Russia, 1918– 1921* (Montreal and Kingston: McGill and Queens University Press, 1992).

Divall, C., 'Down the American Road? Industrial Research on the London, Midland & Scottish Railway, 1923–1947', in Armstrong, J., Bouneau, C. and Vidal Olivares, J. (eds), *Railway Management and its Organisational Structure: Its Impact on and Diffusion into the General Economy* (Seville, 1998), pp. 131–40.

Divall, C., 'Learning from America?', *Railroad History, Millenium Special* (2000): 124–42.

Doherty, J.M., 'Evolution of Internal Combustion Locomotives', *Journal of the Institution of Locomotive Engineers*, 46/251 (1956): 235–68 and discussions 268–91.

Do istorii bil'shovits'kikh organizatsii na ukraini v period pershoi rosiis'koi revoliutsii (1905-1907 rr.) (Kiev, 1955).

Dolzall, G.U. and Dolzall, S.F., *Diesels from Eddystone: the Story of Baldwin Diesels* (Milwaukee: Kalmbach Press, 1984).

Domb, C., 'Besicovitch, Abram Samoilovitch (1891–1970)', in *Oxford Dictionary of National Biography*, vol. 5 (Oxford: Oxford University Press, 2004), pp. 509–11.

Donkov, I. and Nikonov, A., 'Ia.E. Rudzutak (1887–1938): Emu doverial Il'ich', in Proskurin, A. (comp.), *Vozvrashchennye imena: Sbornik publitsisticheskikh statei v 2-kh knigakh* (2 vols, Moscow: Novosti, 1989), vol. 2, pp. 131–47.

Duffy, M.C., *Electric Railways, 1880–1990* (London: Institution of Electrical Engineers, 2003).

Edson, W.D., 'The Russian Decapods', *Railway and Locomotive Historical Society Bulletin*, 124 (1971): 64–75.

Ell, S.O., 'The Testing of Locomotives', in Ransome-Wallis, P. (ed.), *The Concise Encyclopaedia of World Railway Locomotives* (London: Hutchinson, 1959), pp. 386–410.

Emel'ianov, P., *Bor'ba Smolenskikh krest'ian za zamel' v revoliutsii 1905–1907 gg.* (Smolensk: Smolenskoe knizhnoe izdatel'stvo, 1955).

Emmons, T., *The Formation of Political Parties and the First National Elections in Russia* (Cambridge, MA: Harvard University Press, 1983).

Emmons, T., 'Russia's Banquet Campaign', *California Slavic Studies*, 10 (1977): 45–86.

Engelstein, L., *The Keys to Happiness: Sex and the Search for Modernity in Fin-de-Siècle Russia* (Ithaca: Cornell University Press, 1992).

Engelstein, L., 'Weapon of the Weak (Apologies to James Scott): Violence in Russian History', *Kritika*, 4/3 (2003): 679–94.

Erickson, C., *Alexandra, the Last Tsarina* (London: Robinson, 2003).

Erlich, A., *The Soviet Industrialization Debate, 1924–1928* (Cambridge, MA: Harvard University Press, 1960).

Erman, L.K., *Intelligentsiia v pervoi russkoi revoliutsii* (Moscow, 1966).

Erman, L.K., *Uchastie demokraticheskoi intelligentsii v stachechnom i profsoiuznom dvizhenii, 1905–1907 gody* (Moscow, 1955).

Erokhin, A., '"Spetsial'no sledit' za etim delom...": Rasskaz o knigakh iz lichnoi biblioteki V.I. Lenina', *Pravda*, 4 February 1984, p. 3.

Evseev, M., 'Sozdanie teplovoza', *Oktiabr'skaia magistral'*, 1 October 1986.

Fainerman, I.D. et al. (eds), *KPI: 50 let kievskogo ordena Lenina politekhnicheskogo instituta: Sbornik nauchnykh trudov* (Kiev: VNITOMash, kn-zhurn f-ka, 1948).

Ferneyhough, F., *Liverpool and Manchester Railway, 1830–1980* (London: Robert Hale, 1980).

Finkel, S., *On the Ideological Front: The Russian Intelligentsia and the Making of the Soviet Public Sphere* (New Haven: Yale University Press, 2007).

Foglesong, D.S., *America's Secret War Against Bolshevism: US Intervention in the Russian Civil War, 1917–1920* (Chapel Hill: University of North Carolina Press, 1995).

Foglesong, D.S., 'Xenophon Kalamatiano: An American Spy in Revolutionary Russia?', *Intelligence and National Security*, 6/1 (1991): 154–95.

Fufrianskii, N.A. and Dolganov, A.N., *Opytnoe kol'tso Vsesoiuznogo Nauchno-Issledovatel'skogo Instituta Zheleznodorozhnogo Transporta*, 2nd edn (Moscow: Transport, 1977).

Galai, S., *The Liberation Movement in Russia, 1900–1905* (Cambridge: Cambridge University Press, 1973).

Galai, S., 'The Role of the Union of Unions in the Revolution of 1905', *Jahrbücher für Geschichte Osteuropas*, 24/4 (1976): 512–25.

Garaevskaia, I.A., *Petr Pal'chinskii: Biografiia inzhenera na fone voin i revoliutsii* (Moscow: Rossiia molodaia, 1996).

Garmany, J.B., *Southern Pacific Dieselization* (Edmonds: PFM Publishing, 1985).

Gatrell, P., *Russia's First World War: A Social and Economic History* (Harlow: Pearson, 2005).

Gatrell, P., *The Tsarist Economy, 1850–1917* (London: Batsford, 1986).

Geifman, A., 'Psychohistorical Approaches to 1905 Radicalism', in Smele, J.D. and Heywood, A.J. (eds), *The Russian Revolution of 1905: Centenary Perspectives* (London: Routledge, 2005), pp. 13–33.

Geifman, A. (ed.), *Russia under the Last Tsar: Opposition and Subversion, 1894–1917* (Oxford: Blackwell, 1999).

Geifman, A., 'The Russian Intelligentsia, Terrorism, and Revolution', in Brovkin, V.N. (ed.), *The Bolsheviks in Russian Society: The Revolution and the Civil Wars* (New Haven: Yale University Press, 1997).

Geifman, A., *Thou Shalt Kill: Revolutionary Terrorism in Russia, 1894–1917* (Princeton: Princeton University Press, 1993).

Genis, V.L., 'General ot parovozov', *Politicheskii zhurnal*, 36 (4 October 2004): 78–80.

Genis, V.L., *Nevernye slugi rezhima: Pervye sovetskie nevozvrashchentsy (1920–1933): Opyt dokumental'nogo issledovaniia v 2-kh knigakh: Kniga 1: "Bezhal i pereshel v lager' burzhuazii..." (1920–1929)* (Moscow, 2009).

Genis, V.L., 'Nevozvrashchentsy 1920-x–nachala 1930-x godov', *Voprosy istorii*, 1 (2000): 46–63.

Genis, V.L., 'Sergei Dmitrievskii: "Nam nuzhen liberal'nyi tsezarizm"', in Korostelev, O.A. (ed.), *Diaspora: Novye materialy*, vol. 8 (Paris and St Petersburg, 2007), pp. 75–172.

Geyer, M., 'Some Hesitant Observations Concerning "Political Violence"', *Kritika*, 4/3 (2003): 695–708.

Gilchrist, A.O., 'A History of Engineering Research on British Railways: IRSTH Working Paper No.10' (York: National Railway Museum/IRSTH University of York, 2006).

Gill, G., *The Origins of the Stalinist Political System* (Cambridge: Cambridge University Press, 1990).

Glavak, T.V., Babko, Iu.V., Berenshtein, L.E. et al. (eds), *Ocherki istorii kievskikh gorodskoi i oblastnoi partiinykh organizatsii* (Kiev: Izd. politicheskoi literatury, 1981).

Gleason, W., 'Alexander Guchkov and the End of the Russian Empire', in *Transactions of the American Philosophical Society*, 73/3 (1983): 1–90.

Golikov, G.N., Antoniuk, D.I., Mchedlov, M.P. et al. (eds), *Vladimir Il'ich Lenin: Biograficheskaia khronika*, 12 vols (Moscow: Izdatel'stvo politicheskoi literatury, 1970–1982).

Goodstein, J.R., *Millikan's School: A History of the California Institute of Technology* (New York: Norton, 1991).

Gordin, M.D., *A Well-Ordered Thing: Dmitrii Mendeleev and the Shadow of the Periodic Table* (New York: Basic Books, 2004).

Gordin, M.D., Hall, K. and Kojevnikov, A. (eds), *Intelligentsia Science: The Russian Century, 1860–1960 (Osiris*, vol. 23) (Chicago: University of Chicago Press, 2008).

Goriacheva, V.P. and Shelest, P.A., *Aleksei Nesterovich Shelest: Pioner teplovozostroeniia, 1878–1954* (Moscow: Nauka, 1989).

Graham, L.R., *The Ghost of the Executed Engineer: Technology and the Fall of the Soviet Union* (Cambridge, MA: Harvard University Press, 1996).

Graham, L.R., *Science in Russia and the Soviet Union: A Short History* (Cambridge: Cambridge University Press, 1994).

Graham, L.R. (ed.), *Science and the Soviet Social Order* (Cambridge, MA: Harvard University Press, 1990).

Grant, J.A., *Big Business in Russia: The Putilov Company in Late Imperial Russia, 1868–1917* (Pittsburgh: Pittsburgh University Press, 1999).

Gregory, M.S., *History and Development of Engineering* (London: Longman, 1971).

Grishukov, S.M., *Leningradskii ordena Lenina institut inzhenerov zheleznodorozhnogo transporta: kratkii ocherk* (Leningrad, 1959).

Grosser, M., *Diesel: The Man and His Engine* (Newton Abbot: David & Charles, 1980).

Gvishiani, L., *Sovetskaia Rossiia i SShA (1917–1920)* (Moscow: Mezhdunarodnye otnosheniia, 1970).

Haimson, L.H., '"The Problem of Political and Social Stability in Urban Russia on the Eve of War and Revolution" Revisited', *Slavic Review*, 59/4 (Winter 2000): 848–75.

Hall, A.R. and Smith, N. (eds), *History of Technology: First Annual Volume, 1976* (London: Mansell, 1976).

Hamm, M.F., *Kiev: A Portrait, 1800–1917* (Princeton: Princeton University Press, 1993).

Hamm, M.F., 'On the Perimeter of Revolution: Kharkiv's Academic Community, 1905', *Revolutionary Russia*, 15/1 (June 2002): 45–68.

Hasegawa, T., *The February Revolution: Petrograd, 1917* (Seattle: University of Washington Press, 1981).

Hatchen, E.A., 'In Service to Science and Society: Scientists and the Public in Late Nineteenth-Century Russia', *Osiris*, 17 (2002): 171–209.

Hellbeck, J., *Revolution on My Mind: Writing a Diary under Stalin* (Cambridge, MA: Harvard University Press, 2006).

Heywood, A.J., 'The Armstrong Affair and the Making of the Anglo-Soviet Trade Agreement, 1920–1921', *Revolutionary Russia*, 5/1 (June 1992): 53–91.

Heywood, A.J., 'The Baltic Economic "Bridge": Some Early Soviet Perspectives', in Johansson, A., Loit, A., Kangeris, K. et al. (eds), *Emancipation and Interdependence: The Baltic States as New Entities in the International Economy, 1918–1940* (Stockholm: University of Stockholm, 1994), pp. 63–85.

Heywood, A.J., 'Barriers to East–West Technology Transfer: Iu.V. Lomonosov and Diesel Railroad Engineering in the Interwar Period', *The Russian Review* (forthcoming).

Heywood, A.J., 'Breaking the "Window into Europe": A Case-Study of Soviet–Estonian Economic Relations, 1920–24', *Revolutionary Russia*, 12/2 (December 1999): 41–62.

Heywood, A.J., 'Iu.V. Lomonosov and Locomotive Testing in Russia, 1898–1926', in Chester, K.R. (ed.), *Parovozy, Vol. 2: Elements of Locomotive Development in Russia and the USSR* (Skipton: Trackside Publications, [2002]), pp. 4–15.

Heywood, A.J., 'Iu.V. Lomonosov and the Science of Locomotive Testing in Russia: Consolidation, Methodology and Impact, 1908–17', *Transactions of the Newcomen Society*, 72/2 (2000–2001): 269–93.

Heywood, A.J., 'Iu.V. Lomonosov and the Science of Locomotive Testing in Russia: First Steps, 1895–1901', *Transactions of the Newcomen Society*, 72/1 (2000–2001): 1–15.

Heywood, A.J., 'Liberalism, Socialism and "Bourgeois Specialists": The Politics of Iu.V. Lomonosov to 1917', *Revolutionary Russia*, 17/1 (June 2004): 1–30.

Heywood, A.J., *Modernising Lenin's Russia: Economic Reconstruction, Foreign Trade and the Railways* (Cambridge: Cambridge University Press, 1999).

Heywood, A.J., '"The Most Catastrophic Question": Railway Development and Military Strategy in Late Imperial Russia', in Otte, T.G. and Neilson, K. (eds), *Railways and International Politics: Paths of Empire, 1848–1945* (London: Routledge, 2006), pp. 45–67.

Heywood, A.J., 'Russia's Foreign Supply Policy in World War I: Imports of Railway Equipment', *The Journal of European Economic History*, 32/1 (Spring 2003): 77–108.

Heywood, A.J., 'Socialists, Liberals and the Union of Unions in Kyiv during the 1905 Revolution: An Engineer's Perspective', in Smele, J.D. and Heywood, A.J. (eds), *The Russian Revolution of 1905: Centenary Perspectives* (London: Routledge, 2005), pp. 177–95.

Heywood, A.J., 'Trade or Isolation? Soviet Imports of Railway Equipment, 1920–1922', in Hiden, J. and Loit, A. (eds), *Contact or Isolation? Soviet–Western Relations in the Inter-War Period, 1917–1940* (Stockholm: University of Stockholm, 1994), pp. 137–60.

Heywood, A.J., 'War Destruction and Remedial Work in the Early Soviet Economy: Myth and Reality on the Railroads', *The Russian Review*, 64/3 (July 2005): 456–79.

Heywood, A.J. and Button, I.D.C., *Soviet Locomotive Types: The Union Legacy* (Malmo: Stenvall, 1995).

Hiden, J. and Loit, A. (eds), *Contact or Isolation? Soviet–Western Relations in the Inter-War Period, 1917–1940* (Stockholm: University of Stockholm, 1994).

Hildermeier, M., 'The Terrorist Strategies of the Socialist-Revolutionary Party in Russia, 1900–14', in Mommsen, W.J. and Hirschfeld, G. (eds), *Social Protest,*

Violence and Terror in Nineteenth- and Twentieth-century Europe (London: Macmillan, 1982), pp. 80–87.

Hilken, T.J.N., *Engineering at Cambridge University, 1783–1965* (Cambridge: Cambridge University Press, 1967).

Hofsommer, D.L., 'Getting to Know Her', *Railroad History, Millenium Special* (2000): 100–109.

Hofsommer, D.L., *The Southern Pacific, 1901–1985* (College Station: Texas A & M University Press, 1986).

Holquist, P., *Making War, Forging Revolution: Russia's Continuum of Crisis, 1914–1921* (Cambridge, MA: Harvard University Press, 2002).

Holquist, P., 'Violent Russia, Deadly Marxism? Russia in the Epoch of Violence, 1905–21', *Kritika*, 4/3 (2003): 627–52.

Howell, D., 'Brockway, (Archibald) Fenner, Baron Brockway (1888–1988)', in *Oxford Dictionary of National Biography*, vol. 7 (Oxford: Oxford University Press, 2004), pp. 765–6.

Hughes, G., *Sir Nigel Gresley: The Engineer and his Family* (Usk: Oakwood Press, 2001).

Hutchings, R., *Soviet Science, Technology, Design: Interaction and Convergence* (London: Oxford University Press, 1976).

Iakhontov, Iu., 'Tysiacha parovozov dlia Lenina', *Pravda*, 27 January 1969, p. 5.

Iakobson, P.V., *Istoriia teplovoza v SSSR* (Moscow: Transzheldorizdat, 1960).

Ianush, B.V., 'Professor Lomonosov', *Oktiabr'skaia magistral'*, 18 April 1992; 21 April 1992; 22 April 1992, p. 4; 23 April 1992, p. 4; 24 April 1992, p. 3.

Ianush, B.V. and Strekopytov, V.V., *Uchenye Universiteta putei soobshcheniia – sozdateli pervykh lokomotivov* (St Petersburg: PGUPS, 1995).

Ignat'ev, A.V., *Vneshniaia politika vremennogo pravitel'stva* (Moscow: Nauka, 1974).

Igolkin, A.A., 'Algembra: Nefteprovod v nebytie', *Ekonomicheskii zhurnal*, 1 (2001): 5–34.

Igolkin, A.A., 'Leninskii narkom: U istokov sovetskoi korruptsii', *Grazhdanin*, 3 (2004).

Iurchuk, V.I., Babko, Iu.V., Vargatiuk, P.L. et al., *Ocherki istorii kommunisticheskoi partii Ukrainy*, 4th edn (Kiev: Izd. politicheskoi literatury, 1977).

Iurii Vladimirovich Lomonosov (1876–1952): Biobibliografiia (St Petersburg: PGUPS, 1993).

Ivanova, O.V. et al. (eds), *Khto e khto: Dovidnik: Profesori Nats. Tekh. Universiteta Ukraini 'K.P.I.'* (Kyiv: Osvita, 1998).

Johansson, A., Loit, A., Kangeris, K. et al. (eds), *Emancipation and Interdependence: The Baltic States as New Entities in the International Economy, 1918–1940* (Stockholm: University of Stockholm, 1994).

Jones, D.R., 'Nicholas II and the Supreme Command: An Investigation of Motives', *Sbornik*, 11 (1985): 47–83.

Joravsky, D., 'What Do We Ask of the History of Technology?', in Bugliarello, G. and Doner, D.B. (eds), *The History and Philosophy of Technology* (Chicago: University of Illinois, 1979), pp. 128–34.

Kagarlitsky, B., *The Thinking Reed: Intellectuals and the Soviet State, 1917 to the Present* (London: Verso, 1989).

Kalmykov, S.V., Mishenev, Iu.D., Roshchevkin, V.T. and Senin, A.S., *Moskovskoi okruzhnoi zheleznoi doroge 100 let* (Moscow: URSS, 2008).

Kargon, R.H., *The Rise of Robert Millikan: Portrait of a Life in American Science* (Ithaca: Cornell University Press, 1982).

Karpova, R.F., 'Sovetskie otnosheniia so skandinavskimi stranami 1920–1923 gg.', *Skandinavskii sbornik*, 10 (1965): 135–69.

Kassow, S.D., *Students, Professors and the State in Tsarist Russia* (Berkeley and Los Angeles: University of California Press, 1989).

Katkov, G., *Russia 1917: The February Revolution* (London: Fontana, 1969).

Kennan, G.F., *Soviet–American Relations, 1917–1920*, 2 vols (Princeton: Princeton University Press, 1956–1958).

Khlevnyuk, O.V., 'The Objectives of the Great Terror, 1937–1938', in Cooper, J. et al., *Soviet History, 1917–53: Essays in Honour of R.W. Davies* (Basingstoke: Macmillan, 1995), pp. 158–76.

Kievskii politekhnicheskii institut (Kratkii istoricheskii ocherk), 1898–1973 (Kiev: Ministerstvo Vysshego i Srednego Spetsial'nogo Obrazovaniia UkSSR, 1973).

Kim, M.I., *Velikaia oktiabr'skaia sotsialisticheskaia revoliutsiia i stanovlenie sovetskoi kul'tury, 1917–1927* (Moscow: Nauka, 1985).

Kirkland, J.F., *Dawn of the Diesel Age: A History of the Diesel Locomotive in America* (Pasadena: Interurban Press, 1983).

Kirkland, J.F., *The Diesel Builders, Vol. 3: Baldwin Locomotive Works* (Pasadena: Interurban Press, 1994).

Klein, M., 'The Continued Neglect of the Diesel Locomotive', *Railroad History, Millenium Special* (2000): 6–15.

Klimenko, V.I. et al. (eds), *Magistral': Nachal'niki zheleznykh dorog Zapadnoi Sibiri, 1896–2006: Istoricheskie ocherki* (Novosibirsk: Istoricheskoe nasledie Sibiri, 2006).

Koblitz, A.H., *A Convergence of Lives: Sofia Kovalevskaia: Scientist, Writer, Revolutionary* (Cambridge, MA: Birkhäusern, 1983).

Koenker, D.P., Rosenberg, W.G. and Suny, R.G. (eds), *Party, State, and Society in the Russian Civil War: Explorations in Social History* (Bloomington and Indianapolis: Indiana University Press, 1989).

Kolodezhnyi, V. and Teptsov, N., 'A.I. Rykov (1881–1938): Predsedatel' Sovnarkoma', in Proskurin, A. (comp.), *Vozvrashchennye imena: Sbornik publitsisticheskikh statei v 2-kh knigakh* (2 vols, Moscow: Novosti, 1989), vol. 2, pp. 149–75.

Konarev, N.S. et al. (eds), *Zheleznodorozhnyi transport: Entsiklopediia* (Moscow: Bol'shaia Rossiiskaia Entsiklopedia, 1994).

Kondufor, Iu.Iu., Artemenko, I.I., Zgurskii, V.A. et al. (eds), *Istoriia Kieva, tom 2* (Kiev: Naukova dumka, 1983).

Korostelev, O.A. (ed.), *Diaspora: Novye materialy*, vol. 8 (Paris and St Petersburg, 2007).

Kovalevskii, A., 'Fevral' Semnadtsatogo: otrechenie', *Oktiabr'skaia magistral'*, 5 March 1992, pp. 2–3; 6 March 1992, pp. 2–3.

Kraskovskii, E.Ia. and Uzdin, M.M. (eds), *Istoriia zheleznodorozhnogo transporta Rossii, tom 1: 1836–1917 gg.* (St Petersburg: Ivan Fedorov, 1994).

Krementsov, N., *Stalinist Science* (Princeton: Princeton University Press, 1997).

Krivosheev, G.F. (ed.), *Rossiia i SSSR v voinakh XX veka: Poteri vooruzhennykh sil: Statisticheskoe issledovanie* (Moscow: OLMA-Press, 2001).

Kudriakova, E.B., *Rossiiskaia emigratsiia v Velikobritanii v period mezhdu dvumia voinami* (Moscow: INION RAN, 1995).

Kuprienko, O.G., 'Ostorozhno: Shelest! Pora postavit' tochku v istorii o tom, kto sozdal pervyi otechestvennyi teplovoz', *Gudok*, 25 December 1999, p. 2.

Kuprienko, O.G., 'O "svoekorystnoi pomoshchi zagranichnym kapitalistam...": k 80-letiiu nachala deiatel'nosti Rossiiskoi zheleznodorozhnoi missii za granitsei', *Lokomotiv*, 11 (2000): 46–8; and 12 (2000): 38–41.

[Kuprienko, O.G.] Pegurokki, Leon, 'Par ili podkova? O tom, kakaia tiaga luchshe, sporili desiatok let', *Gudok*, 8 February 2001, p. 5.

Kuprienko, O.G., 'Zabytye polpredy Rossii: 80 let nazad nachala rabotu za granitsei nasha zheleznodorozhnaia missiia', *Gudok*, 27 February 2001, p. 5.

Kushch, E. and Glovatskii, M., 'Kievskii Politekhnicheskii Institut nakanune i v period pervoi russkoi revolutsii', in Troitskii, P.N. (ed.), *Kievskii Politekhnicheskii Institut: uchenye zapiski: Trudy kafedry Marksizma–Leninizma* (Kiev: Ministerstvo Vysshego Obrazovaniia SSSR, 1956), pp. 122–42.

Kuz'mich, V.D., 'Liudi i lokomotivy: u istokov teplovozostroeniia: Neskol'ko stranits iz zhizni Iu.V. Lomonosova – sozdatelia pervogo magistral'nogo teplovoza', *Lokomotiv*, 8 (1999): 5–12; 9 (1999): 18–21; and 10 (1999): 40–45.

Kuz'mich, V., 'Pervye sovetskie', *Gudok*, 6 October 1987, p. 4.

Kuz'mich, V., 'Tovarishchu Lomonosovu...', *Gudok*, 31 October 1987, p. 4.

Lamb, J.P., *Evolution of the American Diesel Locomotive* (Bloomington: Indiana University Press, 2007).

Lampert, N., *The Technical Intelligentsia and the Soviet State: A Study of Soviet Managers and Technicians, 1928–1935* (London: Macmillan, 1979).

Landes, D.S., *The Unbound Prometheus: Technological Change and Industrial Development in Western Europe from 1750 to the Present* (Cambridge: Cambridge University Press, 1969).

Lane, A.T. (ed.), *Biographical Dictionary of European Labour Leaders*, 2 vols (Westport: Greenwood, 1995).

Lane, M.R., *The Rendel Connection: A Dynasty of Engineers* (London: Quiller Press, 1989).

Laqueur, W., *Black Hundred: The Rise of the Extreme Right in Russia* (New York: Harper Perennial, 1994).

von Laue, T.H., *Sergei Witte and the Industrialization of Russia* (New York: Columbia University Press, 1963).

Layton, Edwin T. Jr, *The Revolt of the Engineers: Social Responsibility and the American Engineering Profession* (Cleveland: Press of Case Western Reserve University, 1971).

Leonov, M.I., *Partiia sotsialistov-revoliutsionerov v 1905–1907 gg.* (Moscow: Rosspen, 1997).

Lewin, M., *Political Undercurrents in Soviet Economic Debates from Bukharin to the Modern Reformers* (London: Pluto, 1975).

Lewis, R., *Science and Industrialisation in the USSR: Industrial Research and Development, 1917–1940* (London: Macmillan, 1979).

Libbey, J.K., *Russian–American Economic Relations, 1763–1999* (Gulf Breeze: Academic International Press, 1999).

Libbey, J.K., 'Shlikhter, Aleksandr Grigor'evich', in Lane, A.T. (ed.), *Biographical Dictionary of European Labour Leaders*, vol. 2 (Westport: Greenwood, 1995), pp. 888–9.

Lieven, D.C.B., *Russia and the Origins of the First World War* (London: Macmillan, 1983).

Limonov, N., 'L.P. Serebriakov (1888–1937): Ostanetsia navsegda', in Proskurin, A. (comp.), *Vozvrashchennye imena: Sbornik publitsisticheskikh statei v 2-kh knigakh* (2 vols, Moscow: Novosti, 1989), vol. 2, pp. 203–19.

Lincoln, W.B., *Passage through Armageddon: The Russians in War and Revolution, 1914–1918* (Oxford: Oxford University Press, 1994).

Lincoln, W.B., *Petr Petrovich Semenov-Tian-Shanskii: The Life of a Russian Geographer* (Newtonville: Oriental Research Partners, 1980).

Lokshin, A., 'The Bund in the Russian–Jewish Historical Landscape', in Geifman, A. (ed.), *Russia under the Last Tsar: Opposition and Subversion, 1894–1917* (Oxford: Blackwell, 1999), pp. 57–72.

Long, J.W., 'Searching for Sidney Reilly: The Lockhart Plot in Revolutionary Russia, 1918', *Europe–Asia Studies*, 47/7 (1995): 1225–41.

Longo, J. and Van Burkleo, S., 'Abbott, Grace (17 Nov 1878–19 June 1939)', in *American National Biography*, vol. 21 (New York, 1999), pp. 24–6.

Lovtsov, L.V., 'Iz istorii bor'by Kommunisticheskoi partii za sozdanie otechestvennogo teplovozostroeniia', in *Stranitsy velikogo puti: Iz istorii bor'by KPSS za pobedu kommunizma: Sbornik statei aspirantov, Chast' III* (Moscow: Izdatel'stvo Moskovskogo universiteta, 1970), pp. 100–133.

Lovtsov, L.V., 'Leninskii printsip edinstva resheniia khoziaistvennykh i politicheskikh zadach v rukovodstve partii vosstanovleniem zheleznodorozhnogo transporta (1921–1925 gg.)', in *Iz istorii bor'by KPSS za pobedu sotsializma i kommunizma: Sbornik statei* (Moscow: Izdatel'stvo Moskovskogo universiteta, 1970), pp. 159–83.

Lubrano, L. and Solomon, S. (eds), *The Social Context of Soviet Science* (Boulder: Westview Press, 1980).

Luntinen, P., *French Information on the Russian War Plans, 1880–1914* (Helsinki: SHS, 1984).

McCall, J.B., 'Dieselisation on an American Railroad: A Case Study', *Journal of Transport History*, Third Series, 6/2 (September 1985): 1–17.

McClelland, J.C., *Autocrats and Academics: Education, Culture and Society in Tsarist Russia* (Chicago, 1979).

McClelland, J.C., 'The Professoriate in the Russian Civil War', in Koenker, D.P., Rosenberg, W.G. and Suny, R.G. (eds), *Party, State, and Society in the Russian Civil War: Explorations in Social History* (Bloomington and Indianapolis: Indiana University Press, 1989), pp. 243–66.

McFadden, D.W., *Alternative Paths: Soviets and Americans, 1917–1920* (Oxford: Oxford University Press, 1993).

Makarov, L.L., *Parovozy serii E: fotoal'bom* (Moscow: Zheleznodorozhnoe Delo, 2004).

Malle, S., *The Economic Organisation of War Communism, 1918–1921* (Cambridge: Cambridge University Press, 1985).

Marakhov, G.I., *Kievskii universitet v revoliutsionno-demokraticheskom dvizhenii* (Kiev: Vishcha shkola, 1984).

Marsden, B. and Smith, C., *Engineering Empires: A Cultural History of Technology in Nineteenth-Century Britain* (Basingstoke: Palgrave, 2005).

Marshall, A., *The Russian General Staff and Asia, 1800–1917* (Abingdon, 2006).

Marshall, J., *A Biographical Dictionary of Railway Engineers*, 2nd edn (Oxford: RCHS, 2003).

Materialy k biobibliografii deiatelei Instituta, 6 vols (Leningrad: NTB-LIIZhT, 1953–1959).

Mawdsley, E., *The Russian Civil War* (Boston: Allen and Unwin, 1987).

Mazour, A.G., *The First Russian Revolution 1825: The Decembrist Movement* (Stanford: Stanford University Press, 1937).

Medvedev, Zh., *Soviet Science* (Oxford: Oxford University Press, 1979).

Melancon, M., 'Neo-Populism in Early Twentieth-Century Russia: The Socialist Revolutionary Party from 1900 to 1917', in Geifman, A. (ed.), *Russia under the Last Tsar: Opposition and Subversion, 1894–1917* (Oxford: Blackwell, 1999), pp. 73–90.

Melua, A.I., *Inzhenery Sankt-Peterburga: Entsiklopediia* (St Petersburg and Moscow: Izdatel'stvo Mezhdunarodnogo fonda istorii nauki, 1996).

Memorial, 'Zhertvy politicheskogo terrora v SSSR': http//lists.memo.ru (last accessed 30 June 2008).

Mierzejewski, A.C., *The Most Valuable Asset of the Reich: A History of the German National Railway, vol. 1: 1920–1932* (Chapel Hill: University of North Carolina Press, 1999).

Miller, M.A., *The Russian Revolutionary Emigrés, 1825–1870* (Baltimore: Johns Hopkins University Press, 1986).

Mommsen, W.J. and Hirschfeld, G. (eds), *Social Protest, Violence and Terror in Nineteenth- and Twentieth-century Europe* (London: Macmillan, 1982).

Moon, D., *The Abolition of Serfdom in Russia* (Harlow: Pearson, 2001).

Morrissey, S., *Heralds of Revolution: Russian Students and the Mythologies of Radicalism* (New York: Oxford University Press, 1998).

Mosse, W.E., *Alexander II and the Modernization of Russia* (London: I.B. Tauris, 1992).

Nahirny, V.C., *The Russian Intelligentsia: From Torment to Silence* (New Brunswick: Transaction Books, 1983).

Neilson, K., *Britain and the Last Tsar: British Policy and Russia, 1894–1917* (Oxford: Clarendon Press, 1995).

Neilson, K., *Strategy and Supply: The Anglo–Russian Alliance, 1914–1917* (London: George Allen and Unwin, 1984).

Nitske, R. and Wilson, C.M., *Rudolf Diesel: Pioneer of the Age of Power* (Norman: University of Oklahoma Press, 1965).

Norman, E.A., 'Teplovoz professora Lomonosova – pervenets sovetskogo i mirovogo teplovozostroeniia', *Voprosy istorii estestvoznaniia i tekhniki*, 4 (1985): 116–25.

Nove, A., *An Economic History of the USSR* (Harmondsworth: Penguin, 1982).

O'Connor, T.E., *The Engineer of Revolution: L.B. Krasin and the Bolsheviks, 1870–1926* (Boulder: Westview Press, 1992).

Otte, T. and Neilson, K. (eds), *Railways and International Politics: Paths of Empire, 1848–1945* (London: Routledge, 2006).

Ozhegov, S.I., *Slovar' russkogo iazyka* (Moscow: Russkii iazyk, 1975).

Parry, A., *The Russian Scientist* (New York: Macmillan, 1973).

Patrikeeff, F. and Shukman, H., *Railways and the Russo–Japanese War: Transporting War* (Abingdon, 2007).

Pavlov, D.B., 'The Union of October 17', in Geifman, A. (ed.), *Russia under the Last Tsar: Opposition and Subversion, 1894–1917* (Oxford: Blackwell, 1999), pp. 179–98.

Pearson, R., *The Russian Moderates and the Crisis of Tsarism, 1914–1917* (London: Macmillan, 1977).

Pearson, T.S., 'Russian Law and Rural Justice: Activity and Problems of the Russian Justices of the Peace, 1865–1889', *Jahrbücher für Geschichte Osteuropas*, 32/1 (1984): 52–71.

Pearson, T.S., *Russian Officialdom in Crisis: Autocracy and Local Self-Government, 1861–1900* (Cambridge: Cambridge University Press, 1989).

Pegurokki, Leon – see Kuprienko, O.G.

Perrie, M., 'Political and Economic Terror in the Tactics of the Russian Socialist-Revolutionary Party before 1914', in Mommsen, W.J. and Hirschfeld, G. (eds), *Social Protest, Violence and Terror in Nineteenth- and Twentieth-century Europe* (London: Macmillan, 1982), pp. 63–79.

Pethybridge, R.W., 'The Significance of Communications in 1917', *Soviet Studies*, 19/1 (1967–1968): 109–14.

Phillips, H.D., *Between the Revolution and the West: A Political Biography of Maxim M. Litvinov* (Boulder: Westview Press, 1992).

Printed and bound by CPI Group (UK) Ltd, Croydon, CR0 4YY

21/10/2024

01777082-0019

Phillips, H.D., 'Riots, Strikes and Soviet: The City of Tver in 1905', *Revolutionary Russia*, 17/2 (December 2004): 49–66.

Pimlott, B., *Labour and the Left in the 1930s* (Cambridge: Cambridge University Press, 1977).

Pintner, W.M. and Rowney, D.K. (eds), *Russian Officialdom: The Bureaucratization of Russian Society from the Seventeenth to the Twentieth Century* (London: Macmillan, 1980).

Pipes, R. (ed.), *The Russian Intelligentsia* (New York: Columbia University Press, 1961).

Proskurin, A. (comp.), *Vozvrashchennye imena: Sbornik publitsisticheskikh statei v 2-kh knigakh* (2 vols, Moscow: Novosti, 1989).

Raeff, M., *Russia Abroad: A Cultural History of the Russian Emigration, 1919– 1939* (Oxford: Oxford University Press, 1990).

Rakov, V.A., *Lokomotivy otechestvennykh zheleznykh dorog, 1845–1955* (Moscow: Transport, 1995).

Rakov, V.A., *Lokomotivy zheleznykh dorog Sovetskogo Soiuza: Ot pervykh parovozov do sovremennykh lokomotivov* (Moscow: Gosudarstvennoe transportnoe zheleznodorozhnoe izdatel'stvo, 1955).

Ransome Wallis, P. (ed.), *The Concise Encyclopedia of World Railway Locomotives* (London: Hutchinson, 1959).

Raskin, D.I. and Sukhanova, O.P. (comps), *Fondy Rossiiskogo Gosudarstvennogo Istoricheskogo Arkhiva: Kratkii spravochnik* (St Petersburg: RGIA, 1994).

Read, C., *From Tsar to Soviets: The Russian People and their Revolution, 1917– 21* (London: UCL Press, 1996).

Reed, B., 'Lomonossoff: A Diesel Traction Pioneer', *Railroad History*, 128 (1973): 35–49.

van Reenen, P., 'Alexandra Feodorovna's Intervention in Russian Domestic Politics During the First World War, *Slovo*, 10/1–2 (1998): 71–82.

Rees, E.A., *Stalinism and Soviet Rail Transport, 1928–41* (Basingstoke: Macmillan, 1995).

Rees, E.A., *State Control in Soviet Russia: The Rise and Fall of the Workers' and Peasants' Inspectorate, 1920–1934* (Basingstoke: Macmillan, 1987).

Reichman, H., *Railwaymen and Revolution: Russia, 1905* (Berkeley: University of California Press, 1987).

Reutter, M., 'The Great (Motive) Power Struggle: The Pennsylvania Railroad v. General Motors, 1935–1949', *Railroad History*, 170 (Spring 1994): 15–33.

Reutter, M., 'The Revolutionary', *Railroad History, Millenium Special* (2000): 17–21.

Rice, C., *Russian Workers and the Socialist Revolutionary Party through the Revolution of 1905–07* (Basingstoke: Macmillan, 1988).

Rieber, A.J., 'The Rise of Engineers in Russia', *Cahiers du Monde Russe et Soviétique*, 31/4 (1990): 539–68.

Rielage, D.C. *Russian Supply Efforts in America During the First World War* (Jefferson: MacFarland, 2002).

Rigby, T.H., *Lenin's Government: Sovnarkom, 1917–1922* (Cambridge: Cambridge University Press, 1979).

Roberts, A.J., *The American Diesel Locomotive* (South Brunswick and New York: A.S. Barnes, 1977).

Rogers, H.C.B., *Chapelon, Genius of French Steam* (London: Ian Allan, 1972).

Rogers, H.C.B., *Express Steam Locomotive Development in Great Britain and France* (Sparkford: Haynes/Oxford Publishing Company, 1990).

Rogger, H., *Russia in the Age of Modernisation and Revolution, 1881–1917* (London: Longman, 1983).

Rolt, L.T.C., *The Mechanicals: Progress of a Profession* (London: Heinemann, 1967).

Ross, D., *The Steam Locomotive: A History* (Brimscombe Port: Tempus, 2006).

Rovner, O.S. and Korolik, R.Ia., *Oleksandr Grigorovich Shlikhter, 1868–1940: bibliografichnii pokazhchik* (Kiev: AN URSR, 1958).

Rowney, D.K., *Transition to Technocracy: The Structural Origins of the Soviet Administrative State* (Ithaca: Cornell University Press, 1989).

Rutherford, M., '"Export or Die!": British Diesel-Electric Manufacturers and Modernisation, Part One: Roots', *Backtrack*, 22/1 (January 2008): 52–60.

Rutherford, M., 'Failure? Part Two', *Backtrack*, 9/12 (December 1995): 657–65.

Rutherford, M., 'Measurement not Mystification: The British Dynamometer Car', *Backtrack*, 9/8 (August 1995): 436–44.

Rutherford, M., 'Quantification and Railway Development Before the Micro-electronic Revolution with Special Reference to the Great Western Railway and its Successors', in Cossons, N., Patmore, A. and Shorland-Ball, R. (eds), *Perspectives on Railway History and Interpretation: Essays to Celebrate the Work of Dr John Coiley at the National Railway Museum, York* (York: National Railway Museum, 1992), pp. 38–47.

Rutherford, M., 'Sages are not Fixers: Science, Invention and Dr Diesel', *Backtrack*, 9/7 (July 1995): 377–85.

Rutherford, M., 'When Britain was a Contender: Some Pioneer Diesel-Electrics before La Grange', *Backtrack*, 14/6 (June 2000): 351–8; 14/7 (July 2000): 416–21; and 14/8 (August 2000): 479–86.

Sanderson, M., *The Universities and British Industry, 1850–1970* (London: Routledge & Kegan Paul, 1972).

Saul, N.E., *War and Revolution: The United States and Russia, 1914–1921* (Lawrence: University Press of Kansas, 2001)

Saveliev, I.R. and Pestushko, Yu. S., 'Dangerous Rapprochement: Russia and Japan in the First World War, 1914–1916', *Acta Slavica Iaponica*, 18 (2001): 19–41.

Schapiro, L., *The Communist Party of the Soviet Union*, 2nd edn (London: Methuen, 1970).

Senchakova, L.T., *Boevaia rat' revoliutsii: ocherk o boevykh organizatsiiakh RSDRP i rabochikh druzhinakh 1905–1907 gg.* (Moscow: Politicheskaia literatura, 1975).

Senin, A.S., 'Aleksei Ivanovich Rykov', *Voprosy istorii*, 9 (1988): 85–115.

Senin, A.S., *Ministerstvo putei soobshcheniia v 1917 godu*, 2nd edn (Moscow: URSS, 2008).

Senin, A.S., 'Zheleznye dorogi v marte–oktiabre 1917 g.: ot krizisa k khaosu', *Voprosy istorii*, 3 (2004): 32–56.

Serkov, A.I., *Russkoe masonstvo, 1731–2000: Entsiklopedicheskii slovar'* (Moscow: Rossiiskaia politicheskaia entsiklopediia, 2001).

Seroshtan, N.A., Burik, A.I., Voevodin, N.A. et al. (eds), *Istoriia gorodov i sel Ukrainskoi SSR: Khar'kovskaia oblast'* (Kiev: Institut istorii AN UkSSR, 1976).

Service, R., *Lenin: A Political Life*, 3 vols ((Basingstoke: Macmillan, 1985–1995).

Service, R., *Stalin: A Biography* (London: Pan, 2005).

Sharikova, E.S. and Pokrovskaia, L.Iu., 'K istorii teplovozostroeniia v SSSR (Novye materialy ob izobretatel'skoi deiatel'nosti prof.Iu.V. Lomonosova)', *Voprosy istorii estestvoznaniia i tekhniki*, 4 (1986): 186–88.

Shelestov, D., 'L.B. Kamenev (1883–1936): Odin iz vidneishikh bol'shevikov i kommunistov...', in Proskurin, A. (comp.), *Vozvrashchennye imena: Sbornik publitsisticheskikh statei v 2-kh knigakh* (2 vols, Moscow: Novosti, 1989), vol. 1, pp. 211–33.

Shilov, D.I., *Gosudarstvennye deiateli Rossiiskoi Imperii, 1802–1917: Biobibliograficheskii spravochnik* (St Petersburg: Dmitrii Bulanin, 2001).

Shishkin, V.A., *Sovetskoe gosudarstvo i strany zapada v 1917–1923 gg.: Ocherki istorii stanovleniia ekonomicheskikh otnoshenii* (Leningrad: Nauka, 1969).

Shmorgun, P.M., *Bol'shevistskie organizatsii Ukrainy v gody pervoi russkoi revoliutsii 1905–1907* (Moscow: Gosudarstvennoe Izdatel'stvo Politicheskoi Literatury, 1955).

Shoenberg, D., 'Kapitza, Fact and Fiction', *Intelligence and National Security*, 3/4 (1988): 49–61.

Shoenberg, D., 'Piotr Leonidovich Kapitza, 9 July 1894–8 April 1984', *Biographical Memoirs of Fellows of the Royal Society*, 31 (November 1985): 327–74.

Sidorov, A.L., 'Otnosheniia Rossii s soiuznikami i inostrannye postavki vo vremia pervoi mirovoi voiny, 1914–1917 gg.', *Istoricheskie zapiski*, 15 (1945): 128–79.

Sidorov, A.L., 'Zheleznodorozhnyi transport Rossii v pervoi mirovoi voine i obostrenie ekonomicheskogo krizise v strane', *Istoricheskie zapiski*, 26 (1948): 3–64.

Siegel, J., *Endgame, Britain, Russia and the Final Struggle for Central Asia* (London: I.B. Tauris, 2002).

Siegel, K.A.S., *Loans and Legitimacy: The Evolution of Soviet–American Relations, 1919–1933* (Lexington: University Press of Kentucky, 1996).

Siegelbaum, L.H., *The Politics of Industrial Mobilization in Russia, 1914–17: A Study of the War Industries Committees* (London: Macmillan, 1983).

Simmons, J., *The Victorian Railway* (London: Thames and Hudson, 1995).

Simmons, J. and Biddle, G. (eds), *The Oxford Companion to British Railway History* (Oxford: Oxford University Press, 1997).

Simon, E.W., Jr, 'Dawn of the Diesels', *National Railway Bulletin*, 63/5 (1998): 4–27.

Smele, J.D. and Heywood, A.J. (eds), *The Russian Revolution of 1905: Centenary Perspectives* (London: Routledge, 2005).

Smith, C., *The Science of Energy: A Cultural History of Energy Physics in Victorian Britain* (London: Athlone, 1998).

Sochor, Z., 'Soviet Taylorism Revisited', *Soviet Studies*, 33/2 (1981): 246–64.

Solovyova, A.M., 'The Railway System in the Mining Area of Southern Russia in the Late Nineteenth and Early Twentieth Centuries', *Journal of Transport History*, Third Series, 5/1 (March 1984): 66–81.

Spring, D., 'Railways and Economic Development in Turkestan before 1917', in Symons, L. and White, C. (eds), *Russian Transport: An Historical and Geographical Survey* (London: Bell, 1975), pp. 46–74.

Staudenmaier, J.M., *Technology's Storytellers: Reweaving the Human Fabric* (Cambridge, MA: Society for the History of Technology/MIT Press, 1985).

Steinberg, J.W. et al. (eds), *The Russo–Japanese War in Global Perspective: World War Zero*, 2 vols (Leiden, 2005–2006).

Stepanenko, T., 'Kievskaia sotsial-demokraticheskaia organizatsiia nakanune pervoi russkoi revoliutsii 1905–1907 gg.', in Troitskii, P.N. (ed.), *Kievskii Politekhnicheskii Institut: Uchenye zapiski: Trudy kafedry Marksizma-Leninizma* (Kiev: Ministerstvo Vysshego Obrazovaniia SSSR, 1956), pp. 89–109.

Stevenson, D., 'War by Timetable? The Railway Race before 1914', *Past and Present*, 162 (1999): 163–94.

Stockdale, M., 'Liberalism and Democracy: The Constitutional Democratic Party', in Geifman, A. (ed.), *Russia under the Last Tsar: Opposition and Subversion, 1894–1917* (Oxford: Blackwell, 1999), pp. 153–78.

Stockdale, M., *Paul Miliukov and the Quest for a Liberal Russia, 1880–1918* (Ithaca, NY: Cornell University Press, 1996).

Stone, N., *The Eastern Front, 1914–1917* (Abingdon: Purnell, 1976).

Sutton, A.C., *Wall Street and the Bolshevik Revolution* (New Rochelle: Arlington House, 1974).

Swain, G., *Trotsky* (Harlow: Longman, 2006).

Symons, L. and White, C. (eds), *Russian Transport: An Historical and Geographical Survey* (London: Bell, 1975).

Szontagh, G., 'Brotan and Brotan–Dreffner Type Fireboxes and Boilers Applied to Steam Locomotives', *Transactions of the Newcomen Society*, 62 (1990–1991): 21–51.

Taylor, A.J.P., *War by Timetable: How the First World War Began* (London: MacDonald, 1969).

Tester, A., 'An Introduction to Steam Locomotive Testing', parts 1, 2, 2A, 3A, 3B, *Backtrack*, 23/4 (April 2009): 199–203; 23/5 (May 2009): 308–15; 23/9

(September 2009): 564–9; 24/2 (February 2010): 100–7; and 24/4 (April 2010): 242–50.

Thatcher, I.D. (ed.), *Late Imperial Russia: Problems and Prospects* (Manchester: Manchester University Press, 2005).

Thatcher, I.D., *Trotsky* (London: Routledge, 2003).

Thomas, D.E., *Diesel: Technology and Society in Industrial Germany* (Tuscaloosa: University of Alabama Press, 1987).

Timoshenko, S.P., *Engineering Education in Russia* (New York: McGraw-Hill, 1959).

Tiutiukin, S.V. (ed.), *Istoricheskie siluety* (Moscow: Nauka, 1991).

Todes, D.P., *Pavlov's Physiology Factory: Experiment, Interpretation, Laboratory Enterprise* (Baltimore: Johns Hopkins University Press, 2001).

Torrey, G.E., 'Indifference and Mistrust: Russian–Romanian Collaboration in the Camapaign of 1916', *The Journal of Military History*, 57/2 (1993): 279–300.

Torrey, G.E., 'Rumania and the Belligerents 1914–1916', *Journal of Contemporary History*, 1/3 (1966): 171–91.

Torrey, G.E., 'The Rumanian Campaign of 1916: Its Impact on the Belligerents', *Slavic Review*, 39/1 (1980): 27–43.

Trebilcock, C., *The Industrialization of the Continental Powers, 1780–1914* (London: Longman, 1981).

Troitskii, P.N. (ed.), *Kievskii Politekhnicheskii Institut: Uchenye zapiski: Trudy kafedry Marksizma-Leninizma* (Kiev: Ministerstvo Vysshego Obrazovaniia SSSR, 1956).

Tumanova, A.S., 'Vlast' i tekhnicheskii progress: Vozdukhoplavatel'nye i avtomobil'nye obshchestva dorevoliutsionnoi Rossii pod nabliudeniem Departamenta Politsii', *Voprosy istorii estestvoznaniia i tekhniki*, 4 (2002): 787–94.

Turgaev, A.S. (ed.), *Vysshie organy gosudarstvennoi vlasti i upravleniia Rossii, IX–XX vv.: Spravochnik* (St Petersburg: Obrazovanie–Kul'tura, 2000).

Ushakov, K., *Podgotovka voennykh soobshchenii Rossii k mirovoi voine* (Moscow and Leningrad: Gosudarstvennoe izdatel'stvo, otdel voennoi literatury, 1928).

Vasil'ev, N., *Transport Rossii v voine 1914–1918* (Moscow: Voenizdat, 1939).

Vatlin, A. and Malashenko, L. (eds), *Piggy Fox and the Sword of the Revolution: Bolshevik Self-Portraits* (New Haven: Yale University Press, 2006).

Velidov, A.S. et al. (eds), *Feliks Edmundovich Dzerzhinskii: Biografiia*, 3rd edn (Moscow: Politizdat, 1986).

Veviorovskii, I.V., Voronin, M.I., Klauz, P.L. et al. (eds), *Leningradskii ordena Lenina Institut Inzhenerov Zheleznodorozhnogo Transporta imeni akademika V.N. Obraztsova, 1809–1959* (Moscow: Vsesoiuznoe Izdatel'sko-Poligraficheskoe Ob"edinenie MPS, 1960).

Vickers, R.L.,'The Beginnings of Diesel Electric Traction', *Transactions of the Newcomen Society*, 71/1 (1999–2000): 115–27.

Virnyk, D.F., *Aleksandr Grigor'evich Shlikhter* (Kiev: Nauka dumka, 1979).

Virnyk, D.F., *Zhiznennyi put' bol'shevika A.G. Shlikhtera* (Moscow, 1960).

Volobuev, P.V. et al. (eds), *Politicheskie deiateli Rossii 1917: biograficheskii slovar'* (Moscow: Bol'shaia Rossiiskaia entsiklopediia, 1993).

Vucinich, A., *Science in Russian Culture, 1861–1917* (Stanford: Stanford University Press, 1970).

Wade, R., *The Russian Search for Peace, February–October 1917* (Stanford: Stanford University Press, 1969).

Walsh, W.B., 'The Bublikov–Rodzyanko Telegram', *Russian Review*, 30/1 (1971): 69–70.

Warren, K., *Armstrongs of Elswick: Growth in Engineering and Armaments to the Merger with Vickers* (Basingstoke: Macmillan, 1989).

Wartenweiler, D., *Civil Society and Academic Debate in Russia, 1905–1914* (Oxford: Clarendon Press, 1999).

Webb, B., *The British Internal Combustion Locomotive, 1894–1940* (Newton Abbot: David and Charles, 1973).

Weinstein, E.G., 'Andre Chapelon and French Locomotives in the Twentieth Century', *Railroad History*, 176 (Spring 1997): 7–38.

Westwood, J.N., *A History of Russian Railways* (London: George Allen and Unwin, 1964).

Westwood, J.N., *Locomotive Designers in the Age of Steam* (London: Sidgwick and Jackson, 1977).

Westwood, J.N., *Soviet Locomotive Technology During Industrialization, 1928–1952* (London: Macmillan, 1982).

Wheatcroft, S.G. (ed.), *Challenging Traditional Views of Russian History* (Basingstoke: Palgrave, 2002).

White, C.A., *British and American Commercial Relations with Soviet Russia, 1918–1924* (Chapel Hill: University of North Carolina Press, 1992).

Williams, B., '1905: The View from the Provinces', in Smele, J.D. and Heywood, A.J. (eds), *The Russian Revolution of 1905: Centenary Perspectives* (London: Routledge, 2005), pp. 34–54.

Zenzinov, N.A., 'Iu.V. Lomonosov 1876–1952 g.g.', *Locotrans*, 4 (2000): 6–8.

Zenzinov, N.A., 'Lomonosov i ego teplovoz', *Lokomotiv*, 2 (1996): 38–40.

Zenzinov, N.A., 'Vydaiushchiisia rossiiskii zheleznodorozhnyi inzhener', *Zheleznodorozhnyi transport*, 4 (1993): 57–63.

Zenzinov, N.A. and Ryzhak, S.A., *Vydaiushchiesia inzhenery i uchenye zheleznodorozhnogo transporta*, 2nd edn (Moscow: Transport, 1990).

Zhuravlyov, V.V., 'Private Railway Companies in Russia in the Early Twentieth Century', *Journal of Transport History*, Third Series, 4/1 (March 1983): 51–66.

Zuckerman, F.S., 'The Russian Army and American Industry, 1915–17: Globalisation and the Transfer of Technology', in Wheatcroft, S.G. (ed.), *Challenging Traditional Views of Russian History* (Basingstoke: Palgrave, 2002), pp. 55–65.

Zuckerman, F.S., *The Tsarist Secret Police in Russian Society, 1880–1917* (Basingstoke: Macmillan, 1996).

Index